ROWAN UNIVERSITY
CAMPBELL LIBRARY
201 MULLICA HILL RD.
GLASSBORO, NJ 08028-1701

Jack Phillips

General Spatial Involute Gearing

Springer

*Berlin
Heidelberg
New York
Hong Kong
London
Milan
Paris
Tokyo*

Jack Phillips

General Spatial Involute Gearing

With 159 Figures

Springer

Professor Jack Phillips
B.Mech.E. (Melb.), Ph.D. (Melb.),
Honorary Associate, School of Information Technologies, University of Sydney,
Formerly Associate Professor in Theory of Machines, University of Sydney,
Honorary Member IFToMM

Library of Congress Cataloging-in-Publication Data
Phillips, Jack
General spatial involute gearing / Jack Phillips
p.cm.
ISBN 3540442049
1. Gearing, Spur. I. Title
TJ189.P73 2003
621.8'331--dc21   20022042593

ISBN 3-540-44204-9 Springer-Verlag Berlin Heidelberg New York

This work is subject to copyright. All rights are reserved, whether the whole or part of the material is concerned, specifically the rights of translation, reprinting, reuse of illustrations, recitation, broadcasting, reproduction on microfilm or in other ways, and storage in data banks. Duplication of this publication or parts thereof is permitted only under the provisions of the German Copyright Law of September 9, 1965, in its current version, and permission for use must always be obtained from Springer-Verlag. Violations are liable for prosecution act under German Copyright Law.

Springer-Verlag Berlin Heidelberg New York
a member of BertelsmannSpringer Science + Business Media GmbH

http://www.springer.de

© Springer-Verlag Berlin Heidelberg 2003
Printed in Germany

The use of general descriptive names, registered names, trademarks, etc. in this publication does not imply, even in the absence of a specific statement, that such names are exempt from the relevant protective laws and regulations and therefore free for general use.

Typesetting: Camera ready copies from author
Cover-Design: medio Technologies AG, Berlin
Printed on acid-free paper   62/3020 Rw 5 4 3 2 1 0

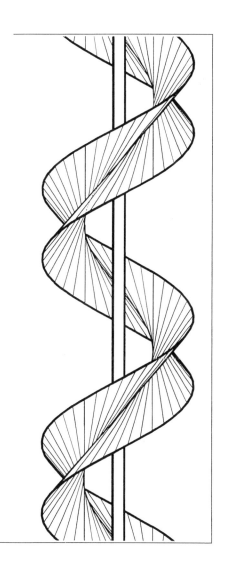

Frontispiece: an involute helicoid: the apparently real rod, whose diameter plays no role in defining the ruled surface shown here, should be imagined to be invisible.

PREFACE

It has been hard for me to escape the imprint of my early, strong, but scattered trains of thought. There was, at the beginning, little to go by; and I saw no clear way to go. This book is accordingly filled with internal tensions that are not, as yet, fully annealed. Subsequent writers may re-present the work, explaining it in a simpler way. Others may simply invert it. I mean by this that, by writing it backwards, from its found ends (practical machinable teeth) to its tentative beginnings (dimly perceived geometrical notions), one might conceivably write a *manual*, not on how to understand these kinds of gears, but on how to make them. Indeed a manual will need to be written. If this gearing is to be further investigated, evaluated and checked for applicability, prototypes will need to be made. I wish to say again however that my somewhat convoluted way of presenting these early ideas has been inevitable. It has simply not been possible to present a tidy set of explanations and rules without exploring first (and in a somewhat backwards-going direction) the complexities of the kinematic geometry. There remains, now in this book, a putting together of primitive geometric intuition, computer aided exploration of certain areas, geometric explanations of the discovered phenomena, and a loose sprinkling of a relevant algebra cementing the parts together. One might, I suppose, have gone on to reduce the work to a very small bundle of tight, algebraic formulae, which would have been 'neat', but, even in the presence of pictures, such a bundle would have been incomprehensible. I have in any event a distinct feeling, not that I have invented this kind of gearing, but that I have discovered it. This book is my account of that discovery.

Jack Phillips, Sydney, June 2002.

ACKNOWLEDGEMENTS

I wish to mention with thanks the contributions of Dr Friedjhof Sticher of UTS in Sydney, a one-time doctoral student of mine, and Michael Killeen, a recently graduated undergraduate student of Sticher. Killeen wrote his own algebraic interpretation of my geometry of the equiangular architecture, he devised an extensive spreadsheet to provide accurate numerical values for the example used for demonstration, and he was responsible for the first working, two-toothed, equiangular prototype described at chapter 11.
The algebraic abilities of Sticher and Killeen eclipse my own, and the skill displayed by Killeen in the design and supervision of the building of his prototype was a pleasure to see. Their separate contributions are foot-noted or otherwise highlighted wherever appropriate. I wish to thank Dr Ian Parkin of the School of Information Technologies at the University of Sydney for his enthusiastic support, and for so effectively nurturing my Honorary Associateship there. I wish to thank Dr Koljo Minkov of the Bulgarian Academy of Sciences for much of my meagre understanding of the wide field of non-involute gearing and for his introduction to, and clarification of, the Russian and Russian-related literature. I thank Friederike Binder of Sydney and Alice Currie of Sydney respectively for their generous help with the less recent parts of the German and French literatures. Yet again I wish to mention the steadfast support, over these many years, of my loving companion Elayne Russell.

SYNTAX

Each paragraphs is identified first by the chapter number and then by a serial number within the chapter. Thus for example the twelfth paragraph of chapter 3 is called §3.12. Any such number inserted suddenly on-line and within rounded brackets, thus (§3.12) for example, is an invitation to turn to that paragraph for parallel reading. These internal references may be casting either forwards to material yet to be encountered, or backwards to material that might require a second reading. Appendices appear at the ends of the chapters. They are lettered A at 1, B at 1, A at 2, B at 2, etc., for easy identification. A list called Notes and References appears at the end of the book. Sendings to this are made by means of consecutive numbers within square brackets, thus [4] for example. I have sometimes enlarged this device to refer to specific paragraphs within a quoted work. This [1] [§12.13] for example refers the reader specifically to §12.13 in reference [1]. The many inter-connected themes of the book cannot be reduced, one-by-one, each to simple number. Developing as they go, they run interwoven throughout. Accordingly the listings at the Subjects Index tell the reader (a) reading down, which chapters deal with the subject matter at issue, and (b) reading across, in which particular paragraphs the developing details may be found.
A straight reading of the contents of such listings may often be fruitful. Authors' names appear as starred items within the Subjects Index. Algebraic symbols are listed separately at the Symbols Index. Generally speaking, upper case Roman characters are reserved for points, lower case Greek characters for angles, lower case Italic Roman for radii, and upper case Italic Roman for lengths. Many of the drawn figures and much of the written argument are variously based upon or bolstered by numerical example. For the reader interested in reproducing the CAD calculations which abound, most numerical values are given either directly on line with the text [but isolated there within square brackets] or, more often, among the Appendices or at the Table of Numerical Results (TNR). This latter appears just before the two, already mentioned Indices at the end of the book.

TABLE OF CONTENTS

PROLOGUE, page 1
SYNOPSIS, page 5

CHAPTER ONE
BASIC KINEMATICS OF THE GEAR SET

Preliminary remarks, 13
Kinematics and the pitch line, 14
On the dichotomy synthesis-v-analysis, 19
Algebraic formulae for the pitch line, 20
On the numbering and significance of these equations, 23
Various graphs telling about the pitch line, 24
On the question of sign, 27
The axodes, 28
The imaginary spline-line teeth of a pair of axodes, 30
Singly-departed naked wheels of the crossed helicals, 32
The twisted surface marked (d) in figure 1.07, 33
Some aspects to be remembered of figure 1.07, 34
Special cases among the radii occurring at P, D and O, 34
Ordinary planar spur and the parallel helical gears, 36
The various kinds of bevel gears, 37
A first look at the question of offset, 37
Conclusion, 40

CHAPTER TWO
THE FUNDAMENTAL LAW OF GEARING

Introduction, 41
The fundamental law, 41
Involute gears, 43
Prognoses, 44
Some assumptions and two disclaimers, 44
An equivalent linkage, 45

Enumerative geometry, 47
In screw theory the fundamental law manifests itself, 48
Proofs of the law by means of the vector algebra, 50
On what there is to follow, 52
On the non-involute hypoids, 52
Design in the absence of the fundamental law, 53
Sketchy account of the (non-involute) hypoid literature, 54
The manufacture of hypoid gearing, 55
Some eclectic theoretical works, 56
Survey of the general spatial involute literature, 57
Other laws, 58
Some observations in respect of all gears, 58
Robert Willis 1838, 61

CHAPTER THREE
MATING INVOLUTE HELICOIDS

Introduction, 63
Circular hyperboloids proliferate, 64
A straight path for the point of contact, 65
Two ruled surfaces intersecting perpendicularly, 66
Imagined milling of the involute tooth profile, 71
The core helix and the twisted line of intersection, 71
The involute helicoid and its whole slip track, 72
The whole geometric construct, 74
Brief look at a special (the planar) case, 75
The triad of lines mutually perpendicular at B, 75
The angles ζ and σ and the curvature at B, 75
A mating pair of non-alike involute helicoids, 77
The unimportance of errors upon assembly, 80
Two cut gear-bodies put anyhow into contact, 81
On the kinematic equivalences, 84
The G-circles intersecting at Q, 84
Circles intersecting in general, 85
The triad of planes at a general Q, 85
Gyratory motion of the triad at Q, 89
Location here of the pitch line, 89
Another triad of lines in and about the polar plane, 90
More about the variable angle δ_{REMOTE}, 91
Special points [Qp], [Qm] and [Qc] in the path of Q, 91
Practical limits to the otherwise endless motion, 92
Degenerate slip track geometry at the two ends of Z_l, 93
The slip tracks unsupported, 97
Theodore Olivier 1842, 98

Intersection of the base hyperboloids, 98
Penultimate remarks, 101
Raoul Bricard, 102
Appendix A at chapter 3, 102
Appendix B at chapter 3, 103
Appendix C at chapter 3, 104
Appendix D at chapter 3, 104
Appendix E at chapter 3, 105

CHAPTER FOUR
KEY ASPECTS OF THE GEOMETRY

Introduction, 107
The hyperboloid and its generators, 108
Theorems about generator-joined hyperboloids, 109
Theorem #1, about circular hyperboloidal axodes, 111
Theorem #2, about the points [Γm], 113
Intersecting circles, 115
Theorem #3 part one, about the points [Γt], 115
Theorem #3 part two, about the point [Γd], 117
An isolated note looking in retrospect, 118
An isolated note looking ahead, 118
Definition of [Qf], 119
Looking now at the figures 4.04, 119
Looking next at figure 4.04(d), 122
Specifically in the case of gearing, 123
The central fortuitous fact now apparent, 125
Legitimate paths, 126
On the distribution of [Qm], 127
Kinematics at [Qm], 129
Statics at [Qm], 130
Strain energy at [Qm], 132
Another look at the gear bodies of figure 3.10, 132
The moving transversal, the FAXL, and the F-surface, 134
Variation of δ-remote with the distance of Q from [Qf], 134
The point [Qs] occurring at the line of striction, 135
Position of [Qs] in the special case of equiangularity, 137
More about the gyratory motion of the triad at Q, 138
The two planes that continuously define j–Q–k, 140
Mechanical significance of [Qf], 140
From the path of Q in 1 to the slip tracks by inversion, 143
The two paths of the two Q, 144
Coaxiality of the circular hyperboloids, 145
The parabolic hyperboloid, 146

Four different rectilinear parabolic hyperboloids, 148
Appendix A at chapter 4, 149
Appendix B at chapter 4, 158

CHAPTER FIVE (A)
THE SIMPLICITY OF EQUIANGULARITY

Transition comment, 159
Prognosis, 159
Synopsis of the chapters 5, 160
The common objectives of the chapters 5B and 6, 160
Two important theorems that need to be studied, 161
Aspects of the pitch helices not well studied as yet, 161
Problems concerning our choice for the travel paths, 162
Radial offset R and the FAXL, 164
The fortuitous aspects of an equiangular architecture, 166
A selected few of the algebraic formulae, 171
Two special values of R, 172
The special case when C is at P, 173
An interim mention of the polyangular, 173
Last remarks and a transition comment, 173
Appendix A at chapter 5A, 174
Appendix B at chapter 5A, 176

CHAPTER FIVE (B)
SYNTHESISING AN EQUIANGULAR SET

Introduction, 177
Naked wheels, 177
The angles τ, κ and λ at the naked wheel, 178
Rolling, rocking and boring, 181
Watershed, 182
Beginning of worked example #1 (part 1), 182
More on the lengths Z_j and related questions, 193
On the angles α and the shapes of the slip tracks, 196
Danger-locations for excessive wear, 198
Adverse circumstances at the crossed helicals, 199
On the sets of wheels combined at figure 5B.06, 200
Beginning of part 2 of worked example #1, 200
Unwrapping the roll of paper, 204
WkEx#1(2) continued, 206
On the results of this synthesis, 207
The phantom rack, 209
The anatomy of a tooth, 212
The very small angles χ at the crossed helicals, 212

On the plane of the E-radii and the radiplan view, 214
The ellipses of obliquity, 218
Beginning of worked example #2, 220
Mutual decapitation versus continuity of action, 227
Transition comment, 228
The special sets with Σ zero, 229
The special sets with R zero, 230
The special sets with C zero, 230
The so-called straight bevels of ordinary practice, 231
More transition comment, 232
Load bearing and lubrication, 233
Conclusion, 233
Appendix A at chapter 5B, 234
Appendix B at chapter 5B, 236

CHAPTER SIX
THE PLAIN POLYANGULAR OPTION

Recapitulation, 237
Introduction to polyangularity, 237
Timely exclusion of an as yet unwanted difficulty, 238
On getting ready to begin, 238
On what happens in the G-space, 239
The plain polyangular and the exotic, 240
Important aspects of the equilateral transversal, 240
An important observation, 243
The figures 6.02, 243
The figure 6.03, 246
Circular cylinder of radius R coaxial with the CDL, 247
Some more-incisive remarks about the F-surface, 249
Location of [Qf], 250
Finally some definitive remarks about the elusive [Qm], 250
Two special plain polyangulars, 251
Again the plain and the exotic, 252
More about axodes and the naked wheels, 252
Geometric relations within the linear array, 254
The velocity vector at F, 255
The parabolic hyperboloid of the rectilinear array, 256
Need for a dedicated algebra, 256
An algebra for the plain polyangular architecture, 257
Law of the speed ratio $k = [a_3 \cos \alpha_3]/[a_2 \cos \alpha_2]$, 260
Beginning of worked example #3, 261
Numerical check on the law of the speed ratio, 270
Putting one of the paths upon the transversal, 270

Beginning of worked example #4, 271
Some ongoing comments in conclusion, 278
Appendix A at chapter 6, 279
Appendix B at chapter 6, 282
Appendix C at chapter 6, 283

CHAPTER SEVEN
WORM DRIVES OFFSET AND NON-OFFSET

Introduction, 285
Recapitulation, 286
Beginning of worked example #5, 287
Looking more generally now at the plane of adjacency, 292
The radial planar section through the flank of a tooth, 294
Beam's axvolute, 294
Visual recognition of properly cut involute worms, 296
Beginning of worked example #6, 297
Load bearing and the small angles χ in worm drives, 307
Appendix A at chapter 7, 308

CHAPTER EIGHT
EXOTICISM BY ACCIDENT OR DESIGN

Preliminary remarks, 311
Overview of exoticism, 311
Analysis of the Merritt figure, 313
The wholly-completed F-surface, 318
Forehanded and backhanded design, 319
Non-intersecting paths, 319
The wider generality of exoticism, 322
On the difficulties surrounding truncation, 322

CHAPTER NINE (A)
HELICALITY TIGHT MESH AND BACKLASH

Screw symmetry at the flanks of teeth, 323
On the misnomer, 325
The fronts and the backs, 326
Thick teeth, 326
Conical blanks, 327
Drive and coast, 327
The fixity of the paths, 328
Axial withdrawal from tight mesh, 328
Shifting of the slip tracks, 329
On mobility, 330

Relative re-location at the axes of the wheels, 331
The flush configuration, 333
Axial adjustment for location, 333
Axial adjustment for backlash, 334
A general observation, 334

CHAPTER NINE (B)
MECHANICS OF THE PHANTOM RACK

Introduction, 335
First look at a puzzling matter, 335
Second look at the same matter, 338
Axodes in general and the ideas of Reuleaux, 339
Another theorem, 340
On deeper consideration, 346
Concluding remarks, 347

CHAPTER TEN
MATHEMATICS OF THE MACHINING

Recapitulation, 349
Some early hints about milling and hobbing, 350
The rack triad, 351
On the nagging question of the 'conicality' of teeth, 351
More on the shapes of teeth, 354
The swivel angle lambda, 354
The angles kappa and lambda taken together, 355
Incidentally some interesting interrelated formulae, 355
Two other important angles, gamma and theta, 356
Matters to do with delta and gamma, 357
The real rack, 358
Length-measured pitches of teeth, 360
The flat-action phenomenon, 364
First algebraic look at gamma in terms of delta, 366
Second algebraic look at gamma in terms of delta, 370
An incidental second look at alpha also, 371
An already established fact not to be overlooked, 372
The helix angles θ at the opposite profiles of a tooth, 372
The various triads, 375
The helix angles in Beveloid gearing, 375
Aspects of the motion of the real rack cutter, 376
Some misconceptions that must be rejected, 379
Rack velocity, 380
The analogous oblique-rolling vehicular wheel, 381
How the rack moves to cut the teeth of a wheel, 382

A mating pair of naked wheels and its phantom rack, 383
Finding precisely the location and pitch of ISA_{24}, 386
Finding precisely the location and pitch of ISA_{34}, 387
Some well known but misleading phenomena, 387
The angles iota, 389
The cylindroid at the meshing, 390
The other, as yet unmentioned cylindroids, 393
Cutting velocity across the flanks of the rack, 393
A confirmation by means of inversion, 395
Application here of the theorems at chapter 9B, 397
Speeds and feeds at the machine tool, 401
Overall strategy of the found mathematics, 402
Appendix A at chapter 10, 404

CHAPTER 11
ASPECTS OF THE PHYSICAL REALITY

Contents, 407
Interference at teeth and among teeth, 407
Kinetostatic evaluation of the result at WkEx#2, 412
Commenting more widely now but inconclusively, 416
Ideas about the envelope, 416
A worked example previously tried but discontinued, 417
The slip track locations matrix, 426
The closed envelope of kinematic possibilities, 427
An exercise for the reader, 430
Early confirmation with a first prototype, 430
Appendix A at chapter 11, 430

CHAPTER 12
THE SCREW SYSTEMS AND GEAR DESIGN

Explanation, 435
Introduction, 435
Intersecting screw systems inhabiting the gear set, 436
The overall design strategy seen now in retrospect, 439

TABLE OF NUMERICAL RESULTS
NOTES AND REFERENCES
SYMBOLS INDEX INCLUDING LIST OF ACRONYMS
SUBJECTS AND AUTHORS INDEX ALPHABETICAL

PROLOGUE

01. This book deals in a wide open area, *the synthesis of involute teeth for general spatial gearing*. The gearing is *spatial* in the sense that the two shafts are skew. It is *general* in the sense that the zone of the meshing is not *at,* but *offset from,* the common perpendicular between the two shaft axes. The teeth are *involute* in the sense that they are not the non-involute teeth that are commonly employed to achieve curved-line contact in the so-called *hypoid* gearing of current spatial practice. The teeth accordingly relate, not merely to the well known planar involute which determines the kinematic foundation for ordinary spur gears, but to the *involute helicoid*. This is a spatial, ruled, developable surface generated by a line moving always tangent to a given helical curve (or *helix*). And the area is *wide open* in the sense that needed new analyses and new methods of synthesis for involute teeth in gears of this kind have only begun to be explored as yet. The term *offset* is here used, not to measure the length of the common perpendicular between the axes of the two shafts, which length is the *centre distance*, but to imply that, whereas the so-called crossed-axes cylindrical helical involute gearing of the well known kind is not offset [1], the involute gearing being explored here is. The meshing is located, as has been said, not at the centre distance line (the CDL), but remote from it. In the broad sense conveyed by logo and otherwise, the gears are commonly described as being *hyperboloidal conical* [76]. But where do we find, exactly, the implied hyperboloids and the associated cones? Finding (as we shall see) that the axis of the involute helicoid at the flank of a tooth remains in all cases collinear with the axis of the relevant shaft (§9.08), there is much more to this than meets the eye. Given the required locations of the two shaft axes, the required angular velocity ratio, and the required *offset* (§5A.15), the forthcoming theory permits the synthesis of conjugate involute teeth that (a) mesh without interference, and (b) are capable of being cut by ordinary machining methods.

02. Whereas I say that most aspects of the present study are relatively new, I must explain that the origins of the general enquiry are very old. Leaving aside the earlier works of Euler (1767) and others [70], great books of the more recent past have been those by Olivier (1842) [40], Willis (1870) [12], Mc Cord (1883) [18], and Grant (1889) [94]. I should mention also Disteli who wrote a series of serious papers between 1899 and 1911 [77]. Next there were papers from Giovannozzi (1947) [66], and Poritsky and Dudley

(1948) [67]. Some recent writing in the arena of the regular text book for students may be found in Merritt (1953) [58]; his sketchy remarks are interesting; see his §10.11; this somewhat hasty, unexplored contribution to the theory underpins a substantial part of my chapter 8. Subsequently (but not consequently) came the more recent work of Beam (1954) [59].

03. The forthcoming theory and its related practice are radically unrelated to their counterparts in the already mentioned, other kind of offset skew gearing, namely the generically called *hypoid gearing*. Hypoid gearing employs offset; but that gearing is *non-involute*. Much of hypoid theory begins however with the valid assumption that the carefully chosen, easy-to-cut, but *non-involute* teeth of the one wheel, and the subsequently cut, *generated* teeth of the other, automatically ensure conjugality in the first instance. By virtue of this there is said to be *curved-line contact* between the teeth but, after modifications to deal with overconstraint, clearances, possible shaft misalignment and other such matters, this curved line is corrupted and shortened, and no claim (kinematically based) is made by designers that transmission with exact constant angular velocity ratio is achieved. Hypoid gearing involves, as I have said, a *mutual tooth generating process,* but the involute gearing dealt with in this book involves *separate generation of the teeth*.

04. The methods employed here spring directly from *screw theory* [1]. Partly for this reason, much of the terminology will appear strange to regular readers of the gear literature. Without [1], however, with its applicable concepts and established terminology, the current study could not have begun as it did, or developed as quickly as it did. The new terminology (along with its consequently new, algebraic symbolism) has been necessary for a number of compelling reasons, and these, of course, will be made clear.

05. It is in the realm of such well used terms as *pitch, pitch-point and pitch-surfaces* that the reader skilled in the areas of hypoids and worms will find some difficulty. In the present work these terms are given new, precise meanings. For example the pitch point in any gear set is here a unique point called P upon the CDL, and the pitch line is a unique line through P in a plane parallel to the pair of planes containing the two shaft axes. There are the two radii r_2 and r_3 from the shaft axes to the pitch point, and the corresponding angles ψ_2 and ψ_3 between the shafts and the pitch line. The two said radii add to the centre distance C, of course, and the two said angles add to the shaft angle Σ. This is not altogether new — see for example Grant [94], Lagutin [23] and Litvin [5] — but what may be new to some is that the pitch line, which is that line about which the two wheels *screw* with respect to one another, has a *pitch*. This pitch, called p, is measured mm/rad.

06. The pitch of a gear set, whether the shaft axes be parallel, intersecting or skew, depends solely upon the centre distance C, the shaft angle Σ, the tooth ratio k and the relative directions of rotation of the wheels. We say for convenience that a screw of the said pitch *inhabits* the pitch line. *Screw* may be used as a verb, or as a noun. We may also say that the pitch line is, con-

tinuously, the *instantaneous screw axis* for the relative motion of the two wheels. We thus see that the pitch at the pitch line of an ordinary spur gear set with parallel axes, or an ordinary bevel-gear set with intersecting axes, will always be zero. The pitch on the other hand of a pair of crossed helicals, or the pitch of a pair of offset skew gears (such as we find for example in hypoid sets), will always be finite, either positive or negative.

07. Concepts in the region of the meshing that currently go by the names for examples of *mean contact point, pitch point, pitch surface and pairs of pitch surfaces* are avoided in the present work. These somewhat unsatisfactory concepts and terms are found to be not only often ill-defined by most accounts, but also inappropriate in the different circumstances of offset skew involute theory. Such points and surfaces might be said to relate to pitch when this term is seen to relate to the various kinds of length or angle measure that might be used between successive teeth of the mating wheels, but they bear no relation to pitch when this is seen as the measure of the fundamental screw inhabiting the pitch line. Parts of this book re-examine some of these older terms, and others, of course, take a careful look at some of the new.

08. I wish to mention here that there is already produced a gearing called by its makers *Beveloid* [59] [60] [61]. This gearing consists essentially of separately cut, separate wheels that are cut to mesh at various unusual angles with existing spur wheels, existing cylindrical helicals, or with one another in such a way that actual movement of the *slip tracks* — this is my term (§5B.26) — can and does occur. With Beveloids, however, or, as they are generically called, *taper gears* [61], the actual production work has been in advance of theory, which is, of course, often the case. It appears to have gone from the found geometry of the phantom rack and newly invented cutting machines towards various *discovered meshings*. It appears not to have gone from (a) a required layout of the shaft axes, and (b) a given location for the meshing, towards calculated details of the necessary cutting. Beveloid gearing, a special case of the *general spatial involute gearing* discussed in this book, is considered in some detail at the chapters 5, 6 and elsewhere.

09. Please be aware that at the time of writing few prototypes involving purposive general examples of the offset skew involute gearing have been built and even fewer tested, and much remains to be done. None of this book is meant to mean that the existing kinds of involute and non-involute gearing is becoming obsolete. It is however meant to mean that, if the theory outlined here is critically examined, corrected if necessary, and further developed, there will be real chances in the future that, after optimization to maximize benign contact and studies to improve our understanding of elastic deformation, strain energy, hydrodynamic lubrication and so on, we can further improve the theory and practice of offset skew gears in general.

SYNOPSIS

01. In gear sets the acronym CDL stands for the centre distance line; this is the line of the shortest distance between the axes of the two rotating shafts. All gears sets of whatever kind include not only those whose axes are skew, but also those wherein the zone of the meshing is remote from the CDL. Chapter 1 deals with those parts of gear theory that apply to all gears of whatever kind. It isolates the main kinematic parameters and relates these with one another. They are C the centre distance namely the shortest distance between the axes of the two rotating shafts, Σ the angle between the said axes, k the angular velocity ratio (k being less than unity), and, derived from these, p the pitch at the pitch line. The pitch line (the axis about which each wheel screws with respect to the other) is explained and located; this line cuts the CDL perpendicularly and thus lies in a plane parallel with the unique pair of parallel planes that contain the two shaft axes. The magnitude of p (mm/rad) is worked out and presented algebraically. Both the pitch line and its pitch are openly held to be (and later shown to be) of fundamental importance in design. The *offset*, which is roughly speaking the radial distance from the CDL to the zone of the meshing, lacks a precise definition at this early stage. It is, however, not yet relevant.

02. Chapter 2 explains and proves, in three dimensions, the fundamental law of gearing [14]. This law appears not to have been known. The law states that, for constant angular velocity ratio k and as we go from tooth to tooth across teeth, the contact normal must at all times pass through the fixed space (namely the space of the gear box) in such a way that $q \tan \phi$ remains a constant, namely p. Refer to figure 2.01. Distance q and angle ϕ are the shortest distance and the angle respectively between the contact normal and the pitch line. A special case of the said law (which is here written for spatial gearing) is already well known in the simple case of planar gearing. It is well known that, for constant angular velocity ratio k and as we go from tooth to tooth across teeth in planar gearing, the contact normal must at all times pass through the pitch point [12]. The implications of the new law for both involute and non-involute spatial gearing are examined.

03. In chapter 3 the said fundamental law is used to prove that pure involute action among pure involute teeth in offset skew gearing is not only possible but practical. The bounded surface at the flank of an involute tooth

is a portion of the shape that is formed by all of the straight lines drawn to be tangent to a given helical curve (a helix), *the axis of the helix being at the axis of the wheel.* We are dealing here not in a narrow way with the so-called crossed helicals (which are of course pairs of involute wheels), but in a much wider way with offset skew involute gears in general. The theory, developed from first principles, deals with the intricacies of the geometry and concludes by establishing that offset skew involute action provides (a) constant angular velocity ratio k as we go from tooth to tooth across teeth, (b) single-point contact at the meshing but with surface curvatures there that are benign, (c) a straight path in the fixed space for each of the points of contact Q, there being separate paths for drive and coast, (d) insensitivity within limits to all kinds of errors at assembly, and (e) an opportunity for conventional methods of manufacture such as NC milling or CNC hobbing.

04. We show by way of lemma in chapter 3 that, when the shaft angle Σ becomes zero, there is a catastrophic change in the conditions at contact: single-point contact suddenly becomes line contact, and many of the fortuitous aspects of the *spatial* action disappear. The kinematics become planar, teeth become spur (or helical), the mechanism becomes overconstrained, and some form of *crowning*, which provides a combination of tip and end relief (at the expense of course of kinematic exactitude), becomes necessary.

05. Given point contact, and a point of contact Q, there are three different paths being taken by Q: in the space (link 1) of the gear box itself; in the space (link 2) of the first wheel; in the space (link 3) of the second. Whereas the path of Q in 1 is a straight line, the paths of Q in 2 and 3 are twisted curves described upon the profiles (the flanks) of the teeth. These twisted paths are called, by me, the *slip tracks.* It must now be said that, among the ordinary, non-offset crossed helicals, we have the well known tendency toward rapid wear. It does not follow from this however that the same tendency will pervade the wider range of the offset skew involute sets. In chapter 3 and the chapters 5 the alleged inevitable tendency towards rapid wear is shown to be a fallacy. There are many factors at work. There is not only (a) the mere values of the curvatures of the mating surfaces at the point of contact, but also (b) the question of whether or not the sliding velocity reverses itself during passage across the contact zone (as it does for example in planar gears), and (c) the question of whether or not the two relevant rulings of the mating involute helicoids and the sliding velocity at the point of contact cross one another satisfactorily (as they do for example in skew gears with adequate offset). I wish to remark here (and I speak very roughly) that, whereas due to elasticity the actual *contact patch* in modified hypoid gearing is somewhat *cucumber shaped in silhouette*, the contact patch in offset skew involute gearing (whether only mildly skew or indeed fully *square*) is more rounded, more *circular shaped in silhouette*. The implications of these phenomena for wear and lubrication warrant study.

06. Chapter 4 clarifies the many questions that abound in connection with the various kinds of pairs of hyperboloids that intersect along a common generator. It deals in the abstract, but provides the guidelines necessary for a smooth transition of the work from the analyses which precede this chapter to the syntheses which follow it. The different kinds of pairs of hyperboloids that share a common generator in gears are (a) the *axodes*, which are innate, (b) the *base hyperboloids*, of which there are two pairs obtained by swinging the two straight paths of Q, one for drive and one for coast, about the shaft axes, and (c) the *naked wheels*, these being selected narrow slices of the so called *doubly departed sheets* whose generators originate in the direction of the sliding velocity at the point of contact. The naked wheels are imaginary, scaffolding-like constructions upon which the actual teeth are mounted.

07. The chapters 5 (A and B) deal with the synthesis of offset skew involute teeth in the relatively straightforward realm of *equiangular architecture*. The E-line is a special line which, parallel with the pitch line, cuts the CDL perpendicularly at a special point D. Whereas the pitch point P divides the centre distance C according to a complicated formula involving both k and Σ, the point D divides C quite simply according to k. Chapter 5A explains that, if we put a selected intersection point [Qx] for the two paths of the two Q anywhere upon the E-line, we establish equiangularity. We find that equiangularity means in effect that the asymptotic cones of the doubly departed sheets are equiangular. Chapter 5B includes two worked examples, WkEx#1 and WkEx#2. *These show that a wide range of machinable equiangular offset involute skew gear sets are available, even to the extent of there being square sets where the shaft angle Σ becomes right angular.* Chapter 5B also opens discussion about the *ellipses of obliquity*. These involve a new concept which allows intelligent choices to be made for the two angles δ of obliquity, These angles are required to be markedly different whenever Σ is chosen large and k is chosen small.

08. Chapter 6 deals with synthesis of offset skew involute teeth in the more difficult realm of *plain polyangular architecture*, where the intersection point of the paths [Qx] may be put, not exclusively upon the E-line, but anywhere upon the so-called F-surface. There is a so-called F-axis line (FAXL) which cuts the CDL at its mid point O, bisects the shaft angle Σ, and resides in a plane through O parallel with the parallel planes containing the two shaft axes. The F-surface (which contains the E-line) is a *parabolic hyperboloid* [69] whose principal axes are the CDL and the FAXL (§4.70). The F-surface contains also the two shaft axes. Two more worked examples, WkEx#3 and WkEx#4, are extensions respectively of those pursued in chapter 5B. Both within and beyond the two shaft axes, the F-surface extends into space as far as we may wish; and it has indeed two branches; but in knowing this we must remember that internal meshing with involute gearing is only possible (except in special cases) when the shaft axes are put to be parallel. The F-surface makes it clear, in any event, that we can take a closer look, whenever we

wish, at square (and nearly-square) polyangular sets, involute spiroid gears and offset involute worms. These are all part of one gigantic family.

09. Chapter 7 deals in effect with the edges of the windows of opportunity. As we go further and further into the realm of polyangularity, that is as we go, in our design work, further and further away from the relative safety of the E-line, we begin to encounter the usual, not unexpected difficulties: we run into the need for changing the angles of obliquity, increasing problems of friction (non back driving), diminishment below unity of the contact ratio unless we are careful, undercut, and pointed teeth. In this chapter, two worked examples deal with worms. At WkEx#5 a conventional worm drive is erected upon the same fundamental principles as in all other involute gears, and, at WkEx#6, a spiroid involute set is found to be interesting, not only in this universal respect, but in its own special way as well.

10. In chapter 8, wider possibilities are mentioned. It is not necessary that the intersection point [Qx] be put exactly upon the F-surface; it is simply relatively safe and convenient to do so. *Nor is it necessary that there be an intersection point [Qx] at all.* It is entirely legitimate (given the fundamental law) to choose two legitimate paths, one for drive and one for coast, almost anywhere in the architectural space, and for these to be, if we choose, skew with one another. If we sailed off in this direction, however, we would sail with primitive charts into uncertain seas. We would encounter *exoticism*. We would almost certainly find the need for two different rack generating procedures for cutting the opposite flanks of teeth, and other uncertainties would abound. Among the interesting exotics that might be found, however, are some of the works of the well remembered Olivier and some of the *ad hoc*, non-straight, but matching modern Beveloids.

11. In chapter 9A we look at the inherent *helicality* of the involute helicoidal teeth which obtains in any event. The active surfaces, namely the flanks (or the profiles) of the teeth, are not only (a) *ruled*, there being a continuum of straight lines upon the flanks, not only (b) generated by a planar involute of fixed base circle screwing steadily about the shaft axis while remaining within a plane moving normal to the that axis, but also (c) *helicoidal.* This latter means that the tooth profiles (however they might be truncated conically) are akin to ordinary screw threads coaxial with the axis of the wheel. The pitches of the flanks at the fronts and the backs of the teeth of a wheel are different however, and this means there are *two* such helicoidal surfaces on any given wheel, each with a different pitch. This phenomenon allows us physically to adjust the relative axial location of the wheels upon their shafts, and thus, in the absence of a variable centre distance, *to adjust the extent of backlash*. This facility is not available among any of the offset skew non-involute gears, the hypoids for example.

12. In chapter 9B we state and prove some very important theorems to do with the mechanical generation of the profiles of the teeth that relate not only to the material of the previous chapter 9A but also and more impor-

tantly to the required procedures for the actual machining — the hobbing, milling, grinding etc. that may ordinarily be employed. We distinguish between the *phantom rack,* a remote geometrical abstraction, and the mere *imaginary rack,* which, although an abstraction also, is somewhat less remote from the actual mechanical reality of the chosen cutting tools and the teeth.

13. Collectively the chapters 9 teach us incidentally to see that (a) the bevel sets of the ordinary Beveloid kind mentioned at §P.08, (b) the better organized offset involute sets studied in chapters 6 and 7 mentioned at §S.08, and (c) the well known parallel cylindrical helicals of ordinary engineering practice, are not qualitatively different in so far as the shapes of the flanks of the teeth are concerned. While the flanks of the teeth in all of these cases remain regular involute helicoids, whose axes coincide as usual with the axes of the wheels, it is the pitches of the helicoids of the opposing flanks (which differ in general in both magnitude and handedness) and the question of which branch of a particular helicoid is accepted as real and which unreal that make the apparent difference. In the first of these, the straight Beveloids where the axes nominally intersect, and in the second where the axes are nominally skew, the above remarks about backlash remain applicable, but in the third, the parallel helicals where the axes are nominally parallel (and these include the ordinary involute spur gears), the remarks about backlash do not apply. Plain involute spur gear sets are clearly special (but accessible) cases of the overall theory being expounded here. *The straight bevel gear sets of ordinary current practice (and their derivatives such as the so called spiral bevels), however, which are non-involute for reasons made clear at §5B.62, have no logical place in this book.*

14. In chapter 10 we use a mixture of methods to deal with the mathematical aspects of the manufacture by machining of general involute gears. In the design and cutting of ordinary planar spur gears we customarily put the two angles of obliquity δ (the pressure angles) to be equal, and we find consequently that the same two angles δ at the two straight sides of the required hob-cutter are equal to the angles δ and to one another. *In the general, offset skew involute gearing, however, these simplicities do not apply. The angles δ at the straight sides of the phantom rack (let these be measured in the plane normal to the rack) do not equal the angles γ measured at the sloping sides of the cut teeth (letting these be measured in the plane normal to the axis of the wheel).* Refer to figure 10.01. If, in design, we set the two angles δ to be equal, the angles γ will not be equal to the angles δ, and will not be equal to one another. Although with equiangular architecture and with equal angles δ there are for the given δ two sets of two equal angles γ, there are with polyangular architecture, both with and without equal angles δ, *four* different angles γ. We find, furthermore, that a similar set of circumstances applies for the helix angles θ at the teeth. These are the helix angles of the helices cut upon the profiles by the cylinder containing the F-circle, which circle is the key circle of the naked wheel.

15. Exploring and explaining such matters, we expose some remarkable relationships within the overall geometry. We employ the angles τ, κ, and λ. Angle τ is the half angle at the apex of the asymptotic cone associated with the hyperboloidal surface of the naked wheel, angle κ is the half angle at the apex of the Wildhaber cone (which comes intact from the hypoid literature), and λ is an important angle relating to the *swivel* of the teeth at the naked wheel; refer to figure 5B.02(b). The angles λ and κ turn out to be the *asymuth* and *altitude* angles associated with the shaft axis and the phantom rack in machining; see figure 10.08(a). Each of these angles reflect themselves in the overall algebra, which algebra — simple, but somehow deceptively so — lies at the root of computer programmes that will need to be written for overall design and fast optimization. *Above all, in chapter 10, we clarify and then confirm that the teeth of both wheels in the case of general spatial involute gearing can be cut by the regular machining methods geometrically summarized by the straight sided imaginary, generating rack.* The rack, however, does not simply roll in a parallel manner upon its fixed cylindrical axode as it does for example in generating the teeth of planar spur gear wheels. Inclined at the mentioned angles λ and κ, it screws (with a calculable pitch) with respect to the cylindrical fixed axode, the ISA remaining as before parallel with the axis of the wheel. Unless the teeth are exotic (chapter 8), in which case we need to employ two racks, the said rack generates both flanks of the teeth at the one single pass.

16. As might be expected overall, the algebra breaks down to that for the simpler cases; it breaks down, for example, step by step and in a satisfactory way to the single helix angle θ which is to be found in ordinary parallel-axes cylindrical helical sets. The mathematics shows moreover why that single angle θ, unlike the two pitch radii in such sets, may take any value. In general the angles δ, γ and θ are intimately related geometrically. They are describable by means of simple expressible formulae; they each relate, of course, to the overall architecture of the set; and the algebra of the interrelationships is intriguing. Especially so is its implications in the area of involute worm sets and spiroids where debate has occurred about the proper shapes of cutters.

17. In chapter 11, under a heading contrived to suit, we look in some detail at a worked example where two troublesome, but not unusual matters are thrown up for consideration. They are the questions of how to avoid (a) the occurrence of contacts ratio less than unity, which is associated with the mutual decapitation of the enclosing flanks of individual teeth, and (b) the occurrence of unavoidable interference between the mating teeth in mesh. These lead to a discussion about an ill-defined *envelope of practical possibilities*, within which we need to keep for fear of physical catastrophe. The envelope divides, mysteriously, that which is kinematically possible from that which is not. These matters are not peculiar to involute gearing; they are inherent in all gearing; and it is with this undisputed fact in mind that this exploratory but inconclusive investigation is undertaken.

SYNOPSIS   §S.18

18. In chapter 11, also, current report is made about the building of an early prototype. The young researcher Michael Killeen, working under the supervision of myself, John Gal and Friedjhof Sticher (who were themselves, earlier, doctoral students of mine), has been of assistance in various ways. As well as making important contributions to the needed algebra (reported and acknowledged within the text at chapters 5 and 6), he has constructed the first purposive prototype. Refer to §11.30 for an account of this, and see the apparatus itself at the figures 11.08(a) and (b).

19. Chapter 12 deals with two difficult matters. Each of them is an overall aspect obtaining wherever we look. Firstly we have the logically consistent, self contained, interwoven collection of geometric concepts that make up the kinetostatics of the involute gear set when seen from the general points of view of the line systems of classical line geometry and the related screw systems of modern screw theory; this is a seamless study that deals in the local space and the outer limits at infinity at the same time. But secondly there is an attempt, in the latter part of chapter 12, to give a coherent account in a few words of the overall *design strategy* of this book.

20. The work contains 175,000 words and 160 computer constructed figures. It cites about 100 references, carries a number of appendices dealing with peripheral matters, and there is a table of numerical results. There is an index of acronyms, another of algebraic symbols, and a subject and author index referring the reader, progressively by means of paragraph number, into the listed aspects. Reference [1], often cited because of its direct relevance, refers to the author's earlier work, *Freedom in Machinery*, in two volumes, Cambridge University Press, 1984, 1990.

CHAPTER ONE
BASIC KINEMATICS
OF THE GEAR SET

Preliminary remarks

01. The term *gear set*, in gears, includes the box (namely the box of the gear set) as well as the two toothed wheels that mesh with one another. Accordingly there are in a gear set, thus defined, three links, the box (or the frame) 1, the driving wheel (or the driver) 2, and the driven wheel (or the follower) 3. This division of the mechanism into three links applies in whatever way the shafts are arranged, whether they be parallel, intersecting or skew, and however complicated the meshing may be. At a first approximation, in mechanism, links are assumed rigid. The three links form in any event a 3-link loop [1] [§1.15], and the *apparent mobility* of the loop is unity [1] [§2.07]. Look, however, at an ordinary spur gear set with ordinary parallel axes and notice that, because there is held to be line-contact between the teeth, the mechanism is not simply *constrained*, but *overconstrained* [1] [§2.18]; and this of course carries the consequence that a high level of accuracy is required in both the manufacture and the assembly of such sets.

02. Nearly always in machinery the function of a gear set is to transmit from one wheel to the other a continuous, steady, rotary motion in such a way (a) that sufficient power flows from wheel to wheel without excessive stress, (b) that there is some known *speed ratio* between the wheels, and (c) that, as the teeth go into and out of mesh with one another, there is *no pulsation, no periodic fluctuation of the instantaneous angular velocity ratio*. While the speed ratio of a gear set is determined wholly and exactly by the ratio of the numbers of teeth upon the wheels, the presence or absence of pulsation is wholly determined (in the absence of elasticity) by the shapes of the profiles of the teeth.

03. In the wide and very complicated area of gear set design we have the mechanics of the steady transmission of power. We have the periodically changing relative motions at the teeth, the changing forces at work at the changing contact patch or patches [2], the mechanical properties of the materials of the teeth, and the problems of achieving adequate hydrodynamic, wedge-film lubrication (HDL). These are some of the critical matters at issue.

Many of them are often outweighed for the designer, however, by the necessity for accepting whatever standard sets of tooth cutters are available, by the wholeness of whole numbers, and by the limited methods available for generating or otherwise cutting the correct shapes for the profiles of the teeth. Overall it is held that, in designing a gear set, we should aim our attention at the three following matters in the listed order of priority: *the likelihood of pitting* — this may occur at the tooth surfaces due to a too heavy concentration of the contact pressures across the relatively small contact patches involved; *the strength of the teeth in bending* — if this is inadequate the teeth will bend or break off; *the occurrence or otherwise of scoring* — scoring at the profiles of the teeth is related with the relative sliding which is inevitable there, and with the absence or otherwise of adequate lubrication.

04. Mentioned only fleetingly above is the sometimes overlooked matter, *the correct shapes of the profiles of the teeth*, and in these few words is hidden, of course, the most important matter. The shapes of the profiles of the teeth must be matched in such a way that they remain, in mating throughout the meshing, *conjugate with one another*. They must be shaped in such a way, in other words, that all pulsation be absent. In conformation with this requirement there are, as we know, the older *cycloidal* and the newer *involute* schemes for achieving conjugality in planar gears, namely in those gears where the shaft axes are parallel; and there are, as we know, the multitudinous other schemes (and these are neither cycloidal nor involute) for achieving approximate conjugality in (a) those gears where the shaft-axes are skew and the meshing offset (as in most of the *hyperboloidals* for example), and (b) those gears where the shaft axes are intersecting and the meshing offset (as in all of the so called *straight-bevels* and *spiral-bevels* for example). All of this, with its extensive literature, points to the overwhelming importance of the fundamental kinematics. We go now to look at that.

Kinematics and the pitch line

05. Refer in the mind's eye to any gear set. Let us assume that the links of the gear set are rigid, that contact is being made at all relevant points within the set, and that we are dealing solely with the circumstance of *drive* [3]. The three links are connected into a single, simple loop by means of three *joints* [1] [§1.11]. The joints are, taken in turn around the loop, an f1-revolute R at the input shaft, an f5-joint with its single point of contact (called Q) between the neighbouring flanks of the mating teeth, and another f1-revolute R at the output shaft [1] [§1.25]. The point of contact Q between the mating teeth is an important point as we shall see, and the implied conclusion here is that there is but a single point at Q. This is a consequence of the simplifying assumptions that have been made. These, which reject the possibility of flexure, are saying, in effect, not only that the links are rigid, but also that, due to the inevitability of irregularity [1] [§6.01], there can be no way to assemble such a mechanism otherwise.

BASIC KINEMATICS OF THE GEAR SET §1.06

06. Refer to figure 1.01. This is a semi-pictorial figure showing the three links or bodies of a generalised offset skew gear set and the joints of its kinematic chain (§1.01). The ends A and B of the unique shortest distance between the two shaft-axes are indicated. The segment (A–B) is called the *centre distance*; its length is called *C*; and the whole of the line itself is called *the centre distance line* (the CDL). A carefully chosen origin of coordinates O bisects the centre distance, and an axis O-z is mounted collinear with it. The body of the gear box (link-1) is represented by (a) the shown, stationary, block-like structure at the left of the figure, and/or (b) the unique pair of parallel planes at A and B which contain respectively the two shaft axes. It will be convenient throughout this book to employ the convention that, while the CDL is held to be *vertical*, the planes at A and B and the two shaft axes,

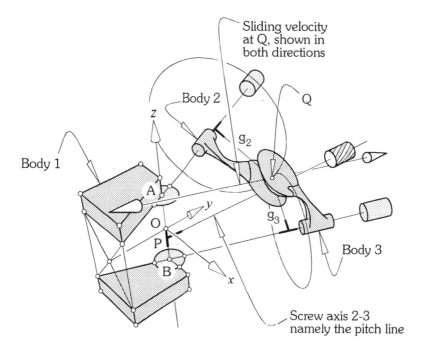

Figure 1.01(§1.06). This figure is not simply a mere sketch. It shows the members of the closed-loop, 3-link, kinematic chain of the offset skew gear set (§1.06), but it shows also, by virtue of its carefully drawn dimensions, two inescapable relationships that exist between the variables. These are (a) that the sliding velocity at Q can occur only within the common tangent plane to the curved profiles touching at Q (§2.16), and (b) that the said velocity (which is indicated and may be determined) must belong as a member of the helicoidal field of velocity vectors surrounding the pitch line (§1.38). We see thereby that the generally disposed profiles mating at Q and the location of and the pitch at the pitch line soon to be mentioned are inextricably interwoven (§2.06). The screwing at the pitch line is left handed. See also figure 1.08; it shows some detail occluded here.

§1.07   GENERAL SPATIAL INVOLUTE GEARING

accordingly, are held to be *horizontal*. Measured axially around the CDL and between the two shaft axes is the *shaft angle* Σ. Whereas origin O bisects the centre distance, axis O-y bisects Σ, and O-x is drawn accordingly. The block-like structure is oriented square with the triad O-xyz. Figure 1.01 (like many others in this book) is generally representing the algebraic relationships at issue, but carefully drawn to scale by computer graphics [25]. Most of the figures are however in perspective, so direct scaling from them is useless. Numerical values so far employed are: $C = 80$ mm, and $\Sigma = 50°$.

07. The contacting flanks of a pair of generalised teeth are shown in figure 1.01. These may be seen as the backs of two smooth, but generally shaped spoons, fixed rigidly and respectively to the shafts. The spoons are in contact back-to-back at Q. The contact normal at Q is not shown; this and other related matters are discussed at §1.13. See also §2.02, and glance in advance at the gear bodies (§3.35). Point Q (which moves of course) is located anywhere in the fixed space, and it remains (generally speaking) remote from the CDL. In figure 1.01 the angular velocity ratio k of the imagined gear set is of course not constant. It depends upon the configuration (namely the relative shapes and locations of the links). In other words the shapes and locations of the spoons (which are arbitrarily chosen) do not automatically lead to conjugality in the circumstances here. At the instant depicted, however, the ratio k is set to be 0.6; the speed ratio, in other words, is 3/5. Notice the two circles intersecting at Q. They are coaxial respectively with the two shaft axes. These will later be called the G-circles at that Q (§3.37), but in the meantime they may be seen as simply representing the pair of gear wheels that might be imagined. The ratio of their radii g_3/g_2 is not equal to k. Please read the caption for further explanation, and be aware that certain unmentioned matters and occluded detail in figure 1.01 are dealt with later at figure 1.08. There are inescapable relationships that exist between the variables, and these require explanation.

08. We have, here, three rigid bodies in relative motion, and the theorem of three axes in kinematics [1] [§13.12] is applicable. The theorem says among other things that, independently of whether the three bodies are connected in any way or are entirely free, the three instantaneous screw axes, which always exist at the instant between the pairs of links 1-2, 2-3 and 3-1, must share the same line as common perpendicular. The three instantaneous screw axes (variously called the ISAs, the screw axes, or, simply, the screws), each with its pitch (mm/rad), must, in other words, all cut the same line perpendicularly. The screw axis 1-2 is by inspection along the fixed axis of the rotating input shaft, and its pitch by inspection is zero. The screw axis 3-1 is similarly along the fixed axis of the output shaft, and its pitch is also zero. Thus we know already by means of the theorem that the screw axis 2-3, namely the axis about which the wheel 3 screws with respect to wheel 2 (or vice versa), must be (a) of non-zero pitch, and (b) cutting the CDL somewhere perpendicularly. Taking $k = 0.6$, this screw has been located by formula (§1.14) and inserted correctly to scale (at the shown point P) into the

BASIC KINEMATICS OF THE GEAR SET §1.09

figure here. Other writers such as Steeds [4] [§189], and for example Litvin [5] [§3.5], have located the same screw axis 2-3, by using other methods. Litvin calls the axis the axis of screw motion [5] [§3.51], remarking in passing that it has not, so far, been used effectively in design. He accordingly speaks instead about the operating pitch surfaces for gears with crossed axes [5] [§3.6], and these are, incidentally, the pitch cones of Wildhaber [6].

09. I say conversely that the said screw axis lies in a useful way at the root of the kinematics. It is a central theoretical concept of overwhelming importance in design. I call it the *pitch line*, and I call the point P, at its intersection with the CDL, the *pitch point* (§1.14). In the figures 1.08, which enlarge upon certain details occluded in figure 1.01, one can see how the perpendicular dropped from Q onto the pitch line meets that line at N. Denoting the directed distance (N–Q) by h (mm), denoting the relative angular velocity ω_{23} by ω (rad/sec), and calling the pitch at the pitch line p (mm/rad), the sliding velocity at Q, v (mm/sec), may be written $v = p\omega + (\omega \times h)$. Find an independent examination of this important relationship at [1] [§5.44 *et seq.*]. This and its widespread ramifications will be examined again (§1.51).

10. The location and pitch of the pitch line depends upon the centre distance C, the angle Σ and the ratio k obtaining at the instant. Formulae relating these variables appear at §1.14. But the ratio k is possessed of sign. If $|k|$ is held, for the sake of argument, to be constant, as it is (generally speaking) in gears, there are, for a given direction of rotation of the driving wheel, two different directions for the driven. These relate to $+k$ and $-k$. Be aware of chapter 22 in [1], and refer in advance to figure 1.02. In the planar case of ordinary planar spur gears, the two signs of k relate, quite simply, to the mutually alternative external and internal meshing, but in the case of skew gears the relationship is more complicated.

11. Figure 1.02 shows some aspects of the geometry of the cylindroid. The cylindroid relates to the theorem of three axes [1]. It may be seen to determine in a clear geometrical way the two alternative screws 2-3, namely the two pitch lines (marked * and †) which in turn reflect the two signs (±) of k. Figure 1.02 is a re-drawing of figure 22.03 in [1], except that for this chapter new dimensions are employed [7]. [Refer just now if you wish to the table of numerical results (TNR) and check the details of the first of the worked exercises (WkEx#1) listed there; its dimensions are here being used for the sake of convenience.] Ratio k is taken, as in [1], to be ±0.6. Looking at figure 1.02, we can see (without a worked calculation) that, in a skew gear set, there are two possibilities for the location of the pitch line and for the pitch of the screw upon it. They are (a) the most commonly used pitch line marked here star (*) where k is positive and where the angular location is within the acute angle Σ and where $|p|$ is relatively low, p being -17.25207 mm/rad (left handed) in this case, and (b) the less commonly used pitch line marked dagger (†) where k is negative, where the angular location is within the obtuse angle Σ, where $|p|$ is relatively high, p being $+62.39999$ mm/rad

§1.12   GENERAL SPATIAL INVOLUTE GEARING

(right handed) in this case, and where the pitch line dagger (†) actually lies beyond the confines of the two shaft axes. For various reasons not explained as yet, but later (§1.23), the pitch line star (*), equipped like the other screws with a knurled knob, is the one that I shall pay attention to exclusively until further notice. I should explain incidentally that, whereas the incipient motion suggested in figure 1.01 is to be seen as being instantaneous, the smooth motion suggested in figure 1.02 is to be seen as being continuous.

12. The knurled knobs in the figures are of the same size. This accentuates the fact that, kinematically speaking, they are of equal importance. The knobs upon the pitch lines, however, are decorated with helical markings. These markings, drawn to scale, indicate the pitches at these pitch lines. Let the knobs in figure 1.01 each be connectable (by means of some suitable, switchable, clutch-like device) to either one of the two links whose relative motion forms the relevant axis. Please accept the following instructions. Physically hold the frame 1 fixed, connect the knob at 1-2 to the wheel 2, and rotate that wheel. Due to the revolute R at 1-2, the wheel 2 under these

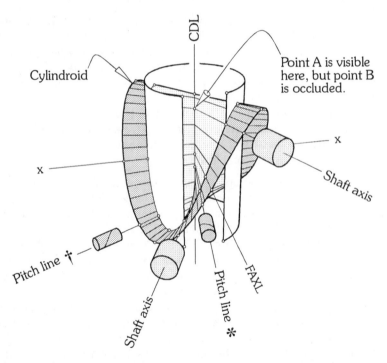

Figure 1.02 (§1.11). The relevance of the cylindroid is illustrated here. Given C and Σ, the fixed cylindroid for the gear set is determined. For a given angular velocity ratio $|k|$, there are two different pitch lines to be found upon the cylindroid. Which one is which depends upon the sign of k (namely the relative directions of the rotations of the wheels). At pitch line (*) the axodes engage in external rolling and the circumstances are benign, but at pitch line dagger (†) practical design becomes more difficult.

BASIC KINEMATICS OF THE GEAR SET §1.13

circumstances will be rotated about the axis 1-2 without axial translation, and the axis itself will, incidentally, remain stationary. This will cause wheel 3 to rotate about the revolute R at the axis 3-1, and the gear set will appear to behave in the usual way, with we the observers attached at link 1. Go now if you wish to figure 5B.01(a); its Σ and its k conform with those of figure 1.01; it may help with the visualization needed here. Next, however, physically hold the wheel 2 fixed, connect (by means of the clutch) the knob at 2-3 to the wheel 3, and rotate that wheel. Due to the kinematic constraints within the loop (whose overall mobility is unity), the wheel 3 under these circumstances will be rotated about the relevant axis 2-3 with axial translation, and the axis itself, incidentally, will not remain stationary. This will cause the frame 1 to rotate with zero pitch about the revolute at axis 1-2, and the gear set will appear to behave unusually. But this is not anomalous. It seems odd simply because we the observers are attached, not as before at the frame 1, but now at the wheel 2. This changing of the frame of reference from link to link in spatial mechanism is called, as it is in planar mechanism, kinematic inversion [8], and all imagined visual phenomena associated with such inversions in the gear set (which is inescapably spatial) need to be though about here [1] [§13.23]. See figures 1.08. See also figure 10.11(a), which is not very clear, but which could be improved, to show the movement of some moving teeth on a moving rack relative to some mating teeth on a fixed wheel. Given the data above, we might note that, for every radian turned by wheel 3 relative to wheel 2, it translates in the direction of the pitch line a distance of minus 17.25207 mm. The negative sign means that wheel 3 screws left handedly with respect to wheel 2, and vice versa. Note the left handed helices drawn at the surface of the pitch-line knob in figure 1.01, and note the compatible direction of the sliding velocity v drawn at Q. See also the same sliding velocity v at the figures 1.08.

On the dichotomy synthesis-v-analysis

13. This paragraph might be omitted upon a first reading. A question arises that must surely have nagged the reader: if we were given complete dimensions for separately building the links of the apparatus depicted in figure 1.01, and if we were either prohibited from, or unable to build the actual apparatus to measure its characteristics physically, how could we calculate (graphically or otherwise) the sliding velocity at Q which is shown in figure 1.01, how could we discover the angular velocity ratio k at the various configurations of the apparatus, and how, correspondingly, could we determine the changing location of the pitch line and its changing pitch as the apparatus engaged in its motion? The trouble is that we have no immediately useful data. How do we begin? The answer is this: we must, first of all, either explicitly or implicitly, write a closure equation [1] [§19.47-51]. To do this we would need to nominate input and output angles, say θ and δ (these symbols, which come from the relevant passage in [1], are otherwise used both in other passages in [1] and in this book), and write the equation $\delta = f(\theta)$,

which equation would employ, of course, the fixed variables (the relevant dimensions) of the apparatus. But this, of course, is not an easy thing to do. The writing of such equations is notoriously difficult, and the algebra involved will not be contemplated here. What we can do here however, and I do it, is to construct (by trial if necessary) an accurate drawing, whereby the closure equation (at least for the instant under consideration) is written automatically albeit implicitly. Were we to construct the actual apparatus (or to animate the drawing), we would construct, indeed, a mechanical computer, which could give input-output information across the whole range of the available motion. But such building of an apparatus would defeat the object of the exercise; for the built apparatus would almost certainly not be the optimum we are seeking. In the present instance (given the accurate drawing) we could (a) locate the point Q, (b) locate the common tangent plane to the mating profiles there, and thus (c) locate the contact normal. Appealing to the basic equation of meshing (§2.32), we could next see that the sliding velocity at Q must be perpendicular to the contact normal. Taking unit angular velocity ω_{12} at the input shaft we could next discover the unit linear velocity v_{1Q2} at Q [1] [§12.08], and construct its component along the contact normal. The component of the corresponding velocity v_{1Q3} along the same contact normal must be equal (for otherwise contact would be lost or the profiles would dig-in), and by the same token we could thus find the whole value of the vector v_{1Q3}. Thence we could determine the angular velocity ω_{31} at the output shaft, and thus the ratio k. Using this k and the appropriate formulae selected from the list at §1.14 (which were obtained, of course, by analysis), we could locate the pitch line. In figure 1.01, I did all of this in reverse. I selected a value for k ($k = 0.6$), located the pitch line by using the formulae, located a point Q somewhere in the fixed space, located the polar plane (§3.47) that must exist at Q, and thus (given the pitch p at the pitch line) the direction of the sliding velocity at Q. Then I drew the spoons. We also have in the figure, if we wish, the sliding velocity by vector difference, there being redundant pieces of information.

Algebraic formulae for the pitch line

14. The formulae that follow relate to pairs of generalised mating gear teeth as illustrated in figure 1.01. They are thus quite general. Accordingly they apply to pairs of actual gear wheels, where, as usual, by one means or another, a constant k has been pursued. The theorem of three axes [1] [§13.12] is a most convenient key (but not the only key) to the necessary calculations concerning the pitch line and its pitch. Although it has been employed in those calculations here that come directly from [1], it is not an overt feature of the whole paragraph. The three relatively moving links, as has been said, are (a) the frame 1, (b) the driver 2, and (c) the driven 3. The radii r_2 and r_3 of the wheels measured along the CDL from the shaft axes to the pitch line sum to C, but the ratio r_2/r_3 is found to be a function not only of k but of Σ also. The angular location of the pitch line in the plane parallel with

BASIC KINEMATICS OF THE GEAR SET   §1.14

the parallel planes containing the shaft axes is governed by the simply drawn, triangular polygon of angular velocities [1] [§12.12], but the ratio ψ_2/ψ_3 of the relevant angles which sum to Σ is similarly found to bear no obvious relation to k. I refer now to some algebraic formulae proved in [1] and extracted from there. They are listed here, and commented upon at §1.15. These groups of formulae for locating the pitch line were separately arrived at by their respective authors, Phillips [1], Konstantinov et al. [9], and Steeds [4]. They involve redundancies and are much less formidable than they look. Not surprisingly they reconcile with one another [10]. The terminology was made uniform by the present writer and the groups are transcribed exactly from [1] [§22.24].

---

(1) $\theta = \Sigma/2 - \arctan[(k \sin \Sigma)/(1 + k \cos \Sigma)]$
(2) $r_2 = C/2\,[1 - \operatorname{cosec} \Sigma \sin 2\theta]$
(3) $r_3 = C - r_2$

Phillips [1]

---

(4) $r_2 = C[(k^2 + k \cos \Sigma)/(1 + 2k \cos \Sigma + k^2)]$
(5) $r_3 = C[(1 + k \cos \Sigma)/(1 + 2k \cos \Sigma + k^2)]$

Konstantinov [9]

---

(6) $\Sigma = \psi_2 + \psi_3$
(7) $k = (\sin \psi_2)/(\sin \psi_3)$
(8) $r_2/r_3 = (\tan \psi_2)/(\tan \psi_3)$

Steeds [4]

In the now present context of a known first law (§2.06), however, we need, next, explicit formulations for r_2 (and/or r_3), ψ_2 (and/or ψ_3) and p, employing none other than the three main variables C, Σ and k. For r_2 and r_3 we have equations (4) and (5). For ψ_2 and ψ_3, and p, we have nothing direct from [1]; but for the angles ψ we have equations (6), (7) and (8), and for p we have equation (9), drawn newly from [1], along with (1). The angle θ in equation (1) relates to the axes of the cylindroid [1] [§22.20], where it plays an important role, but it may be eliminated here. So, having (6), (7), (8), we can, first,

---

(9) $p = (C/2) \cot \Sigma - (C/2) \operatorname{cosec} \Sigma \cos 2\theta$
(1) $\theta = \Sigma/2 - \arctan[(k \sin \Sigma)/(1 + k \cos \Sigma)]$

Phillips [1]

from equations (6) and (7), find (10) and (11), and then, from equations (9) and (1), find (12). The complete list of required equations is, accordingly,

$$
\begin{aligned}
(4)\quad & r_2 = C[(k^2 + k\cos\Sigma)/(1 + 2k\cos\Sigma + k^2)] \\
(5)\quad & r_3 = C[(1 + k\cos\Sigma)/(1 + 2k\cos\Sigma + k^2)] \\
(10)\quad & \psi_2 = \arctan[(k\sin\Sigma)/(1 + k\cos\Sigma)] \\
(11)\quad & \psi_3 = \arctan[(\sin\Sigma)/(k + \cos\Sigma)] \\
(12)\quad & p = -C[(k\sin\Sigma)/(1 + 2k\cos\Sigma + k^2)]
\end{aligned}
$$

The numbers put within square brackets and following here are telling the relevant numerical values obtained with the current working example. This working example is being used not only for its own sake but also for constructing to scale the various figures. Even though questions of accuracy and the actual machining of teeth will need to be dealt with later, the numbers of decimal places to which these numbers are carried here are not at issue yet.

[r_2 = 52.001194 mm]
[r_3 = 27.988806 mm]
[ψ_2 = 31.650 deg]
[ψ_3 = 18.350 deg]
[p = –17.25207 mm/rad]

Notice in (12) that, when Σ is zero, p is zero, and that when C is zero, p is again zero. Often in skew gears, but not always [11], Σ is taken $\pi/2$. This results in those sets that might be called *square*. In this special arena of the square skew sets, the above list of equations reduces to

$$
\begin{aligned}
(13)\quad & r_2 = C[k^2/(1 + k^2)] \\
(14)\quad & r_3 = C[1/(1 + k^2)] \\
(15)\quad & \psi_2 = \arctan k \\
(16)\quad & \psi_3 = \arctan(1/k) \\
(17)\quad & p = -C[k/(1 + k^2)]
\end{aligned}
$$

which is a simpler list. I wish to say here with regard to worms and hypoids (whether they be square or not) that there is no generic difference between (a) the worm wheel and the worm of a worm-reduction set (whether it be offset or not), and (b) the crown wheel and the pinion of the decidedly offset, hypoid-reduction set. These different kinds of skew gear set are conceived, designed, drawn and manufactured differently, and they are in fact different. Both kinds of set, like gear sets of whatever kind, submit to the same equa-

BASIC KINEMATICS OF THE GEAR SET   §1.15

tions listed above however. If a constant ratio k is required, sets must moreover obey the fundamental law (§2.06). As we pass from tooth to tooth in a gear set, the location of the pitch line will not remain fixed unless the ratio k remains constant. Unless k remains constant, in other words, and as we move from tooth to tooth across teeth, the location of the pitch line will move; it will move in an oscillatory manner; and the frequency obtaining will be, as might be expected, tooth frequency.

On the numbering and significance of these equations

15. Unlike other equations of this book, which, appearing later, are referenced usually by means of paragraph number, the equations of this chapter are unique in the sense that they are referenced by consecutive numbers in rounded brackets in the well accepted manner. The reason for this is that the numbers here are chosen to coincide with the corresponding numbers at chapter 22 in [1], which work is, as might be expected, a companion piece. Speaking more about matters of substance now, one could relate directly to θ in equation (1), where θ is measured from the x-axis of the relevant cylindroid (§1.30) [1] [§15.13], or one could, where the angles ψ are not only the helix angles at helical teeth in ordinary crossed helical gears but also the angles between the shaft axes and the pitch line, relate to equations (6), (7) and (8) which are due to Steeds [4]. Equations (4) and (5) are also equations about the location of the pitch line that come from Konstantinov *et al.* [6]. These workers all knew of the pitch line (but not by that name), and they found its location with their different formulae. They were seeking, however, not its pitch, but the least velocity of crosswise sliding to be found there, none of them paying particular attention to pitch. The pitch at the pitch line may be obtained from equation (12). This pitch p, as has been said, is the pitch of that screwing which occurs relatively between the wheels, and it might be seen, overall, as an important quantity (a signed length according to left or right handedness and measured mm/rad) which is characteristic of the gear set. Pitch p is a function solely of C, Σ and k. It is independent of the shapes of the teeth (provided those shapes maintain a constant k of course); and it conjures up for us the system of helices of the same pitch p that may be drawn coaxial with the pitch line, such helices filling the whole of space. See figure 9.01 in [1]. Litvin [5] has given a short account of these helices. The pitch at the pitch line (and thus the pitch of all of the helices) might be right handed positive or left handed negative. In figure 2.01, for example (and this is an exercise in advance for the reader), the pitch is positive, while in figure 2.02 it is negative. The pitch in ordinary parallel spur gears and cylindrical helicals (and in bevels) as we know is always zero. Finally here it might be argued that the above formulae need to be — as, indeed, they lend themselves to be — graphically displayed to facilitate qualitative design. It is for this reason that the following material is presented.

§1.16   GENERAL SPATIAL INVOLUTE GEARING

Various graphs telling about the pitch line

16. Refer to the figures 1.03, 1.04 and 1.05. They comprise a set of three interrelated Cartesian 3-D plots, each with k and Σ at the abscissae. The plots apply to all gearing. The first two are reproduced 60% the size of the third and are somewhat simplified, but all three share the same perspective point of view and are thereby directly comparable. The plots deal, in turn, (a) with the so called RADIUS, the linear location of the pitch point P upon the CDL, (b) with the so called ANGLE, the angular location in the plane normal to the CDL of the pitch line with respect to the shaft axes, and (c) with the PITCH, the signed magnitude (mm/rad) of the pitch of the screwing at the pitch line. We take k to vary from –1 to +1 (thus excluding the reciprocals $1/k$ of k), and we take Σ to vary from zero to 2π. Figures 1.03 and 1.04 reveal, (a) one of the radii r that add to C, and (b) one of the angles ψ that add to Σ, both of which are required to locate the pitch line. Figure 1.05 reveals, (c) and most importantly, the pitch at the pitch line.

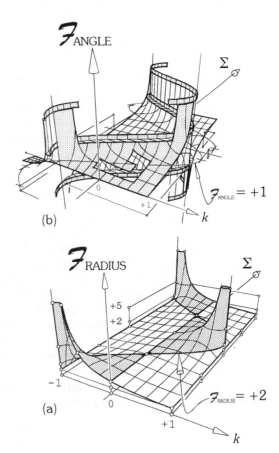

Figures 1.03 and 1.04 (§1.18 and §1.19). These figures, (a) below, and (b) above respectively, are grouped together here because, in comparison with the forthcoming figure 1.05, they are relatively unimportant. They do however go together to make figure 1.05 up, and for that reason they need to be shown here. They illustrate, against the fixed variables k and Σ of a gear set, the variation firstly of the radius (distance from O to the pitch line measured vertically along the CDL), and secondly the angle to the pitch line (measured from OY horizontally). Refer for more detail to appendix A at the end of this chapter.

BASIC KINEMATICS OF THE GEAR SET §1.17

17. We seek, in the graphical figures, r_3 (mm), ψ_3 (degrees), and p (mm/rad). To normalize the disparate dimensions, however, and to take account globally of a changing C, we plot the surfaces by using some simple scaling devices. For radius r_3, we plot (figure 1.03) a contrived function of the radius, $\mathcal{F}_{\text{RADIUS}} = +(2/C)r_3$; for angle ψ_3, we plot (figure 1.04) a contrived function of the angle, $\mathcal{F}_{\text{ANGLE}} = \tan\psi_3$; and for pitch p, we plot (figure 1.05) a contrived function of the pitch, $\mathcal{F}_{\text{PITCH}} = -(2/C)p$. Equations (5), (11) and (12) reveal that these three functions $\mathcal{F}$ have the double advantage of being given by an elegant set of expressions involving k and Σ alone, and of being all dimensionless numbers.

18. *Notes on figure 1.03.* These notes and this figure may be ignored upon a first reading. Contours for constant $\mathcal{F}_{\text{RADIUS}}$ appear at values 1, 2, 5 and 12. If r_3 becomes greater than twice half the centre distance $2(C/2)$, $\mathcal{F}_{\text{RADIUS}}$ becomes greater than $+2$, internal (as distinct from external) rolling of the axodes begins and very large values of r_3 occur as $k \to -1$ when Σ is near zero, or as $k \to +1$ when Σ is near π. This internal rolling occurs within the D-shaped regions delineated by (a) the straight line $k = 0$ at $\mathcal{F}_{\text{RADIUS}} = +2$, and (b) the shown sinusoid at $\mathcal{F}_{\text{RADIUS}} = +2$ whose equation is $k = -\cos\Sigma$. Internal rolling of the axodes, although a phenomenon fraught with some difficulty in design, does not necessarily require internal meshing at the teeth of the corresponding real wheels. (§1.29-30).]

19. *Notes on figure 1.04.* These notes and this figure may also be ignored upon a first reading. Although we have avoided the problems of multivalued angles by plotting not ψ_3 but $\tan\psi_3$, the variation of ψ_3 with k and Σ is difficult to portray. Be aware however that when $\mathcal{F}_{\text{ANGLE}} \to \pm\infty$, $\psi_3 \to 90°$ and $r_2 \to 0$. The vertical rulings mark out the relevant asymptote. The equation there is again $k = -\Sigma$. Strange transition cases occur at the asymptote, see for example figure 22.10 in [1]. Notice that, when $\mathcal{F}_{\text{ANGLE}} = \pm 1$, ψ_3 is 45°, but that, except that they clarify the figure, the shown contours at $\mathcal{F}_{\text{ANGLE}} = \pm 1$ are not important borderlines. They have the equations, however, $k = -\cos\Sigma \pm \sin\Sigma$.

20. *Notes on figure 1.05.* The important behaviour of $\mathcal{F}_{\text{PITCH}}$ is shown here. Because we have shown one complete cycle of Σ, confusing questions of sign are made clear. We remember of course that $p = -(C/2)\mathcal{F}_{\text{PITCH}}$. This means incidentally that, as we go down the holes, $|p|$ becomes progressively larger right handed positive, while as we go up the peaks $|p|$ becomes progressively larger left handed negative. Whenever $|p|$ is high either positive or negative we have a strong adverse affect upon (namely an increase in) the sliding velocities at the real teeth in real gears (§1.09, §1.51), so generally speaking the troughs and peaks are best avoided in conventional design. Contours for constant $\mathcal{F}_{\text{PITCH}}$ appear at values 1, 2, 5, and 12, both positive and negative. See also the non-continuous twisted curve running down the ridges and into the troughs. The straight line $k = 0$ at $\mathcal{F}_{\text{PITCH}} = 0$ and this twisted curve are the imprints here of the same borderlines delineating the

§1.20 GENERAL SPATIAL INVOLUTE GEARING

D-shaped regions of internal rolling at the axodes as before. Equations to the twisted curve are $k = -\cos \Sigma$, and $p = -2 \cot \Sigma$. Let me remark again that, although internal rolling at the axodes is a phenomenon fraught with some difficulty in design, it does not necessarily require that internal meshing must also occur at the actual teeth on the corresponding real wheels (§1.29-30). Note the two pairs of plotted spots marked star (*) and dagger (†) (§1.30).

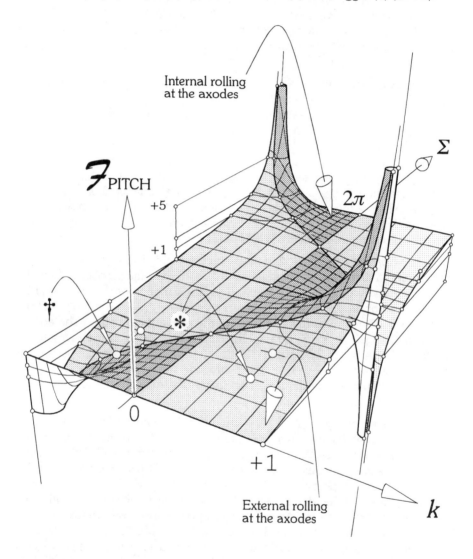

Figure 1.05 (§1.20). For a given C we see here the variation of pitch p at the pitch line with k and Σ. For the sake of uniformity the scaling factor $\mathcal{F}_{\text{PITCH}}$ is plotted, where $p = -(C/2)\,\mathcal{F}_{\text{PITCH}}$. Notice the blebs at star (*) and dagger (†). These refer to the two alternative pitch lines discussed at §1.11.

21. *Notes on the nature of k.* Everywhere among the plots, k may be taken beyond its shown limits without loss of meaning. It can be taken as k or as its reciprocal $1/k$ without fazing the formulae. The shapes of the plots beyond the shown limits of k have not been shown. Our comprehension of these shapes (in figures 1.03 and 1.05) must be colored by our knowledge of the circumstance that the reciprocal of zero is infinity and that as we go to from $k = -1$ to $k = -\infty$, and form $k = +1$ to $k = +\infty$, we need to examine not r_3 and ψ_3, but r_2 and ψ_2.

22. The plots show in general that, while the fixing of sign conventions in the realm of real skew gearing is a confusing business, the effort required to do so is not overly worthwhile when seen against the wider scheme of things. What we really need to know is that, if we come (in design) too close to the steeply sloping parts of figures 1.03 or 1.04, and thus to the steep spikes and deep holes of figure 1.05, we should do so with foresight and caution. The plots apply to all gearing. Every conceivable gear set has its own set of coordinates (k, Σ) at the same place in plan view on each of the plotted surfaces. The universality of the relationships depicted here, especially of those depicted at figure 1.05 which relate so directly to pitch and the fundamental law (§2.06), justify their reproduction here. The plots have shown, among other things, that the allegedly fundamental distinctions commonly made between planar and spatial gears and the different kinds of skew gear set (spiroids, hypoids, worms etc.) are in most geometrical respects illusory.

On the questions of sign

23. Troubles with sign pervade the issues of gear design in such a confusing way that they deserve special consideration here. Firstly I wish to mention that the somewhat inelegant minus sign appearing at equations (12) and (17) above is due to a convention adopted by the author, not stated but implied in [1], with respect to the sign of k. And in this connection it should be stressed that, in all of the formulae listed, substitution first of $+k$, then of $-k$, will yield two, quite different solutions for the location of and the pitch at the pitch line. This reflects the fact that, given the direction of rotation of a given driver and a non-signed ratio k, a driver and a driven may be found to give rotation of the latter in either of the two directions. Please recall the star (*) and dagger (†) of this chapter, refer to §22.20 in [1], and study the photo-frontispiece in [1] at chapter 22.

24. If k be taken between 0 and +1 anywhere at §1.14, and Σ is taken acute, the equations there will show that the pitch line is bracketed within the acute Σ, that pitch is always negative, but that $|p|$ is never greater than half the centre distance, all of which is interesting, usual and benign. Taking Σ to be negative under the same circumstances will afford right handed pitches incidentally. If k be taken between 0 and -1, on the other hand, again with Σ acute, however, we lead ourselves into that region where the pitch line is bracketed within the obtuse S, where the axodes intersect except at P (§1.26-

27, §4.11-15), where the pitch is always positive, but where $|p|$ may reach very high values. In this latter, less benign region, which may, nevertheless, be the region of choice, very high levels of sliding at the teeth may occur even when the meshing itself is close to the pitch line. As $k \to -1$ and $\Sigma \to 0$, the shafts in the limit become parallel, the angular velocities become to be of the same magnitude (namely zero) and in the same direction (not opposite), the meshing is internal (which is not practicable except at the limit), and the wheels become straight racks.

25. There are, in other words (and please refer to figure 1.05), two distinct regions for design activity, (a) the benign region, the usual one where the pitch line is bracketed within the acute Σ and $|p|$ is always less than $(C/2)$, and (b) the much more unforgiving region, already discussed above, where the pitch line is within the obtuse Σ, and where difficulties occur (§1.11, §8.12). Note that, when $\Sigma = \pi/2$ exactly (the case for many ordinary worms and hypoids), r_3 can never exceed $2(C/2)$, internal meshing cannot occur, and events in the two regions are mirror images of one another. Going to the new hypoids of [11], however, where Σ is other than $\pi/2$ exactly, events in the two regions do differ, and designers will be careful.

The axodes

26. Having established and discussed the pitch line in some detail, we can now discuss the axodes for the relative motion of the wheels. I wish to begin by making the following quasi-general remarks. In the study of the general relative motion of two bodies in *planar* motion, we (a) nominate two rigid *laminae* (representing the bodies) moving relatively in the plane, (b) find at each instant the unique *instantaneous centre of rotation* calling that point the *pole*, (c) plot the two separate paths of the pole upon the two separate laminae calling those paths the *polodes*, (d) find that the said polodes share a common point namely the *pole*, and (e) find that the polodes roll upon one another at the pole (which is in general itself moving in the plane) without slip. In space, in general and correspondingly, we (a) take the two bodies, (b) locate at each instant the unique instantaneous screw axis calling this often by some other name (and here in this special case of gears I have called it the pitch line), (c) plot the ruled surfaces traced by the instantaneous screw axis in each of the bodies calling them the axodes, (d) find that the two axodes share a common generator namely the said screw axis, and (e) find that, being tangential, they roll upon one another at the said generator, slide with respect to one another along the said generator, but sometimes intersect one another in a way that is, to say the least, somewhat mysterious.

27. What I have said here may sound either blotchy in its content or unnecessarily complicated in the relatively special context of the gear set, where the two shaft axes are fixed in space (namely in the frame 1), and where the pitch line (the common generator shared by the axodes) also remains fixed. The axodes for the relative motion of the wheels in this very

BASIC KINEMATICS OF THE GEAR SET §1.28

special case of spatial three-body motion are simply circular hyperboloids as shown in figure 1.06, and their common generator is simply the said pitch line. Figure 1.06 also shows that the axodes for the benign example chosen are non-intersecting along the common generator namely the pitch line. Given an increasing Σ, however, the axodes will (a) 'snug into' one another exactly when Σ becomes $\pi/2$, then (b), while remaining tangential, intersect.

28. Following figure 22.12(a) in [1], whose content refers to a general relative movement in space (and whose caption is somewhat confusingly written unfortunately), figure 22.12(b) comes closer to explaining this special phenomenon of the intersecting axodes that we need to understand in gears. In gears, we have the special circumstance that the wheels rotate respectively about the two shaft axes fixed in frame. The axodes are thus both mere circular hyperboloids concentric with these fixed axes, and the hyperboloids share a common generator (the fixed pitch line) in a relatively simple way. The axodes remain tangential along the whole (infinite) length of the common generator but, as fully explained at [1] [§22.64], and as briefly explained again, above, however, the axodes intersect if Σ is greater than $\pi/2$.

29. Despite these somewhat difficult though unavoidable discussions about the geometry of axodes, their rolling, axial sliding, intersections and so on, it is important to see that in real gears (as in all real machinery) the axodal surfaces are mere geometrical abstractions that in no way inhibit the in-

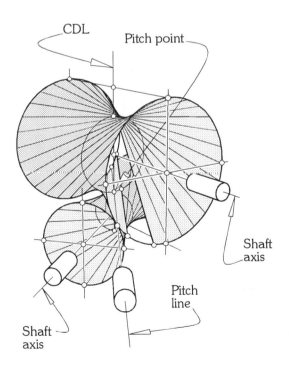

Figure 1.06 (§1.27). Axodes, which roll circumferentially yet slide axially with respect to one another, either touch or intersect along the pitch line (§1.27). Although in general the meshing of the real wheels is such that the meshing does not coincide with the pitch line (§5B.03), the relative motion of the axodes reproduces exactly the relative motion of the wheels.

§1.30 GENERAL SPATIAL INVOLUTE GEARING

visible motions of one another or the seen physical motions of the real bodies. There is a clear distinction between the surfaces for example of axodes that roll, slide, and interpenetrate ghostlike in space, and the surfaces of real bodies such as gear wheels that work in actual, physical, surface contact with one another, teeth to teeth. These latter, the physical surfaces of the real bodies in machinery, have the solid material of the relevant body on the one side (namely the inside) and the absence of such material on the other. Real bodies cannot interpenetrate; and if, in gear design, they are held to do, they undercut. The axodal surfaces, and other such ghostlike surfaces yet to be mentioned, do not undercut in this physical way. They operate smoothly and quietly; they sometimes interpenetrate; they do not interfere with the physical motion; they simply, accurately and usefully, represent the relative motion.

30. They do this, however, in often puzzling ways. A striking example of such a puzzle is the circumstance that sometimes exists in offset skew involute gearing whereby the axodes are engaged in internal rolling (with one of the mating hyperboloids rolling inside the other) while the actual wheels they represent are engaged in external meshing. See the deeper point marked (†) on the surface of the plot in figure 1.05. Keep such matters as this in mind while looking at figure 1.02, and see the significance of what I have just been saying exemplified in chapter 6.

The imaginary spline-line teeth of a pair of axodes

31. Refer to figure 1.07. Much of its contents is a summary of the matters discussed so far. The surfaces marked (a) and (b) are the axodes of figure 1.06 severely truncated. The remaining short portions of these surfaces (which are of course infinitely long measured axially) evenly straddle the throat circles. The throat circles intersect at P. Thus the surfaces (a) and (b), the axodes, take on the vague appearance of being an actual pair of straight-cut meshing gears, both evenly truncated to straddle the gorge. The surface marked (c) is the cylindroid. It governs the location of the pitch line according to the value of k [1] [§22.20]. The surface marked (d) is yet to be mentioned.

32. The surfaces marked (a) and (b) are not actual gears of course; they are simply the axodal surfaces which result directly from the given k, or, as I wish to call them now, the non-departed sheets. While the meaning of the adjective non-departed will be clarified soon, the meaning of the word *sheet* is borrowed here from the phrase hyperboloid of one sheet. This latter is used sometimes to describe such ruled hyperboloidal surfaces and to distinguish them from the other, two-branched hyperboloidal surfaces which are not ruled and said to be of two sheets. These two-sheeted hyperboloids make no appearance in this book. Anyway the single-sheeted surfaces which do — and they are, just here, the axodes — are marked with appropriate sets of generators. Please imagine however that all of the generators are drawn, and that, with no space between them, they cover the surfaces. Please imagine also that, as we go to the limit of reducing the space between the successive

BASIC KINEMATICS OF THE GEAR SET §1.32

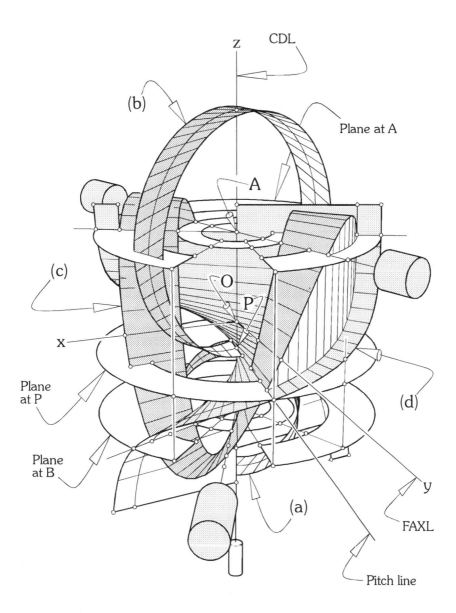

Figure 1.07 (§1.31). This omnibus picture graphically summarizes some of the main aspects of chapter 1. It relates exclusively to the crossed helicals. I shows at (a) and (b) the pair of axodes for the prevailing k, which are also the naked wheels for the special set based on P, at (c) the cylindroid which relates the speed ratio k with the location of the pitch line, and at (d) the parabolic hyperboloid which tells about the singly departed sheets which contain the naked wheels.

generators to zero, we maintain the ratio k between the numbers of them on the respective axodes.

33. The equations (7) and (8) at §1.14 may now be used to show that, as we approach the said limit, the spaces (the shortest distances) between successive generators of the two axodes tend towards equality. What this says is that the infinitesimally small, shortest distances between successive generators will be the same on both of the mating axodes. This fundamental property of a mating pair of axodes makes clear what our meaning is when we speak not only about pure axial sliding at each separate pair of mating generators as we go, but also about pure circumferential rolling as we step from each pair of mating generators to the next. In any event these geometrical facts permit an early glimpse of my idea of the imaginary, spine-line teeth of an axode. Please imagine each generator of an axode to consist of a long, yet infinitesimally thin, spline, a line that is smooth in the direction of its length but rough in all directions perpendicular to it. The pair of mating spline-line teeth of a pair of mating axodes might thus be seen to be able to slide upon one another in the mutual axial direction of their lengths, while at the same time to be able to roll upon one another without circumferential slip. It may thus be imagined that, by virtue of their shapes and by virtue of their spline-line teeth, a pair of axodes have the capacity to drive one another and to reproduce thereby the relative motion of the two actual gear wheels. After some further experience I will speak some more about axodes at §6.29.

Singly-departed naked wheels of the crossed helicals

34. There is the well known class of gears that might be called, with some precision, *cylindrical crossed helical involute gears*. The class is more often called by the simpler name, *crossed helicals*, and sometimes, indeed, by the older and somewhat absurd name, *spiral gears*. The class appears to enjoy its existence more by accident than design. We appear to take the view that we might look for some odd yet matching pair of ready-made, ordinary, cylindrical, involute helical wheels, of helix angles θ of opposite sign (§10.22), separately designed for other, parallel drives, and simply *cross* them. The class is said to suffer tooth-wear rapidly and is said to be, for this and other reasons, unsatisfactory. Using my earlier definition of the term offset (§0.01), we might say that all of the gears of chapter 22 in [1], the said, ordinary crossed helicals, are non-offset.

35. Speaking about the kinematics (and speaking loosely), it is well known in this class of gear that the region of the meshing of the real teeth, while straddling the CDL, does not in general coincide with the location of the pitch point. If we wish (keeping the centre distance C and the ratio k both constant) to change the so-called pitch diameters of the wheels (a misnomer that I try to avoid), all we need to do is to change the relevant radii (I call these radii c_2 and c_3 which, like r_2 and r_3, add to C), then change the

helix angles of the teeth to suit. This is well explained in Steeds [4], where he writes some relevant equations. These equations were lifted by me into [1], and from there into §1.14. From these it can be deduced very easily that, as we depart (either upwards or downwards) from the pitch line, where the axodes namely the non-departed sheets exist, we can erect new rulings at suitable helix angles as we go [1] [§22.40] (§1.36). Refer to figure 22.08 in [1]. By rotation, we may produce a whole range of pairs of singly-departed sheets. Suitably located slices of finite thickness taken from these departed sheets, evenly to straddle the gorge, become the naked wheels. Thus each pair of naked wheels is obtained by sweeping the relevant departed ruling about the two shaft axes in turn. A naked wheel is naked in the sense that, while it is equipped with rulings, it is not (as I wish to argue) equipped with spline-line teeth (§6.29). We begin to see that, for a given C, Σ and k, there exists a continuum of pairs of naked wheels existing, all of which are legitimate. Each pair will be seen to have its own radius ratio c_2/c_3.

36. Please notice incidentally that there is only one pair of radii and thus only one pair of throat circle circumferences having radius ratio k, and that this pair occurs not at P but at D. There is thus something as yet unexplained about the question of the rolling of these naked wheels upon one another. See §1.41 and again later, at §6.29. The naked wheels under consideration here have been called singly-departed because, as we shall see, they will need to be distinguished later from the doubly-departed naked wheels which are a feature of the general, spatial involute gears. Refer again to [1] and see that some aspects of the argument here are missing there.

The twisted surface marked (d) in figure 1.07

37. Plotting the mentioned directions of these singly departed rulings (which are mounted along the CDL and perpendicular to it), we obtain the ruled surface marked (d) at figure 1.07. Based on the relationship, $\Delta r = p \tan \Delta \psi$, between the so-called departure Δr and swing $\Delta \psi$ of the teeth, which is equation (11) derived in [1] [§22.37], the surface is a parabolic hyperboloid with its central point at P. One principal axis is collinear with the CDL as shown; the other is at the pitch line; and the surface itself intersects the other lines and surfaces as shown. Note that the surface does not contain the shaft axes at the extremities of the centre distance C, but that it does contain the other pair of generators of the cylindroid which are located there. Consistent with the fixed points A and B at the two ends of the centre distance C, I take a general point upon the CDL called C. At this point (which should not be confused with C which is a length), the following formulae obtain:

(18)	$c_2 = r_2 \pm \Delta r$
(19)	$c_3 = r_3 \mp \Delta r$
(20)	$c_2 + c_3 = C$

These are explained more carefully in [1]. There was however no occasion there for me to speak about both singly and doubly departed sheets. The idea of double departure is new. It comes only with this book (§5B.02).

Some aspects to be remembered of figure 1.07

38. In figure 1.07, the cylindroid has its relevance all the way from its central axis outwards. For given C and Σ, it relates the speed ratio k to the linear and angular location of the pitch line. The parabolic hyperboloid, however, applies only for that k which is extant, and only within the immediate neighbourhood of the CDL. As explained in [1] [§22.50] and as shown in figure 22.08 in [1], the shown generators of this surface tell about the layout of the rulings upon that limited range of different pairs of naked wheels which have, not only the same k, but also their rulings evenly straddling the CDL. It applies in other words only to these limited ranges of gears known as the crossed helicals. Figure 1.07 says nothing about offset skew gears.

Special cases among the radii occurring at P, D and O

39. As we have seen, the crossed helicals are characterized by the occurrence of singly departed sheets, and it is the throat radii of these sheets, c_2 and c_3, that have been defined at §1.35. Two important special cases occur among a range of crossed helicals with given C, Σ and k. The first occurs when c_2 and c_3, which are of course variable (provided they add to C), coincide exactly with r_2 and r_3, in which case the meeting point of the throat circles is at the pitch point P.

40. In this first special case, when C is at P, (a) the naked wheels are at the axodes and are thereby non-departed (Δr is zero), (b) the path of Q not only cuts the CDL but cuts it at the pitch point, (c) the circumferences of the contacting throat circles of the naked wheels are not proportioned according to k, so the said circles must, it seems, be 'slipping' somehow, and (d) the helix angles at the spline-line teeth of the mating naked wheels are different.

41. A second special case occurs when the ratio c_2/c_3 becomes exactly k. This occurs, not at P, but at another point upon the CDL that I wish to call D. Point D can be seen at figure 5A.01, but this figure is, as yet, premature. In this second special case, when C is at D, (a) the naked wheels are no longer at the axodes and are thereby departed (Δr is non-zero), (b) the path of Q cuts the CDL not at P but at this other point D, (c) the circumferences of the contacting throat circles of the naked wheels are proportioned according to k, so the said circles must, it would appear, be 'rolling' somehow, and (d) the helix angles at the smooth generating lines of the mating naked wheels are equal.

42. It is important to understand, in any event, that these two special cases characterized by the points P and D are quite different. Accordingly I wish to reserve for these second, special-case radii (those occurring when

$c_2/c_3 = k$) the terms d_2 and d_3. The logic of this terminology will become evident when I begin to define and employ, not only the C-radii (c_2 and c_3) and the D-radii (d_2 and d_3) as I have done here, but also E-radii, F-radii, and (as might be expected from figure 1.01) G-radii as well.

43. As well as the points P and D, however, there is another important point along the CDL, namely O, the mid-point of the centre distance. Although O is clearly relevant in the case when k is unity (in which case both P and D are at O), it should also be seen that there exists at O a unique line perpendicular to the CDL and bisecting there the acute, shaft angle Σ. This line I call the FAXL. Mentioned in more detail in the text to follow, it is collinear with the algebraic axis O-y (§1.06). Whereas the positions of P and D upon the CDL, for given C and Σ, depend upon the value of k, the position of O upon the CDL and the location of the FAXL are both independent of k.

44. The kinematic significance of the FAXL, especially in the general event of involute action, will be explained later. For now, however, let the FAXL simply be seen to be one of the principal axes of a doubly-ruled surface called the F-surface; see figure 5B.01; the FAXL will sometimes be called the F-axis line; but I will call it most often by the pronounceable acronym FAXL. Both the CDL and the FAXL are indicated by means of labels at figure 1.07. They occupy there the axes O-z and O-y of the cylindroid [1] [§15.13].

45. Please notice also in figure 1.07 that, whereas the pitch point P and the mid-point O are both marked and indicated, the point D is neither marked nor indicated. Point D is omitted for the sake of clarity. Despite the fact that each of P and D has its own separate significance, each of which is crucial for a clear understanding of involute action, the points P and D are often (but not always) quite close to one another. That P and D are different is taken up again at §5A.20

46. Readers in advance of the argument here will already have seen that there is surely more to all of this than meets the eye. Risking that I use here terms not yet properly defined, I say the following. The 'helix angles' of the 'teeth' of the naked wheels are not necessarily the helix angles θ of the flanks of real teeth occurring in offset involute gears (§10.22). They are however special cases of the angles γ shown at figures 5B.02(b) and 6.07. These figures are drawn respectively for equiangular and polyangular offset naked wheels. If we allow, in the said figures, the key circles of the naked wheels to migrate towards the throats of the departed sheets, whereupon they tend towards intersection with one another, the wheels become, in the limit, the ordinary crossed helicals under discussion here. The angles λ become the angles τ; and the angles τ, as we can see, are the half angles at the vertices of the asymptotic cones of the naked wheels. When C is at D, the wheels are equiangular (§5A.23), with zero offset; when C is elsewhere along the CDL, the wheels are polyangular (§6.02), again with zero offset. As has been said, the meeting point C of the key circles of the naked wheels might be either at or remote from the axode. Let the implied questions surrounding the as yet

vaguely mentioned matters of rolling and/or slipping (sliding) at the mating of the naked wheels remain for the meantime unresolved. They require careful attention.

Ordinary planar spur and the parallel helical gears

47. Any general theory of gearing that could not deal with its own special cases would of course be useless, so it behoves the author now to examine the special case where the shaft angle Σ becomes zero, the special case where, in other words, the gears become planar spur or helical gears with parallel axes. If we simply put Σ to zero in the equations (1) (2) and (3) at §1.14, the equations break down, revealing only that r_2 and r_3 are indeterminate. The Steeds equations (6) (7) and (8) break down also. The alternative formulations of Konstantinov *et al.* at (4) and (5), however, allow the radii r_2 and r_3 for a spur gear set to be determined. We can find from these that $r_2 = Ck/(k + 1)$, and that $r_3 = C/(k + 1)$. We can find, moreover, intelligible results by substituting in these two equations values of k both positive and negative; we can find, for example, that the same speed ratio $|k|$ may be made available on going from external to internal meshing.

48. These simple substitutions do not explain, however, why it is that, *with spur gears, there is a unique result for the two radii.* This is contrary to the circumstance we have been studying above (with the crossed helicals) whereby the two radii c_2 and c_3 may take any pair of values, provided the pair adds to the centre distance C. This is not by any means an easy matter to understand; the algebraic absurdities apparent here are bound up with the *geometric singularities* that occur among the screw systems that are involved; the reader might refer to [1] [§22.30 *et seq.*] for a much more detailed account of this somewhat difficult matter.

49. The fundamental trouble is that, as we move from the crossed helicals into the special case of the parallel helicals, we move from an open circumstance of mobility unity into a closed circumstance of overconstraint [1] [§2.06]. Given involute action, we suddenly have straight line contact between the teeth. In this special case, it is strange for us also to find that, provided of course they add to Σ (which is zero), the helix angles θ at the teeth (which must accordingly be of opposite hand) may take any value. Please be warned here that the angles θ — not in this special case but in general when the gears are not only skew but also offset — are difficult to define and somewhat difficult to understand; go for this to §10.26. The said paradox, however, of the ordinary parallel cylindrical helicals (which includes of course the ordinary planar spur gears as a doubly special case), lies in the nature of the special linear complex [1] [§9.41]. Go next to [1] [§22.38] for an explanation of the corresponding doubly special fact that, in both the parallel helicals and the ordinary spur gears, the radii c_2 and c_3 (and thus the position upon the CDL of C) depend inescapably upon the value of k. For a given k, the position of C is fixed; it is fixed at D, which is at P.

The various kinds of bevel gears

50. These gears appear when the centre distance C becomes zero, while the radial offset R, which is not defined yet (but see §5A.15), either (a) remains finite or (b) becomes zero. The so called straight bevel gears of regular industrial practice (along with their related fellows the so called spiral bevel gears) both belong inescapably to the difficult and, dare I say, somewhat obscure field of hypoid gearing (§0.03). These straight ones can be manufactured in various ways to give straight line contact between teeth, but with only approximate conjugality, and the thus cut teeth can be physically modified (by suitable crowning) to alleviate the overconstraint [79]. These gears are not, however, nor are they ever likely to be, involute gears. If we insist, however, upon a sincere application of the found involute geometry as it presents itself in this book, to the extent of allowing without fear not only (a) the centre distance C to become zero, but also (b) the radial offset R to become zero, we arrive as follows. In the simpler case (a) we arrive inescapably at the Beveloid gears first mentioned at §0.08 but not to be mentioned again until §5B.61; and in the more difficult case (b) we arrive at a definitive explanation of the so called 'spherical' involute action often mentioned, but not often well explained, in the current literature. See Steeds [4] [§175-§183] for a welcome, well written exception. The necessary argument for the introduction of this new idea namely (b), however, is not available yet; so an answer to these somewhat persistent questions about the various bevels and the Beveloids can and should, accordingly, be put off, just now, until later. Following a few more inconsequential remarks (§2.06, §5A.30, §5B.61), the next serious mention of this somewhat puzzling matter of the conventional straight bevels of the one and, and the bevels that I would wish to call *legitimate involute bevels* on the other, occurs at §5B.62.

A first look at the question of offset

51. Although this paragraph is an ill disguised re-run of §1.09, it will do no harm to look again at the relevant crucial circumstances of the matter. In figure 1.01 we are led to imagine the mentioned field of helices of the same pitch which coaxially surrounds the pitch line and fills the whole of space. Associated with this is the helicoidal field of linear velocity vectors [1] [§5.44-51]. At each point upon each helix and tangential with it there is a linear velocity vector representing either the linear velocity of that point in wheel 3 relative to wheel 2, or the equal and opposite linear velocity of that point in wheel 2 relative to wheel 3. Consider that particular helix of the set of helices that passes through the point of contact Q between the spoons; and think about the sliding velocities occurring there. They are v_{2Q3} and v_{3Q2} [1] [§12.08]. The figures 1.08(a)(b) now show some important detail occluded in figure 1.01. Let h be the length of the perpendicular dropped from the general point of contact Q onto the pitch line. Let **h**, accordingly, be the directed distance from the pitch line to that Q, let ω be the relative angular velocity

§1.52 GENERAL SPATIAL INVOLUTE GEARING

between the wheels (this particular ω is ω_{23}), and let v be the said linear velocity. We can now write more confidently than we wrote before at §1.09 that

$$(25) \quad v = p\omega + (w \times h)$$

which is, perhaps, the best known basic equation of screw theory. It relates exclusively to the relative motion of two bodies. A particular point in question, in this case Q, resides fixed within one of the bodies while the frame of reference resides fixed within the other.

52. In the special case of the crossed helicals, and at the CDL, h is collinear with the CDL, and v is the rubbing velocity occurring at the point of contact Q when that Q is at the CDL. The vectors v for various positions of C find themselves arranged in figure 1.07 coinciding with the generators of the surface marked (d). Each is perpendicular with the CDL, and all have their basepoints upon the CDL. Refer to figure 22.08 in [1] for a translation of this into terms of the naked wheels for a continuous range of crossed helicals having a fixed C and Σ, a moving point C, changing radii c, and a given k.

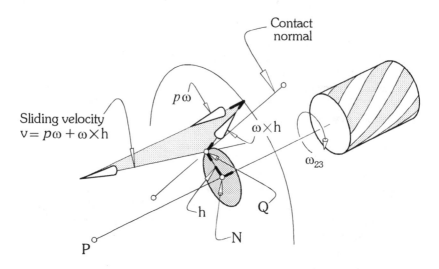

Figure 1.08(a) (§1.51). In reference to figure 1.01, and looking again from that same point of view, here are the mentioned occluded details of the vectors at and about the pitch line. The line marked contact normal is normal to the common tangent plane to the mating surfaces of the spoons at Q. It is somewhere perpendicular to the sliding velocity. In this view, however, the overall picture is not overwhelmingly clear. The reader might accordingly go to figure 1.08(b), where matching orthogonal views of the tangent plane at the point of contact and the participating vectors there might help to clarify the underlying nature of this instantaneous screw motion.

BASIC KINEMATICS OF THE GEAR SET

§1.52

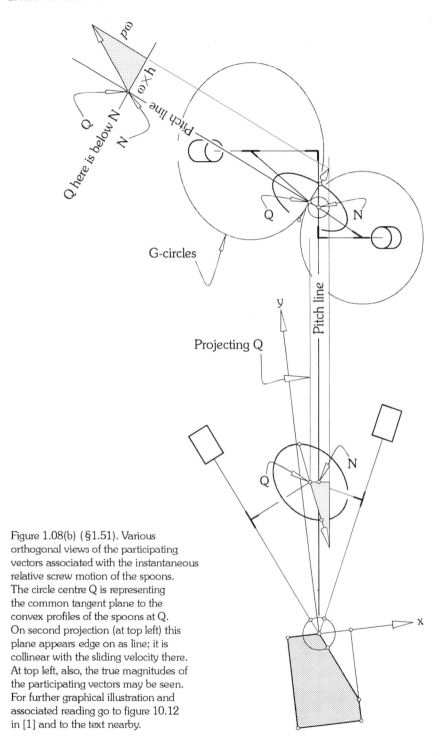

Figure 1.08(b) (§1.51). Various orthogonal views of the participating vectors associated with the instantaneous relative screw motion of the spoons. The circle centre Q is representing the common tangent plane to the convex profiles of the spoons at Q. On second projection (at top left) this plane appears edge on as line; it is collinear with the sliding velocity there. At top left, also, the true magnitudes of the participating vectors may be seen. For further graphical illustration and associated reading go to figure 10.12 in [1] and to the text nearby.

53. Now see figure 5B.02(a) for a more general case, where Q is not necessarily at the CDL and where, indeed, Q might never be at the CDL . It is in this sense that the offset skew gear sets is the main subject matter of this book. Steeds [4] cites Mac Cord [18], and speaks highly of his work, be he uses the unadorned term *skew* to describe those gears that I and others, following Mac Cord, might wish to call (as I have called them here) the non-offset, crossed helicals. See Steeds [4] [§184-189]. Offset skew gear sets were well delineated by Mac Cord. He explains his view in 1889 (and I agree with it) that his genuinely skew, skew gear sets have their being when the meshing is put, not at the CDL as it is with the relatively simple crossed helicals, but somewhere remote from the CDL. See Mac Cord [18]. See Rankine [55].

54. For these reasons I wish to employ the terms (a) *offset skew*, which is general, to describe those gears where the meshing occurs remote from the CDL, and (b) *non-offset skew*, which is special, to describe the kinds of gears like the relatively simple crossed helicals discussed above, where the departed sheets are singly departed, where the naked wheels straddle the gorges of the relevant hyperboloids, and where the meshing straddles the CDL. Skewness in gears is a matter of semantics of course, but seen this way, we can argue with logic that the crossed helicals are a limited, special, non-offset sub-set of the offset skew gears. Whereas in offset skew involute gearing (yet to be studied) the straight path of the point of contact never intersects the CDL, in the crossed helicals as here defined it always does.

Conclusion

55. We have dealt in this chapter with the most basic of the kinematics associated with, and applicable to, all gear sets of whatever kind. There has, however, been no kinematics here of the interactions among real gear teeth, let alone the interactions among real involute teeth. We have, moreover, not yet enlarged upon the complications associated with offset skew gears, where the meshing occurs in the open space between the shaft axes and remote from the CDL. Figure 1.01 did illustrate the general possibility, and figure 1.08 pursued the matter a little further; but, by and large, the question of offset has so far been avoided. We have mentioned the mechanics of the pitch line and dealt with some of its ramifications, we have taken a preliminary look at the idea naked wheel, we have defined in a rough way the idea of the spline-line teeth of axodes, and we have looked preliminarily at the simple crossed helicals, but we have, in fact, done very little else. We have not yet clarified, even, the mechanics of the rolling and/or slipping of the singly departed naked wheels. Truly skew, namely general, offset skew, involute action among real spatial involute teeth is the main subject matter of this book. That subject begins to reveal itself soon, but not before the beginning of chapter 3.

CHAPTER TWO
THE FUNDAMENTAL
LAW OF GEARING

Introduction

01. The fundamental law itself is yet to be enunciated; see §2.02; but the following needs to be said by way of introduction. The word *conjugality*, in gear parlance, means, in effect, *constancy of angular velocity ratio*. Having said that, we can say this: *among mating sets of accurately cut teeth, the wheels being unloaded and the teeth being in contact, the only requirement for conjugality is accurate compliance with the fundamental law.* In the machine-cutting of offset skew gears such as the well known *hypoids*, the conveniently chosen shapes of the cut teeth of the wheel are imposed upon the mating cut teeth of the pinion by means of an imaginary, concave-convex exchange, followed by a generating process which is designed precisely to cut with this particular end in view — exact conjugate action [50]. One might accordingly say that, in the absence of clearances introduced for needed backlash, and without manufacturing inaccuracies and the usual minor modifications made to avoid the need for accurate assembly, hypoid gears and their kin do automatically obey the fundamental law. This happens by virtue of their construction. See however §2.25. Properly constructed *involute* gears, on the other hand, obey the fundamental law by virtue of the shapes of the profiles of their *separately cut* sets of teeth. The shapes are determined by the underlying geometry of the required, smooth transmission of motion, and are thus describable by discoverable algebraic formulae. The shapes may be cut by generating processes which apply equally and independently at each of the two wheels — not unequally and interdependently at the said wheels. This latter gearing, namely *involute gearing*, insensitive both to introduced clearances allowing the necessary backlash, and to related but other accidental errors at assembly, is the main subject matter of this book.

The fundamental law

02. The fundamental law of gearing, stated by the author [14], applies to all circular, shaft driven gearing of whatever kind; the teeth may be invo-

lute or non-involute. Without sufficient explanation as yet, it may be stated as follows. Refer to figure 2.01. *For constant angular velocity ratio in gears, the contact normal must at all stages of the meshing be located in such a way that q tan φ remains a constant namely p.* The parameters q and ϕ are the shortest distance and the angle, respectively, between the contact normal and the pitch line. The pitch line is by definition the axis of the relative screwing of the mating wheels (§1.09); and the parameter p — this depends solely upon C, Σ and k — is the pitch of the said screwing. This pitch (an overall characteristic of the gear set) is measured mm/rad (§1.14).

03. By way of confirmation of the law it may be pointed out that, when Σ becomes zero, the gears become parallel-axes gears, the gears being then *planar*; p is zero, ϕ is $\pi/2$; and the law breaks down to stating, merely, that q must always be zero. This is the special, well known, planar case of the fundamental law which, when seen within the confines of the so-called *reference plane*, states this: *that the contact normal must at all stages of the meshing pass through the pitch point.* Refer to Robert Willis [12] and to the brief historical remarks I have put together at §2.44. Those readers ready enough for a direct, vector proof of the law in space may go in advance to §2.23.

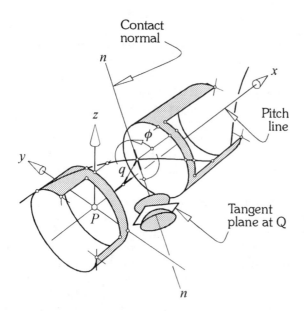

Figure 2.01 (§ 2.02). Illustrating the fundamental law of gearing. The contact normal n-n is seen here at distance q from the pitch line and inclined at angle ϕ to it. Across time, q and ϕ may both vary, but for constant k across time, we need $q \tan\phi$ to remain a constant, namely p. This is an algebraic statement of the law which, otherwise speaking, requires that, for constant k, the contact normal must at all times be a member of the linear complex of lines surrounding the pitch line. The axis system O-xyz is local.

Involute gears

04. This chapter like chapter 1 deals with gearing in general. It is however convenient to make the following particular remarks at this early stage. Just like other kinds of offset skew gearing, involute skew gearing requires the provision of inter-tooth clearance to allow backlash, but *unlike* other kinds of skew gearing, involute skew gearing is characterised by (a) an exact constancy of angular velocity ratio as the action passes from tooth to tooth across teeth, (b) a single point of contact between teeth, whose flanks are ruled surfaces easily machined, (c) a straight line path for the point of contact, along which path the point of contact travels with uniform velocity, (d) a continuous smooth rolling and sliding upon one another of the said flanks, (e) the capacity within limits to sustain errors at assembly in respect not only of the centre distance, but also of shaft angle and the axial locations of the wheels upon the shafts, and thus (f) the capacity to function properly even after a change of configuration due to a flexing of the shafts or of the frame.

05. Applications of offset skew involute gearing might conceivably be found in (a) machinery where the maintenance of accuracy in the relative angular locations of shafting might be a problem, (b) machinery with great angular inertia at output, where accuracy of angular velocity ratio across teeth is important, (c) flexible machinery designed for operation in the absence of gravity, (d) off-road and extra-terrestrial vehicular machinery, where gear sets with skew axes might be needed, (e) ordinary engine driven road vehicles where flexibility at the complex joint between the transmission shafting and the wheels might allow simpler suspension systems, (f) special drives where the shaft axes are set intentionally but only slightly off parallel to diminish the severity of overconstraint due to full line contact, (g) gear sets where easy, axial, accurate adjustment for backlash might be a valued feature, (h) bicycles, motor cycles, boats and human powered flying machines where light, flexible, skew gearboxes might be used, (i) flexible robots, (j) helicopters, and (k) selected drives in ordinary NC or CNC milling and/or hobbing machines which might in the future be dedicated to cutting these involute skew gear teeth.

06. Looking in general not only at offset skew involute gearing but at all involute gearing, we might legitimately see planar involute spur and helical gearing with parallel axes, and the corresponding bevel gears with intersecting axes, as the somewhat troublesome special cases. Single point contact has given way in these special cases to straight line contact, and kinematic overconstraint has become a problem [1]. Line contact in general (leading in practice to elongated areas of contact) helps to improve the lateral distribution of load of course, but the need for great accuracy in the mounting of all line-contact gears is a difficulty. So great is the difficulty that *crowning*, a material-removing, 'shaving' process which bulges somewhat the otherwise correct profiles of the teeth [51], is often resorted to. Straight bevel gears of the

ordinary, non-Beveloid kind have been mentioned only in passing here (and this happened also at §1.50); they come up for further consideration later, first preliminarily at §4.31 and §5B.30, then more seriously at §5B.62.

07. For an early treatment of *planar* involute action (with straight line contact) go to [12], for subsequent developments see for example [13], and for later works go to the recent literature. The mechanics of *spatial* involute action, on the other hand, is barely understood. Even less understood is the likely practical consequences of *offset spatial* involute gears, with their openly avowed single points of contact [58] [59] [60] [61] [62] (§3.29). Little is known of the materials or lubrication that might be suitable for the somewhat more rounded contact patches clearly involved, or of the kinds and rates of wear, or of power transmission. *But what we will say with certainty at the end of this book is that the path of the point of contact in properly constructed, offset skew involute gearing is always a straight line and that, at all stages of the meshing, the tangent plane at the point of contact (the plane of adjacency) is normal to that line.* This augurs well for the absence of vibration in such gears. The spatial rolling, oblique sliding and benign curvatures at the meshing augur well for lubrication also.

Prognoses

08. The fundamental law of gearing (§2.02) and its ramifications for all gearing are to be dealt with in this chapter; the kinematics of offset skew involute action is to be analysed in chapter 3; and actual skew involute teeth are to be synthesised in the chapters 5 and 6. Synthesis is not the mere reversal of analysis, however, so ample geometrical foundations need to be laid in this and the next chapter for ultimate use in the following chapters. This chapter 2, solely about the fundamental law and its ramifications for all gearing, is, accordingly, much longer than it needs to be for simply stating the law and proving it. The matters at issue in the design of gear teeth are more complicated than a mere algebraic statement of the law and a quick proof of it. Accordingly the said law (§2.02) is not proven immediately here, but somewhat soon (§2.15). Some intervening comments need to be made.

Some assumptions and two disclaimers

09. *Assumptions*. I assume that all participating bodies are rigid, and I make the following other assumptions in my forthcoming proofs and analyses. For mobility unity, single-point contact needs to be seen at Q. This is not to deny that curved line contact, in the presence of adequate accuracy, cannot be envisioned to occur (§2.25). For smooth transfer of motion as we step from tooth to tooth in any event, several points Q need to occur at several teeth in tandem. In tight mesh all teeth remain in contact, front to front and back to back. The shapes of the profiles (there being two front profiles and two back profiles) will be, in general, each different from the other. There being no friction, all rotational motions are reversible. I assume, in short, that

10. *Disclaimer.* The well known *pressure angle*, well entrenched in planar gear theory, is quite other than my ϕ appearing in the above statement of the fundamental law (see also figure 1.01). My ϕ here comes directly from earlier work. See for example [1] [§19.08] and figure 19.01 in [1] where ϕ, already well established in screw theory, refers to the angle between a pair of reciprocal screws [1] [§19.09] (§2.28). The pressure angle of planar gearing may be seen to have its counterpart in spatial involute gearing (§5B.15), and it may be seen to have a precise meaning (§4.60-61, §10.11-12), but I wish to call it now the a*ngle of obliquity,* or the *disability angle,* both of which were earlier names [4] [§147], and to use for it the symbol δ. The term pressure angle can at last be seen to be a misnomer in the new circumstances of space, and I wish to let it go. The law makes no mention of course of δ, which relates more to questions of physical clearances between the bounded bodies of the teeth and statics, than to questions of kinematics and the constancy of angular velocity ratio. As we have seen the fundamental law reduces in the case of planar motion to the well known fundamental law of planar gearing, namely that the contact normal must at all times pass through the pitch point. For planar gearing p is zero, q is zero, and ϕ is $\pi/2$. But δ, as we know, may be set to any value consistent with practicality.

11. *Disclaimer.* Despite some possible, perceived indications to the contrary in what there is to follow, I wish to say that the phenomenon of friction, the consequent tendency for the inter-tooth force at the point of contact between real teeth during incipient action and subsequent motion to lean in the direction of the rubbing velocity away from collinearity with the contact normal, and the fact that some gear sets might thus be rendered irreversible due to friction, are matters which are irrelevant here. Given rigidity and accuracy, the constancy or otherwise of angular velocity ratio in gear sets is a purely geometrical matter, unrelated to questions of force.

An equivalent linkage

12. Figure 2.02 shows two hinged contacting *spoons,* 2 and 3, mounted back to back (§1.07). Looking ahead, these spoons may be seen to be related to the *gear bodies* back to back at §3.35. We see however that, whereas the gear bodies of figure 3.10 are of carefully defined shapes and known dimensions, together producing a straight path for the point of contact Q in the frame 1 of the mechanism and a fixed k, the spoons of figure 2.02 are, as yet, quite generally shaped. At figure 2.02 all three paths of Q — they are traced in the links 1, 2 and 3 of the mechanism (§0.13) — are curved in general, and are of unknown shape; and, with the spoons, k is not constant. The shapes of the spoons themselves are relevant only in the given configuration and at the given instant. Looking back, we note that figure 2.02 relates directly to figure 1.01. See also in figure 2.02 the revolutes R at the frame link 1. This 3-link mechanism may be seen to represent in general a pair of gear

§2.12 GENERAL SPATIAL INVOLUTE GEARING

teeth in contact at Q along with the relevant shaft axes. Point Q is eclipsed in the figure but the tangent plane at Q, called also the *plane of adjacency* there, is clear to see. *Now the fact is that the mechanism of the contacting spoons may be replaced, equivalently for all small displacements and their first derivatives with respect to time namely velocities, by the 4-link loop RSSR as shown.* The straight line through Q normal to the plane of adjacency at Q is called, universally, the *contact normal*; and the rod (S–S), link 4 of the equivalent loop, is arranged collinear with this. The rod may be of any length but, without loss of generality here, it has been cut to such a length that the transmission angles at the S-joints are both 90° at the instant. Links 2 and 3 of the RSSR and the shown cylinders tangential with the rod (S–S) have radii a_2 and a_3. See also the corresponding angles α_2 and α_3 between the shown velocity vectors and the rod. In a design process aimed at finding properly shaped teeth for a skew gear set of given k, a proposed Q may be chosen, on first trial, to be anywhere. Next a proposed contact normal at Q, a rod (S–S), may be drawn to be anywhere through Q, provided it is tangent to two circular cylinders coaxial respectively with the two shaft axes (see figure 2.02) in such a way that the ratio $(a_2 \cos \alpha_2)/(a_3 \cos \alpha_3)$ equates with k. This latter, characteristic of the RSSR, is a well known relationship. It is independent of the length of the rod. This length will later be designated Z_j.

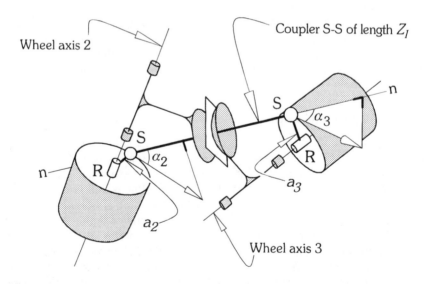

Figure 2.02 (§2.12). An equivalent 4-bar (RSSR) loop for the offset skew gear set. Link SS is collinear with the contact normal, the revolutes R are at the shaft axes, and the links a_2 and a_3 are the common perpendiculars as shown. Ratio k is $(a_2 \cos \alpha_2)/(a_3 \cos \alpha_3)$.

In practice Q will be chosen within the as yet vaguely defined yet limited region of the meshing wherein the shapes of the teeth are as yet unknown. The direction of the proposed contact normal will be within some reasonable limits too, due mainly to problems of interference and considerations of force. The region chosen for the meshing may be quite remote from the pitch line of course, as it is for example in worms and hypoids [6].

13. Figure 2.02 requires further explanation. Let the backs of the spoons be the *front* of a driving tooth and the *front* of a driven tooth in contact at Q. The shaft axes to which the spoons are rigidly attached are shown, and the contact normal n-n at Q is containing the coupler S-S of the instantaneously equivalent RSSR. It is important to notice, however, that in real gears there is a difference between (a) a *reversal of function*, when the driver becomes the driven (and vice versa) and the teeth remain in contact at their fronts, and (b) a *reversal of travel*, when the force at the contacting fronts of the teeth becomes zero, and mechanical catastrophe is averted by the immediate presence nearby of the supporting *backs*. We thus speak about *drive* (when the torque is in the one direction) and *coast* (when the torque is in the other).

14. As has been said, the RSSR of figure 2.02 will be instantaneously transmitting at the required angular velocity ratio k provided $(a_2 \cos \alpha_2)/(a_3 \cos \alpha_3) = k$, *but we must remember that the imaginary rod S-S of figure 2.02 cannot sustain tension*. We need to see in other words that the imaginary S-joints there have sockets less than hemispherical, and that, for change from drive to coast, we need a similar but different equivalent linkage. Another way of appreciating this is to see that we are dealing here with a pair of gears which have working profiles on one flank only of their teeth and can thus sustain and transmit torque in one direction only. An understanding of this will become important later.

Enumerative geometry

15. There follows now a commentary based on the figure 2.02. It uses the language of enunmerative geometry [1] [§4.01]. Although the results will be instructive, this whole paragraph may be omitted upon a first reading. Take any two cylinders coaxial respectively with their shaft axes, and consider the number of common tangents that may be drawn. We may choose any point on any one of the cylinders (say S) and observe that there is containing S a single tangent plane to that cylinder, that that plane intersects the other cylinder in an ellipse, and accordingly that, through S, two distinct tangents may be drawn to the ellipse and thus to the other cylinder. This duality reflects the double valued nature of the skew gear problem; having chosen a direction of rotation for one of the wheels, the other may be chosen to rotate in either of the available directions; thus we become aware of the acute and the obtuse angles Σ; this is not the same phenomenon as the one of the two kinds of reversals discussed above incidentally; it is another; see [1] [§22.20]

and the photo-frontispiece to chapter 22 in [1]. But on each cylinder there is an ∞^2 of points S and so, without counting each line an ∞ of times too often, and disregarding the number 2 because twice ∞ is simply ∞, there is an ∞^2 of common tangents that may be drawn to the pair of cylinders. Now let each of the cylinders be of any radius, and thereby see that the number of common tangents that may be drawn to all possible pairs of cylinders is ∞^4, namely the number of all of the lines in space. We shall see however that only a tiny fraction of these, namely only an ∞^3, is the somewhat limited number of common tangents that may be taken as legitimate rods SS and thus legitimate contact normals among the actual teeth of the gear set, for given k. All this suggests a certain layout of the legitimate contact normals, namely that of a *linear complex* [1][§9.12, §9.26, §11.30], and, suspecting that, we might try to show next that, through any chosen Q, for a given k of course, only a single planar pencil of contact normals may be drawn. One may show by computer graphics, as the author has done [25], (a) that the perpendicular dropped from Q onto the pitch line is a possible rod S-S and thus a legitimate contact normal, (b) that a second line through Q, perpendicular at Q to the line just mentioned, but inclined to the pitch line at such an angle ϕ that $\phi = \arctan(p/q)$, is similarly a legitimate contact normal, (c) that all lines through Q coplanar with the two just found are legitimate contact normals, but (d) that no other line through Q is a legitimate contact normal. The reader might also show to his or her satisfaction that, upon a generally disposed plane, a single infinity of legitimate contact normals lie which are concurrent at a single point [1] [§11.35].

In screw theory the fundamental law manifests itself

16. Confirmations and proofs of the fundamental law could emerge directly from the above remarks. Independently of that possibility, however, the said law may be seen to manifest itself directly from within the *principle of reciprocity* that arises in the field of screw theory and kinetostatics [1] [§14.01]. Two such manifestations, by means of which, incidentally, the author first saw the truth of the law, follow. For the sake of those who might wish to avoid the kinetostatical arguments to be offered next (§2.17-20), two other arguments, in the terms of two different kinds of vector algebra, are offered later (§2.21-23). Refer just now however to figure 2.01. The axes P-x and P-z are the pitch line and the CDL respectively, P (at the origin) is the pitch point, and the rest of the figure is self explanatory.

17. *First manifestation.* The joint at the actual meshing in any gear set may be replaced equivalently at the instant by a screw joint connecting the wheels; this (call it H) must be coaxial with the pitch line and its pitch must be p. Thus the 3-link loop of an actual gear set may be seen to become, at the instant, a 3-link RHR loop, overconstrained and with zero travel, but with apparent mobility unity. Joint H will exhibit its 1-system of motion screws, namely the single motion screw of the two real wheels. But within H there will exist, as well, the reciprocal 5-system of action screws upon whose sepa-

rate screws may reside the wrenches that could be transmitted in the absence of motion (or in the presence motion if the joint were workless) by the joint. This 5-system contains an ∞^4 of screws upon an ∞^1 of coaxial complexes, each of which contains an ∞^3 of screws of the same pitch [1] [§23.77]. The particular complex having screws of zero pitch is the complex in question here, because we have, not up to as many as five points of contact within H and thus the transmission of wrenches under consideration, but only a single point of contact and a single contact normal within the real meshing. The mentioned complex is the locus of all contacts normal at points Q within the meshing; and this is the complex of lines consisting of all lines inclined at angle ϕ to the pitch line where $\phi = \arctan(p/q)$. It may accordingly be said that $p = q \tan \phi$ gives the locus of the limited number of lines at the meshing which are legitimate contact normal lines, the limited number being only ∞^3, there being an ∞^4 of lines in space from which to choose them.

18. I have not ignored here the severality in tandem of points Q. Provided the law is obeyed at all points Q, the severalty of Q does no more than render the mechanism overconstrained, which demands, in turn, no more than mere accuracy of construction, already assumed. In gears, the number of teeth in contact at any one time changes as the teeth continue to let go and re-engage but, if, as is commonly done in the simple case of planar involute gearing, we measure the effective length of the straight path of Q (along which path the speed of Q is constant if the angular velocities of the wheels are constant) and divide this by the base pitch of the teeth, we obtain the so called *contact ratio*. This ratio clearly needs to be greater than unity, and it may, according to kinematic circumstance, be found to be as high as 2 or even 3. Switching, I wish to write now a second manifestation of the main law, not to be repetitive, but to show that the law is ubiquitous. The following may be seen to be a needlessly exotic exercise for the imagination, and bizarre but, in so far as one can ever offer a proof of a law, this is a proof; it may be seen to come to one, mysteriously.

19. *Second manifestation.* Put the massless, frictionless, 3-link mechanism of some skew gear set into a black bag [1] [§1.31]; let two rigid material extensions occur, like bat handles, one protruding from each wheel of the gear set, and let these handles protrude beyond the bag; take one handle in each hand and investigate, blindly, by means of the handles only, the now-mysterious 3-link joint inside the bag. Discover by kinesthetic sense that the freedom of the complex joint inside the bag is unity [1] [§1.26, §14.23]. Become aware also that the pitch of the single motion screw available is finite, and either positive or negative; it is of course our *p*. Grasp the handles firmly, twist and push or pull them towards or away from one another in one of the many ways that results in no motion at the joint (there is an ∞^4 of ways). Relax the condition that zero backlash must obtain; thus see that all transmission forces within the joint will be acting along contact normals at their points of contact, and that this will apply to the single force at the single point of contact *Q* between the teeth. Unless the direction of the contact

normal at Q is such that the force there (residing in its screw of pitch p_a which is zero) can be reciprocal to the incipient rate of twisting at the pitch line (residing in its screw of pitch p_b which is non-zero), there will be a movement; the pitch line will change location to suit the wrong k. After rest is achieved, we can apply Ball's geometrical requirement for reciprocity at the simple joint at Q, namely that $[(p_a + p_b)\cos\phi - q\sin\phi]$ must be zero [26]; we have $p_a = 0$ and $p_b = p$; thus we have for a fixed pitch line, namely for a constant k, that $q\tan\phi$ must be p.

20 The reader may go to [1] [§14.09] for a near direct account of the first manifestation. See also [1] [§14.23] where a mythical gear box is mentioned. Begin at [1] [§11.30] for an account of the linear complex, and at [1] [§23.38] for an account of the general 5-system. At the time of that writing, the writer remained unmindful of the theory of gear teeth, and this of course now surprises him. Refer in advance to §5A.15.

Proofs of the law by means of the vector algebra

21. Without implying that manifestations are not proofs, three other derivations (or proofs) of the law have since been given using the vector algebra. A somewhat laborious proof, worked out however upon the basis of [14] solely, and published independently by Karsai [28], is (to the best of my knowledge) the first proof of the law to appear in print. A second and a third proof follow [27]. In each of these the original work of its author has been paraphrased and transcribed by me to match the terminology of [1] [§12.08]. The second, with its idiosyncratic finite-screw terminology, employs to begin the 6-vector algebra of Plücker [1] [§19.28], while the third is a very short, elegant exercise in ordinary 3-vector algebra.

22. *Second proof (by courtesy Ian Parkin, July 1994).* Refer [27]. The screw axes containing points A and B at the shafts of two gear wheels having angular velocities ω_{12} and ω_{13} are, in Plücker coordinates, the screws, $\Omega_{12} = (\omega_{12};\ A \times \omega_{12})$, $\Omega_{13} = (\omega_{13};\ B \times \omega_{13})$, both of zero pitch. At a generally disposed point of contact Q between teeth, the wheels respectively produce the linear velocities $v_{1Q2} = \omega_{12} \times (Q - A)$, and $v_{1Q3} = \omega_{13} \times (Q - B)$. A requirement that the angular velocity ratio k be constant is simply that the points Q_2 and Q_3 instantaneously coincident at Q should neither separate nor dig in. This requires that we can find a direction at Q, such that the projections of v_{1Q2} and v_{1Q3} upon that direction are equal. We can find such a direction (there being a whole polar plane of them at Q), and we must adopt one of them for the contact normal there. While the difference of the velocities projected upon the tangent plane at Q will be evidenced by the sliding there, the difference of the velocities projected upon the contact normal might be set to zero

THE FUNDAMENTAL LAW OF GEARING §2.23

to establish the criterion for constant k. The criterion is, accordingly

$$
\begin{aligned}
0 &= [v_{1Q2} - v_{1Q3}] \\
&= [\omega_{12} \times (Q - A) - \omega_{13} \times (Q - B)] \\
&= [(\omega_{12} - \omega_{13}) \times Q - (\omega_{12} \times A) - (\omega_{13} \times B)] \\
&= (\omega_{12} - \omega_{13}) \times Q + [(A \times \omega_{12}) - (B \times \omega_{13})],
\end{aligned}
$$

which is just the criterion for reciprocity [26] between (a) the line (n; Q × n) of the contact normal, having direction n and passing through Q, and (b) the screw $[\omega_{12} - \omega_{13}; (A \times \omega_{12}) - (B \times \omega_{13})]$, which is the difference-screw of the screws at the wheel-shafts, namely the screw at the pitch line.

23. *Third proof (by courtesy Karl Wohlhart, August 1994)*. Refer [27]. Let u and n be unit vectors in the directions of the pitch line and the contact normal respectively. Let q and ϕ be the shortest distance and the angle respectively between these two. Taking origin O on the pitch line at the foot of q (figure 2.01), erect a new O-x along q, a new O-y along the pitch line, and a new O-z right handedly. Let position vector r originating at O locate the point of contact Q. With link 2 fixed (3 may move), erect the velocity vector v_{2Q3} at Q. The orthogonality condition, namely that $v_{2Q3} \circ n$ be zero, demands that v_{2Q3} (namely v) be drawn perpendicular to the contact normal. By definition, angular velocity ω_{23} (namely ω) is in the direction u, and by virtue of rigidity [1] [§5.49], the linear velocity of a generally chosen point in body 3 (namely v) has its component in the direction u to be $p\omega$. So we have

$$v = p\omega + \omega \times r,$$
and
$$v \circ n = 0,$$

which together give

$$
\begin{aligned}
0 &= [(p\omega + \omega \times x) \circ n]/|\omega|. \\
&= (pu + u \times x) \circ n \\
&= p \cos \phi + (u \times x) \circ n \\
&= p \cos \phi + (n \times u) \circ x \\
&= p \cos \phi + u \sin \phi \circ x \\
&= p \cos \phi - q \sin \phi
\end{aligned}
$$

which is the same as

$$q \tan \phi = p.$$

This compact analysis involves not only the orthogonality condition. It employs as well the pitch line and the general architecture of the gear set. If k were not constant, the analysis would be invalid.

On what there is to follow

24. Matters arise in this chapter now as follows. First I comment qualitatively on the non-involute hypoids. These are of course well known to many; and I count myself among this many. They are however not well understood by many; and nor by me. I try nevertheless to write a short account of what the problems must surely be with hypoids in the absence of a fundamental law. I give a brief and probably inadequate survey of the hypoid literature, followed by a much more detailed survey of the less copious, but also less empirical, general spatial involute literature. I go on to mention some other laws, make some comments about gearing in general and, at the end, I write an appreciative account of the pioneering work of Robert Willis [12]. The applicability of the fundamental law in the processes of design and inspection is of course important, especially in the area of involute gearing as we shall see, but methods of application are of secondary importance at the present stage. Here I deal with the law itself. In later chapters I deal with its application.

On the non-involute hypoids

25. I have been vague so far about the contact geometry in present-day skew gearing. Most such gearing is openly seen, and said to be *non-involute*. This remark applies to both its conception geometrically and its methods of design and manufacture. I actively leave aside here all those kinds of gearing which are based on no genuine theory of conjugality at all; these are often presented quite frankly as being kinematically wrong, while yet, somehow, as being satisfactory. The hypoids that have followed Wildhaber [6], however, *are* based upon the idea of conjugality (§0.03); and, assuming kinematic rigidity, there is said to be a *curved line of contact* between the moving mating teeth and a *curved surface of action* traced out in space by the said curved line [15] [16]. It is also said, speaking in terms of flexibility, that the curvature of the profiles meeting at the curved line are carefully examined for their elongated patches of contact, the likely hydrodynamic lubrication (HDL), and thus for the power to weight ratio. It must however be a fact in such gearing that, if contact along the curved line and the constancy of k across teeth are both to be maintained, the contacts normal at all points along the line must, at all stages of the meshing, *each obey the fundamental law*. Otherwise trouble will occur [56]. It is not possible to trace the separate paths of separate points of contact in such non-involute gearing unless (a) the ideal, overconstrained conditions of curved line contact are not in fact met, which is likely, or (b) the teeth after first trial are modified by means of crowning, tip-relief or otherwise to ensure what boils down to being a single point of contact [80]. In light of this, possible paths of single points of contact are, in general, twisted paths with no easily definable shape [81]; and the imagined single force being transmitted, as the imagined single point of contact moves, must continuously vary in magnitude, direction, location, and speed at the point of

application. It must moreover be said that, since line contact is actively pursued in design and often almost achieved, accuracy of assembly (involving centre distance C, shaft angle Σ, and the separate axial locations of the two wheels) is of paramount importance. It is therefore not surprising that much of our present-day skew gearing emits noise, exhibits discernible vibration at tooth frequency, is difficult to design, complicated to manufacture, and expensive. It might be said very roughly that efficiency drops from near unity (sometimes greater than 0.98) in well cut involute spur gears down to 0.85 (± 0.05) in hypoids.

Design in the absence of the fundamental law

26. It is possible with involute, planar, spur and helical gears to engage in the various well regulated processes of tooth design without thinking about the fundamental law. Thanks to de la Hire, Euler, the cycloids, the involute, Willis (§2.46) and the others, the fundamental kinematics of planar gearing for the maintenance of constant angular velocity ratio is no longer a trouble; the law is (in its planar form) obeyed; conjugality is inherent in the undistorted shapes of the teeth; the matter is quite settled for involute teeth in the plane.

27. On looking first into space, at the well known *crossed helicals*, we can again overlook the fundamental kinematics, even though unexplained phenomena abound. Whatever the tooth numbers, we do know that a crossed pair will mesh satisfactorily and afford constant angular velocity ratio provided both wheels derive from the same basic rack. But where, now, have the contact normal and the pitch line gone? Calculated wheel radii taken directly from the mean angular velocity ratio (the tooth ratio) appear to bear no relationship with the physical location of the meshing, and we loose sight of the well known planar maxim that the contact normal must pass through the pitch point. But this latter is, as has been pointed out (§2.07), only a special case of the law and quite inadequate here. These deceptively simple, crossed helicals are more complicated (and somewhat less satisfactory) than they look (§5B.60).

28. On stepping fully and openly into space, where what I wish to call the *offset skew gear sets* operate, and leaving aside for the meantime my remarks at §2.08, we meet our difficult problems in design almost entirely without the fundamental law. In 1970 and again more recently [23] [24] geometrical statements have been made which do bear upon the conditions that need to be met for the maintenance of constant angular velocity ratio; but otherwise Mac Cord [18], Buckingham [13], Wildhaber [6], Baxter [15] [16], Dudley [16], Dyson [19], Shtipelman [20], Litvin [5], Minkov [56] and Minkov, Abadjiev and Petrova [21], with the exception of Lagutin [46], have avoided the central issue of the missing law. They all rely on the often unspoken fact that the gears are, in the first instance, conjugate by virtue of their method of manufacture (§0.02, §2.01, §2.25), and they reveal the same on-

going disjunction in the general thread of argument that I wish to mention shortly. The disjunction exists, in my view, because writers have not been separating properly the questions that need to be asked in the kinematics of skew gearing: and these questions are, firstly, whether or not the methods of manufacture and assembly accurately ensure the conjugality that they allegedly offer; secondly, which are the relevant relatively moving links, where are the relative instantaneous screw axes, what are the pitches at those axes, and why not study the kinetostatics.

Sketchy account of the (non-involute) hypoid literature

29. In his paper about basic geometry and tooth contact in hypoid gears, which summarises and clarifies important parts of [6] to be mentioned later, Baxter [15] explains (a) that, as the designer of hypoid teeth shifts his attention from his first intelligently guessed *mean point* M in the middle, as it were, of a tooth in the middle of its changing relationship with its mating tooth, backwards to the beginning and forwards the end of that relationship, little attention can be paid to the question of constant angular velocity ratio, and (b) that, having effected a trial design for that tooth and its hopefully conjugate mate (paying attention of course to the need for a practical method of manufacture), the designer must next smooth out the transition which occurs at the *transfer point* between the end of that tooth relationship and the beginning of the next. Thus Baxter speaks in effect of a bumpiness in the angular velocity ratio, the bumpiness reflecting the passage of teeth. He says next that by careful first choice of M and the subsequent reduction of errors to unspecified but minuscule proportions, consequent disturbances can be made so small that they are absorbed by various elastic deformations and the lubricant.

30. In the arena of worms, exact details of the various theories of design are hard to find. In Buckingham [13] and generally in Dudley [16], however, we do see the importance of curvatures, firstly of the principal curvatures at the points of contact (or at the curved lines of contact), and secondly of the relative curvatures there in consideration of improved (namely enlarged) patches of contact, lubrication, heat generated, force transmitted, and thus of power transmission.

31. Dyson [19] discusses gearing mathematically and comes close in some places to some of the issues being addressed by me here. He identifies the same screw axis as I do (the pitch line with its pitch as I openly call it), and he explains that Wildhaber [6], who was the first to write about hypoids, also did. Shtipelman [20], writing a clear work about design and manufacture, also mentions the pitch line and its ramifications usefully and unambiguously. Litvin [5] identifies the same axis, locates it, speaks about the system of helices surrounding it, and writes an equation like mine for its pitch. But all three of these writers fail to exploit the significance of what they have done, addressing themselves mainly to questions of contact along the curved lines

and the curvatures there. Study some recent work at reference [63].

32. Dyson, Shtipelman and Litvin, and, more recently, Minkov [56], Minkov et al. [21] and Abadjiev [101], which latter also deal with aspects of hypoid and spiroid gearing, each stress however that the dot product of the sliding velocity at the point of contact and a unit vector in the direction of the contact normal at the point of contact must be zero, thus stating in effect that the velocity of sliding must occur in the tangent plane at the point of contact or, conversely, that the contact normal must be normal to the tangent plane. Many workers have called this the *basic equation of meshing* [22]. The equation is of course correct, and it must be attended to in design, but, in the harsh context of gear law, especially when carefully generated teeth (§2.33) are subsequently modified to overcome the otherwise inevitable overconstraint in the gear set, it is not enough. The matter requiring consideration, namely the disjunction of which I speak (§2.28), is that the contact normal needs to be located legitimately, *not only within its local environment at the continuously changing point of contact, but also within the overall architecture of the two shaft axes and the pitch line.* Conjugality, which may be inbuilt by virtue of generation in the manufacturing method, accurately or otherwise (§2.08) or not at all (§2.25), is an important feature, and it must be kept in mind, but modifications made to properly generated shapes will always lead to trouble. The following further references and anecdotal information might be found to be interesting: there is a book by Pismanik, K.M., *Hypoid Gearing*, Maschinostroenie, Moscow 1964 (in Russian); Shtipelman [20] (now in the US) was a student of Pismanik; Minkov-Petrov (Bulgaria) was a student of Litwin (then in Leningrad). There was recently a book by Nieman, G. and Winter, H., *Maschinen Elemente*, Springer-Verlag, circa. 1980-85.

The manufacture of hypoid gearing

33. This paragraph is a very sketchy account indeed. Its contents come from my own limited knowledge of current hypoid gearing manufacture, and might, therefore, be largely discounted as a non-definitive document. Much of the modern theory associated with accurately cut hypoid gearing is implicit in the special machinery used for its manufacture. *Dudley's Gear Handbook Second Edition (1991)* is a valuable source of information. Chapter 20 of this compendium is the collective work of Hotchkiss, McVea, and Kitchen [50], all three of the Gleason Works, Rochester New York. Paying attention as might be expected mainly to American practice, these authors deal with the details of the machine manufacture of hypoid gearing in a comprehensive way. Chapter 21 of the same compendium, whose author is Shoulders [74], of Reliance Electric, Columbus Indiana, deals in a similar way with the manufacture of worm drives and spiroids. There exist, no doubt, corresponding sources of information for the German and Russian industrial practices, which probably differ from one another. On standards, there is the German DIN. In Germany there is the Sigma Group (a combination of Klingelnberg, Oliken of Switzerland, Lorrenz, and Hurth), and the well known Zhanrad

Fabrik, ZF. Gleason use indexing, one tooth at a time with circular fly cutter for their wheels, while Sigma, non-indexing, use a continuous cutting action, going steadily around the wheel. Very recently Simon [80] wrote a succinct, general account of design and production methods, and clearly explains how transmission error (expressed as a very small angle) can be measured.

Some eclectic theoretical works

34. I mean by *eclectic* here that the mentioned authors have a foot in both camps. They are in the camps of (a) the non-involute hypoid workers, and (b) the non-hypoid involute workers. In any skew gear set there is the unique pair of parallel planes containing the two shaft axes. Korestilev [23] wrote (and I paraphrase): *the two points pierced in these planes by a legitimate contact normal will have their distances from the shaft axes in proportion to the angular velocity ratio.* Korestilev did work against the background of the general architecture (§2.32, §5A.19), and he did take an important step towards establishing the fundamental law as I study it here; but the exact location of his contact normal was hidden implicit in his statement and not explicit in relation to the pitch line. He neither defined the pitch line nor located his contact normal in relation to it.

35. Lagutin [46], writing later, picks up the work of Korestilev, mentions it, and puts it properly into context within his own study of the architecture, the meshing space, and the pitch line. Paying careful attention to the relevant line geometry, he touches upon such crucial indestructibles as the cylindroid, the parabolic hyperboloid, and the linear line complex, all of which continue to figure in the problems of gearing today.

36. Dooner and Seireg [24] dealt preliminarily in that paper with what they called non-circular gears, namely pairs of screw-pivoted, single-toothed *gear bodies* in contact. These might be seen to resemble the contents of my figures 3.09 and 3.10. They made an isolated assertion there that appears to run in parallel with what I have said here. They treated a single pair of gear bodies (with its fixed frame-link) as 3-link function generating mechanism where constancy of angular velocity ratio was generally not being sought. Expanding their symbols into my words for transcription here, they said (on page 87), "it is conceivable that candidate points for tooth contact may exist anywhere in space provided the line coincident with the tooth's surface normal is reciprocal to the instantaneous screw axis determined by the screw pivots at the shaft axes and the angular velocities." But this is tautological. Given the stated circumstances of the two bodies (namely their freedom), the contact normal actually determines the ratio of the angular velocities and the screw axis. The said contact normal and the said screw axis will of course be reciprocal whatever happens. The question at issue is whether the screw axis is freely determined as I have here explained, or whether it is predetermined by the intention of the designer to construct a pair of multi-tooth, circular, skew wheels for continuous meshing, with a chosen tooth ratio and thus with

a chosen, hopefully constant, angular velocity ratio as well. Writing more recently in their book [49], they clarify their view significantly. They say there among other things that there are *three* fundamental laws (none of them mine), *and that the linear complex of legitimate contact normals does indeed lie at the root of the problem.* I can agree with that. The book however deals in general with that kind of gearing where the teeth of one arbitrarily cut wheel generate the teeth of the other, where curved line contact is accordingly the norm, and where detailed machine design and likely methods of manufacture are not discussed extensively.

Survey of the general spatial involute literature

37. I wish to turn now to that somewhat remote part of the gear literature that relates most directly to the main subject matter of this book. I mention, in chronological order, Olivier, Mac Cord, Disteli, Poritsky, Giovanozzi, Beam, Merrit, Mitome, Smith, and Innocenti. These authors either mention as a possibility, or imply it, that offset skew involute gearing as explained in this book is likely, somehow, to be found possible. Olivier (1842) [40], with an eye to straight line contact between teeth, laid important foundations and produced some isolated special cases (§3.61); Mac Cord (1883) [18] enlarged upon Olivier, set out the likely shapes of real teeth, but made some crucial mistakes (§3.61); Disteli (1901-1911) [77] struggled successfully to set down and to clarify his new screw geometry; Giovanozzi (1947) [66] and Poritsky (1948) [67] independently studied the interaction of mating involute helicoids and thus introduced the idea (indeed the inevitability) of point contact; Merrit (1954) [58] spoke of *conical* gearing as an overall desirable concept, producing his figure 10.11 (which is my figure 9.01); Beam (1954) [59] wrote an incisive mathematical paper about the geometry of *tapered* gearing — or *conical* gearing, using Merritt's term — which was known at the time by the name Beveloid. Beveloid is not a generic term; it was then (and it still is) the propriety name for the kinds of wheel produced by the Invincible Gear Company, Livonia, Michigan, USA; Mitome (1981) [60] discussed at some length the machining (by hobbing) of Beveloid teeth; Smith (1989) [61] described the actual commercial manufacture of Beveloid gears at Invincible; and Innocenti (1997) [62] has written convincingly about the phenomena surrounding the adjustment for backlash in such gearing. See also [103]. All of these writers, however, except for Mac Cord, Merritt, and Innocenti, appear to have regarded the *conical, the taper, the beveloid,* gear as somehow being a single isolated wheel, a somewhat independent entity designed to mesh in unspecified ways — except for the known, phantom rack and the regular Beveloid set with intersecting axes — with an existing cylindrical helical gear, an ordinary straight bevel gear, another similar Beveloid, or a real rack. Nowhere in this special corner of the literature, however, with the exception perhaps of the works of Olivier, Mac Cord, and Disteli, do we find a direct attack on what might be seen to be the general central question. The general central question might be seen to be: *given a pair of shaft axes (which are skew),*

given the centre distance C (which is collinear with the CDL), given the required angular velocity ratio (which must be a vulgar fraction), given the location in space of the meshing (which is not at, but somewhat remote from, the CDL), and given that the teeth might be, conical-tapered-beveloid on both of the wheels, how can we construct to the best advantage the shapes of the teeth? In this connection it is important to be aware that Bricard (1926) [82] proved for planar gearing the reverse argument (the argument in reverse of the usual argument) that, if we require variation in centre distance without loss of constancy of the angular velocity ratio, *the well known planar involute shape for the profiles of the teeth is the only solution that can be found; the planar involute shape, in other words, is for adjustable planar gearing inevitable.* No one to date, however, appears to have discovered the likely, spatial analogy of this, Bricard's reverse argument for the general spatial case. I say now (but I say it with some timidity because of the said missing analogy) that this book makes another attack upon the general central question.

Other laws

38. There are of course other laws and/or theorems concerning the kinematics of gear teeth. Refer for example to figure 1-10 by Baxter in Dudley [16], read section 1-6 and find, from the second law stated there for planar gearing, that any completed body of law for gears in space will need to have a second (or some other) law dealing (as this law does) with continuity of the motion and the curvatures. The Euler-Savary geometry concerning the relative motion of two laminae in the plane is twice applied in figure 1-10 by Baxter, firstly for motion between the first wheel and the basic rack, and secondly for motion between the basic rack and the second wheel. Addition of the two resulting E-S equations will reveal his given formulae. Refer [29]. Refer also Skreiner [30] for an incisive study of the spatial analogue of the Euler-Savary geometry; this may well be useful. Read also Dooner and Seireg [49] where they give what they call the first, second, and third laws of gearing according to their ways of analysis. *But please remain aware overall that so far as I am concerned there is only one first, fundamental law, which alone expresses the main matters comprehensively.* It has already been the central subject matter of this chapter.

Some observations in respect of all gears

39. The fundamental law taken in isolation falls a long way short of what we need for effective syntheses of actual gear teeth. It becomes however a sure guide in the processes of design, especially with unfamiliar k and Σ, and it can be seen to have its clear role in the checking of already cut, non-involute teeth. Refer to figure 2.01. The fundamental law may be seen to be insisting that, under all circumstances, the contact normal (which emerges as it were from the unseen mysteries of the meshing) must run on to cut one of the mentioned helices surrounding the pitch line in such a way that the fol-

lowing three lines at the point of intersection, (a) the tangent to the helix, (b) the radius of the helix, and (c) the contact normal itself, are mutually perpendicular. The contact normal must be binormal to one of the helices in other words, there being, as we know, always one (and only one) of the helices available for this [1] [§9.26].

40. This brings us to remember that, in elementary texts about ordinary planar gearing — which texts might deal, for example, not only with involute action but also with the older, cycloidal action — a figure like figure 2.03 might sometimes be found. This figure is intended to convey of course that the fundamental law, although pictured in this way instantaneously in the plane, must be applicable continuously and smoothly across all configurations of the set if the gear-tooth profiles are to be satisfactory. Versions of figure 2.03 (which is planar) are difficult enough to draw convincingly, but refer to figure 2.04. This figure, non-existent here as yet, is left as an exercise for the reader. It will recreate in the fullness of three dimensions the same impression, *namely that there are no restrictions laid by the fundamental law upon the shape of the path of Q.* Both q and ϕ may continuously change as the spatial action proceeds but, for a constant angular velocity ratio in all gears of whatever kind, *the law must continuously be obeyed*.

41. All of this brings into focus the great importance not only of the fundamental law, but also of the pitch line. *It is vital for any valid theory of skew gearing (a) that some unique line (the same line for all kinds of gears and gear teeth), possessed of some clear kinematic function and definable unambiguously, be available as a central theoretical concept, and (b) that its definition be consistent with that for the pitch line as we currently understand the term in ordinary gears with parallel axes.* The definition given at §1.14,

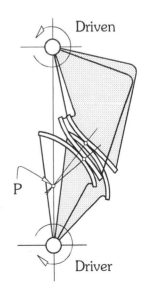

Figure 2.03 (§2.40). Typical text book illustration of the fundamental law as it applies in the special case of planar gearing. The law is well known [12]. For constant angular velocity ratio k, the contact normal must at all stages of the meshing pass through a fixed point, namely the pitch point (called P), on the CDL. It is moreover inescapable that P divides the CDL exactly according to k.

along with the relevant remarks at various places elsewhere in this book, fulfil these criteria. The pitch line is well established here. It is important, precise in its definition, and thus amply warrants the figures 1.03, 1.04 and 1.05 which have been presented.

42. The pitch line and the fundamental law can effectively guide the processes of synthesis and analysis, all of which involve, at the present time, both iterative computing and sophisticated metrology; and without the law, it must be said, these processes will continue to be essentially directionless. It might be mentioned again and parenthetically that skew gears with offset involute action are being written about. See §2.37, then chapters 3, 4 and 5 of this book. The geometric gist of their spatial involute action arises directly from the pitch line and the fundamental law. Notwithstanding that, the following further remarks about the pitch line and the law and their ramifications need to be made in respect of all gears.

43. As soon as the overall configuration of a proposed gear set has been determined, or, in the case of an existing set, discovered by inspection (we may measure C and Σ directly and count the teeth for k), the important quantity p may be calculated immediately; see equation (12) at §1.14. With an existing set, we can check the handedness namely the sign of p with certainty and its magnitude approximately by manual manipulation. To estimate p, we dissemble the set, fix the bigger wheel, engage the teeth and slowly roll the smaller wheel through one radian by hand; we detect (as we roll) the linear motion in the direction of the pitch line (which will itself be moving) and we note not only this motion, but also the magnitude p (mm/rad) of the movement. See figure 13.06 in [1].

44. See [31] for a recent guide to some present methods available in practical CAD. There are clear possibilities nowadays for numerical syntheses of new gear teeth where the importance of accuracy in the maintenance of constant angular velocity ratio might well outweigh cost. For we have, now, iterative computer methods and NC machine tools. Paper [32] addresses itself to a few early obvious steps that might be taken in non-involutes in the new light of the fundamental law.

45. The practical business of checking the inexactitudes of already-cut skew gear teeth (which inexactitudes are due of course not only to the absence, hitherto, of a fundamental law but also to the necessarily approximate methods of manufacture and subsequent adjustments to deal with elasticity, vibration, noise, wear, lubrication etc.) needs to be discussed elsewhere. These and other problems of design, manufacture and inspection certainly obtrude, but, for constant angular velocity ratio k, and given the usual preliminarily accepted conditions of accuracy and rigidity, we may now do the following: *we may (a) locate the contact normal at Q, (b) look along the pitch line to see the true length of q, (c) look along q to see the true size of ϕ, and then (d) know that q tan ϕ must always be p.*

Robert Willis 1838

46. In chapter 4 of Dudley's *Gear Handbook Second Edition*, 1991, p. 4.8, under the sub heading *Conjugate action*, Dudley himself writes, 'The fundamental requirements governing the shapes that any pair of these curves may have is summarised in Buckingham's "basic law of gearing," (*sic*) which states, "normals to the profiles of mating teeth must, at all points of contact, pass through a fixed point located on the line of centres."'. He makes immediate reference to Buckingham, E., *Analytical mechanics of gears*, McGraw-Hill, New York 1949, thus, apparently, attributing the law to Buckingham [13]. Since its discovery more than 150 years ago, this law has been variously named, but I wish to name it here, for the sake of argument, *the Willis fundamental law of gearing for planar gearing*. Robert Willis (1838, 1841, 1870) wrote long ago and at least three times about the law. The law is mooted in his paper "On the teeth of wheels", *Trans. Civ. Eng.*, vol ii, 1838, pp. [find these, I have them], and well mentioned in both the first and second editions of his famous book, *Principles of Mechanism*, 1841, 1870. See the 2nd edition, chapter V, §116 (second paragraph down), to find that Willis said there (and I quote exactly) the following:

> *Any convenient curve being assumed for the edge of one revolving piece, if we can assign such a form of another revolving piece that the common normal of the two curves shall divide the line of centres in a fixed point in all positions of contact, then will these curves preserve a constant angular velocity ratio when one is made to move the other by sliding contact.*

Willis in a footnote at §164 of the same book mentions his main mentor Leonard Euler [70]. Euler proposed in his paper of 1767 the involute curve for the profiles of gear teeth, but he appears to have had no clear idea of how to generate (namely to *make*) the shape by appropriate mechanical processes. Willis knew about (and probably discovered) how to generate the involute profiles of teeth by means of reciprocating straight-sided rack, and he wrote — see his §164 *et seq.*, and especially his §171 (p. 126) — extensively about that. For more about Willis see [12]. He was a major contributor to the theory of gear teeth, and his *fundamental law of gearing for planar gearing*, as I have chosen here to call it, was the direct forerunner of my first paper, since [1], in gearing [14].

CHAPTER THREE
MATING INVOLUTE
HELICOIDS

Introduction

01. It has long been a matter for conjecture that a rational theory might be written for the constancy of angular velocity ratio among offset skew gear teeth. Speaking in terms of rigidity (in the absence of elasticity, vibration, wear and so on), the attainment of exact constant angular velocity ratio in spur and helical gears by means of involute action has been an exploitable and exploited possibility for 150 years [12] [13]. Matters have been at least explicable, also, among the so-called crossed helicals [33]. Within the wider range of general spatial gears, however, the geometry of an acceptable, involute transmission with exact angular velocity ratio has been seen, as yet, as being perhaps discoverable [59] [61], but mysterious [48]. During the recent forties, two important papers appeared on the subject of mating involute helicoids. They came from Giovanozzi [66], and Poritsky and Dudley [67]. Each of them showed that any two involute helicoids, axially mounted upon two shafts generally disposed, and in contact a single point, would under all circumstances (a) transmit rotary motion with uniform angular velocity, and (b) do so with the point of contact moving in a straight line. Various editions of a regular text by Giovanozzi [68] have since consolidated the basic material. This means that some of the aspects of this chapter 3 have been known before by others, but the fresh treatment here exposes new material that can be used effectively in mission-oriented design. I mention again Beam and the Beveloids [59] [61]. Since Beam, the Beveloids have been known, generically, as *taper (or conical) gears*. They were doubtlessly discovered during the testing of previously untested 'taper' methods of machining, but they are, on the other hand, best explained, as we shall see, by means of the 'cylindrical' principles outlined in this book (§8.02).

02. In chapter 1 of this book, following [1], we (a) laid a foundation for the kinetostatics of skew gears in screw theory, and (b) established the pitch line as a pivotal aspect in the overall architecture of gear sets. In chapter 2 we (c) formulated the fundamental law of gearing. In this chapter 3 (and in those to follow) we go on to show (d) how offset involute action among skew gear

teeth may be achieved, (e) how suitably shaped teeth for such action may be synthesised for actual multi-toothed wheels, and (f) how simple machine generation of such teeth by means for example of ordinary NC milling machine or by special hobbing may be achieved. The involute theory that follows applies whatever the centre distance C, the shaft angle Σ, and the velocity ratio k. Everything explained here applies in other words to all involute gear sets of whatever kind, all skew crossed-axes sets (whatever their ratios k and whatever their layouts), and by virtue of their being unsurprising special cases, planar spur gear sets and ordinary cylindrical helicals with parallel axes as well. Although the presentation makes its analyses with an attempt at rigour, it also departs from this from time to time. It sometimes goes into unsubstantiated hypotheses in the realms of gear synthesis and tries to deal in advance with untested, real gear technology.

03. This chapter 3 may be summarised, preliminarily, as follows. The curved line of intersection between a circular hyperboloid and its related involute helicoid is explained. The line of intersection is called the *slip track* because it is destined, ultimately, to become the path traced upon the flanks of involute teeth by the point of contact between teeth in involute skew gearing. Two involute helicoids, erected in space upon two skew axes, are put into contact at a single point Q and caused to drive one another. We study the straight line traced by Q, the curvatures of the mating profiles at Q, and the constancy of the angular velocity ratio k. We thus begin to uncover the kinetostatics of offset involute action. We locate an important triad of planes intersecting orthogonally at Q, *namely the transnormal, the polar plane, and the best facet*. Once the helicoids here are themselves established, the constant k of their interaction is found to remain independent of the distance C and the angle Σ between their axes. This means that (a) *transmission error*, an angular velocity error occurring at tooth frequency in non-involute gears [33], and (b) various difficulties upon assembly in such gears, namely those arising from accuracy and overconstraint [34], might now be avoided by means of involute teeth. The involute helicoid of an involute tooth profile is screw-symmetric about is own axis (which axis is the shaft axis in actual gearing), and the pitch of the involute helicoid is the pitch of its *core helix* which is also coaxial with the shaft. Thus we draw attention to the possibility of machining such involute teeth by ordinary NC milling machine. The even more striking possibility of being able to hob such teeth becomes apparent later, in the chapters 5.

Circular hyperboloids proliferate

04. The hyperboloid is one of the so-called quadric surfaces whose Cartesian point-coordinate equation may be written in terms of the second order algebra [1] [§11.13]. In the event of its being a ruled surface, it has an elliptical throat. There are two systems or families of rulings or families of generators [1] [§11.13]. In general the inclination of a generator measured against the plane of the throat varies as we move around the periphery. In gearing,

however, the ruled hyperboloids commonly met are circular hyperboloids; they are mere surfaces of revolution; and the said angle of inclination remains a constant. Refer to figure 3.01 and see there the distinction I wish to make between (a) the fixed angle of inclination to the plane of the circular throat of the generators of a circular hyperboloid, which angle I wish to call the inclination angle α of the hyperboloid, and (b) the fixed angle between the generators and the direction of the central axis, which angle I wish to call the twist angle τ of the hyperboloid. In this and the following chapters we meet several different circular hyperboloids having quite different significances, and it will be wise by then to have clarified this. Here the angles α and τ are clearly complementary; they sum to $\pi/2$; but please read note [90].

A straight path for the point of contact

05. In classical gear-involute theory (which is based very largely upon planar geometry), there is in the reference plane (the plane of the paper) a single point of contact Q between the profiles of mating teeth, and the path of this point is well known to be a straight line in the reference plane. In the actual physical event of real spur gear teeth with axial width, of course, contact takes place along a straight line of points Q normal to the reference plane and this line of points traces out a series of parallel paths forming a *plane of action* in the space. Correspondingly in the classical theory of non-involute gearing, which naturally involves the spatial geometry of conjugate action, there is seen to be a *curved line* of contact between teeth which traces out a *curved surface of action* in the space. In non-involute gearing (and this includes, just now, all crossed-axes gearing except the crossed helicals), the path of a single point of contact Q is only a vague notion because there is,

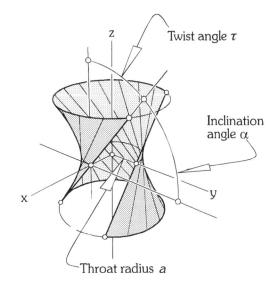

Figure 3.01 (§3.04). Only two parameters, throat radius a and inclination angle α, are enough to define a circular hyperboloid. The twist angle τ is also shown here. Angles α and τ are complementary; but please read note [90].

due to the curved line of contact, a multiplicity of points Q and a confusion of their separate curved paths with one another as the relative motion takes place. Nor are straight lines for the paths of these Q envisioned or intended. Designers of offset skew gearing aim, in the first instance, at achieving full curved-line contact between the flanks of mating teeth. *If this aim were achieved, however, the fundamental law (§2.06), with given p, would need to apply at all Q. Otherwise trouble would occur* (§2.08). Keeping in mind the known questions of flexibility, local compression and gross tooth movements due to bending etc., designers however next modify their discovered flanks by crowning or otherwise, (a) to ameliorate troubles due to the inevitable mislocation of shafts in the otherwise overconstrained circumstance of line contact, curved or otherwise, (b) to permit the use of available machining processes, and thus, incidentally, (c) to converge upon a discoverable Q with a definable path. Litvin and Gutman [81] appear to have studied such paths.

06. *It is, on the other hand, and speaking in terms of rigidity, an aim of this chapter to introduce the notion that a straight path for a single point of contact in skew gears may be actively pursued in the first instance.* A straight path with a continuously collinear contact normal (§3.07) is a first requirement towards the banishment of vibration at tooth frequency, consequent noise, and the sensitivity of loaded teeth to fracture [35] [36] [83].

07. To clarify the continuously instantaneous nature of the fundamental law and to accentuate its applicability under all circumstances, I avoided discussing in detail in chapter 2 the various generally curved paths that might be taken by the point of contact Q in gears. I did speak about the contact normal (a straight line through Q by definition) and, for constant k, its correct though generally changing location with respect to the pitch line. But I did not speak about another straight line, *the line travelled by Q* in involute gears. I thereby avoided explaining that, when the path of Q is a straight line, as it is with involute gears, there is the somewhat confusing, doubly special circumstance that *the contact normal remains always collinear with the said path*. Before going on with the analysis of skew involute action and the proper design of involute skew gear teeth, however, I wish to demonstrate some kinematical certitudes. They are new to gearing, geometrically exact, and germane.

Two ruled surfaces intersecting perpendicularly

08. Refer to figures 3.02 and 3.03. Take upon a central axis o-o a circular hyperboloid of throat radius a, the generators of whose doubly ruled surface are inclined at angles α to the plane containing the throat circle. *I have, incidentally, not overlooked the fact that the symbols a and α used here are the same as those that were used at the RSSR in figure 2.03.* Next choose an initial point A on the said throat circle at some angular displacement Ω about o-o from some origin. Take a point B upon one of the two generators intersecting at A, thus nominating a segment of line (A−B) upon that generator

MATING INVOLUTE HELICOIDS   §3.09

whose length is hereby designated H [37]. Having chosen a generator we now choose to work with that family only of the generators. Next mount a plane at B normal to the segment (A–B), and allow the said segment together with the said plane to move successively from generator to generator in such a way that A remains at the throat circle, that (A–B) occupies successive generators, but that B traces a smooth though twisted curve which cuts every generator perpendicularly. Let the path of B, in other words, take an *orthogonal trajectory* across the ruled surface. The moving plane at B will generate an envelope, and this, because each successive plane cuts each next plane in a straight line, will be a ruled surface. Clearly, the rulings of the envelope will cut the relevant rulings (but not the other rulings) of the hyperboloid at all points along the path of B perpendicularly.

09. For this to happen as described, however, H must vary smoothly in some specified way with Ω. We might, accordingly, first try to write an expression for $dH/d\Omega$ and thence, by integration, to obtain an equation connecting the mentioned variables in the form, $H = f(\Omega)$. Refer to figure 3.02. We see the throat circle of radius a (fixed), the point A (movable), and the angle α (movable with A but fixed in magnitude). We see an initial location of the segment (A–B) marked with its length H, and, following a small change $d\Omega$ in Ω, the new location (C–D) of the initial (A–B). The segment (A–B) drawn here is not the same as the segment (A–B) at figure 1.01 incidentally; this is due, quite simply, to a shortage of symbols. See at figure 3.04 the two corresponding locations of the plane; the locations are represented there by facets; the facets are circular portions of the planes with centres conveniently at B [1] [§7.20]. Point A has moved a distance $a\,d\Omega$ along the circumference of the throat circle to its new location C; B has moved to its new location D; the new generator CE has been drawn; and the angle BDC has been constructed a right angle. Point E is that point on the new generator that B would have reached if length H had not been obliged by the rules to change. Useful as mere stepping-stones in the present calculation, the pair of equal radii from o-o to B and E, and the angle EBD at B, are designated s and η respectively; see these at figure 3.02. The facets shown at figure 3.04 may be seen to intersect in a line j-j which, in the limit, as $d\Omega \to$ zero, becomes the mentioned ruling at B of the mentioned envelope. Refer to appendix A at 3 (§3.75) for an algebraic argument which shows that H varies directly with Ω. It is shown that $H = c_H \Omega$, where $c_H = a\cos\alpha$. *This means among other things that the length of the intercept measured along any chosen generator of the hyperboloid between any two consecutive turns of the said envelope, which envelope is indeed the ruled surface we already know by the name involute helicoid, is a constant.* The constant is [$2\pi\,a\cos\alpha$]. See a representative intercept of this length indicated at figure 3.03.

10. See, in figure 3.02, the right angled triangle WAG. As Ω varies, the said triangle remains similar in shape and orientation. Its side (W–A) remains parallel with o–o. Because (A–B) is by definition the perpendicular dropped from vertex A onto the side (W–G) of the said triangle, the smaller triangle

§3.10 GENERAL SPATIAL INVOLUTE GEARING

WBA also remains right angular as Ω varies. And, as Ω varies, B travels the slip track. These relationships permit the construction in figure 3.03 of the successive facets (each one seen in the limit as $d\Omega \to 0$) of the ruled helicoidal surface shown. They permit the construction also of the slip track. The ruled surface is the envelope of the succession of planes at the moving B. One of the facets at the moving B is shown touching the surface at the repre-

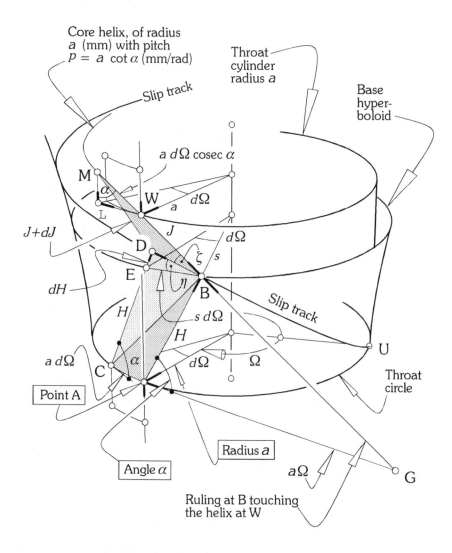

Figure 3.02 (§3.08). Omnibus mathematical sketch showing the layout of the problem and the parameters needed for calculation. The slip track and the helix intersect and are tangential at U. Refer to the text and appendices for details of the calculus.

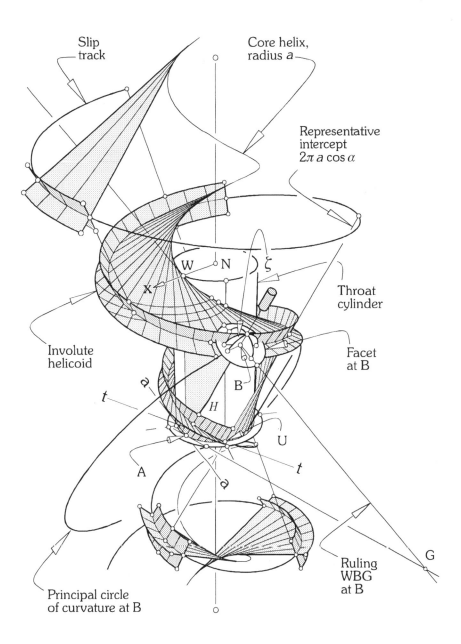

Figure 3.03 (§ 3.08). Omnibus geometrical figure showing among other things the slip track formed at the intersection of the base hyperboloid and the involute helicoid. See the generating plane of the surface of the involute helicoid — the surface is a developable — occupied by the so-called facet at B, and be aware that the surface could be cut by a saucer-shaped milling cutter suitable controlled, or by a hobbing cutter, as we shall see.

sentative B. The facet contains the ruling at B and its axis of symmetry (shown with a knurled knob) is collinear with the line AB. I wish to introduce just here (and I speak in advance of the general argument) an activity for the imagination. *Please observe that, as the rolled-up 'taper-cut paper' WAG is 'unwrapped' from the fixed circular cylindrical core, the line of the taut, tapered edge (W–G) of the paper generates the involute helicoid.* See in advance figure 5B.08 at §5B.31. I ask the reader to be aware, as I point this out, that we are working here in the open context of general spatial involute gearing; we are not being restricted to the mere special case of the parallel cylindrical helicals, where such an argument about the unwrapping of paper is commonly and usefully employed. The point I am trying to make is that, as we unwrap this paper in the imagination, we should see the scene as being useful *for application in general.* We will wonder, in due course, why we should find such a simple fact to be surprising. Go in advance, for a thought provoking example, to §5B.61; but do not proceed to §5B.62 at this early stage; §5B.62 will probably appear, just now, to be unintelligible.

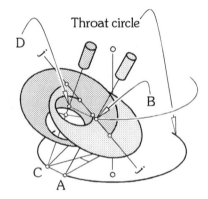

Figure 3.04 (§3.09). Two successively drawn facets at B (see figure 3.03) intersecting along a ruling j–j of the involute helicoid.

Figure 3.05 (§3.09). The triad of planes intersecting at B, (i) the tangent plane to the involute at B, (ii) the plane normal to the ruling at B, and (iii) the plane containing the said ruling and the relevant ruling of the hyperboloid. We see also (iv) the plane containg B that is not a member of the triad; it is the tangent plane to the hyperboloid at B. We meet this figure again at §3.12.

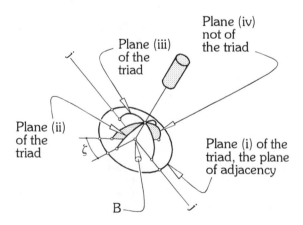

Imagined milling of the involute tooth profile

11. Be aware that the facet at B represents the whole expanse of the plane at B. That the circular facet drawn at B has had B for its centre is of no special significance. The facet may with equal significance have been put (with its axis) anywhere upon the plane it defines. This plane, the tangent plane at B, touches the ruled surface not only at B but at all points along the ruling at B, namely the infinitely long line FBG. The ruled surface is, accordingly, *developable*. Refer to the two successive, intersecting facets at figure 3.04 and see that these remarks (naive as yet) are intended to convey some early hints about the cutting of involute teeth by means of milling machine.

The core helix and the twisted line of intersection

12. We have seen that each ruling of the just-constructed ruled surface resides (a) perpendicular to its corresponding relevant generator of the hyperboloid, (b) in the plane of its corresponding angle α, and thus (c) in a plane tangent to the throat circle and parallel with the central axis o-o. Seeing this, it might be said that the involute helicoid is wholly characterized by its *a* and its α, namely its (a, α). Each of the planes WAG thus defined is therefore tangent to a circular cylinder coaxial with o-o whose radius is *a*. We can next find easily that each ruling resides (d) tangent to the same single helix upon the said cylinder. I wish to call this helix the *core helix* of the involute helicoid, and to note in passing that its helix angle is none other than α. *We see that the ruled surface (a, α) comprises, quite simply, all of the tangents to the said helix, the pitch (mm/rad) of the helix being a cot α.* See figure 3.06 which shows, in a series of three orthogonal views taken against the central axis how, for a given throat radius *a* of an hyperboloid, the pitch $a \cot \alpha$ of the core helix increases more and more rapidly towards infinity as the inclination angle α of the hyperboloid diminishes steadily towards zero. The implications of the different degenerate geometries of this figure, which occur when $\alpha \to 0$ and $\alpha \to \pi/2$, will need to be (and will be) investigated later.

13. We have seen that H varies directly with Ω. $H = c_H \Omega$, and $c_H = a \cos \alpha$. The length of the spiralling path of B (which is shaped like a bedspring) is infinite. There is an infinite number of turns, and the distance between turns measured along a generator of the hyperboloid is a constant, namely $2\pi c_H$, namely $2\pi a \cos \alpha$. Notice that the smoothness of the twisted path of B is not interrupted as we cross the throat circle (where the radius of the path is least), but that the path appears to undergo some special changes there. The path (the twisted path of B) is tangential with the core helix at the throat circle. See the line of tangency marked t-t at U in figure 3.03.

14. Unlike the turns of the twisted path of B which differ, as we have seen, *all turns of the ruled surface itself are the same*. Helicoidal, the system of rulings repeats itself endlessly in the direction o-o and, because all of the rulings are tangential to the core helix and thus to the mentioned core cylin-

§3.15   GENERAL SPATIAL INVOLUTE GEARING

der, the core cylinder encloses an infinitely long, coaxial, circular hole among the rulings. *Unlike the circular hyperboloid which in this analysis gave birth to it, the involute helicoid, screw symmetric about its own axis, has no throat.*

The involute helicoid and its whole slip track

15. In figure 3.03 a projecting portion the *throat plane* is drawn to look like a collar; it cuts the indicated *throat cylinder* orthogonally at the *throat circle*. Refer to those upper parts of the figure which appear above the said throat plane, and project the rulings there past the point of tangency with the helix. See for example the projected ruling at W. Find that the ruled surface has two branches, one of which is the one shown there, while the other is the one shown in the lower parts of the figure below the said throat plane. At the risk of confusion, refer in advance to the figures 3.13. The ruled surface is a known developable; it has a cuspidal edge (also called a line of regression) at the core helix. It is known in both the gear and the pure mathematical literature by the same name, *involute helicoid*. Whereas in the one, however, it is seen to be obtained by 'twisting' the plain involute flank of a spur tooth to produce the well known flank of the so called helical tooth on a cylindrical

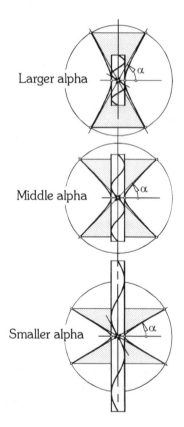

Figure 3.06 (§3.12). Longitudinal squashing-in of the base hyperboloid leads to a longitudinal squirting-out of the core helix. For a given a, the configuration depends entirely on α. It is instructive to study what it might mean, for gear teeth in the limit, when alpha tends to zero or to $\pi/2$. If, at one of these limits, planar spur teeth are portended, what is portended at the other?

72

blank, see for example [13], it is seen in the other to be defined directly in terms of the helix itself, see for example [38]. For brevity I will sometimes call (a) the involute helicoid just introduced the *involute to its base hyperboloid*, (b) the core cylinder of radius a about axis o-o the *throat cylinder or the base cylinder of the base hyperboloid*, and (c) the intersection of the two surfaces namely the path of B, for reasons yet to be made clear, *the slip track upon the involute*.

16. Figure 3.03 shows the slip track upon the involute. It is the bold, openly drawn, twisted line running from top to bottom of the figure. It marks the intersection of the two surfaces and is shaped, as has been said, like a bed spring. It is important to see that, at $\Omega = 0$, the slip track and the core helix are tangential with one another at the point marked U. See the straight line t-t in the figure which is tangential to both curves at U. The core helix is, indeed, as has been said, the aforesaid cuspidal edge. See in advance the figures 3.13. Are there, on account of the two branches, two slip tracks? Do we somehow switch from one to the other as we cross the throat? While on the one hand this question relates to certain benign aspects of planar gearing, it helps to explain on the other certain self destructive aspects of the crossed helicals. I return to this somewhat difficult matter at §3.59.

17. Whereas the segment (A−B) namely the distance along the generator of the hyperboloid from A to B has been called H, the segment (A−W) namely the distance along the throat cylinder from A to the point of tangency W might be called I, and the segment (W−B) namely the distance from the said W to B along the ruling, might be called J. Thus we can see by simple trigonometry that the following equations apply:

$$H = a\,\Omega \cos\alpha = c_H\,\Omega$$
$$I = a\,\Omega \cot\alpha = c_I\,\Omega$$
$$J = a\,\Omega \sin\alpha = c_J\,\Omega$$

The rulings are defined by these equations. For any one involute helicoid, both a and α remain fixed while Ω varies. It will thus be seen that, as Ω increases, the right-angled triangle ABW (with its right angle remaining at B) remains with its proportions intact as B 'unwinds' itself from the throat cylinder (§3.10). The side (A−W), namely I, of the triangle moves circumferentially with Ω and grows in length with Ω, but remains a generator of the throat cylinder as the unwinding takes place. The pitch of the core helix, which is the pitch of the whole ruled surface, is $a \cot\alpha$ (mm/rad). Notice also that $dI/d\Omega = a \cot\alpha$, which is not surprising.

18. I wish to mention parenthetically that any planar section through an involute helicoid taken normal to the axis o-o will reveal a planar involute curve (of the well known planar sort) emerging from its planar base circle of radius a. A few of these are shown in figure 3.03. The shape of the said planar involute depends solely upon the radius a of the throat circle, is independent of the pitch of the helix, and, by virtue of the prevailing axi-symmetry, independent of the location along the axis of the sectioning plane. Its

equation, in terms of the length A of its tangent, is $A = a\,\Omega$. See this clearly in the planar involute spur and helical gears. In general, however, the slip track, when projected along o-o onto a plane normal to o-o, becomes a *planar spiral* ranged about the throat circle and defined by the equation for the length B of its tangent, $B = a\,\Omega\cos^2\alpha$. Compare this with the corresponding equation for the *planar involute* just mentioned. These planar curves are both spirals; but neither of them is an Archimedean spiral; Archimedean spirals involve radial distances from a point; the spirals here might better be called *involute* spirals because they involve tangential distances from a circle.

The whole geometric construct

19. In summary we may write as follows. Let Ω define the angular location of the rotating throat radius a about an axis o-o of a circular base hyperboloid, and let α be the angle between the generators of one family of that hyperboloid and the plane of its throat. Let a system of lines (the rulings of the related involute helicoid), comprising the tangents to a helix of pitch $a\cot\alpha$ coaxial about o-o and defined by the same a and α, intersect the hyperboloid in a twisted curve. This curve, the slip track, is shaped like a bed spring, and at all points along it the generating planes of the envelope, which envelope is the said involute, are normal to the generators of the hyperboloid. These planes occur at distances H from the throat circle such that $dH/d\Omega$ is a constant c_H, where $c_H = a\cos\alpha$. Thus we have, ignoring the constant of integration which merely defines the origin for Ω, $H = c_H\Omega$. This statement permits the involute and its slip track to be seen as being either (a) the helicoidal system of its generators tangent to the helix and the curved line of intersection between that system and the base hyperboloid, or (b) the path traced by the point B upon the hyperboloid along with the envelope of rulings generated by the moving plane normal to the generator at B whose distance H from the throat circle varies according to $H = c_H\Omega$. *More embracingly, we should see the whole combination of parts (the hyperboloid, the helix, the involute helicoid and the intersection of the surfaces namely the slip track) as a unified, whole geometric construct defined entirely by its sole parameters a and α.* The involute can be constructed seemingly independently of the hyperboloid upon which it depends, and the path of B upon the involute may, accordingly, be begun at any point upon its surface; but the facts of these interdependencies (or apparent lack of them) are hidden in the values of the constants c_H, c_I and c_J as we know. These remarks thus circularly stated will be useful when we come to the practical business of synthesising actual skew gear teeth. Notice that the varying angles ζ and σ obtrude, and that the various curvatures of the surface at B might eventually be important. Look again at figure 3.06 and be aware that, whereas in this chapter 3 we are using a relatively large angle α for the sake of clarity in these demonstrations of the basic geometry, we most often use in real gears a much smaller angle α. See chapter 4 and the chapters 5. But see also §7.06 *et seq.*, where we begin to look at *worm drives*.

Brief look at a special (the planar) case

20. To see what I have in mind by speaking as I did above about the base cylinder and its implications, take the special case of the hyperboloid and its involute, defined as explained by a and α, which occurs in the limit (a remaining finite) when α goes to zero. The generators collapse into the plane of the throat circle, all of them becoming tangential to that circle; the general equation reduces to $H = a\Omega$; the throat circle becomes the so-called base circle in planar gear terminology; and the path of B becomes the well known, planar involute to its base circle. Notice that the whole construct in space becomes in the limit, as α goes to zero, a parallel set of rulings, the rulings all running parallel with the axis of the base cylinder and accordingly normal to the plane of the throat circle namely the base circle in gear terminology. The twisted curve of the path of B blooms [1] [§11.03] into becoming a surface which is the same surface as the one traced out by the parallel rulings. Now consider the flanks of ordinary planar involute spur gear teeth, and begin to suspect what there is to follow soon.

The triad of planes mutually perpendicular at B

21. Please go back to figure 3.05. At any point B along a slip track, we have a mutually perpendicular triad of planes. First there is (i) the tangent plane to the involute which contains the ruling at B; this is plainly shown as the largest facet at B. Next we have (ii) the plane normal to the said ruling, and (iii) the plane containing the said ruling and the relevant ruling upon the hyperboloid. For the sake of future reference (§10.28) I wish to call this triad of planes *the curvature triad* (§3.24). Please note next that the hyperboloid, unlike the involute helicoid, is not a developable and, as we go from point to point along any chosen ruling of an hyperboloid the tangent plane at the point changes; it continues to contain the chosen ruling, but it rotates as it goes so as to contain at each new point the other ruling as well. In figure 3.05 at B there is shown a fourth plane (iv), not a member of the triad. This plane contains not only the relevant one but both generators of the hyperboloid at B; it swivels about the line of H within the triad as B moves; it osculates to the slip track; and it is inclined at angle ζ to the triad as shown.

The angles ζ and σ and the curvatures at B

22. This and the following two paragraphs may be omitted upon a first reading. Between the slip track and the ruling at B we have the angle ζ. It may be shown that $\tan \zeta = \Omega \cos \alpha \cot \alpha$. This means that, for a given α, $\tan \zeta$ varies directly with Ω, which is tidy. Refer to the inset at figure 3.07 for a sketched variation of ζ against Ω, and to appendix B at 3 (§3.76) for the associated algebra. For reasons associated with lubrication in real gears (if for no other reason), it would seem that small angles ζ should be avoided. This result is a beneficial one for skew involute gearing, for we can (as we know) always put the point of contact and the travel path somewhat remote from

§3.22 GENERAL SPATIAL INVOLUTE GEARING

the CDL, thus avoiding the smaller angles ζ. It should also be noted that as the point Q in its path (see later) approaches the throat circle of either of the base hyperboloids, that is, should the point of contact be allowed to run on until it approaches the end of its allotted allowable span Z_l, the slip track approaches tangency with the ruling at Q (§3.58 *et seq.*).

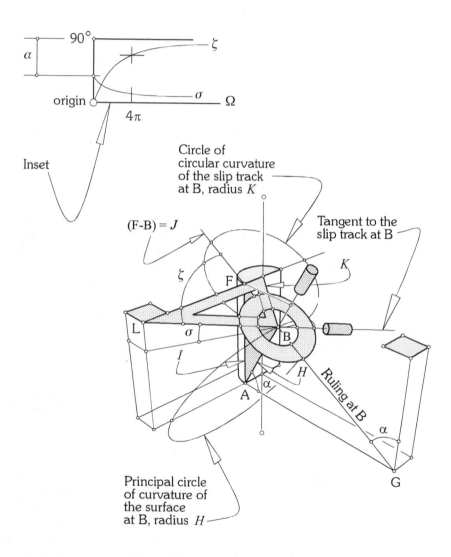

Figure 3.07 (§3.22). The geometry of ζ and σ, and (at inset) the two sketched curves of thier variation with Ω. These angles relate to friction, lubrication, non-reversibility and so on.

76

23. Taking o—o vertical, angle σ is the slope of the slip track; see figure 3.07 for σ; this σ is not the angle η of figure 3.02. Angle σ resides in a vertical plane parallel with o-o whereas angle η does not. Angle σ is somewhat smaller than η. By easy argument in figure 3.07 we can see that $\sin \sigma = \cos \alpha \cos \zeta$. When $\Omega = 0$, $\zeta = 0$, $\sin \sigma = \cos \alpha$, and $\sigma = \arcsin (\cos \alpha) = \pi/2 - \alpha$, which checks correctly with the known geometry of the slip track at the throat of the base hyperboloid. Angle σ has to do with either free transmission of motion or jamming due to friction, non-reversibility, and so on. See the variation of σ with Ω sketched at figure 3.07.

24. The smallest of the two principal radii of curvature of the helicoidal surface at B (the largest of the two being infinitely large) is none other than H. See appendix C at 3 (§3.77). Refer also Bär [53], a careful examination of which might confirm this. The relevant circle radius H may be swung (about the axis a-A-a) in the plane (ii) of the triad at B, which plane is normal to the ruling upon the surface as is shown in figure 3.03. It may also be shown by the same token that the radius of curvature of the slip track at B is none other than the line-segment (B—N). We have had, for the sake of continuity, H, I and J, so let us call (B—N) by the name K, and see that the relevant circle of curvature radius K resides in the plane (i) of the triad, namely in the plane of the triangle NBL. See figure 3.06 for both of the radii K and H. The shown circles of curvature there are not tangential with one another at B; they reside in planes that are perpendicular but, because their centres are not both in the plane (iii), they intersect at B.

A mating pair of non-alike involute helicoids

25. Refer to figure 3.08. This explains the geometry of an imagined kinematical arrangement being newly presented here. Two different involute helicoids (each with its own base hyperboloid, core helix and slip track) have had their geometric axes mounted in space upon two skew, rotational axes o-o. Each of them is in rolling and sliding contact with the other across a common tangent plane *at a single point of contact Q*. The common tangent plane at Q has already been illustrated at figure 2.01. This plane will sometimes be called, from here on, the plane of tangency at the point of contact, or, more often (and indeed more accurately), *the plane of adjacency at Q* (§2.12). Following convention the number 1 has been used for the fixed, or the frame link of the relevant 3-link mechanism; and this link might be said to be, indeed, the *gear box*. Both the gear box itself and its number 1 are omitted, while the mating helicoids have been given the next available numbers namely 2 and 3.

26. At this stage of the argument I need to make a subtle switch from symbol B to symbol Q. It could be said that along any slip track freely traversed in the imagination by its generating point B there is a special home-point for B namely Q, where contact is currently being made with a mating slip track. A similar home-point may under certain special circumstances be

§3.26 GENERAL SPATIAL INVOLUTE GEARING

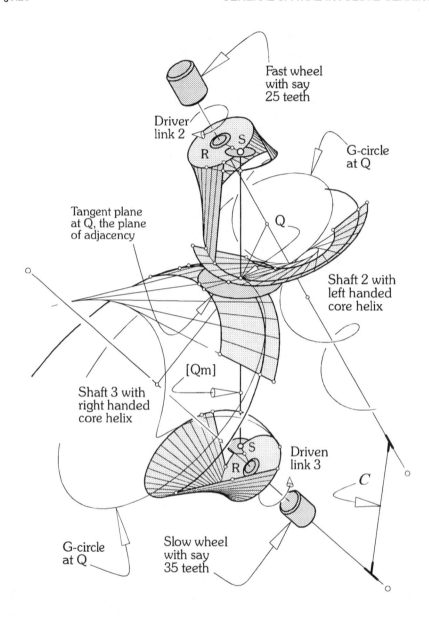

Figure 3.08 (§3.25). The mating of two non-alike geometric constructs. Their involute helicoids are put into contact, both of them oriented convex inwards, at a generally chosen point of contact Q. That done, the axes of rotation of the constructs (wherever thay happen to be) are fixed in the fixed space. Following convention, the fixed space becomes link 1 of the resulting mechanism, while the constructs become links 2 and 3. Provided contact is maintained at the moving Q, either one of the involute helicoids may now drive the other.

MATING INVOLUTE HELICOIDS §3.27

provided for the travelling Q on its geographical path in the fixed space incidentally, but more of that later.

27. Observe in figure 3.08 that a line-segment (S—S) of arbitrary length Z_l has been chosen and mounted, without loss of generality, vertically in space. The whole geometric construct of figure 3.03, its rigid involute helicoid, hyperboloid, helix and slip track (§3.19), has been transferred bodily into figure 3.08 and designated, there, link 2. It has been attached with its point A_2 put at the top of Z_l, its H_2 put collinear with Z_l, and, accordingly, its point B_2 (namely Q_2) put within the length Z_l. Normal to Z_l there is mounted at Q_2 the unique plane which contains the ruling upon, and which is tangential to, the helicoidal surface there. This plane is to become, as we shall see, the plane of adjacency there. We may say that its point Q_2 is coincident with point Q_1 at Q [1] [§12.10].

28. Let us next (a) take any *other* whole geometric construct, whose a_3 and α_3 will differ from a_2 and α_2, this to become link 3 of the mechanism, (b) find its sole particular segment H_3 whose length matches the open gap (namely Z_l minus H_2), and (c) install it with its point A_3 at the other, the bottom end of Z_l and its point B_3 (namely Q_3) at the point Q already established. Notice that, while the shown throat radius a_3 will be perpendicular to Z_l (in the same way as the shown a_2 already is), and while angle α_3 will determine the angle between its o-o and Z_l, *the shown a_3 may be put into any angular relationship (about axis Z_l) with the already installed, shown a_2*. Notice that, although the relevant rulings of the two involutes intersect at Q, they are both touching the tangent plane (the plane of adjacency) which is common to them both. Notice also the planes drawn at the extremities of Z_l parallel to the plane of adjacency at Q. They contain the mentioned radii a_2 and a_3. These radii may now be easily seen in the imagination as the input and output links of the RSSR of §2.12. Recognise (a) the omitted link 1 as the link RR of the instantaneously equivalent RSSR, (b) segment Z_l as the rod SS of the RSSR, and (c) the throat radii a_2 and a_3, each of which resides in a plane normal to Z_l and both of which are accordingly parallel with the plane of adjacency at Q, as the input and output links of the same RSSR.

29. Implicit in this is the fact that, as the involute helicoids continue to mate in this manner explained, *their base hyperboloids are locked into contact with one another along the generator common to them both, namely the straight line Z_l*. The actual mechanics of the contact (which is not pure rolling) must be left until later (chapter 4). *Implicit in this also is a proof of the proposition that Q travels a straight line in space. The straight travel path is along Z_l, and Q (as it goes) marks out upon the mating surfaces the two slip tracks which are shown*. See also in figure 3.08 that, as H_2 becomes longer at its rate $(dH/d\Omega)_2$ as shaft 2 (the driver) rotates in its only possible direction, H_3 becomes shorter at the same rate $(dH/d\Omega)_3$ while shaft 3 (the driven) rotates in its only possible direction. *Thus we see also that, if the apparatus steadily operates with constant angular velocities ω_{12} and ω_{13}, the point Q*

travels at constant linear velocity v_{1Q1} along its straight-line path in link 1. Refer to [1] [§12.08-13]. Note also that all of this is not all together new [62] [66] [67] [68].

30. Notice the shortest distance C drawn between the two axes o-o, and see the directions of rotation at the axes. These directions would be reversed if the functions of the driver and the driven were interchanged, *but, seen as an imaginary gear set, this apparatus cannot operate in reverse.* We must see in the imagination that the imaginary ball joints at the points SS have sockets less than hemi-spherical (§2.14). If the point of contact at Q remains maintained, however, and because as we know $Z_I = H_2 + H_3$, the sum of the radii of curvature at Q will remain a constant, namely Z_I, and the two opposing radii will be evenly matched when Q is at the middle of Z_I.

31. Notice here, and in the work of Innocenti [62], that, from the practical point of view in figure 3.08, the transmission angles within the RSSR are poor [1] [§18.08]. This means that the rubbing velocity at Q would be high in a corresponding physical situation. This along with any significant friction at Q would jam the sliding — in the one direction of coast at any rate. For reversible action (along the other slip track), given friction, we would need smaller angles α, and thus larger angles σ. The question of reversibility in gears is directly associated with these angles α, which angles depend among other things upon k and the angles δ (§2.10, §5B.17, §7.07). The kinematic principles of the motion as described in figure 3.08 apply correctly nevertheless, and they are directly applicable, as we shall see, in the synthesis of gear teeth. For technical reference, the actual numerical data associated with figure 3.08 and subsequent related figures of this chapter (all of which are computer-drawn) are listed at appendix D part (a) at 3 (§3.78).

32. Please become aware from the lower parts of figure 3.03 that special circumstances obtain when Q goes beyond the limits of Z_I. Geometrically speaking the helicoidal involutes can mesh correctly there, but we must see them then as being infinitesimally thin surfaces capable of double-sided action, which action is not possible in real gears. Action beyond the outer limits of Z_I corresponds with internal meshing. This can occur in ordinary practice only with planar gears. For real (as distinct from imaginary) meshing in skew gearing, we must operate with our point of contact Q in the range between the two throats, namely between the two shown, imaginary, ball joints SS, namely within that limited ideal zone in the line of action which is not only (a) called Z_I, but also (b) of length Z_I.

The unimportance of errors upon assembly

33. Look back at figure 2.02. Having erected the kinematic construction at figure 3.08, and having seen that it is a viable 3-link, cam-like mechanism with a fixed frame and two relatively moving, rigid links, we note from the caption at figure 2.02 that the input-output angular velocity ratio may be written $k = (a_3 \cos \alpha_3)/(a_2 \cos \alpha_2)$. But now see this: *given the links 2 and 3*

both fixed in shape with their revolute connections to frame at one of their two extremities and their involute helicoidal surfaces in contact at the other, k remains independent of errors (or intentional variations) made both in the length of Z_l and/or in the angle between the input and output links a_2 and a_3. It thus appears that not only C and Σ but also the *axial offsets* (the locations of the surfaces along their axes) are, without having any affect on k, freely adjustable. We can worry about the origins of and the actual reference points within the surfaces for these axial offsets (namely these distances from the CDL) later (§5B.11, §5B.17 etc.).

34. Let me present here a proof of these hypotheses. See in each of the involute helicoids (each entirely defined by its own characterising a and α) that, due to the prevailing screw symmetry, *the surface contains not only one but an infinity of slip tracks.* The slip tracks cover the whole surface of the involute helicoid, each one of them cutting its own point at the throat circle. Take a generally disposed point upon any involute 2 and another such point on another involute 3, and mate the two points by bringing them tangentially together. With the already determined H_2 and H_3 thus brought collinear, they will set up their own intercept Z_l. With Z_l thus fixed in length, we can erect the rest of the apparatus anyhow, and still be left with the unchanging ratio k. We can even slide one of the involutes along its own axis (so as to separate the surfaces), then rotate it about that same axis to take up the slack. Given the above-said formula for k, which is independent of Z_l, all of this could not be otherwise. There are of course, in gears, the practical matters of tight mesh, backlash and contact ratio to consider. *But these results obtained here point to that favourable property of spatial involute gearing whereby involute action with its openly avowed single-point contact between teeth allows errors of whatever kind to occur (within limits) at the assembly of two properly cut wheels without causing damage to the constancy of angular velocity ratio.*

Two cut gear-bodies put anyhow into contact

35. Speaking more physically now I wish to employ the term *gear-body*. I borrow the term from Dooner and Seireg [24] [49]. The two pieces illustrated at figure 3.09 are gear-bodies, and they suggest, more clearly than words could do, any two generally chosen physical versions of the two mating *whole geometrical constructs* of figure 3.08. The ruled profiles machined upon the said pieces are generally chosen but correctly cut involute helicoids. Each profile has been based upon its own core helix (drawn coaxial with the shaft axis), each helix having been ascribed its own radius a and helix angle α. The rulings have been cut tangential to the respective helices of course, and the helices were chosen to be of opposite hand. The gear-bodies, which shall here be known as body 2 and body 3 (§1.01), differ in their basic dimensions both from the geometrical constructs of figure 3.08 and from one another. See appendix D at 3, part (b), (§3.78) for numerical values. The gear-bodies have been provided, already, with points Q_2 and Q_3 chosen anywhere (but roughly centrally) upon them, and through each of these points

§3.36

has been plotted the relevant slip track (§3.19). At each of them has been drawn the plane of adjacency and the normal to the profile there.

36. Referring to figure 3.10 (and looking again at figure 2.02), we can say that, if the gear-bodies of figure 3.09 are next put into contact with one another in such a way that Q_2 and Q_3 become coincident, the said tangent planes will coalesce into the common plane of adjacency at the point of contact and the said normals will coalesce into the contact normal. Having achieved contact, the gear-bodies will be free to adopt any angular relationship with respect to one another about the contact normal, and, what's more, be able to alter (within the truncated edges of the cut profiles by sliding and/or rolling) the location of the point of contact. Next we can predict that, if the shown, free shaft journals are secured with suitable journal bearings suitably fixed in space, the combination will become a viable cam set with a fixed angular velocity ratio determined solely by the aforesaid expression $(a_2 \cos \alpha_2)/(a_3 \cos \alpha_3)$. Thus, in other words, the angular velocity ratio k for the pair of gear-bodies shown in figure 3.10 is independent of the choice of the point of contact Q and independent of where the two shafts axes might

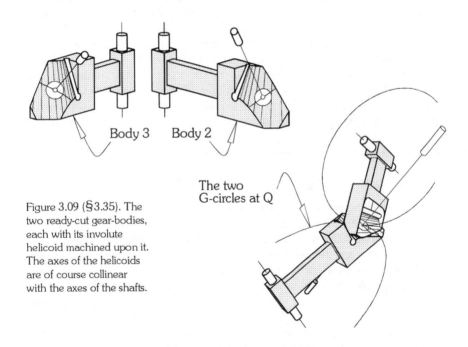

Figure 3.09 (§3.35). The two ready-cut gear-bodies, each with its involute helicoid machined upon it. The axes of the helicoids are of course collinear with the axes of the shafts.

Figure 3.10 (§3.36). The two gear-bodies interacting. Wherever (within bearings) the shaft axes might be located, and provided the profiles remain in contact, the predetermined angular velocity ratio remains constant throughout the whole range of a movement.

MATING INVOLUTE HELICOIDS   §3.37

be located. Seen thus, the facts of the matter are (a) surprising, and (b) of profound significance for involute gearing. It might be mentioned that in figure 3.10 the relative movements of the contacting gear bodies has been graphically checked to confirm, in this generally chosen assembly of two generally chosen skew involute gear bodies, (a) that the path of the point of contact is a straight line as predicted, and (b) that the angular velocity ratio remains constant as predicted. Refer to appendix D at 3 for further numerical details. Refer also to chapter 11 for the construction and testing of prototypical gear bodies derived from an example of real gear design.

37. In figure 3.10 the two circles shown are drawn coaxial respectively with the two shaft axes and they intersect at the point of contact Q. The planes of the circles are thus normal to the axes. They are located in no special way however with respect to the flat physical sides or straight edges of the gear-bodies. We need to study in a general way the geometry of such general pairs of intersecting circles, and we do that soon (§3.41). First, however, I ask the reader to stay with figure 3.10 and to carry out the following investigation. Provide by means of your own imagination a rigid gear-box for the two gear bodies, which bodies are otherwise left floating in the fixed space. Thus provide the 3-link loop of §1.01 with its missing link, link-1. Take the resulting closed-loop apparatus, note the generality of the underlying geometry, and be aware that the straight-line path of Q_1 drawn in link-1 (the path of Q) will pierce somewhere the open-ended region between the two shaft axes. Maintain contact at Q, and ask yourself the following double-barrelled, apparently simple, but very tricky question: *is there a point that we can call the middle of the path of Q and, if so, where is that point?; or, putting the same question in other words, where does some significant, central, definable event occur, as Q passes through the region of the meshing?*

38. The bare geometry, as we have it here, consists in (a) three straight lines fixed in the fixed space, (b) a moving point Q upon one of the lines (which line, the path of Q, is legitimate according to the first law), and (c) the radii of the two circles drawn at figure 3.10, namely the lengths of the two perpendiculars dropped from Q onto the other two lines namely the shaft axes. I might mention that legitimacy, here, means that the rubbing velocity at Q will remains always perpendicular to the path (§5A.08). There is no one single answer to the above question. One might, as a first thought, consider the line of striction around the regulus defined by the three lines [1] [§11.09, §7.57] (§4.60). One might similarly consider the corresponding throat ellipse [1] [§11.13]. One might look for certain conditions that might, at certain locations only of Q, occur in respect of the transversal at Q. One might swing the path respectively about the two shaft axes, thus obtaining the two base hyperboloids for consideration. One might swing the two shaft axes about the path, thus producing two coaxial, circular hyperboloids for consideration. Alternatively, one might naively say that the sum of the two shown circle-radii will arrive at a minimum at some point, and that that point might be taken as the middle point.

39. If we chose this latter and called the thus-discovered point [Qm], we would be locating a point of some significance. For we can see that, for viable through-going action, *and in the absence of other teeth,* the physical edges of the cut profiles of the gear-bodies must be sufficiently spaced to accommodate actual contact at this [Qm]. What this means in the language of [1] [§12.10], is that, if the gear-bodies are to have this through-going action, they must have points $[Qm]_2$ and $[Qm]_3$ fixed upon the real parts of their respective profiles which must, in due course and before the interaction is complete, become instantaneously coincident with one another at $[Qm]_1$. What this means in the special, simple case, incidentally, of ordinary planar (parallel axes) gears is that the so-called addenda and dedenda (in the absence of other teeth) must both be positive. But what it means in the generality of the spatial case is a far more elusive matter. We shall be returning again and again to this complicated question of the location of this and other significant points in the path of Q as Q 'passes through the meshing'. The question is not only complicated but also ambiguous. We cannot say in general what we might mean by the vague phrase 'viable through-going action', unless we specify exactly what the circumstance is that might, in our view, need to be 'gone through' in this way. I wish to mention obliquely here, not the well known Euler-Savary geometry for the general relative motion two rigid laminae in the plane, but its analogue for two rigid bodies in space which was worked out to some extent and discussed by Skreiner [30]. Despite these meanderings, I wish to foreshadow a definitive answer here by remarking obliquely that we will, in due course, be paying attention to an important theorem (as yet unmentioned) which, in brief, might be stated as follows: *there is a unique, other point in the legitimate path of any Q where the transversal at Q, which will, in general, be skew with the FAXL, cuts the FAXL perpendicularly* (§4.27, §4.38).

On the kinematic equivalences

40. Because the action here is involute action, the contact normal at Q and the straight-line path of Q are collinear (§3.07). We see in this connection that, whereas the RSSR at figure 2.02 is only *instantaneously equivalent* to the two-spoon mechanism there, the corresponding RSSR at figures 3.09 and 3.10 are *continuously equivalent* to the mechanisms there. It further needs to be mentioned that, whereas continuous equivalence occurs without limit at figure 3.08, it occurs within limits at figure 3.10. The working profiles of the gear-bodies at figure 3.10 have *edges,* and these edges, in real gears, cannot afford to be forgotten.

The G-circles intersecting at Q

41. In figure 3.08, coaxial respectively with the two shaft axes, are drawn the two circles that pass through the point of contact Q; they are neither coplanar nor tangential; they are intersecting at Q. Refer to figure 3.11(a) which, without loss of generality, is drawn directly from figure 3.08.

MATING INVOLUTE HELICOIDS §3.42

The figures are dimensionally the same and may be superimposed. The circles in figure 3.11(a), obtained by dropping perpendiculars from Q onto the two shaft axes, are the same pair of circles. As implied by the labels, I wish to call these circles, *the G-circles at Q*. Let their respective radii be g_2 and g_3, not forgetting that the magnitudes of these will always depend entirely upon the location of Q, which may, among the architecture, be anywhere. *Certain progressively nominated sub-sets of these, first the special F-circles, and then the doubly special E-circles to be introduced at §5A.21, are significant circles, but these general G-circles being dealt with here are of no special significance. We are dealing here, in other words, with a Q that is, just now, not at any special point in its path.*

42. Looking at figure 3.11(a), see constructed at Q the tangent to each of the G-circles, and see that these tangents (which themselves intersect at Q) define a plane containing Q. I wish to call this plane the *transnormal* at Q. Next we can show by simple geometry — see appendix E at 3 (§3.79) — that the normal to this plane at Q cuts the circle-axes at points j and k as shown. I wish to call this normal the *transversal* through Q. *Notice here that the transnormal at Q is not the same plane as the previously mentioned plane of adjacency at Q which appears both here and in all of the previous figures.* The transnormal at Q is a different and, in many respects, a more important plane than the plane of adjacency at Q.

Circles intersecting in general

43. The matter just explained is a basic piece of pure geometry. One may experimentally examine the facts of the matter by taking three circular cardboard coasters (take beer mats at the bar). Grip one of them (seen as the transnormal) between the generally angled edges of the other two. Ignoring thicknesses, there will be a common point of contact belonging mutually to all three of the mats. See how sensitive the location of the transnormal is to the relative angularity in space of the gripping circular edges of the other two. Become aware most importantly of the fact that the transversal at the point of contact remains normal to the transnormal and together the two make up an axi-symmetric, unchanging geometric entity whose location depends not on one or other of the two gripping mats but upon both of them. But what needs to be noticed is this: *the line of the transversal always cuts both of the axes of the circular edges of the gripping mats.* It must be said that the solidity of the actual beer mats in this experiment prevent the otherwise open circles from choosing to pierce the planes of one another, but in the imagination we must allow for this as an even 50/50 geometrical possibility. The next reference to this important matter occurs at §4.20.

The triad of planes at a general Q

44. At Q, which is not fixed but steadily travels along its straight path as the kinematic action occurs, there is to be found not only a single plane and

§3.45   GENERAL SPATIAL INVOLUTE GEARING

its axis, the above defined transnormal (the plane) and its transversal, *but a triad of planes and axes mutually perpendicular.* Refer to the figures 3.11(a) and (b), the latter being an enlargement of the central portion of the former. The said triad consists of (a) the already mentioned transnormal with its transversal (which depends upon the geometry of the pair of G-circles at Q as we have seen), and (b) two other planes and their axes which depend upon the velocity ratio k. This triad at Q is not the triad at B illustrated previously (§3.21). Nor is it the triad at B when B is at Q, there being two different points B that could be at Q in any case. The planes and axes of the triad at Q, see figures 3.11(a) and (b), are as follows.

45. *Plane #1 of the triad at Q, the transnormal.* For a given pair of involutes in mesh and contacting at Q, this plane is defined solely by the G-circles as explained above. It is also true of course that, through Q, only one line (the mentioned transversal) can be drawn to cut both of the shaft axes, and it is normal to this line at Q that the transnormal lies. I say again that the

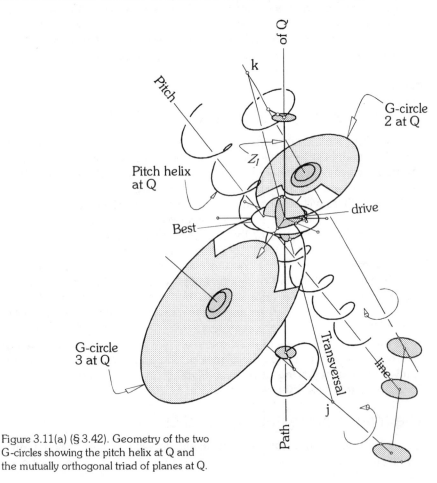

Figure 3.11(a) (§3.42). Geometry of the two G-circles showing the pitch helix at Q and the mutually orthogonal triad of planes at Q.

G-circles are of no special significance: they depend entirely upon the location of Q upon its path. Because the transnormal contains the two tangents to the G-circles at Q as explained, it contains the velocity vectors v_{1Q2} and v_{1Q3} [1] [§12.10]; it thus contains, also, the vector differences v_{Q2Q3} or v_{Q3Q2}, namely v_{2Q3} in one direction and v_{3Q2} in the other, both of which equal and opposite rubbing velocities occur in the direction of the helitangent (§1.41). So the transnormal can be seen to contain not only the vector triangle that determines the rubbing velocity but also the rubbing velocity itself [39]. We need to know here the ratio k of course, which, in the case of the exercise used for the figures here, is 5/7. The rubbing velocity is by physical definition within the plane of adjacency at Q. The plane of adjacency at Q and the transnormal at Q are different planes, but they do intersect in the helitangent (as is shown), and they both, thereby, contain the line of the rubbing velocity. *Normal to the transnormal, as has been said, is the transversal.*

46. As Q moves along its straight travel-path in figure 3.08 the transversal traces a regulus upon an hyperboloid [1] [§11.05]. This transversal has been called by Baxter and others (when relating exclusively, however, to their mysteriously located *middle point* M) the *pitch vertical* [15]; but, because it is neither vertical in any proper sense nor in any clear way associated

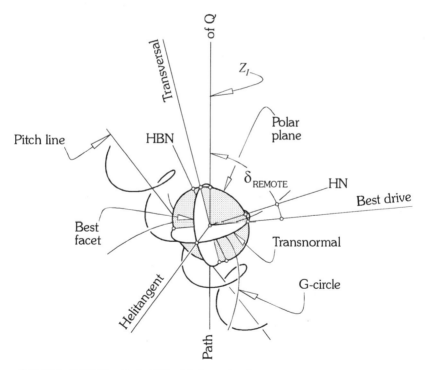

Figure 3.11(b) (§3.44). The triad at Q in detail, naming the planes and intersections. The planes are mutually orthogonal; thus each intersection is normal to one of the planes.

with pitch, I have given it the name I have. At all positions of Q along its travel path, the transversal (the pitch vertical) will be found to be, not surprisingly, perpendicular to the helitangent namely the direction of the rubbing velocity at Q. This first plane of the triad, the transnormal, is also the same plane as that called by Baxter and others the *pitch plane* when Q is at their 'middle point' M [15]. This plane cuts the shaft axes at the apices of the *pitch cones*, but these cones of the Wildhaber geometry hold no special place in the present theory at this stage of the argument because they are here (as they were there) different cones for each point of contact Q, and there is (as yet at least) no intelligibly chosen 'central point' for Q.

47. *Plane #2 of the triad at Q, the polar plane.* For a given pair of involute helicoids in contact, k will be known or calculable, so the direction of the rubbing velocity at Q may always be found. This occurs in the direction of the helitangent as mentioned (but not explained) in §3.45 above. Normal to the helitangent we can erect (at Q) the polar plane. This plane (which depends not only upon the G-circles but upon the ratio k and thus the helitangent) contains the travel path. The polar plane is called by that name by me in the context here because it is, in fact, a polar plane of the linear complex of lines surrounding the pitch line (§1.14) [1] [§9.27, §11.30]; otherwise the travel path contained within it would not be legitimate (§2.24). *Normal to the polar plane, as has been said, is the helitangent.*

48. *Plane #3 of the triad at Q, the best facet.* Defined by the said helitangent and the said transversal this third plane of the triad completes the orthogonally arranged set. The line of intersection of the polar plane and the transnormal is that line in the transnormal along which the components of the velocity vectors v_{1Q2} and v_{1Q3} are equal and in the same direction. This line is where the transmission force at a matching pair of facets (which pair must necessarily contain the helitangent) would be least. Accordingly I call the said line, for the want of a better name, the *best drive*. The matter has to do with kinetostatics [1] [§10.01-75]. This line would also be the best path for Q if it were not for the unavoidable (but vaguely defined as yet) disability angle δ. This is a likely question: where does the best drive go among the helices? And this is an answer: although by definition the helitangent is tangent to the helix at Q, and the best drive is perpendicular to the helitangent there, the best drive is neither the normal (namely the radius) nor the bi-normal to the helix at Q. It is simply some one of the generators of the polar plane at Q. *Normal to the best facet, as has been said, is the best drive.*

49. *Summary of the triad at Q.* The central part of figure 3.11(a) is enlarged at figure 3.11(b). The circular facet indicating the plane of adjacency at Q is there omitted, firstly to reveal details of the triad which would otherwise be hidden by it, and secondly to diminish the relative importance of this already-mentioned plane which is not (in general) a member of the triad. The three planes of the triad mutually perpendicular at Q are #1 the *transnormal*, #2 the *polar plane*, and #3 the *best facet*. The three lines of the

triad at Q are as follows: normal to #1 the transnormal is the *transversal,* to #2 the polar plane is the *helitangent,* and to #3 the best facet is the *best drive*. Each of these three mutually perpendicular lines at Q is accordingly and alternatively defined by its own pair of intersecting planes.

Gyratory motion of the triad at Q.

50. The above-described structure of the triad at Q deserves our careful attention. It should firstly be said that the triad does not remain a fixed thing in the fixed space. It is important to understand that, while retaining its *structure* unaltered, the triad continuously changes its *orientation* with respect to the fixed space as it travels (pivoting spherically about Q) along the path of Q. The helitangent at Q, the transversal at Q, and the best drive at Q all continuously change direction as Q moves; *and, in particular, the angle I wish to call δ_{REMOTE} between the path of Q and the best drive continuously changes magnitude*. The said angle is shown at figure 3.11(b). The symbol δ carries the subscript REMOTE firstly to indicate that, as Q moves, the angle varies, but secondly and more importantly to indicate that, as Q moves, Q is, most often, 'not at home'. I have already given an early hint about the meaning of this remark at the end of §3.39. We return to this and the related matter of an easily definable, unique angle δ among the spatial architecture, *the angle of obliquity,* namely the spatial analogue of the angle called by many in the planar realm *the pressure angle*, soon and again later (§3.54, §4.59, §4.63).

Location here of the pitch line

51. The triad at Q, as we shall see, is crucial to the construction of involute gearing. But, because it changes its orientation with respect to link-1 as Q travels its travel path, we need to look for some geometrical framework to hang the matter on. We might, accordingly, now recall *the pitch line and its pitch helices*. We have as yet no gear teeth among the figures here, but we have our 3-link single-loop mechanism whose three links are the frame 1 and the two pivoted involute helicoids 2 and 3, and the freedoms at the three joints in the single loop are f1, f1 and f5, which sum to seven [1] [§1.46], so the mobility is unity. Please see the two spoons at figure 2.02. Due to the carefully selected shapes of the spoons here at figures 3.08 and 3.10, however, where the spoons are indeed the involute helicoids, namely the profiles of yet to be located mating teeth, there is a constant angular velocity ratio k obtaining [66] [67]. The ratio k in the main series of figures of this chapter has been set for the sake of example at 5/7; this was done at §3.28 where the numerical data was decided; look for complete details of these (but only if you need them) at appendix D at 3, part (b), (§3.78). So the pitch line for the kinematic action is fixed in space in this instance, and it can be located (§1.13). See now in figure 3.11(a) the pitch line located there. The particular pitch helix that passes through Q (at the pictured instant in the travel of Q) is also shown, and from this we can see that the pitch at the pitch line is left handed in this example. In any event the pattern of the pitch helices pierced

by the travel path will determine the distribution of the helitangents as Q moves, thus indicating in some small way the complexity of the gyrations of the triad. For more about these gyrations of the triad, go to §4.60 *et seq.*

Another triad of lines in and about the polar plane

52. There are two other lines upon the polar plane appearing in figure 3.11(b), and I wish to deal with them now. They are the helinormal (HN) and the helibinormal (HBN). They are perpendicular with one another. I now refer to the *heliradius* at Q. This is a *segment of line* whose length h is the radius of the helix through Q. It is the perpendicular (Q−N) dropped from Q onto the pitch line (§1.51), and it is, not surprisingly, collinear with the said helinormal in the polar plane at Q [1] [§9.46]. See figure 3.11(c) for the three, mutually perpendicular, *endless* lines through Q which I have called here, and elsewhere [1], the *helinormal,* the *helibinormal* and the *helitangent.* See also in figure 3.11(c) the mentioned *segment of line*, the *heliradius at Q,* which is marked (N−Q) and is of length h. Please refer in retrospect to the figures 9.02 and 9.03 in [1], examine the context there, and become aware that, in this pursuance of the particular problems of involute gearing, we have come, from some early statements in classical, differential geometry, full circle back to first principles in screw theory.

53. It is well known, and already explained at §1.51, that the sliding velocity v at Q is given, in general, by the vector sum of the following two velocities: $p\omega_{23}$ due to the relative sliding of the bodies along the pitch line, and

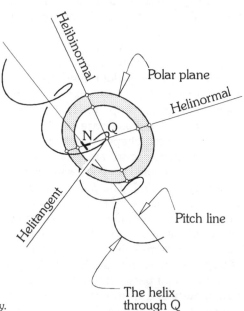

Figure 3.11(c) (§3.52). The polar plane in detail, showing the helinormal containing the heliradius (N-Q) and the helitangent containing the sliding velocity.

(ω_{23} × h) due to the relative rotation at radius h about the pitch line. The sliding velocity is, accordingly, given by v = $p\omega_{23}$ + (ω_{23} × h). It is instructive to notice here that, in involute gearing where the path of Q remains continuously collinear with the contact normal, the minimum value of h occurs when $h = q$, q being the shortest distance between the contact normal and the pitch line; and this, of course, is the same q as the one that is the length-variable appearing in the fundamental law (§2.02).

More about the variable angle δ_{REMOTE}

54. Next in the polar plane at figure 3.11(b) please see the path of Q, marked Z_j, which is inclined at angle δ_{REMOTE} to the plane of the transnormal. As we shall see (§4.58, §10.11), a certain value of this variable angle δ_{REMOTE} is the spatial analogue of the well known 'pressure angle' in planar involute gears. I will be calling that certain value, *both in the special, planar case where δ_{REMOTE} happens to remain a constant, and here in general where it does not, the angle δ of obliquity, or the disability angle of the gear set* (§2.10). The details cannot be dealt with here however. As already said, we have not yet found in the path of Q any point that can legitimately be called the 'central point' of the path; and herein lies the difficulty. I do wish to mention here, however, the *size* of the δ_{REMOTE} appearing in figure 3.11(b). It appears to be big but, as we shall see at §4.58 and subsequently, this is no indication that a gear set based upon the example being taken here might exhibit an equally big angle δ of obliquity. See for example the gears at chapter 7.

Special points [Qp], [Qm] and [Qc] in the path of Q

55. Erecting the shortest distance (namely q) between the travel path and the pitch line, we locate that unique point [Qp] in the travel path closest to the pitch line, where the heliradius h is least (namely where h is q), where the sliding velocity (which always occurs along the helitangent) is least, and where the travel path is not merely some general one of the radial lines of the polar plane at Q, as indicated in figures 3.11(a) and (b), but a special radial line in that plane namely the binormal to the helix at [Qp]. Figure 2.01 conveniently illustrates this. That such a point [Qp] exists in the path of any Q is not remarkable. By definition a right line is a straight line of points in a moving body where the velocities of the points are all perpendicular to the line [1] [§3.01]. Accordingly our straight travel path here is a right line in both of the moving bodies 2 and 3 as they move relatively to one another and, along any right line in a body screwing with respect to another body, there is always a least velocity. Refer to the photo-frontispiece at chapter 3 in [1]. This least velocity occurs, in the case here, when Q is at [Qp]. When Q (in the course of its travel) passes [Qp], the triad at Q becomes, at that moment, specially oriented. The helibinormal at Q becomes collinear with the path. By inspection we can see incidentally that, in the case of external meshing (when both of the mating surfaces are convex towards Q), the point [Qp] lies within the confines of the limited segment of line whose length is Z_j.

56. A second special point [Qm] occurs in the travel path. It is that point (or pair of points) already mentioned at the end of §3.38 where the sum (the algebraic sum?) of the changing radii g_2 and g_3 arrives at a local minimum. The newly inserted question at issue here is whether or not we entertain the possibility of internal as well as external meshing. This point (or pair of points) has yet to be fully discussed.

57. A third special point [Qc] occurs in the travel path. I wish to define [Qc] as that point midway along the segment Z_I where the distances H_2 and H_3 are equal and the principal circular curvatures of the profiles meeting at Q are thus the same (§3.24). This point [Qc] will be of relevance when we come to questions of lubrication. There are the two planes of the two circles of circular curvature at [Qc]; the angle between these may bear some examination; this angle will also be relevant in connection with lubrication.

Practical limits to the otherwise endless motion

58. We have at figure 3.08 two axially pivoted involute helicoids in tangential contact with one another at a continuously moving point Q, and I wish to distinguish now between (a) *general hypothetical* Q, which occur at all points of tangency whether the surfaces be convex or concave towards the point of contact, (b) *possibly physical* Q, which occur at points of tangency which are occurring 'convex to convex' at Q, and (c) *actual physical* Q, which occur at points of contact between the flanks of real involute gear teeth. Clearly the two surfaces of figure 3.08 can interact with one another indefinitely, rotating continuously in one or other of the two directions while maintaining hypothetical contact at the endlessly moving Q. As we know, however, it is always the convex sides of the surfaces that need to be exposed for the physical contacting action, and it is only while Q is within the confines of its allocated segment Z_I that the mating sides of the two involutes are both convex. In the event of a steadily working pair of general, spatial, involute, real gear wheels, the relevant *actual physical* Q — or the relevant *duplicated* actual physical Q running in tandem (§2.18) — will be travelling along their straight-line path in the same direction steadily. Let us envisage upon each shaft a coaxially arranged set of evenly spaced and identical involute helicoids whose number is the number of teeth; and let us be able to cut the physical teeth in such a way that the ongoing physical action can step from one helicoid to the next without allowing collisions to occur at the edges of the teeth or the contact ratio to fall below unity (§2.18). Our study of the gear bodies at §3.35-39 has helped to clarify the nature of this problem. Somewhat similarly we can look later at Z_A, a portion only of the path of a possibly physical Q, which, due to real mechanical truncations of real teeth, does not extend for the whole of the length Z_I. *We need to see that the zone of actual contact Z_A is contained within Z_I; Z_A is a limited portion only of Z_I; see this later at §5B.23.* But what I wish to study now is not the immediate practical limits of Z_A, but the more fundamental, end limits of Z_I. I wish to study the question of what happens to the two sets of rulings and the two slip

tracks as we go across the start-point and/or the end-point of a given segment Z_l. And in this respect we need to understand, among other things, what might happen if either (a) an incoming hypothetical Q is allowed to become actual before the incoming end of its Z_l, or (b) an already travelling, actual Q is allowed to remain actual until after the outgoing end of the same Z_l.

Degenerate slip-track geometry at the two ends of Z_l

59. I made an earlier reference to this paragraph at §3.16. To tackle the question, let us look first at the planar case. Refer to figure 3.12(a). The segment Z_l in this case lies collinear with the common tangent to the two base circles, its length extending in the reference plane between the shown two points of tangency T_2 and T_3 (§3.05, §3.20). Generally speaking in the plane, any involute to a circle has two branches. A few such double-branched involutes are drawn upon the base circle 2. Only one part of only one of the branches is normally used for the profiles of real teeth however. Please follow the path of an incoming hypothetical Q at the lower RHS of figure 3.12(a) and notice that, with its participating involutes, this Q has been taken straight through T_2 and into the physical region of Z_l without hesitation at T_2. Within the physical region of Z_l a participating pair of real teeth at the physical Q are shown diagrammatically. Notice that, as we passed T_2, the leading branch of the involute on circle 2 fell out of play, and the trailing branch of the same involute came into play. Despite the cusp, smooth geometrical motion was continuous at the change-over. Notice, however, that at the changeover we suddenly went from drive to coast (§5A.10), so the action (so far as the directions of the torques were concerned) was discontinuous there. Refer to figure 3.12(b) which is, in the context here, self explanatory.

60. The figures 3.12(a) and (b) depict the special planar case only of the general spatial circumstance which occurs with skew involute gears when, travelling along a twisted slip track upon its twisted tooth, we cross the throat circle of the base hyperboloid. Refer again to figure 3.03 and see that, at that point (the throat-crossing point, which is, incidentally, at the origin of Ω in figure 3.02), the angle ζ of §3.22 becomes zero. *At the throat-crossing point, in other words, the tangent t–t to the slip track becomes collinear with the ruling upon the helicoid.* In this connection we might define here the *base ruling* of the helicoid. *The base ruling is precisely that ruling which, tangent to the base helix at S, straddles the cuspidal edge of the helicoid.* This is further explained at §3.62.

61. In retrospect, however, we see in the figures 3.12(a)(b) that the mentioned slip track is degenerate. It is lying flat in the plane of the paper; it is identical there with the branches of the involute; it is thus a planar curve. At the cusp where the tangent t–t may be seen to be in any direction, the said tangent may be seen to be normal to the plane of the paper; and this direction is, as we know, the direction of the rulings upon the actual helicoid, all of which are parallel (§3.20).

§3.62   GENERAL SPATIAL INVOLUTE GEARING

62. What happens in general is depicted at the figures 3.13. Here we have, at figure 3.13(a), a central axis (mounted vertical) surrounded by a base hyperboloid (a, α) whose throat radius is a, and whose generators are inclined at the relatively small angle α to the plane of the throat circle. The value of α in the figure is not 60° as it was at figure 3.03 but only 15°. To pass through an arbitrarily chosen point say S upon the throat circle (see the points SS at §2.12), the relevant core helix has been mounted whose helix angle is, of course, the same α (§3.16). Two branches of the resulting involute helicoid have been constructed. They are shown in the figure truncated by two parallel planes parallel with and evenly straddling the plane of the throat circle. The slip track (which is drawn bold) was plotted by finding the intersection of the shown hyperboloid with the discovered involute helicoid. It is not clear to see but, as the slip track approaches the throat, it turns more and more severely until, quite suddenly, ζ drops to zero and the slip track

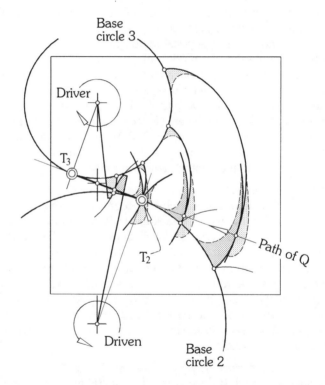

Figure 3.12(a) (§3.59). The special planar case of the trouble at the ends of the ideal zones of contact. In accord with the planarity prevailing here and the nature of the argument involving ordinary tangents, the imaginary rod SS in the figure here is written T_2T_3. In the general spatial arena the lengths of the rods SS are written Z_I, meaning zone ideal.
The lesser length of an actual zone of contact, on the other hand, which is determined by truncations of real teeth to avoid mechanical interferences, is written Z_A.

MATING INVOLUTE HELICOIDS   §3.63

touches, at the throat, the *base ruling* there. *Let me define the base ruling at S as the tangent to the core helix at that same S.* Refer to the sketched graph of ζ against Ω at figure 3.07.

63. What we have at figure 3.13(a) is a purely geometrical picture. We have only one involute helicoid with both of its branches shown. If on the other hand we had had two different helicoidal surfaces touching (and thereby mating) at a point of contact Q (see figure 3.08 for example), we would take only one branch of each of the two surfaces (these destined to become in mechanical reality the real flanks of two contacting teeth), and the matter would be a different matter. Two slip tracks would cross in that case at the point of contact Q and, while one of them would become degenerate at one end S of the relevant Z_I, the other would do so at the *other* end S of the same Z_I (§2.12, §3.27). When cogitating this matter of mounting two different involute helicoids one upon the other to touch at a point of contact Q, be aware that the locations of the shaft axes are to an extent determined by this and that the path of Q will automatically always become legitimate (§3.29). I wish to repeat that in figure 3.13(a) the shown two branches of the same involute helicoid is not a picture of two mating flanks of teeth in mesh; it is a picture of what happens to the shape of an isolated involute helicoid as its generator crosses the throat circle of its parent hyperboloid.

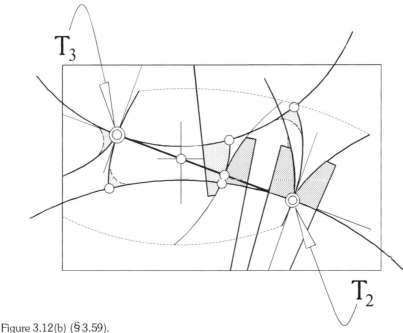

Figure 3.12(b) (§3.59).
This figure is clearly in the
circumstances self evident.

§3.64   GENERAL SPATIAL INVOLUTE GEARING

64. Physically speaking at figure 3.13(a) we are dealing with a single flank on one side only of a real tooth, *and it is important to understand in the context of gears that the axis of the said involute helicoid is always collinear with the axis of the wheel.* That said, we can now go on to say the following. Among the orthogonal projections thrown at figure 3.13(b) we are looking in the various directions as follows: (i) towards +y, namely inwards along the radius a towards point S; (ii) towards +z, namely along the central axis; in this view the planar involutes of §3.18 appear in true shape; (iii) towards −x, namely along the tangent to the throat circle; in this view the helical cuspidal edge of the surface becomes evident; (iv) along the binormal to the core helix (namely tangential to the throat cylinder and orthogonally the cuspidal edge) at S; (v) directly along the base ruling; in this view the slip track gives the appearance of having a cusp at S; as we know, however, it does not really have a cusp there. Both branches of the slip track are shown bold in all five views, and a study of the shapes is instructive. In (v) a part of the surface appears below the horizon of the base cylinder. This is due to the helical shape of the cuspidal edge. Near to the base ruling, the cuspidal edge follows the core helix, not the straight line of the base ruling, and the detail of this is important for those cases in practice where the path of Q is allowed to go all the way to the degenerate limit.

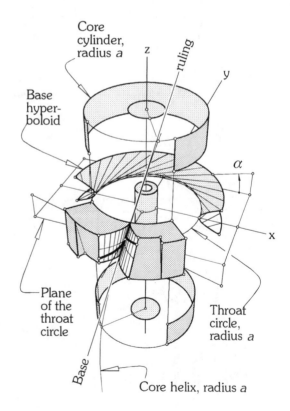

Figure 3.13(a) (§3.62). This is not a picture of two teeth in mesh. It is a picture of the two branches of an involute helicoid. The base ruling, tangent to the core helix at the U-turn of the slip track (at the throat of the relevant base hyperboloid) straddles the cuspidal edge. Compare this figure with figure 3.03

96

The slip tracks unsupported

65. Be aware that the slip tracks alone, each seen as a mere curved, rigid wire of zero diameter attached at the periphery of its own throat circle, could perform in contact the entire kinematic action as accurately and continuously as could the contacting surfaces. The surfaces do not touch except at the slip tracks, and they do not undercut one another (except at distant locations). This latter is an easily seen geometrical property of the construction: each involute is everywhere convex on its convex side: the said involutes can touch one another at (and only at) the single point of contact Q. Alternatively the unsupported slip tracks may be seen as being supported only at the truncated edges of their respective base hyperboloids, the hyperboloids running (as it were) edge to edge, and with their edges crossed at Q; see figure 3.08 where the hyperboloids are, indeed, so truncated. This notion of the unsupported slip tracks was, in a limited way, foreseen by Olivier [40] in his first paper on the kinds of involute gears that might be able work between shafts that were *non situés dans un même plan* (skew) in 1839 [101] (§3.66).

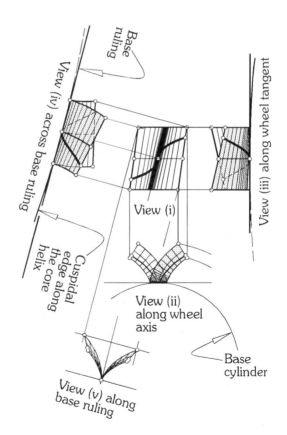

Figure 3.13(b) (§3.64). Here we see in projected detail the shape of the particular slip track illustrated at figure 3.13(a). The shape might be seen as being typical of those shapes arising in the actual gear sets synthesised in the forthcoming chapters 5B and 6. In real gears each slip track will have (a) its real branch (which may be imagined to be inscribed upon the actual flank of the real tooth, and (b) its unreal branch (which should be seen as being inscribed in the open space of the gear wheel, namely that space outside the enclosed space containing the physical material of the real tooth).

Théodore Olivier 1842

66. This brings me to where I must, *pro tem*, speak about Olivier [40]. It seems to me that writers and speakers over the years have implied that Olivier in his book foresaw, then dismissed as impractical, all attempts to solve the skew gear problem by means of involute geometry. Olivier deserves ongoing mention for the clear distinction that he made very early between those gears he called (1) *engranage de force*, gears for *strength*, and those he called (2) *engranage de précision*, gears for *accuracy* [40] [pp. 4, 24]. This distinction, predicated upon the occurrence between teeth of either (1) *curved line contact (which may of course be straight)*, or (2) *single point contact*, was a geometric distinction which remains applicable today. It applies across all forms of gear teeth whether involute or non-involute. It depends upon the kinematic strategy that might be employed for the generation of teeth for conjugality, there being (1) *the first general approach to the problem*, involving direct generation from wheel to wheel resulting in line contact, and (2) *the second general approach to the problem*, involving the separated generation of teeth through some double-sided, thin, intermediate agency resulting in point contact. The intermediate agency is called by some (and by me) *the phantom rack*. Olivier's attack on the problem of skew involute gearing was in my view however, although inspired, flawed. He failed on the one hand to find by his methods my solution to the problem, which is the subject matter of this book, but he did not, on the other, announce that my solution to the problem could not be found. The matter of Olivier's work and his legacy is quite complicated. I will accordingly delay until the end of chapter 4 my critical survey of his literature; see appendix A at chapter 4. This survey will thus find itself placed to appear after the laying of further theoretical groundwork, and before the new material that follows it. I wish to mention in the meantime however that, whereas Olivier in his involute work was overtly pursuing (1) his *gears-for-strength solution*, my work overtly pursues (2) his *gears-for-precision solution*. Our paths openly diverge. I hasten to say however that, while my kind of spatial involute teeth are *precise*, in the precise meaning of Olivier's word, they are certainly not, as a consequence of that precision, not strong.

Intersection of the base hyperboloids

67. At figure 3.08 we have seen (a) the two shaft axes, (b) the two involute helicoids (coaxial respectively with the said axes and in contact at Q), and (c) the whole length Z_l of the possible path of Q (Z_l being the sum of H_2 and H_3). Please remain aware that figure 3.08 relates to drive, there being no coast as yet. In the case of coast, the torques are reversed and a different Q is involved with a different path (§2.14, §5A.10). Now if we swing the infinitely long path of Q in figure 3.08 about the two shaft axes in turn, we sweep out the two base hyperboloids which are already partly shown at figure 3.08 but are now more fully shown at figure 3.15(a). These hyperboloids are at the

root of the kinematic action. They intersect along a common generator namely the path of Q. Their throats are separated from one another along the path by the distance Z_f as can be seen (with difficulty), and the throats are in general of different diameters. Look again at the linkage RSSR sketched at figure 3.08 and see how this same linkage fits in at figure 3.14(a). The links RS and RS are the two throat radii a_2 and a_3, the rod SS is of length Z_f, and the two angles of inclination α of the hyperboloids are the angles α_2 and α_3 of figure 2.02. Please read note [90]. In general (and here generality does obtain), $a_2 \neq a_3$, and $\alpha_2 \neq \alpha_3$.

68. When we come in due course to the work of synthesis, it will be clear from the geometry of this chapter, the shaft axes having been located, and a straight-line path for Q having been chosen, that the throat radii a_2 and a_3 and the inclination angles α_2 and α_3 of the base hyperboloids will already have been determined. The geometrical constructions required in this chapter for the inward-looking, illustrative analysis just completed were easily executed by computer graphics and the numerical values easily measured similarly. See these values listed for those that might be interested at appendix D at 3. But in general we shall require for gear-set syntheses, namely for actual, outward-looking gear design work, where trial by numerical iteration will always be an important feature, that the relevant algebra be thoroughly available for the equations that will be needed to underlie a computer-driven spreadsheet. See chapters 5 and 6 for the beginnings of such an algebra, and see chapter 10 some more algebraic material.

69. The base hyperboloids of figures 3.08 and 3.14(a) do not simply roll and slide upon one another as pairs of axodes do. Firstly they are not throat-to-throat as axodes always are; and secondly they intersect (and interact) not along the fixed pitch line, but along a different fixed line namely the path of Q. In general they may be seen as two circular hyperboloids generally disposed in space, except that they share a common generator namely the path of Q. And I say it again, they are not axodes. Their relative motion, which is, as we know, a continuous screwing motion with a known pitch p about the pitch line (§1.14), which motion is, as we also know, exactly represented by the rolling and sliding of the axodes, is certainly not a simple rolling motion at the said common generator. An animated film could be made to illustrate the relative motion, continuously showing the unbroken whole of both hyperboloids as depicted in figure 3.14(a), but due to the intersection of the surfaces a working model would be impossible to construct unless the hyperboloids were truncated somehow, for example at the slip tracks. They are thus truncated at figure 3.08 incidentally. Even at [Qm], which has now been tentatively defined (§3.38), the interaction is not pure rolling but a rolling with slip, which occurs not only axially (in the direction of the path) but also circumferentially. The point [Qm] has been located in figure 3.08, by means of a poorly conditioned construction not shown, and marked-in there. It is, however, unfortunately occluded at figure 3.14(a). The importance (or otherwise) of this [Qm] is still not fully clarified.

70. Please notice now that in figure 3.14(a) the intersection between the two hyperboloids has been revealed to have two branches, neither one of which is clearly visible. First there is of course the straight line of the common generator itself, but second there is a twisted curve which intersects the common generator at at least one point. This one point (and there may be more than one) is also unfortunately occluded. The twisted curve clearly goes off to infinity in both directions, but its exact shape and its location with respect to the two throats and [Qm] are not very clear to see.

71. Refer to figure 3.14(b). Holding the common generator at figure 3.14(a) fixed, I have here rotated the whole contents of that figure about that line until we are able to see deeply into the throat of hyperboloid 3. This I have done for two reasons, (a) to show more clearly the point [Qm] in its local environment between the surfaces, and (b) to look more carefully at the point of intersection between the mentioned two branches of the intersection between the surfaces. The point [Qm] is visible as expected, and the said point of intersection — I have called it [Qt] — may also be seen. Although the points [Qm] and [Qt] have been found in this case to be close to one another (which invites the conjecture that the two points are indeed the same point, one or both of them having been inaccurately located), the two points are, as we shall see, distinctly separate.

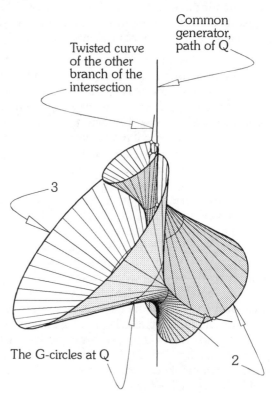

Figure 3.14(a) (§3.67). The two base hyperboloids swung in full, with an indication of their intersection. Thier intersection is a quartic curve having two, infinitely long, separated branches. One of these (being the common generator) is of course a straight line.

MATING INVOLUTE HELICOIDS   §3.72

72. There is enough new material discovered here to warrant a more thorough examination of the phenomena, not by way of an arbitrarily chosen example as has been the case in this chapter hitherto, but by way of a more general approach. See chapter 4. The reader might be warned however that chapter 4 is written more for the sake of geometrical explanation than for pressing the otherwise immediate issue of mechanical design. Readers may prefer to jump, after these next remarks, from here to the start of chapter 5A.

Penultimate remarks

73. Assuming accuracy and rigidity and the absence of friction, we may conclude from this chapter that the flanks of involute teeth as described will mesh properly with one another, and that they will, within limits, be insensitive to all errors upon assembly. This has been an analytical study to discover the possibilities for synthesis. No attempt has been made to simulate an actual practical example here, with the result that, although the fundamental geometry has been thus made clear, many of the angles appearing are absurd when looked at from a practical point of view. When we come to the actual synthesis of real gear teeth on real gear wheels, we will need not merely

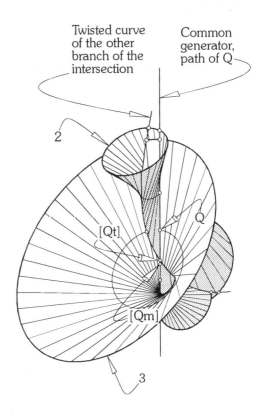

Figure 3.14(b) (§3.71). Whereas figure 3.15(a) was proportioned and oriented to match figure 3.08 exactly, this figure is a rotated version. Figure 3.15(a) has been rotated (clockwise looking upwards) about the vertical line of the common generator through a suitable angle. This was done simply to get a better view of the special points [Qt] and [Qm] which were otherwise occluded.

to find the triad of planes at Q by analysis among the existing geometry as we have done here, but to establish by synthetic methods its likely best orientation at a freely chosen point Q (or at some special point Q), and to locate the travel paths for both drive and coast (namely for both the fronts and the backs of the teeth), legitimately according to the fundamental law (§2.06), and appropriately according to circumstances.

Raoul Bricard

74. As mentioned already (§2.37), Bricard [82] showed in 1926 for planar gearing that, if we demand both conjugality and the freedom to alter centre distance, there is only one solution for the shapes of the profiles of the teeth, the planar involute. To identify the phenomenon, I wish to call the involutes of the statement *Bricard's obligatory involutes for adjustable planar gearing*. Now the parallel rulings upon the involute teeth of ordinary spur gears may be seen to derive from a special case of the spatial involute helicoid as we have learned to know it (§3.15). We are now moreover anticipating the discovery that, if we use for synthesis in spatial gearing the spatial involute helicoid as we have learned to know it, we can get both conjugality and the freedom to alter not only the centre distance, but also all of the other dimensions required to locate the parts of the set (§3.33). *Having demanded in space both conjugality and the said freedom to alter the said dimensions, however, we do not know whether the said spatial involute helicoid for the shapes of the profiles of the teeth is the only solution.* Should it exist, in other words, the spatial analogy of Bricard's important statement is unknown as yet. I wish to mention that a modernised rewrite of Bricard's work in the plane, namely [82] above, has recently been undertaken by Fayet [98], and that this very recent paper (2001) might encourage ongoing workers to examine this overlooked field in the totality of three dimensions.

Appendix A at chapter 3

75. This appendix originates from §3.09. It offers proof of the fact that, at figure 3.03, the distance H varies directly with angle Ω. Referring to figure 3.02, we have in triangle ABC, by the cosine rule, that

$$(B-C)^2 = H^2 + (a\, d\Omega)^2 - 2H(a\, d\Omega) \cos \alpha,$$

and in triangle CBE similarly, where angle CEB $= (\varepsilon + \pi/2)$, that

$$\sin \varepsilon = [(s\, d\Omega)^2 + H^2 - (B-C)^2]/[2\,(s\, d\Omega)\, H].$$

By substitution, simplification, and the removal of second order small quantities, we can find that

$$\sin \varepsilon = (a \cos \alpha)/s.$$

In triangle BDE, where BDE is a right angle, we can write that

$$(E-D) = (E-B) \sin \varepsilon;$$

MATING INVOLUTE HELICOIDS   §3.76

but the intercept (E−D) is dH, and the intercept (E−B) is $s\,d\Omega$,
so we can find without further ado that

$dH/d\Omega = a\cos\alpha$, namely that $dH/d\Omega$ is constant.

Integrating with respect to Ω and ignoring the added constant, we obtain

$H = c_H\,\Omega$, where c_H is a constant for the hyperboloid.

The constant c_H, of course, is the already mentioned $a\cos\alpha$.
This result completes the required proof.

Appendix B at chapter 3

76. This appendix originates from §3.22. It is a sketchy account of the requested algebra for the variation of ζ with Ω. Referring to figure 3.02, we can see first of all that

$s = \sqrt{(a^2 + H^2\cos^2\alpha)} = a\sqrt{(1 + \Omega^2\cos^4\alpha)}$;
(E−B) $= s\,d\Omega = a\,d\Omega\,\sqrt{(1 + \Omega^2\cos^4\alpha)}$; and
(E−D) $= dH = a\,d\Omega\cos\alpha$.

Next, in triangle BDE, we have a right angle at D, so

$(D-B)^2 = (E-B)^2 - (E-D)^2$
$= a^2(d\Omega)^2(1 + \Omega^2\cos^4\alpha) - a^2(d\Omega)^2\cos^2\alpha$; and
$(D-B) = a\,d\Omega\,\sqrt{(\sin^2\alpha + \Omega^2\cos^4\alpha)}$.

Solving for ζ in triangle MBD:

(B−M) $= J + (F-M)$
$= a\Omega\cos\alpha\cot\alpha + a\,d\Omega\,\mathrm{cosec}\,\alpha$
$= a\cos\alpha\cot\alpha\,(\Omega + d\Omega\sec^2\alpha)$; and

(M−D) $= J + dJ$
$= a\cos\alpha\cot\alpha\,(\Omega + d\Omega)$; and
(B−D) $= a\,d\Omega\,\sqrt{(\sin^2\alpha + \Omega^2\cos^4\alpha)}$.

Employing the cosine rule we have next, etc., which leads, after simplification and the removal of all multiples of $(d\Omega)^2$, which are small, to

$\cos\zeta = (\sin\alpha)/\sqrt{(\sin^2\alpha + \Omega^2\cos^2\alpha)}$, which is equivalent to
$\tan\zeta = \Omega\cos\alpha\cot\alpha$.

We thus see that, for a given a and α, $\tan\zeta$ varies directly with Ω.

But going now to figure 3.07 which is extracted directly from figure 3.03, we can more easily get the same result. The circle coaxial with o-o containing F has centre N. I hypothesise that the centre of curvature of the slip track at B is at N precisely. See the pair of coplanar right angled triangles GAF and FBA characterised by angle α as already mentioned, see the ruling FBG at B, and then see that the tangent BL to the slip track at B cuts the line NF in L. Perceive now that we have a second set of coplanar right angled triangles

LBN and BFN characterised by the angle ζ. Thus see that

$\tan \zeta = J/a = \Omega \cos \alpha \cot \alpha$,

as before.

Appendix C at chapter 3

77. This is not yet written but it will give I hope some reasoned explanations in terms of curvature theory for what I have said at §3.24.

Appendix D at chapter 3

78. The numerical values of this chapter are of no practical importance; but for readers interested, details follow. The details are approximations, of course, having come from the computer aided graphics. This graphics was done with care, but there might, independently of that, be ordinary mistakes of transcription for which the author takes no responsibility.

Part (a) pertaining to the ongoing main example of this chapter. Speed ratio $k = 5/7$, thus $k = 0.71428$. Centre distance $C = 254.84$ mm. Pitch p (by formula) $= -14.2422$ mm/rad. The smaller r_3 (by formula) $= 105.9535$ mm, the bigger r_2 (by difference) $= 148.93$ mm. Shaft angle $\Sigma = 13.12°$. The smaller ψ_3 (by formula) $= 5.46°$, the bigger ψ_2 (by difference) $= 7.66°$. [The two radii a and the two angles α are not yet reported here.] The angle of rotation taking us from figure 3.17 to figure 3.18 was 25°.

Part (b) pertaining to the figures 3.09 and 3.10. These involve a quite different example (§3.35). Author used (or found) the following. For gear body 2, a = 65.0987, $\alpha = 33.4933°$, pitch a cot α of the core helix = 98.3784 mm/rad right handed, a cos $\alpha = 54.2891$. For gear body 3, $a = 45.0000$, $\alpha = 25.0000°$, pitch a cot α of the core helix = -96.5028 mm/rad left handed, a cos $\alpha = 40.7839$. For the combination, speed ratio k of the gear-body set = (a cos α)/(a cos α) = 40.7839/54.2891 = 0.7512. The G-circle radii, chosen by open choice, are (2) 80.0000 mm, and (3) 55.6143 mm.

The dimensions of the actual metal material pieces. The actual gear bodies are, at the pivots, 10 × 10 mm square, and 6 mm diameter at the shafting. At the gear-body heads, the material is 30 × 14 rectangular; and along the arms 14 × 6 rectangular.

Looking to chapters 5 and 6, it is evident that these gear bodies when assembled (figure 3.10) will not form an equiangular set. We can see this by seeing that the path of Q will not in general cut the E-line (whose location depends upon our choice for the angle Σ) and that the two angles α are not equal. The set will be polyangular in the sense that the path of Q will in general cut the F-surface somewhere other than at the E-line. There is no point in mentioning [Qx] in connection with a pair of gear bodies because we have only one path, not yet two: the teeth have only fronts, not yet backs.

Appendix E at chapter 3

79. This appendix comes from §3.42. I give here a proof for the mentioned theorem about two circles intersecting one another at a single point in space. Given two circles intersecting at Q, the two tangents at Q and the corresponding circles axes are, taken circle by circle, skew but perpendicular with one another. The corresponding circle-radii are the common perpendiculars respectively between the tangents and the axes. Any line through Q which cuts either axis somewhere is clearly perpendicular with the corresponding tangent at Q. It follows that any line through Q which cuts both of the axes somewhere is perpendicular to both of the tangents at Q. There is only one such line. This line is the *transversal* at Q which is, thereby, normal to the plane of the tangents at Q. The plane of the tangents at Q has already been called the *transnormal* at Q.

CHAPTER FOUR
KEY ASPECTS
OF THE GEOMETRY

Introduction

01. This chapter appears precisely here and separately, not only because it is a timely intervention just now, but also because its contents form a crucial, well-delineated part of the overall theory as currently being worked out and presented by the author. It is a buffer, for him, between his efforts at analysis on the one hand (only partly completed as yet) and his forthcoming problems of synthesis on the other. The chapter directs its relevance both backwards to explain more clearly some of the difficulties already encountered, and forwards to underpin the contents of the chapters 5 and 6. It begins by presenting some general properties of the hyperboloid, and goes on to discuss some theorems about the intersection of two circular hyperboloids sharing a common generator. *It deals among other things with three general questions concerning the following: (#1) the special geometry of the hyperboloids when they constitute the axodes of a gear set; (#2) the smallest sum of the distances from a point on the said common generator to the axes of the hyperboloids when the question is a general one; and (#3), in two parts, (part one) the common tangent planes existing between the hyperboloids at no more than two isolated points along the common generator when the question is a general one, and (part two) the perpendicularly intersecting tangent planes existing between the hyperboloids at no more than one point along the same common generator when again the question is a general one.* When applied in the kinetostatical arena, and in the special case of general spatial involute gearing, these purely geometrical questions relate to the kinds of practical questions we might naturally ask: (#1) about the shapes of the axodes of the gear sets; (#2) about the whereabouts of, and the characterizing features of, the 'middle' of a path of a moving point of contact Q; and (#3) about whether or not the *concept of the naked wheels* is valid. It deals, also, with the special question of the *parabolic* hyperboloid [69]. The chapter deals with the geometric aspects of the various equivalent linkages that crop up. In particular it defines more clearly than before the various special points along the path of any Q. It ends by discussing some of the implications for

general spatial involute gearing of these remarks and theorems, and ventures (albeit somewhat timidly) into the open field of the significance within gear sets of classical line geometry and modern screw theory. This whole chapter might be omitted upon a first reading.

The hyperboloid and its generators

02. The smooth surface of the hyperboloid (when in its standard or reference location within its standard set of Cartesian axes) is defined by the well known equation to be found in most elementary texts on the subject of point coordinate geometry. It runs: $x^2/a^2 + y^2/b^2 - z^2/c^2 = 1$. This equation yields, according to its coefficients, either a smooth surface of two sheets, in which case there are no straight lines to be ruled upon the two mirror-imaged, inwardly convex, separate parts of the one surface, or a *continuously smooth surface of one sheet,* in which case the surface is the double-trumpet shaped, ruled surface with the elliptical throat, the circular-throated special case of which has already become familiar in this book and generally in the area of gears. In respect of these ruled hyperboloids (namely those of one sheet), we know that: (a) any three lines define a regulus [1] [§11.04]; (b) any regulus sweeps out its own hyperboloidal surface; (c) upon any hyperboloidal surface thus swept out, there may be drawn the opposite regulus [1] [§11.06]. We have distinguished here between the hyperboloidal surface itself and the *two reguli,* otherwise known as the *two families of generators,* that belong upon the surface. There are other equations, quite distinct from the Cartesian equation, that deal not with the surface as such, but with the generators and the reguli. There are the parametric equations that deal more directly with the generators themselves [1] [§11.13]. There are also the Plücker equations which are of assistance with of screws [1] [§11.51]. The following other properties of the reguli on the surface of the hyperboloid are listed here

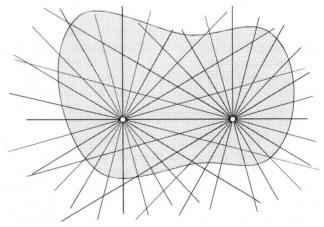

Figure 4.01(§4.02). Orthogonal view of an hyperboloid looking along one (and thereby two) of its generators.

KEY ASPECTS OF THE GEOMETRY   §4.03

because they are directly relevant to what we are about to do. We know also that: (d) every generator cuts all generators of the opposite family, and thus that at every point upon the hyperboloidal surface two generators intersect; (e) for every generator there is one generator of the opposite family which is parallel with it; (f) such pairs of parallel generators cut the throat ellipse in points that are diametrally opposite. These properties explain why, when we look orthogonally at any hyperboloid in the direction of one of its generators, the two reguli together take on the appearance of figure 4.01. From such a figure, taken alone and not truncated in some known way, we cannot tell the shape of the hyperboloid. We cannot tell whether it be *squat* or *tall*, namely whether it has generally small or generally large angles α of inclination of the generators (see figure 3.01). Nor can we tell the proportions of its throat or the shape of its asymptotic cone. We cannot tell from a view such as the one depicted in figure 4.01, in other words, what the coefficients are of the Cartesian equation; but we can use such views to advantage in our work.

Theorems about generator-joined hyperboloids

03. The three propositions listed as #1, #2, and #3, each of which is separately inset under this sub-heading here, are each a carefully selected portion of a more comprehensive, single proposition. The single proposition could be fully worked out and presented with all of its special cases subsequently examined but, for the sake of simplicity and in the circumstances, I prefer to work from the particular to the general. The separate propositions relate, in turn as we shall see, to the different sets and styles of generator-joined hyperboloids we will meet as we go, step by step, through the processes of synthesis in chapters 5B and 6. The listed propositions are severely shortened here for compactness, but they are next enlarged upon, in turn, under the next three sub-headings due to follow soon, namely Theorem #1, Theorem #2, and Theorem #3.

04. First let there be two rotational axes (screws of zero pitch) generally disposed and let the intercept determined by the common perpendicular between them be (A–B). Refer to figure 1.01. Take the *cylindroid* determined by the two said rotational axes [1] [§15.49], and choose any one of the screws of that cylindroid. Let the chosen screw intersect (A–B) in P. Now rotate the line of the chosen screw in turn about the two rotational axes, thus sweeping out two circular hyperboloids which will then share, of course, the line of the chosen screw as common generator. Refer to figure 1.06. Let Γ be a movable point upon the common generator. For unacknowledged reasons as yet, I hypothesize as follows:

(#1) (a) that for each position of Γ along the common generator there will be two radii g obtained by dropping perpendiculars from Γ onto the two rotational axes; that the lengths of these will vary as Γ moves; that the sum (or the difference) of them will be at a minimum (or a maximum) when Γ is at P; and (b) that the two hyperboloids will, in general, be throat-to-throat, and either intersect one another along the whole line of the common generator or not do so, as the case may be, but be tangential with one another along the whole length of this same line nevertheless.

We need to see here that the swung generator of the cylindroid, the common generator of the two hyperboloids, might be either 'between' the two rotational axes or 'outside' of them (§1.11), and thence that part (a) of the above hypothesis becomes self evident in either event. The implications of (b) are more opaque. Given a parallel reading of [1] [§22.56-67], however, this will be less so, and a reading of §4.10 *et seq.* will clarify the matter.

05. Now, for the propositions #2 and #3, let there be as before two lines generally disposed in space, and let us call them both, for convenience, by the same name *Axis*. Let there be this time, however, a third line *also* generally disposed, and let us call this by the name *Line*. Let us swing the line called Line about the lines called Axis in turn, thus producing two circular hyperboloids which share a common generator namely Line. It is important to notice that the circles at the throats of the two circular hyperboloids will not now cut Line at the same point: *the points of intersection T of the throat circles with Line will be, in general, separated from one another*. Let there be as before a movable point Γ that inhabits the common generator.

06. Having accepted the names Axis, Axis and Line as defined in the paragraph above, having noted the two circular hyperboloids with their throats laterally displaced from one another along their common line of intersection namely Line, and having taken the term g as again applying to the already mentioned radii, I hypothesize as follows:

(#2) that there are in general two special points Γ in Line (other than the said separated points T at the two circular throats) where for one of them the sum of the radii g is at a minimum, while for the other the difference of the said radii is at a maximum. The special points in question will be called by me [Γm], the suffix m standing for 'mid' or 'middle'.

What his means, intuitively speaking, is that we could, at these points [Γm], but at no other point along Line, drop perpendiculars onto the Axes, whereupon at one of them the sum of the lengths of the perpendiculars would be at a minimum, while at the other the difference between the lengths of the perpendiculars (taking suitable account of sign) would be at a maximum.

07. Taking again the contents of §4.05 as the basis for argument, having, as we had there, Axis, Axis and Line (where Line is any line), I wish to hypothesize next as follows:

(#3) (part one) that there are, in general, two isolated points somewhere in Line, other than the said separated points T at the said circular throats, and other than the two points [Γm] mentioned above, *where the two hyperboloids not only intersect one* another along Line (as we know by definition that they will), but also be tangential with one another there. The two special points in question will be called by me [Γt], the suffix t standing for 'tangent' or 'tangential'.

What this means, intuitively speaking, is that we could, at either of these two hypothesized points [Γt], *but at no other point along the common generator*, draw a coplanar set of intersecting lines without piercing the surface of either of the hyperboloids, the said planar set delineating the *common tangent plane* to the surfaces at that point. At the special points [Γt] the lines of the said planar set either miss the hyperboloidal surfaces altogether or are collinear with certain of the generators of the hyperboloids. There are four (only four) generators involved at any one point along the common generator; see the truth of this at §4.02, item (d). Two of the said generators at the given point being collinear in any event, the hyperboloids being by definition 'generator joined', all four become coplanar when the moving Γ arrives at one or other of the points [Γt].

08. Taking again the contents of §4.05 as the basis for argument, I wish to hypothesize finally as follows:

(#3) (part two) that there is in general another special point in Line, other than the said separated points T at the two circular throats, other than the said points [Γm] and other than the said points [Γt], *where the tangent planes to the hyperboloids are perpendicular with one another.* This special point will be called by me [Γd], the suffix 'd' coming from perpen-*d*-icular.

What this means, intuitively speaking, is that, at the special point [Γd], the two hyperboloidal surfaces will be cutting one another perpendicularly. Because each tangent plane at the said point is defined by the relevant pair of intersecting generators upon the relevant hyperboloid, there will be at the point [Γd] two generators collinear and two others in planes which mutually intersect in these collinear lines perpendicularly.

09. We now set out under the sub headings that follow either to prove outright the truth of these three hypotheses or to comment effectively upon them. Because they will soon be shown to be self evident and/or otherwise proved, I will refer to them already as theorems: Theorem #1, Theorem #2, and Theorem #3

Theorem #1, about circular hyperboloidal axodes.

10. We are dealing here quite frankly with the simple, circular hyperboloidal axodes of a gear set which are generated by swinging the pitch line (§1.09) in turn about the two rotational axes. The geometry of this particular kind of pair of generator-joined hyperboloids is highly special. Refer to figure 1.06. The relevant text appears at §1.26 *et seq.*

11. The first part of theorem #1, which states that the sum or the difference of the two radii g is at a minimum or a maximum when Γ is at P is self evident. It does need explaining however that, according to the gear ratio k and the directions of rotation (§1.11), the pitch line might be 'inside' (A–B) or 'outside' of it. This in turn brings up the question of whether we have 'external rolling' of the axodes as shown at figure 1.06, or 'internal rolling' as can be imagined. Germane to this are the passages §1.26-30 and, with respect to the special matter of external or internal rolling, please pay careful attention to §1.29-30.

12. With regard to the second part of theorem #1, concerning the matters of tangency, I now intend to show that, with the special conditions prevailing, the two swung hyperboloids will be tangential with one another *at all points* along the common generator namely the said pitch line. By means of a theorem due to Plücker which relates to certain special transversals of the generators of a cylindroid [1] [§15.36-39], it is clear that any line drawn to cut (a) the said pitch line perpendicularly, and (b) either one of the said rotational axes, will cut (c) the other rotational axis also. See figure 22.13 in [1] which shows that this special circumstance applies, not just to any line that cuts the CDL perpendicularly (there being as many as an ∞^2 of these), but only to the ∞^1 of lines that are generators of the cylindroid. Given that the speed ratio k might be open to choice, these latter are the only lines that might conceivably be pitch lines. Appealing next to the proven geometry of two circles intersecting at a single point (§3.41-43), we can see that the criterion for tangency of the hyperboloidal surfaces is fulfilled at all points along the pitch line. At all points, in other words, the two intersecting generators of the hyperboloids mating there and the common generator are coplanar.

13. This kind of tangency that exists along the common generator of a pair of mating axodes is special. It is special in that the surfaces are more intimately in contact with one another than in the case of ordinary *first order* tangency. In the case of axodes in general, there is firstly the required condition that, for any two infinitesimally close pair of mating generators, the *parameter of distribution* must be the same, and next it is necessary that the *lines of striction* upon the surfaces intersect at the *central point* [1] [§22.61]. In the special case that has come to hand here, that of *continuously meshing circular gears,* however, the parameter of distribution along the generators of the axodes is fixed and the lines of striction are simply the throat circles of the circular hyperboloids. But all such *second and higher order* questions are beyond the scope of the present discussion in any event, and all we need to conclude is that, although the axodes may intersect (§1.26), they will be tangential with one another at all points along the pitch line. Refer Skreiner [30] and [1] [§22.61], each of which refers in turn to earlier accounts of these more difficult, second-order matters.

14. Generally with axodes, cases of intersection and consequent interpenetration of the mating surfaces clearly occur. Take for example the inter-

KEY ASPECTS OF THE GEOMETRY   §4.15

secting axodes of figure 22.12(b) in [1] [§22.64]; these are a pair of general, intersecting axodes relating to some smooth, unknown, finite relative movement between two bodies; and in this general case the ISA moves as the relative motion occurs. Take for another example those at the right in the photo-frontispiece at chapter 22 in [1]; these are special circular axodes pertaining to continuously rotating circular gears; and in this special case the ISA does not move as the relative motion occurs. In each of these cases the axodes, although tangential along the whole length of the common generator (the ISA), are intersecting. A physical model of them could not be a *working model* without provision being made for the continuous mutual mutilation of the rolling and axially sliding surfaces. Now take for a third example the already mentioned generator-joined axodal hyperboloids of figure 1.06 where conventional rolling (with axial sliding) can occur at all points along the common generator in the total absence of intersection and consequent interpenetration. We *could* make a working model of figure 1.06 because the axodes here are not intersecting. Please recall the contents of §1.23-30, and go to [1] [§22.61-65]. I wish to repeat: *axodes are always tangential with one another along the whole length of the common generator, which, in the case of gears, is the ISA_{23} namely the pitch line; but they will often intersect as well.* I should also mention that intersecting axodes *in general* will always be tangential at some single isolated point; this point is called the *central point* of the ISA (§4.14) [1] [§22.61]. At this point in gears, at the pitch point, where the circular lines of striction of the axodes intersect, there is an isolated occurrence of a higher-order tangency; and the common tangent plane there is normal to the CDL. Having dealt in this somewhat slovenly way with these axodal matters, we have thus removed from further consideration, just here, this kind #1 of generator joined hyperboloids. The axodes are, just now, for the sake of this book at least, sufficiently well understood.

Theorem #2, about the points [Γm]

15. A convenient way to come to terms with the gist of this second theorem (§4.06) is to imagine first that the one line called Line and the two lines called Axis are smooth thin rigid rods. Next imagine a pair of frictionless cylindrical sliders (thin tubes of short but finite length) free to locate themselves respectively and as required upon the rods called Axis. Next imagine the said sliders to be connected by a thin flexible elastic frictionless cord which passes over the rod called Line in such a way that the cord is strung to be taut and equilibrium among the various forces has established itself. Following the principle of least strain energy the cord will have adopted its minimum possible total length. The tension in the two straight portions of the cord will be equal due to the absence of friction, the angles between the two said portions of the cord and the rod called Line will also be equal due to the fact that the components in the direction of the said smooth rod of the forces due to the tension in the cord will be equal and opposite, and the angles between the cord and the axes of the tubular sliders at its two ends will both be

90°. It is clear to see that the equilibrium configuration of this imaginary apparatus coincides exactly with the first required circumstance of the theorem, namely that the radii be additive. The equivalent apparatus for the case when the radii are subtractive is somewhat more fantastic, but I will come back to that later; this latter, second aspect of the matter is somewhat elusive.

16. From this it may be concluded that, at any genuine 'additive' point [Γm] in Line, the bisector of the angle between the two straight portions of the cord (namely between the perpendiculars dropped from a likely [Γm] in Line onto the axes called Axis) will be perpendicular to Line.

> This fact, this perpendicularity of the said bisector with Line at a genuine additive [Γm], can accordingly be chosen as a convenient criterion to be used for a well conditioned graphical construction for locating the additive [Γm] in the common generator. For the 'subtractive' [Γm] we need to take, not the bisector of the *internal* angle between the two radii, but that of the *external* angle between the two. In this alternative case, however, this more 'elusive' case, the construction to find the point of perpendicularity is equally well conditioned.

See figure 4.04(a) for a preliminary look at the main aspects of the present matter, and see figure 4.04(b) for some of the finer details. At figure 4.04(b) the point $[Qm]_{add've}$ may be seen defined by the smooth flexible elastic cord and the frictionless tubular sliders (§4.16). The same figure shows the position only of the point $[Qm]_{sub've}$. I will discuss all four of the figures 4.04(a)(b)(c)(d) in due course; they need to be seen together as a group. The said figures have been constructed on the basis of carefully chosen but non-special data; they are designed to present a general case that is visually intelligible. They are moreover drawn by means of computer graphics; so they may be relied upon to give an accurate impression of the geometry.

17. This theorem #2 appears to be, just yet, curiously incomplete. I have in the general geometrical circumstance no concomitant occurrence to report, no other kinematical event that happens at the same time as does the right angularity of the mentioned angle. I have found (as yet) nothing of special interest to say, for example, about the *transversal* though [Γm], namely the unique line through [Γm] that cuts both of the lines called Axis; nor, for example, have I found any special relationships occurring with respect to the throat ellipse or the line of striction surrounding the *girdling hyperboloid* (§4.21). I make the conjecture however that, in the actual mechanics (the kinetostatics) of real gearing, we will find, in the path of a moving Q, a mechanical significance for the two points [Qm] which will not be *kinematical,* as might be expected, but (in a strange way) *statical* (§6.25, §4.50).

18. Going again to figure 4.04(b), I wish to mention the two pairs of circles, coaxial respectively with the lines called Axis, which intersect respectively at the two [Γm]. These circles, for the sake of clarity, are not shown, but they are clearly special pairs of circles. They might turn out to be impor-

tant. The first pair, the 'additive' ones, are, quite exclusively, the E-circles of chapter 5A; but this is a red herring; it is a special case. Notice incidentally that the additive [Γm] occurs *between* the two points TT which are at the throat circles respectively of the two participating hyperboloids. Under the right circumstances (when Line is the legitimate path of a Q) you may read this also as, 'between the spherical joints SS in figure 3.08'. The subtractive [Γm] occurs on the other hand *beyond* the intercept between the said two points. This may be shown to have occurred, not by virtue of some accident of the generally chosen data for the figures 4, but by virtue of a general principle; *the said 'inside-outside' phenomenon always occurs.*

Intersecting circles

19. I might need to say some more about the mentioned circles. I might need to speak, indeed, about these kinds of intersecting circles in general. Please recall the contents of §3.43. I might need to ask in the context of real gears, for example, whether such pairs of intersecting circles — they are, in general, intersecting at a single point — may, in certain circumstances, cut the enclosed circular laminae of one another, or whether they may not.

Theorem #3 part one, about the points [Γt]

20. I intend to show under this sub-heading that the said pair of isolated common tangent planes to the surfaces do in general exist along the line called Line, and that we can, in geometric practice, by means of an effective criterion yet to be enunciated, accurately locate the actual points of tangency. Refer in advance to figure 4.04(c) and recall in retrospect the contents of §3.41-43. Let Γ be a point generally disposed upon Line, and let there be circles drawn intersecting at Γ which are coaxial respectively with the two axes. As might be expected, I will call these circles the *G-circles* at Γ. Let the tangents (the two tangent-lines) to these G-circles be drawn intersecting at Γ. Now an obvious criterion for tangency of the two hyperboloidal surfaces at Γ is that the plane containing the two said tangents contains Line also. The plane containing the two said tangents might be called the *transnormal* at Γ. Now take the line through Γ normal to the plane containing the two tangents at Γ and find that under all circumstances this line, which we may call the *transversal* at Γ, cuts both of the axes called Axis. This has already been proved in the general discussion at §3.41 *et seq*.

> It follows that the criterion for locating points of tangency [Γt] between the participating hyperboloids is that the transversal at Γ cuts the line called Line (namely the straight line of intersection of the hyperboloids), perpendicularly.

This is an important result. It is of great significance in the realm of correct meshing. Incidentally the said transversal is not an *equilateral* one; it stands proud of the F-surface. The F-surface was first mentioned at §1.43; it is to be better studied soon (§4.26, §4.58); and it figures largely of course in chapter

6. Anyway please compare this statement for the points [Γt] with the corresponding statement for the points [Γm], and be aware that the distinction between the two is absolute.

21. We can prove and thereby clarify the contents of theorem #3 part one as follows. Please go back to my earlier remarks beginning at §4.03, and newly take into consideration another hyperboloid, the unique, *girdling hyperboloid* as I wish to call it, defined by Axis, Axis and Line. We can construct this by drawing all of the lines that cut all three of these, Axis, Axis and Line, namely all the generators of the opposite family (§4.02). The said girdling hyperboloid will have, in general, the appearance of figure 4.02. Its throat will be, as shown in the figure, not circular but elliptical. Can we also construct this girdling hyperboloid otherwise? Refer to [1] [§11.61] for some general information, or (better) the lack of it, then go to Zhang and Xu [84].

22. Let us next plot the locus of those points upon this girdling hyperboloid where generators of the opposite families cut one another perpendicularly. Refer to figure 4.02. This locus is well explained in the classical literature, see for example Salmon [45] [§186]. In general it may be seen to be shaped as shown. The locus resides upon a sphere whose centre is at the centre of the hyperboloid, and the radius of the sphere is calculable [45] [§186]. There are two branches in the case of the shown locus; they are mirror images of one another across the local xz-plane. The two branches may be coalesced in various ways. If the hyperboloid is more nearly circular the

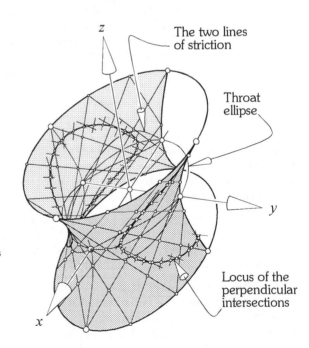

Figure 4.02 (§4.21). An elliptical hyperboloid. Such an hyperboloid contains (girdles) the three lines Axis, Axis and Line of §4.21. See here the locus of those points where intersecting generators cut one another perpendicularly.

KEY ASPECTS OF THE GEOMETRY §4.23

branches may for example surround the hyperboloid like two drooping necklaces, symmetric with one another about the xy-plane. Indeed, by symmetry, the locus is always plane symmetric about all three of the Cartesian planes. If the hyperboloid becomes circular, and the angle of inclination α is less than $\pi/4$ (see figure 3.01 for α), the branches become circular necklaces. If the circular hyperboloid becomes moreover possessed of an inclination angle α of exactly $\pi/4$ the two circular branches of the locus become coalesced into a single circle, namely the throat circle. And if the angle α is greater than $\pi/4$ there is no real locus at all. And finally there is the chance, as we shall see, that the girdling hyperboloid will be not only *parabolic* [69], but also *rectilinear* (§4.78), in which special case the said locus will have degenerated into two intersecting straight lines (§4.23).

23. Now perceive that any generator of this girdling hyperboloid, which might be representing the line called Line at §4.03, may (a) cut this locus twice thus causing two real points of intersection, (b) be tangent to the locus in which case the two real points of intersection will have become coalesced into a double point at the point of tangency, or (c) not cut the locus at all in which case the two points of intersection will have become unreal. There are in general no other possibilities. Thus we have shown that there are in general *two points* along the path of Γ where the transversal at Γ is perpendicular to the said path namely the line called Line. *I wish to call these points, which are points fixed in the common generator namely the path of Γ, namely Line, by the name [Γt]*. Refer to figure 4.04(c), see the two points [Γt] there marked in, and be aware that the four figures 4.04(a)(b)(c)(d) as a group have yet to be explained. I might remark that in the special case of the axodes at theorem #1, Line and both of the two axes called Axis all cut the same line (the CDL) perpendicularly, so the girdling hyperboloid in that particular case is *rectilinear parabolic* (§4.77, §4.78) [69].

Theorem #3 part two, about the point [Γd]

24. It will be clear from the text so far and from figure 4.04(c) that, as the movable point Γ moves from one [Γt] to the other, the two participating tangent planes rotate with respect to one another (rotating about the common generator) through an angle of 180°. It is accordingly clear by inspection that there will be an intermediate point where the two tangent planes are *perpendicular* with one another. I have already called this point [Γd] (§4.09). A graphical construction by trial to find [Γd] might be seen to go as follows:

> Erect a plane normal to the common generator at Γ and let it move with the moving Γ until the point-traces say A' and B' of the two Axes in the said plane are so located that angle A'TB' is a right angle.

This construction could be used in the figures 4.04 to locate the point [Γd] in the general case being considered there. One might also think about the circumstances at infinity along the common generator in each direction. Some phenomenon occurring there might provide good reason for looking more carefully at this [Γd]. It might turn out to be a 'middle' point of some significance, as yet unseen by me.

An isolated note looking in retrospect

25. Lest figure 4.03 be overlooked in the ongoing argument, I wish to comment now and in retrospect as follows. It is easy to see with CAD that the twisted line-locus of the perpendicular intersection explained at §4.22, drawn at figure 4.02, and employed in connection with theorem #3 (part one) above, can be independently defined by two intersecting elliptical cylinders. See figure 4.03. Each of the shown cylinders intersects the hyperboloid in the said locus and so, of course, they intersect one another in the same locus. There is clearly an algebraic relationship between the coefficients of the Cartesian equation for the hyperboloid and the principal dimensions of the two cylinders. As explained indeed in Salmon [45], under the chapter heading *Foci and confocal quadrics*, both loci exist upon the same *sphere* [45] [§186]. Writing in 1928, Salmon [45] refers in retrospect to Chasles (1835) and MacCullach, both of whom are said by him to have made separate, but overlapping, major contributions to this pure mathematical matter; but this and other such general questions are pursued no further here.

An isolated note looking ahead

26. Refer in advance to figure 6.01. Like figure 1.01 it shows the two shaft axes of a gear set, the common perpendicular between them namely the *centre distance* (A–B), and the pitch line for a certain k ($k = +0.6$). Equidistant from A and B are the points K and J on the axes, and the line JK is a

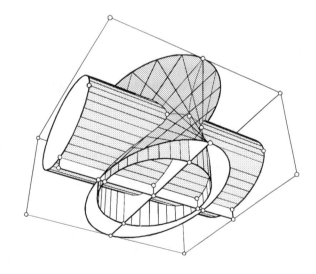

Figure 4.03 (§4.25). Two elliptical cylinders intersecting the hyperboloid of figure 4.02. They cut the hyperboloid itself, a centrally located sphere, and one another, in the same twisted curve, namely the locus of those points where generators intersect perpendicularly. These not-unknown facts might be found to be relevant in the study of gearing.

KEY ASPECTS OF THE GEOMETRY §4.27

prominent feature of the figure. The line JK is said to be a *transversal* of the two axes and, because (B–J) = (A–K), I call JK an *equilateral* transversal. Now, recalling *inversion* (§1.12), and by that token holding one of the wheels to be fixed, and speaking in terms of [1], JK is a *right line in the moving body*. It follows that the relative linear velocities (the rubbing velocities) at all points in the line JK are perpendicular to JK, and the representative vectors are proportioned with respect to one another and distributed along the line JK as shown. The overall gist of figure 6.01 is moreover independent of k and thus of the location of the pitch line: *the linear velocity vectors belonging to points distributed along the line JK (and along all such lines) are, for all k, perpendicular to the said line (or lines)*. This does not dispute the fact all of the linear velocity vectors at the points in a moving body are tangential to the relevant helix of pitch p coaxial with the pitch line (§5A.08); this fact is simply a sub-set of that fact. The sub-set is, however, within itself a spectacular fact; and it leads in turn to an easily defined and relatively simple, *ruled surface* of spectacular importance in design. The surface, generated by all of the equilateral transversals of the two shaft axes, is a *rectilinear parabolic hyperboloid* [69], whose principal axes are none other than the CDL and the FAXL. The FAXL was first mentioned and defined at §1.43, and the said ruled surface, as we shall see, will become to be known, in this book, as the *F-surface*.

Definition of [Qf]

27. Go back to §1.41 for an explanation (given in terms of gearing) of the FAXL. Apply this in the general arena of the two axes called Axis. Now it may be said in terms of the aforesaid Axes, Line and Γ that, as the moving Γ arrives at that unique point in Line where the transversal cuts, *not Line, but the FAXL perpendicularly,* the point in question needs a name. It might be called [Γf]. Accordingly I take the opportunity here to make a definition:

> In general the straight path of a point of contact Q in spatial involute gearing will cut the F-surface at some point, and I wish to call that point (which belongs, of course, not the F-surface, but to the path) the point [Qf].

Looking now at the figures 4.04

28. The figures 4.04(a), (b), (c) and (d) are drawn to illustrate both of the theorems #2 and #3. The dimensions are the same for each of the figures and the figures are chosen in such a way that the relevant curves and points are made visible. The hyperboloids are shown intersecting one another at figure 4.04(a) along the straight line traversed by Γ, namely the said common generator, and, at figure 4.04(c), along that and another (twisted) part of the intersection which is suitably labeled. Together the two branches of the intersection form a quartic. See appendix B at 4 for my comment upon Sticher's algebraic analysis of this quartic; and see [42]. The evident physical shape of the twisted curve of intersection, seen most clearly in figure 4.04(c),

§4.29 GENERAL SPATIAL INVOLUTE GEARING

has been obtained by me by means of computer graphics. Note the two clearly separated, separate circular throats of the two intersecting hyperboloids which are clearest to see at figure 4.04(a).

29. Searching in figure 4.04(a) for the points [Γm], we don't see them, but they are clearly shown at the enlarged figure 4.04(b). Note the thin cord riding over Line and the two sliding tubes at the Axes in the case of the addi-

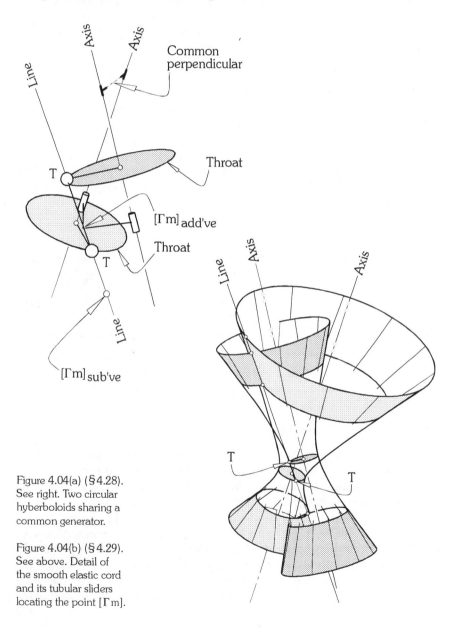

Figure 4.04(a) (§4.28). See right. Two circular hyberboloids sharing a common generator.

Figure 4.04(b) (§4.29). See above. Detail of the smooth elastic cord and its tubular sliders locating the point [Γm].

120

KEY ASPECTS OF THE GEOMETRY   §4.30

tive point [Γm]. The location by construction of this and the subtractive [Γm] have followed the double barreled instructions in the box at §4.17. Figure 4.04(b) shows more clearly, also, how the three common perpendiculars between the three lines (Line, Axis and Axis) come into play. The two between Line and the two Axes determine the shown points T in line, which are in turn indicate the locations of the throat circles of the two hyperboloids. The third, between Axis and Axis, recalls the intercept (A–B) at figure 1.01.

30. Figure 4.04(c) shows more clearly than any of the figures of chapter 3 the shape of the complete intersection of the two surfaces. The two intersections of the two branches of the intersection occur, of course, at the two

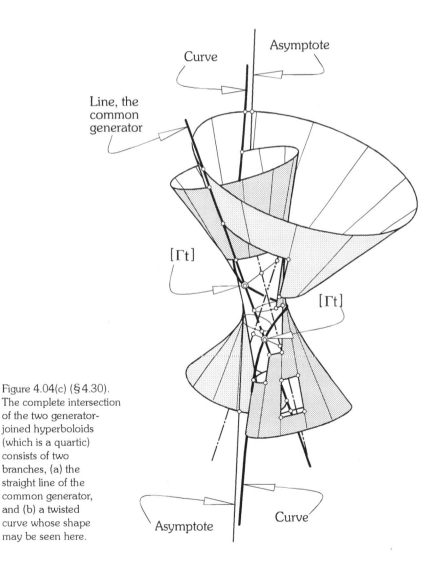

Figure 4.04(c) (§4.30). The complete intersection of the two generator-joined hyperboloids (which is a quartic) consists of two branches, (a) the straight line of the common generator, and (b) a twisted curve whose shape may be seen here.

points [Γt] of tangency of the hyperboloids, for this is what the theorem (theorem #3 part one) demands. See appendix B at 4. Notice in respect of these two points of tangency that, while the lower construction of the transversal at Γ according to §4.20 indicates external rolling at the lower [Γt], the upper construction of the transversal at Γ according to §4.20 illustrates internal rolling at the upper [Γt].

31. Looking next, again in figure 4.04(c), at the single, straight line asymptote towards which the twisted curve tends as it goes in both directions towards infinity, we should perceive that something of this nature was to be expected. If we the observers go off towards infinity looking at the whole immediate scene from further and further away, the two circular hyperboloids tend to become two circular cones sharing the same vertex (they are the asymptotic cones of the hyperboloids), the two throats tend to become two points together at the same vertex, and the two cones will be seen in the limit to be intersecting along two straight lines which, intersecting at an (accurately imagined) point of intersection, are indeed collinear. One of these lines is indeed a straight line (because we defined it to be), while the other is straight except for a minuscule 'twiddle' in the neighbourhood of the vertices taking it through the two points [Γt], both of which are infinitesimally close to the said (accurately imagined) point of intersection at the vertices [1] [§15.51].

Looking next at figure 4.04(d)

32. The figures 4.04(a)(b)(c) are orthogonal projections, and so also is the figure 4.04(d). This latter is showing the same contents as did the previous three, but it looks in a new direction. The new direction is unique. Keeping in mind the material of §4.02, please be aware that the new direction is taken precisely along the straight line asymptote just now mentioned (§4.31). The asymptote naturally appears here as a single point (which is indicated), and the twisted curve associated with it is also shown and suitably indicated. The twisted curve appears in the figure to be wholly elliptical. I refer again to Sticher [42] at appendix B at 4. Sticher has shown among other things that the said apparent ellipse is indeed elliptical. Note that the ellipse in figure 4.04(d) contains the already mentioned point which is the asymptote seen end-on [42]. *What this means is that the twisted curve is escribed upon an elliptical cylinder which has for one of its generators the asymptote itself.*

33. We see moreover in figure 4.04(d) that, parallel with the mentioned asymptote (which belongs to one of the families of generators of one of the hyperboloids), there is not only the generator of the opposite family of that hyperboloid, but two more generators belonging respectively to the two families of the other hyperboloid. Thus *four points* appear in the figure, all of which are the kind of point shown in the special figure 4.01. Each of these represents a generator. Thus we see that not only one of them, but both hyperboloids are being viewed along one of its generators, and that there are, when we look in this unique direction, *four parallel generators to be seen.*

KEY ASPECTS OF THE GEOMETRY §4.34

Three of them (the asymptote and two others) are coplanar, and two of them (one each from the two hyperboloids) are not only coplanar but also intersected by the common generator. For an exercise, explain why this is so.

34. The figures 4.04(a)(b)(c)(d) are together showing, in general, the important points, lines and curves to be expected when we come to deal in gears with the various occurrences of hyperboloids intersecting along a common generator. Please note with respect to the matter of generality, however, that the figures have been dealing exclusively with the case where the common generator (namely Line) goes 'between' the two axes. What this means, more correctly speaking, is that the saddle surface associated with the acute Σ is the one exclusively under consideration in the figure here. If Line were taken 'beyond' the two axes it would mean that the obtuse angle Σ would be the relevant angle. Refer to chapter 8 (§8.15 *et seq.*).

Specifically in the case of gearing

35. Given the shaft axes and the ratio k, we have, in all gearing, (a) the pitch line (§1.05, §1.14). Given the requirements for the legitimacy of a path (§3.07, §5A.08), and given that we have, so far, considered only one-sided teeth, we have in involute gearing, (b) a chosen legitimate path. Each of these lines in an involute gear set – (a) the pitch line, and (b) the path – has

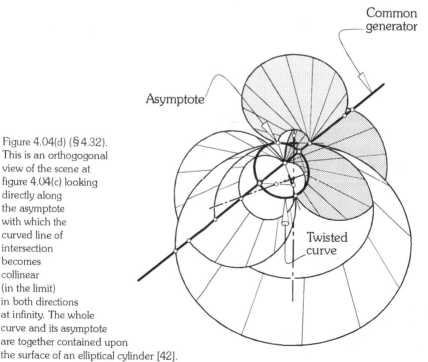

Figure 4.04(d) (§4.32). This is an orthogogonal view of the scene at figure 4.04(c) looking directly along the asymptote with which the curved line of intersection becomes collinear (in the limit) in both directions at infinity. The whole curve and its asymptote are together contained upon the surface of an elliptical cylinder [42].

123

its own significance. When they are swung respectively about the shaft axes, we get two pairs of hyperboloids. Swinging (a) gives (#1) *the axodes*, the pair of rolling and axially sliding hyperboloids which are associated with the relative screwing of the wheels (§4.10-14); and swinging (b) gives (#2) *the base hyperboloids*, which intersect and are associated with the meshing (§3.29, §3.67). Whereas the axodes are always throat-to-throat and always tangential with one another at all points along their common generator (§1.26), the base hyperboloids are not throat-to-throat in general and are not tangential with one another (except at the two special points [Qt] which come from theorem #3 part one). The path of Q has incidentally now a whole ragged collection of various special points along it — [Qp] (§3.55), [Qm] (mentioned already at §3.55 and now again in connection with theorem #2), [Qc] (§3.55), [Qt] and [Qd] from theorem #3, and [Qf] (§4.27) — all of which are yet to be studied intensively. Thus for one pair of wheels, we might say to begin that there are (#1) the axodal pair of intersecting hyperboloids whose throat radii have been designated r (§1.13), and, for one side only of the teeth, there are (#2) the pair of intersecting base hyperboloids whose throat radii have been designated a (§3.08), and whose common generator (the path of Q) contains the said ragged collection of points. Kindly note that the mentioned radii a are in no way related to the coefficient a among the coefficients a, b, and c of the Cartesian equation seen at §4.02.

36. Also in gears, however, we have (c) another kind (#3) of pairs of intersecting hyperboloids. They have only been mooted hitherto (§1.34-36). Associated with the actual shapes of the mating teeth, these are intimately connected with the *Cones of Wildhaber* [6] which, as I intend to explain, they 'hug'. These pairs of hyperboloids (#3) are yet to be better explained, but they might for the meanwhile be described as follows. *At a point of contact Q, when Q in its path is at the F-surface, namely at [Qf] in its path, there is a corresponding value of the rubbing velocity which must occur in the plane of adjacency there and which is inescapably determined by the helicoidal field of linear velocity vectors surrounding the pitch line.* Go to figure 1.01 and recall the helices explained at §1.09 and $1.51, to the singly departed sheets at §1.34, to the rubbing velocities in the directions of the helitangents mentioned at §3.47, to the helices mentioned again at §5A.08, and the definition of [Qf] at §4.27. These references to material already studied (or about to be studied) should conjure up an image of the ruled but smooth hyperboloidal surfaces, only mooted hitherto, which appear as soon as we swing the helitangent at [Qf] about the two shaft axes in turn. We thereby obtain (#3) this third kind of pair of hyperboloids. They touch one another at [Qf]; they slide upon one another circumferentially; they touch, like frilly garters, the cones of Wildhaber at circles; see figures 5B.02 and refer in advance to §5B.13. I have already called them *the cone hugging hyperboloids*, or, more descriptively, *the cone hugging naked wheels*. Refer in retrospect to §1.31-33 for (#1) the axodes, to §3.67 for (#2) the base hyperboloids, and in advance to §5B.02 and §6.29 for more about (#3) the naked wheels.

37. So we have in gears three different kinds of pairs of hyperboloids. They are: firstly kind #1, the unique pair of axodes whose throat radii have already been designated r; secondly kind #2, pairs of base hyperboloids whose throat radii have already been designated a; and thirdly kind #3, pairs of cone hugging naked wheels (which are pairs of severely truncated, doubly departed sheets) whose throat radii are hereby designated t.

The central fortuitous fact now apparent

38. Any legitimate path of a point Q cuts the F-surface at that unique point in the path we call [Qf]. When we erect, as we do, the two naked wheels whose smooth generators in mesh are set collinear with the sliding velocity at [Qf], the hyperboloidal surfaces of the said wheels intersect. *Naked wheels intersect.* The following is however true. *Precisely at [Qf], the naked wheels are tangential.* This could be said in other words as follows. *If H be a general point free to roam along the helitangent at [Qf], one of the two points [Ht] in the common generator of the intersecting naked wheels is at [Qf].* The reason for this is that, because the helitangent at [Qf] cuts the equilateral transversal there perpendicularly (§4.26), and because the sliding velocity (which is along the helitangent) is always perpendicular to the path ($3.47), conditions for the criterion in the box at §4.20 have been fulfilled. It is further a fact, but not especially fortuitous, that the relevant pair of mating F-circles intersecting at [Qf] roll on one another there, *not without, but with circumferential slip*. Even in the case of equiangularity where the F-circles become E-circles at [Qf] at E, there is circumferential slipping at the naked wheels (§5A.19, §6.29).

39. The gist of the argument as taken thus far, it might be said, is this. First, in kind #1 (the axodes), the relevant [Γm] is at P (the other [Γm] is unreal), and the two [Γt] have either coalesced at P or bloomed to fill the whole length of the common generator. Second, in kind #2 (the base hyperboloids), the relevant [Γm], namely [Qm], is at its designated position wherever that might be, and [Γf] is at that point [Qf] in the path where the path cuts the F-surface. Thirdly, in kind #3 (the naked wheels), the pair of smooth generators collinear at the meshing — these constitute the line called Line — together cut the path of Q at the said [Qf].

> This ensures that the relevant [Γt] namely the relevant one of the points [Ht] will also be at the said [Qf]. This needs to be understood. If the were not so, the naked wheels (which are of course ghostly) could not roll and slide upon one another in a proper manner, and the present theory would collapse.

These remarks are applicable to each of the two paths; and I say here also that the same three kinds of sets of generator-joined hyperboloids are evident in the special cases of planar gears and the crossed helicals; but they appear in those special cases in special, degenerate forms of course. These special cases will be discussed more fully when next convenient.

Legitimate paths

40. It is instructive to notice that all of the argument so far has been pursued in the absence of any appeal to the *legitimacy* or otherwise of the paths of possible points of contact Q. Although this has safely resulted in the important conclusion drawn at §4.39, it has left us with no idea about the whereabouts in relation to the F-surface of the points [Qm] on the legitimate path of a Q — and no idea, either, of the corresponding whereabouts of the points [Qt] and [Qd]. *So I wish to speak now about the inescapable requirement obtaining in spatial involute gearing, namely that the path of a Q cannot be just anywhere, but must be legitimate according to the fundamental law* (§2.02). In doing this, however, I will restrict the discussion to paths that cut a single infinity of points [Qf] on the F-surface which inhabits some single, generally chosen equilateral transversal. I make this restriction to clarify the argument; for to deal simultaneously with the whole population (the ∞^2) of points on the F-surface would be too unwieldy.

41. Please refer to figure 6.01. It shows (for a given k and at points along a chosen equilateral transversal) the distribution of the sliding (or the rubbing) velocity. The greater the distance from the pitch line to a point of contact, the greater is the sliding velocity there; recall the equation (25) at §1.51. *Let any point residing within the F-surface be called F.* Please see in the figure that, *because the sliding velocity (which occurs along the helitangent) must at all points of a path be perpendicular to the path (§5.29), it follows that, at all points F along any one equilateral transversal, the legitimate paths must be perpendicular to the helitangent.* This corresponds with, and leads directly into, those parts of the chapters 5 and 6 which deal with the possible paths of points of contact in the real circumstances of practical involute gearing. The chapters 5 and 6 lead independently to the same conclusions as the ones arrived at here, but by a more circuitous route which takes into account, as we go, considerations of real gear design. Those chapters and this that follows should, indeed, be read in parallel.

42. Referring next to figure 4.05, which is a special view of figure 6.01 looking along the helitangent at a chosen F and having the effect that the polar plane there is in the plane of the paper, we can see that, at any point F along the shown equilateral transversal, there may be constructed, not only the helitangent there (for the given k), which shows the direction of the sliding velocity at that point F (for the same given k), but also the *best drive line* (the BDL). At each point F along the equilateral transversal a mutually perpendicular triad of lines presents itself. It consists of (a) the shown equilateral transversal itself, (b) the helitangent (appearing in figure 4.01 as a point), and (c) the BDL, which latter is drawn at all points F. Recall §3.44. At a generally chosen one of the points F, the whole planar array of the possible paths through that particular F has been indicated. *There is a such a planar array (a polar plane) of possible paths at each single point F; and, as we know, together they are a sub-set of all of the polar planes of the linear complex of lines surrounding the pitch line* (§3.47, §5B.14) [1] [§9.12]. Seen as said, this

KEY ASPECTS OF THE GEOMETRY   §4.43

is simply a restatement of the second law (§2.06, §2.17, §2.19). In design, of course, only one or other of these paths in the polar plane needs to be chosen for drive and another for coast, the two choices, often conflicting of course, together constituting the essential difficulty in design; but in the investigative analysis being undertaken here we can look collectively at all of them.

On the distribution of [Qm]

43. Nowhere yet in this book has it been mentioned that a distinction is soon to be made between the so-called *equiangular* and the *polyangular*. This matter is introduced for the first time early in chapter 5A, and the relevant discussion continues, on and off, well into chapter 6. The following material relates directly to equiangularity and the E-line (yet to be mentioned) of the chapters 5 and to polyangularity and the F-surface (yet to be mentioned) of chapter 6. The author believes in this connection that, in the minds of many gear workers — as well as in mine — either lurks or has lurked the intuitive notion that the points [Qm] are somehow important kinematically or, at the very least, significant in some mechanical respect. What follows in connection with figure 4.05 might be seen, just yet, as a mere exploratory exercise; and it is just that. It is poorly executed and poorly presented to boot, but it will be of value perhaps to those researchers who might in the future wish to study [Qm] intensively. Look now in advance at §6.25-26 which has some more to say about what we are doing here, preferably having preliminarily perused the whole of the chapters 5 and 6. The reader will find at §6.26 incidentally a back-reference to the forthcoming §4.50, which probably captures all by itself the real nub of the matter.

44. Please re-read the contents of §3.35-36, reconsider the awkward questions put at §3.37, and look again at the related evasive answers given somewhat expansively at §3.38-39. Questions relating to [Qm], its definition and its mechanical significance, remain. Taking then our currently ongoing definition for [Qm] from the last sentence at §3.38 and from the inset material at §4.16, let us plot, in figure 4.05, by repeatedly using the construction explained at §4.16, the distribution of the population of the additive [Qm] among parts of the population of the legitimate paths, beginning for convenience with the distribution along the transversal of the points F and the available BDLs that display themselves there, all of which latter although impractical as paths are at least legitimate. Let us moreover not forget that intersecting at each point F along the transversal there is not only the BDL that belongs there, but the whole of the polar plane of possible legitimate paths that also belong there. In figure 4.05 the orthogonal view is taken along the helitangent at the representative point F, so the polar plane at F is in the plane of the paper. So also, therefore, are the transversal itself and the BDL at that F in the plane of the paper. The single infinity of the BDLs of the array appear of course to be parallel in this view, but only one of them (the one at the chosen representative F) is in the plane of the paper.

45. Recall that the angle within a given polar plane between the best drive and a chosen path has already been called by me the *angle of obliquity* δ. This angle is otherwise defined and otherwise called, by others, the *pressure angle*. Please see some representative angles δ marked in at figure 4.05. They are quite clear to see there because, in figure 4.05, as has been said, the polar plane at F is in the plane of the paper.

46. Laboriously by computer graphics in figure 4.05, additive [Qm] have been located (a) along each of the best drives which, although in fact not parallel, appear there as being parallel, and (b) around the polar pane of legitimate paths at F across a limited range of δ extending in both directions away from the BDL at F. Loci of the [Qm] have been sketched in: while the locus (a) of [Qm] upon the ruled surface of the array of the BDL is a twisted curve, the locus (b) of [Qm] upon the surface of the polar plane is of course a planar curve. Already from this one begins to see that all points [Qm] are path dependent. What I mean by this is better explained at §6.25.

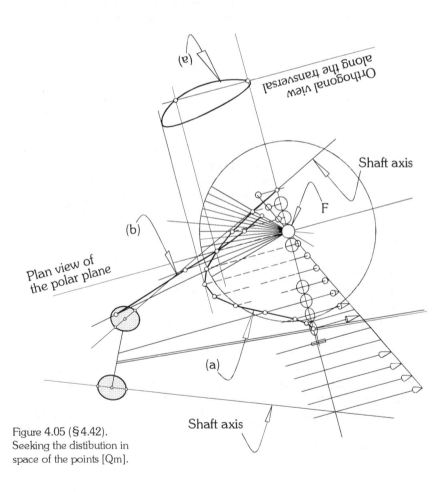

Figure 4.05 (§4.42).
Seeking the distibution in space of the points [Qm].

47. An important thing to notice is that it is only at J and K (see figure 6.01), at the special point E (see chapter 5A), and at certain other isolated points (which happen by accident), that the additive [Qm] of a path falls coincident with the [Qf] of that path. The contents of figure 4.05 (and other investigative graphical work not shown) make it abundantly clear that the additive and/or subtractive point [Qm] and the point [Qf] of a path are, in general, quite different points. *So we can say in general that [Qm] is not at [Qf].* When [Γm] was not at [Γt] in the figures 4.04 — see also the same phenomenon at figure 3.15(b) — we concluded quite naturally that this might be due to the generality, but here it has occurred even with legitimate paths, which clinches the matter. [Qm] is definitely not at [Qf].

48. After more graphical work, the results of which are not shown here, the next thing noticeable was that, whereas the additive [Qm] were found to occur between the two common perpendiculars (the shortest distances a erected between the path and the shaft axes in each particular case, *the subtractive [Qm] were found to occur beyond these limits*. This coincides with what we found in the general case using the line called Line and the two lines called Axes (§4.19). Now consider this: in chapter 2 we looked at figure 2.02 and found the rod (S–S) of the equivalent RSSR; in chapter 3 we looked at figure 3.08 and drew attention to the same rod; and at figure 5B.04 of chapter 5B we study that part of the path of a Q that all researchers call the *line of action* within the path. The line of action is bounded by the points TT in figures 3.12(a)(b), which figures are drawn for the planar case (the next figures 3.13(a)(b) being for the spatial); and the points TT in these figures, when seen in the broader kinematic flow of *spatial* events, correspond exactly with the centres SS of the spherical joints of the RSSR. See for example the two sets of spherical joints SS within their linkages RSSR (one for coast and one for drive) illustrated bold at figure 5B.04. We seem have found something of real mechanical significance here.

Kinematics at [Qm]

49. From the above we can see (albeit somewhat dimly) that, whereas the additive [Qm] is relevant for external meshing (and within the line of action for that kind of meshing), the subtractive [Qm] has its relevance in more remote, 'unreal' portions of the path. These latter, although on the one hand real mathematically (where hard inter-actable two-sided surfaces without material backup might be held to operate), are on the other impossible mechanically (where only single-sided surfaces that have material backup can exist). The subtractive [Qm] applies either in those portions of the path which are beyond the limits SS of the line of action, where undercut interferes with the surfaces and destroys the action (§3.57 *et seq.*), or in that kind of gearing where internal meshing is the general rule. This latter however cannot occur unless the action is wholly planar (§3.32). *We might conclude, therefore, that only one of the two [Qm], the additive one, is worth considering seriously.*

§4.50   GENERAL SPATIAL INVOLUTE GEARING

The line called Line in figure 4.04(b) is not a legitimate path as seen in the context of gears; it is simply a chosen one of the three lines generally arranged in that figure; the additive [Qm] in gears is nevertheless the one represented by the point [Γm] located there by the action of the location-seeking smooth elastic cord and the two cylindrical sliders.

Statics at [Qm]

50. Following §4.18, and still with [Qm], let us examine now the *statics*. Let us do this, however, in some imagined real gear set, where the line of action is legitimate. Under steady conditions and in the absence of friction, power is being transmitted from shaft to shaft without loss. The torques suffered at the respective shafts are inversely as the ratio of the numbers of teeth, namely as the angular velocity ratio k. Ignoring the occasional severality of Q (§2.18), the transmission force at the single point of contact between teeth is a force fixed in magnitude and direction whose magnitude is determined by the steady-state kinetostatics and whose point of application is determined by the current position of Q. If **T** be the torque at one of the shafts, and if (a, α) be as before, the following may be written: **T** = **F** $a \cos \alpha$, where **F** is the force at Q. This force causes an equal and opposite reaction force to occur at the moving Q against the mating tooth. Considering each of the two wheels as an isolated, free, rigid body, the sum of the external forces and couples upon each of them will be zero. Ignoring gravity, these forces and couples may be seen to be, for either one of the wheels, (a) the said transmission force acting along the line of action at Q, and (b) the whole collection of other forces and couples acting reactively at the bearings supporting the relevant shaft. Drop a perpendicular from Q (wherever it may happen to be) onto the axis of the shaft. The component of **F** in the direction of the dropped perpendicular is the radial load upon the shaft applied at the boss of the wheel; and this component, somewhat less than **F**, *depends upon the position of Q*.

51. Refer to figure 5B.04. This shows, not only one of the lines of action in a gear set (our having considered only one hitherto), but both of them, there being one for drive and the other for coast. Both segments (S–S) are shown, each occupying a portion of its own path. Depending upon the direction of motion, each of them 'begins' at the throat of its 'own' hyperboloid, terminating at the throat of another, there being, in all, four hyperboloids. At all four points S the angle between the relevant segment (S–S) and the relevant throat radius a is a right angle. Let me say by way of definition that a point Q remains *real* when within its (S–S), becoming *unreal* when beyond it. Let me say also that a segment (S–S) might be called *the ideal zone of contact*, whose length is Z_I. Within Z_I, and remembering the reality of truncation, let the even more limited zone, *the zone of actual contact*, be called by that name, and let its length be called Z_A. Let me now say by way of further definition that within Z_A the really real, *physical* Q exist (§3.56).

52. Let us imagine that, at each position of a point of contact Q, as that Q travels the length its ideal zone of contact, from its starting-point S where the relevant radius *a* is perpendicular to the path, to its ending-point S where the relevant radius *a* is not thus perpendicular to the path, a perpendicular is dropped onto both of the two shaft axes. Let us see first of all (a) that each of the two resulting populations of dropped perpendiculars forms a rectilinear parabolic hyperboloid whose principal axes are the axis of the relevant shaft and the relevant radius *a*, and (b) that the said rectilinear parabolic hyperboloids intersect along the straight line of action of the force namely the straight line SS of the path (§4.77). Next we can say (c) that the component of **F** in the direction of the dropped perpendicular at any position of Q at the instant is the radial load upon the relevant shaft. The locations of these components and their magnitudes will of course depend upon the position in its path of the traveling Q; but it will moreover be clear that, while one of the shafts is increasing its suffered radial load in this respect, the other is decreasing it, and *vice versa*. There is, in other words, during the passage of any Q, a build-up from zero to maximum of the compressive radial load in the driving (or the driven) member accompanied by a diminishment to zero of the corresponding compressive load in the driven (or the driving) member, *and at [Qm] the said radial loads on the shafts are equal.*

53. The above ingenuous conclusion, that at [Qm] a definitive mechanical event occurs, *namely that the radial loads upon the shafts become equal,* is not only somewhat undramatic, but also based upon the dismissal of two important phenomena that need to be noted. Firstly we have ignored the axial loads upon the shafts which are resisted by the thrust bearings at the walls of the gear box. But secondly there is that, due to the severality of Q (§2.18), the statics becomes sometimes indeterminate. In this respect it needs to be seen however that, unless the contact ratio is greater than 2, the episodes of 'double Q contact' occur only at the beginning and end of the passage of a Q through its zone of actual contact Z_A (§4.51). It might be instructive to argue that, within this already restricted zone Z_A, there is a even further restricted zone that might be called, *the zone of single point contact.*

54. Please also notice that, at §4.52, there is the unstudied implication that, due to the continuously changing relationship between the internal forces, there is, between the shafts, a spatial, oscillatory disturbance occurring at tooth frequency. In the presence of accuracy and rigidity (§1.05) [1] [§6.06], there is no vibratory response to this, but in the presence of elasticity (which is the reality) there will be a response, and the complexities are bewildering. It is certainly true to say that, although with idealized involute action we have a continuous equilibrium existing between the forces and a steady, external kinetostatic state, we have, within the gear box, locked up 'internal' forces continuously changing periodically at tooth frequency, and the phenomena associated with these I choose not to study here.

Strain energy at [Qm]

55. Taking yet another view (and this is a comprehensive straightforward one), let us consider the effect (given elasticity) of the constant transmission force **F** upon each of the two participating teeth. Each of them may be seen as a tapered cantilever in continuous, elastic, loaded, 'oblique sliding' contact with the other. Assume that, with a constant torque (say **T**) applied, we move from configuration to configuration across the range of single-point-of-contact movement (§4.53). As Q_2 traces its slip track from root to tip on the driver 2, Q_3 traces its slip track from tip to root on the driven 3. The elastic deflections at Q are somewhat less than 'mid' when Q is at the root of a tooth and very much greater than 'mid' when Q is at the tip. *This line of argument would appear to confirm, in any event, that [Qm] and its immediate neighborhood is not a local 'torque resisting hump' to be met and overcome, but a local 'dip in resistance' to be 'driven through'.* It is surely possible thus to argue (with certain defensive reservations) that the configuration of least total strain energy occurs when Q is at (or near) [Qm].

56. Please go to §11.30 and see explained there the apparatus shown in the photograph nearby. A mating, movable pair of gear bodies is being held in contact by means of the pulleys and the suspended counterweights. Please ignore all mass in the apparatus other than that of the counterweights which, for equilibrium, are proportioned inversely to the angular velocity ratio k. Let there be a total absence of friction. Using this product of the imagination, and by way of further explanation of the question of strain energy under consideration here, the following might be said. In the presence of rigidity and geometric accuracy the equilibrium among the moving parts of the apparatus remains of course in a state of neutral stability across the range of single point contact. It might however be argued that, in the presence of elasticity and the absence of friction (and the presence of course of a steady **T** applied), the contacting pair of involute teeth, nominally 'at rest', will migrate towards a state of 'super-stable' equilibrium where their Q is at [Qm]. This is a wild statement, poorly supported, for the distribution in space of the elastic material of the teeth is somewhat complicated here; but in a corresponding planar case, where the geometry would lend itself to simple experiment, and where the mathematics might not be intractable, it will surely be possible to make some intelligent analyses. Refer to Routh [85] who speaks about *metastability*, and refer to my valedictory remarks about [Qm] at §6.25.

Another look at the gear bodies of figure 3.10

57. Following §3.35 *et seq.* and these most recent discussions of [Qm] and [Qf], I wish to comment further now about the gear bodies of figure 3.10. Their defining parameters (their core helix radii a and their angles of inclination α) were chosen arbitrarily. The gear bodies were then imagined to be constructed accurately, but assembled arbitrarily, for that was the crux of the exercise. We wished to display (by means of the CAD) the fact that cor-

rect involute action would occur however careless we chose the assembly to be. It is therefore not surprising, firstly, to find that the gear-body set of figure 3.10 displays polyangularity; we find, by CAD, that the path of Q does not cut the E-line. We find moreover that the length of the effective portion of the path of Q is even less than Z_A because the profiles are unnaturally truncated. The profiles have precipitate *edges*. We even find in the figure that the zone of actual contact does not overlap the point [Qf]. These considerations in no way invalidate the exercise however. They simply mean that the gear bodies of figure 3.10, *in the absence of other teeth*, cannot 'follow through'. With their Q traveling towards the F-surface, in other words, the gear bodies cannot propel that Q onwards to its point of piercement, and thence beyond, before the traveling Q meets catastrophe, an edge of profile. But could a pair of gears with teeth like this actually work? It would depend upon the *contact ratio* I think, not upon whether or not [Qm] or [Qf] is embraced within Z_A. The teeth would always be 'pulling' or 'pushing' one another, never a happy combination of the two where the changing stored energies are diminished in magnitude and distributed more evenly on either side of an extant [Qm].

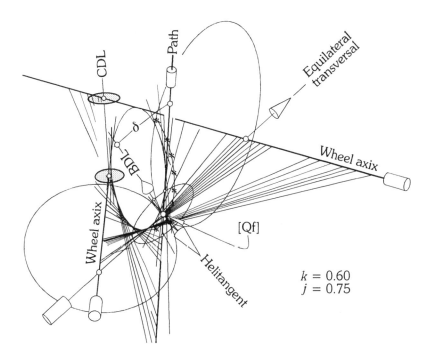

Figure 4.06(a) (§4.58). The moving transversal j-Q-k is here tracing out its regulus, the complete shape of which is only enough sketched-in to indicate the totality of the girdling hyperboloid. The seen 'hole' is is not showing the true elliptical shape of the throat because we are not looking here parallel with the axis of the hyperboloid. See figure 4.06(c) for this. A short portion of one of the lines of striction is plotted here with the barbed wire.

The moving transversal, the FAXL, and the F-surface

58. The FAXL is prominently mentioned and illustrated in the chapters that follow. It is one of the two principal axes of the so-called F-surface, namely the parabolic hyperboloid defined by the CDL, the FAXL itself, and the two shaft axes. Any legitimate path of any Q (in any kind of involute gear set whether equiangular or polyangular) will pierce the F-surface at some point, and I call that point [Qf]. At any *general* position of Q in its path, the moving transversal at Q does not cut the FAXL; it misses the FAXL; it simply cuts the two shaft axes. At [Qf], however, the transversal does cut the FAXL, and it cuts it perpendicularly (§4.27). Also at [Qf], the transversal cuts the two shaft axes equiangularly (see the equal angles ξ at figure 5.01), and the length (j–k) of the changing transversal arrives there at a local minimum. To make matters clear, I write the transversal at [Qf], not j–Q–k, but J–Q–K. At [Qf], as has been said, the transversal is perpendicular to the FAXL, the plane of the transnormal is thereby parallel with the FAXL, and the sliding velocity, because it must always be perpendicular to the transversal (§3.49), must always be within that plane. Figure 4.06(a) relates to a plain polyangular set where [Qf] is somewhat remote from E; whereas $k = 0.6$, $j = 0.75$. As Q travels its path, the transversal j–Q–k at Q is tracing out a regulus (§3.46). I have drawn this regulus for a short way (enough to straddle the F-surface) by locating a series of transversals j–Q–k to the three given lines, namely the two shaft axes and the one given path [1] [§11.05]. The resulting incomplete hyperboloid seen in the figure is, thereby, an example of the girdling hyperboloid of §4.21. But we are here looking in some unknown direction oblique to the principal axis of the hyperboloid; the apparent shape at the throat is not the true shape. See the *barbed wire* marking the line of striction. Note also that, as Q passes [Qf], the generator of the swept regulus becomes contained within the F-surface; it becomes indeed the equilateral transversal; this should not be surprising; while the two shaft axes and the path are members of the one regulus, the equilateral transversal is a member of the opposite regulus.

Variation of δ-remote with the distance of Q from [Qf]

59. In this paragraph and the two that follow I take it for granted that the reader has already perused the chapters 5 and 6. By CAD and by using the architectural data of figure 5A.01, where $k = 0.6$, I have located a polyangular [Qf] with poly-ratio $j = 0.75$; and I use this as an example here. See figure 4.06(a). In the polar plane at the located [Qf] I have set up a legitimate path taking δ to be 20°. I have next plotted a series of locations of the transversal jQk as Q travels along this path from one side of the F-surface to the other. See the plotted regulus which contains of course the unique, equilateral transversal that cuts the FAXL perpendicularly at [Qf] (§4.58). To get at the angles δ_{REMOTE} at a series of points along the path of Q, I have measured the angle between the path and the transversal at each point, and plotted the *complement* of this angle (which is the angle δ_{REMOTE} of §3.54) against the dis-

KEY ASPECTS OF THE GEOMETRY   §4.60

tance of Q from [Qf]. The graph appears at figure 4.06(b). Note the length and location of the segment Z_I in the line of action, and note that a maximum value of δ_{REMOTE} occurs somewhere within it. The maximum value is greater than δ itself, namely greater than 20°. The curve however is asymmetrical about the intercept Z_I, asymmetrical about [Qf], and asymmetrical about the location of its own maximum; and the question is, where (when speaking geometrically) does this maximum occur?

The point [Qs] occurring at the line of striction

60. The maximum value δ_{REMOTE}, found by CAD for the chosen numerical example at §4.59, seems to occur at a well conditioned point accurately determined by what appears to be a tidy theorem. Having not yet proven the theorem, I hypothesize as yet as follows: *as a point goes from one end to the other of any generator of a regulus, the angle at that point between the said generator and the generator of the opposite regulus varies in such a way that a maximum (or a minimum) of the said angle occurs at the line of striction upon the said opposite regulus (§4.02) [1] [§11.09]; whether or not a maxi-*

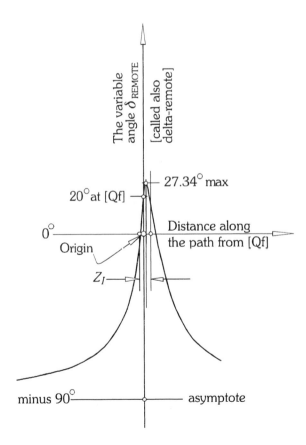

Figure 4.06(b) (§4.59). This figure shows that, whereas the angle of obliquity δ was set, at figure 4.06(a), at 20° exactly, the angle delta-remote has been found to vary considerably within the range of Z_I and beyond. In planar gearing this angle delta-remote remains constant at a fixed value, namely delta.

mum or a minimum occurs depends (a) upon the simple question of what sign convention is adopted for the participating angles, (b) upon whether or not there exists a locus of points on the surface of the hyperboloid where generators cut one another perpendicularly, and (c) whether or not the originally chosen generator happens to cut this locus. Refer to figure 4.02; see there (a) a picture of the said locus, and (b) both of the lines of striction upon the surface, one drawn for each of the two reguli [1] [§11.09].

61. Refer to figure 4.06(b). The angle δ_{REMOTE} in the present numerical example, which is the complement of the angle of the hypothesis, takes the value $[\pi/2]$ at $\pm\infty$, takes its given value namely $[+20°]$ at [Qf], *and reaches its discovered maximum at the said line of striction*. The value at the discovered maximum was here found to be 27.34° approximately. Figures 4.06(a) and 4.06(b) accordingly begin, by way of example, to answer the question of how δ_{REMOTE} varies with the distance of Q from [Qf], and of how δ itself, namely the angle of obliquity, might be unambiguously defined. They help moreover to answer the question of how the triad at Q alters its orientation (gyrates) as Q moves (§3.50, §4.63). Terminologically, I wish to call that point on the path of Q where δ_{REMOTE} is at its local maximum by the name [Qs], the suffix 's' coming from the word *s*-triction.

62. We don't yet know the dimensions of the regulus traced by the moving j–Q–k at figure 4.06(a). We know neither the position of its centre nor the orientation of its principal axes. It is however the *girdling hyperboloid for the given path* (§4.21), so at least we understand its origin in our imagination and something about its general significance. Given its three given generators (we have called them Axis, Axis, and Path), there are theorems already available for making some direct constructional progress in the matter of its dimensions and orientation [1] [§11.08, §11.10], and I have used the first of these to find the centre. But it is not yet possible for me to find by direct methods in my CAD the associated principal axes. What I have done at figure 4.06(c), which purports to look directly along the longitudinal axis of the said hyperboloid, and which takes its clue from [1] [§11.62], is to draw, for the given figure, generators of the asymptotic cone centred at the found centre of the hyperboloid. These cut the surface of a unit sphere whose centre is at the found centre and, by looking repeatedly by trial, until a direction is found whereat the two twisted curves of intersection upon the sphere appear to coincide, I have determined approximately the direction of the central axis. Alternatively I could have located by trial the four principal vertices of the hyperboloid which occur at the intersections of the two lines of striction (see figure 4.02), and proceed by means of CAD from there. Read some old remarks about the problem of the axes at [1] [§11.62], and refer to the recent paper by Zhang and Xu [75], which authors have definitively solved this difficult problem algebraically; their paper deals with three separately given *screws* which define a *three-system* [1] [§6C.10]; but in our much simpler problem here the three screws are simply lines, namely screws of zero pitch.

KEY ASPECTS OF THE GEOMETRY   §4.63

Position of [Qs] in the special case of equiangularity

63. In figure 4.06(c), where polyangularity prevails, and where the matter has had a first, somewhat rough investigation only, the centre-point of the girdling hyperboloid appears to be, but is probably not, exactly at the FAXL. Simpler, more manageable forms of the girdling hyperboloid probably occur in special cases, when for example $k = 1$ and/or when equiangularity obtains. Given equiangularity, we find (a) that the said centre-point is somewhere upon the plane perpendicularly bisecting the centre distance, and (b) that the lines of striction (of which there are two) intersect at E. In this special case, one of the vertices of the girdling hyperboloid is at E and the maximum value of δ-remote (namely δ itself) occurs at [Qx], which is also at E.

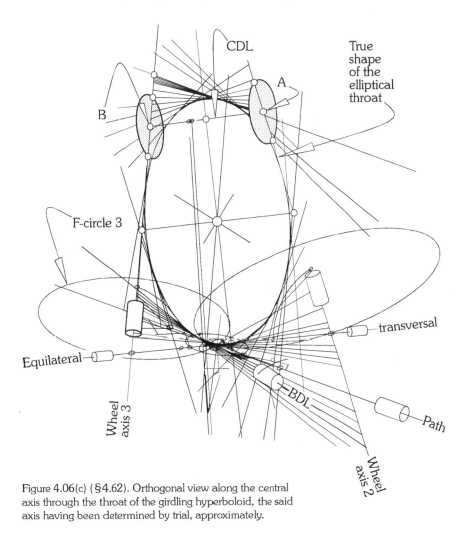

Figure 4.06(c) (§4.62). Orthogonal view along the central axis through the throat of the girdling hyperboloid, the said axis having been determined by trial, approximately.

More about the gyratory motion of the triad at Q.

64. The figures 4.06(a) and (c) are different projections of the same gear set; $k = 0.6$, and $j = 0.75$; and they are both orthographic projections. Figure 4.06(d), however, while again of the same gear set, is drawn in perspective. The central point is at [Qf]; the viewer is looking directly against the plane defined by the FAXL and the equilateral transversal; these latter lines (intersecting in Y) are in the plane of the paper. Angle δ is 20° so the shown path of Q is inclined at this angle to the principal transnormal at [Qf], which latter appears in the figure as a line. The sliding velocities at points Q along the path of any Q will form a rectilinear array (§6.12) [1] [§3.19]. Whereas however at figure 6.01 the velocities are perpendicular to the equilateral transversal, here at figure 4.06(d) the velocities are perpendicular to the path; see figure 3.11(b) at §3.44; this is drawn not for the particular case where Q is at [Qf] but for the general case when Q is somewhere remote from [Qf]. Whereas we are dealing at figure 6.01 with the velocities at [Qf] for various values of a changing j, here at figure 4.06(d) we are dealing with the velocities at various values of the distance from [Qf] to Q for a fixed j. To find the array, we could apply directly the basic equation (25) at §1.42; but, having already from figure 4.06(c) the direction of the transversal j–Q–j at each Q, we can, by appeal to the rectilinearity of the triad at Q (§3.44), locate the triad at each Q. Representative points Q are distributed at equal intervals along the path of Q to reflect the steady speed of Q across time. At each of these, with [Qf] arranged midway, I have established the local transnormal normal to the local transversal; next I have stepped off a distance along the path from the particular Q and dropped from that point a normal onto the said transnormal; thus I have been able to measure δ-remote which, by definition (§10.13) is between the path and the plane of said transnormal. The local helitangent at Q must by virtue of the mechanics be (a) in the plane of the local transnormal, and (b) perpendicular to the path; so I have been able to find the helitangent at each Q and thus the whole triad at each Q. Each triad is drawn in the form of a cubical box, the three joins of the centres of the opposite faces representing the three directions of the axes. The gyratory motion of the triad (§3.50) can now be more clearly seen. See in the figure the spectacular 'double-tumble', not of the helitangent alone whose relatively simple motion we understand (see figure 6.03), but of the triad as a whole, as Q traverses the segment (S–S). It is within the segment (S–S), of course, that the real action occurs. When δ–remote is zero the best drive is collinear with the path; so it might well be interesting to locate the locus of points upon the surface of the plotted hyperboloid where perpendicular intersections of generators occur, and thus to locate the two points [Qt] on the path of Q where the angle δ-remote becomes zero; see figure 4.02 at §4.21; it is precisely at the said locus that δ-remote changes through zero from positive to negative. I do not propose follow such matters further here, but they might productively respond to further study.

KEY ASPECTS OF THE GEOMETRY §4.64

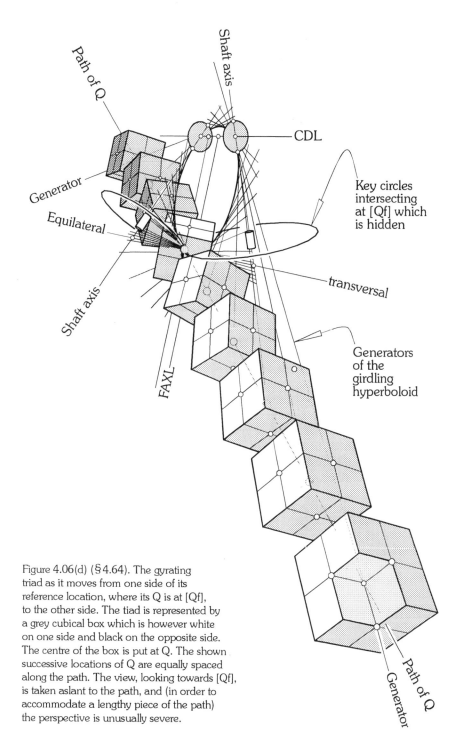

Figure 4.06(d) (§4.64). The gyrating triad as it moves from one side of its reference location, where its Q is at [Qf], to the other side. The tiad is represented by a grey cubical box which is however white on one side and black on the opposite side. The centre of the box is put at Q. The shown successive locations of Q are equally spaced along the path. The view, looking towards [Qf], is taken aslant to the path, and (in order to accommodate a lengthy piece of the path) the perspective is unusually severe.

The two planes that continuously define j–Q–k.

65. The two planes defined by the two shaft axes and the moving point of contact Q continuously intersect in the transversal j–Q–k at that Q. Peruse retrospectively §0.13, §2.12, §3.25, §3.29 and §3.65 to gather the various scraps of information already given in passing about this and the three different paths of Q which occur separately but simultaneously in (or upon) the three links 1, 2 and 3 of a gear set. Find a consistent nomenclature for clearly delineating such different paths at [1] [§12.08]. In a working involute gear set, at any moving point Q_1 in its straight path in 1, the two planes defined by (a) the fixed shaft-axis 2 and the moving point Q_2, and (b) the fixed shaft-axis 3 and the moving point Q_3, intersect in a line which is the transversal at that Q. The locus (as Q moves) of the said transversal constitutes one of the two reguli of the girdling hyperboloid girdling the two shaft axes and the path. Such an hyperboloid appears in each of the figures 4.06(c) and 4.06(d). At [Qf], where the transversal suddenly becomes the *equilateral* transversal, its variable length (j–k) arrives at its local minimum namely (J–K), which length is, as we have said, embedded within the F-surface (§4.58). *That moving intersection of the planes, that changing transversal, is a motional right line in each of the moving bodies 2 and 3 when the bodies are looked at relatively to one another, the said bodies being of course the two wheels of the gear set* (§4.26) [1] [§10.49]. The sliding velocity (when found from the pitch line and the pitch *p*) will be perpendicular to both the best drive and the transversal at all points Q along the path of Q which is, as has been said, a right line in both of the moving bodies; and by virtue of this the relative velocities v_{2Q3} and v_{3Q2} (which together might be seen as equal and opposite aspects of the 'rubbing' velocity) are collinear and perpendicular also to the path. Thus we have been able to find the triad that smoothly moves along the path and through [Qf] continuously gyrating as it goes; see figure 4.06(d). A special case of the said pair of intersecting planes appears at 5A.04.

Mechanical significance of [Qf]

66. Take some *planar* gear-set. Take some generally chosen position for one of the Q in its path (which path is in the reference plane), and imagine an ordinary hinge (a revolute) introduced there with its axis normal to the reference plane, to join the wheels at this Q. It will be evident that the mechanism, now a triangular, 3-link planar loop with three parallel revolutes, has been rendered rigidly immobile. If on the other hand we did the same thing when Q was at [Qf], namely at the pitch point P in this planar case, the mechanism would be seen to remain, at least, transitorily mobile [1] [§2.08]. The axes of all three revolutes would be coplanar, and we would be able to wobble the wheels to and fro within a fairly wide range of available movement, even in the presence of fairly tight clearances at the three revolutes. The point [Qf] might be seen here as that point in the path of Q where rigid immobility (due to the substituted revolute) suddenly gives way to transient

KEY ASPECTS OF THE GEOMETRY   §4.67

mobility. Notice that the F-surface (which is shown in figure 5.01) does not in general contain the pitch line, but that, in the special case of a planar set being discussed here, where the F-surface is collapsed into the plane of the two shaft axes and the CDL, it does contain the pitch line.

67. In general spatial involute gearing, the corresponding argument might run as follows. If, at a general point Q in the path of some Q, we (a) discover the direction of the sliding velocity there, then (b) substitute a ball-in-tube joint in such a way that the centre of the ball is at Q and the centre line of the tube is aligned with the rubbing velocity, we would render the mechanism rigidly immobile. Please read about zero mobility and modes of assembly at [1] [§1.40-45]. There would be no point in arranging the joint carefully as suggested, because, with the joint dissembled, the hyperboloid traced out by the axis of the free tube is intersected at some non-zero angle twice by the circle traced out by the centre of the free ball. If on the other hand we inserted the said substitute joint when Q was at [Qf], the mechanism would be, not rigidly immobile, but transitorily mobile. See figures 4.07 and 4.08, and refer [1] [§2.08]. Please read about transitory mobility in chapter 2 of [1], and study the worked exercises at [1] [§2.45-46]. As the mechanism wobbles, with the substitute joint at [Qf]), and within its soggy (*sic*) range of available movement [1] [§17.31, §19.55], the rate of change of length of transversal remains at zero. The correct direction of the tube-axis accommodates the sliding, and we are at the instant at that important point in the path of Q where this can happen. At this location, moreover, we could, at the instant, re-substitute, for the ball-in-tube, a mating pair of naked teeth (§6.29).

68. Thus we begin to see the mechanical significance of [Qf] and that of the naked wheels and the continuous, isolated tangency of these with one another at [Ht] which is at [Qf]. Thus also we begin to see the ways in which the synthesized involute profiles of the chapters 5 and 6 are inextricably rooted in these phenomena.

69. By way of further explanation, the following remarks are relevant. For unit mobility in a 3-link loop, there needs to be, at the three joints at the instant, three available screws which belong to the same 2-system [1] [§20.51]. In the present event the relevant 2-system is the unchanging cylindroid of figures 1.07 and 5B.01. This cylindroid (characterized by C and Σ) is wholly determined by the two screws of zero pitch inhabiting the two shaft axes and, among the ∞^3 of screws at the f4-joint (the ball-in-tube) [1] [§7.34], there is an available screw which is a member of this 2-system. It is none other than the screw of pitch p inhabiting the pitch line, which line is of course a generator of the said cylindroid. I clarify as follows. Figure 4.07 is of the equivalent 3-bar linkage with Q at [Qf] in the case of an equiangular set (see figure 5A.01 at §5A.17). Figure 4.08 is of the corresponding 3-bar linkage for a related polyangular set where the ratio j is not the same as k, but different, j being not 6/10 namely 0.6, but its reciprocal namely 10/6 (see figure 6.01 at §6.10). In the figures, the ratio k is the same, *but the vectors* h

141

§4.70   GENERAL SPATIAL INVOLUTE GEARING

measured radially away from the pitch line and the directions of the axes of the tubes are different. In each figure, indeed, *the direction of the axis of the tube is consistent with the direction of the sliding velocity at Q which is tangential to the helix coaxial with the pitch line.* In both figures, also, *the direction of the axis of the tube is perpendicular to the transversal.* These phenomena confirm that the required available screw inhabits the pitch line. Another reason for concentrating thus upon the transversal that cuts the FAXL is that, *if [Qf] for a related range of gears (the range having the same R) is moved to and fro along the said transversal, the sliding velocity at [Qf] traces out a parabolic hyperboloid* (§6.11). For the relatively simple case of the crossed helicals, refer to figure 22.08 in [1], and for the case of the general, offset gears being presented here, refer for example to figure 6.01 at §6.10.

70. *Three interrelated remarks.* Remark #1. It is now clear, after this consideration, *that the point [Qf] on the path of a given Q does not depend upon the path*; the said point is of course a point *of* the path; it is however defined, not by the location and orientation of the path, but entirely by the fixed presence of the F-surface; *the F-surface is fixed in the fixed space; it is*

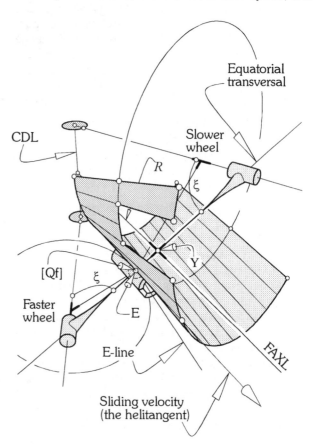

Figure 4.07 (§4.67). This transitorily mobile link mechanism (where the overall mobility f is only 6) is an equivalent mechanism for the gear set in the special event of eqiuiangularity. In this case here, $k = 0.6$, and j, of course, is the same.

KEY ASPECTS OF THE GEOMETRY   §4.71

clear to visualize, easy to handle algebraically, and mechanically significant.
Remark #2. Looking back on the other hand at [Qm], it may be seen that [Qm] *does* depend upon the path; a plot of all [Qm] in the whole of space is not some clear-cut surface among the architecture but some vague volume; see figure 4.05 (§4.42); given the two shaft axes fixed, and a k that might be varied, moreover, it is true to say that any chosen point in the whole of the fixed space could be the [Qm] for an ∞^2 of legitimate paths. Remark #3. I wish to mention here but mainly for the sake of the record that I have briefly studied the angle between the sliding velocity and the radiplane; I give no details here but in general this angle passes through the value zero at some point other than [Qf]; in the special case of equiangularity, however, it does pass through the value zero when Q is at [Qf], which [Qf] is then at [Qx], which [Qx] is then of course at E (§5B.38).

From the path of Q in 1 to the slip tracks by inversion

71. Recognize here and now among the current figures that we might, by inversion, be able to plot the shapes of the two slip tracks. Using our j and k, and having chosen our δ, we have fixed the path of Q. Having divided the

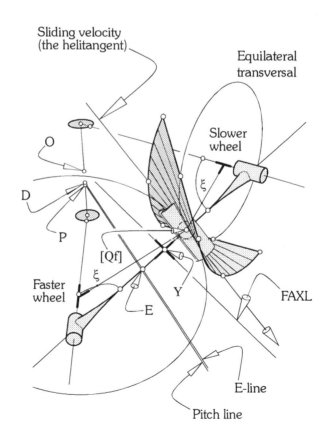

$k = 6/10$
$j = 10/6$

Figure 4.08 (§4.69). This, other things remaining equal, is the set at figure 4.07 with its value for j increased from 0.6 (namely 6/10) to the higher value 10/6. The set has thus become polyangular.

said path into conveniently chosen equal intervals, we might proceed by inversion to obtain the slip tracks as follows. Strategically speaking, what we need to do is to hold a chosen one of the wheels fixed, then to rotate the whole frame and its points Q_1, one at a time, through their respective original angular displacements away from the origin at [Qf]; *but this requires of course an independent construction to find two key angular displacements of the two wheels which correspond to some key location of the moving Q.* If this could be done, all would be well, because we would know from there on that the distance traveled by Q varies directly with the angular displacement of either one of the wheels (§5A.08). Is there however some as yet missing material needed here? Is it that the profiles of the teeth *must* be involute helicoids? Or may they not be? I suspect that an aspect of the yet to be proven, spatial version of Bricard's reverse proposition is lurking here (§2.37, §3.73). Examine the veracity of my remarks at §5B.15 where I speak about the decision-making process being inexorable once the paths are chosen.

The two paths of the two Q

72. Nowhere in chapter 3 is it mentioned that we will need in real gear design to choose two different paths for the two different Q, namely that, in the case of real gear teeth, there will need to be the front-profiles which go front-to-front for drive, and the back-profiles which go back-to-back for coast. In explaining the underlying geometry of mating profiles, chapter 3 deals with one pair of profiles only. Now the trouble is that, with skew gearing, the two sets of profiles will always be obliged to be, as we shall see (§5A.10), different. This difficult matter was mooted at §2.14, where I remarked about our need in the long run to envisage not only one but two different equivalent linkages RSSR, each with less-than-hemispherical S-joints; but the remark there was somewhat oblique in the circumstances, and may not have conveyed its underlying meaning clearly. It was hinting that the two different rods (S–S) would almost certainly not be collinear, would probably be skew, introducing *exoticism* (to be dealt with in chapter 9), but might be intersecting; and this is the matter I wish to bring into focus now.

73. *It needs to be made very clear that the fundamental principles of the involute action of the pairs of profiles on the backs and the fronts of the teeth on a given pair of meshing wheels are in no way vitiated by a circumstance that the two paths for the two Q might not intersect.* Provided both paths are legitimate according to the fundamental law, the paths for the two Q may pass anywhere through the region of the meshing. *They need not intersect.* As we shall see in the chapters 5A, 5B and 6, however, where the fortuitous aspects of equiangular architecture and the plain polyangular option are explained, we sustain severe pressure as designers to choose at least in the first instance that the two chosen paths for the two Q do intersect, and that they do so upon the F-surface. Refer to figure 5B.04. It is clearly much more convenient to choose in this way than not to do so. Accordingly, all the way from here to the beginning of chapter 8 (where we finally take a look at

non-intersection), I assume that the paths have been chosen to intersect. Another special location for Q on each of the paths is thereby (and hereby) the point of intersection [Qx] (§5B.15). Next, when we do put this [Qx] upon the F-surface, we find quite naturally that, *for each pair of real wheels to be designed, only one pair of naked wheels is sufficient.*

Coaxiality of the circular hyperboloids

74. It is worth noting that, while #1 the axodes, and #3 the single pair of cone hugging naked wheels may be seen as being phenomena associated with all gearing of whatever kind, #2 the pairs of base hyperboloids are phenomena peculiar to involute gearing only. In involute gearing we can observe there to be three different kinds of pairs of circular hyperboloids participating in the action, there being indeed four pairs of such hyperboloids altogether. The kinds of pairs have already been listed (§4.37) as follows: kind #1 the one pair of axodes, kind #2 the *two* pairs of base hyperboloids (one pair for drive and the other for coast), and kind #3 the one pair of naked wheels, the conventional decision having been made that the two paths of the two points Q will intersect. Thus we have established that, unless the paths are put to be skew, there will be a [Qx] on both of the paths, the two [Qx] being of course coincident. Having clarified this we can next see that each of the four pairs of

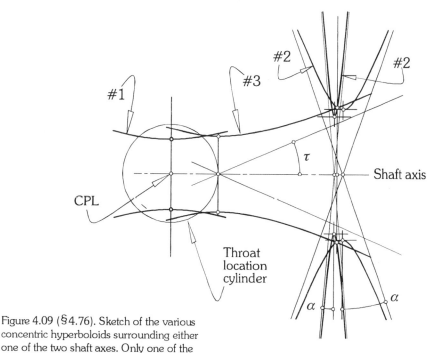

Figure 4.09 (§4.76). Sketch of the various concentric hyperboloids surrounding either one of the two shaft axes. Only one of the two shaft axes is shown. The other is inclined, in this view along the CDL, at angle Σ to it.

circular hyperboloids have for their axes of rotation the same pair of shaft axes; and this means that each of the two shaft axes has, coaxially arranged upon it, four different hyperboloids that can now be examined.

75. Before going on to figure 4.09, I wish to remind the reader that, whereas in chapter 3 we chose for the sake of geometrical clarity large angles α and δ, in this chapter and subsequently — until we get to the worms and spiroids of chapter 7 — we choose for the sake of mechanical practicability much smaller angles α and δ. This switch dramatically changes the appearances of the figures.

76. Refer to figure 4.09 which is not a careful drawing but a sketch. It represents a single shaft axis (drawn across the page in the plane of the paper). The horizontal (sic) edges of the four mentioned circular hyperboloids germane to the single wheel are coaxial upon it. A plane section is taken containing the said shaft axis and the double point [Qx]. In orthogonal view, the horizonal edges are, of course, hyperbolas. See kind #1, the axode with its throat at the CDL. See kind #2, the two base hyperboloids, with their small (but different) angles α; these two intersect in a circle at [Qx], the plane of the circle being normal to the shaft axis. See also kind #3, the single naked wheel upon its doubly departed sheet; note the half apex-angle τ of its asymptotic cone. This single naked wheel also cuts the same circle of intersection. Accordingly I will sometimes call this circle, otherwise called in the chapters 5 and 6 the E-circle or the F-circle, the *key circle* of the wheel. The single naked wheel is the geometric basis for both profiles of the teeth of the single wheel on this single shaft, each tooth having its drive-side profile and its coast-side profile. *It is useful to see (but this is not to be seen in this particular figure) that, because the helitangent at [Qx], which is at [Qf], is by definition perpendicular to the paths of both Q, the generators of both base hyperboloids cut the generators of the naked wheels perpendicularly.* I repeat that, should the two paths for the two Q been chosen not to intersect, there would be no [Qx] in the figure here, and there would have been not only one but two naked wheels coaxial upon the shaft. For skew paths, see chapter 8.

The parabolic hyperboloid

77. This paragraph is an isolated note; it is not part of an ongoing argument; it stands here for the sake of back reference; it is followed however by some accounts of examples relating to it. At [1] [§11.25], I failed to explain that the *parabolic hyperboloid* as I called it there, the so called *saddle surface*, is the same thing as the *hyperbolic paraboloid* of the classical literature. It is thus called in that literature for important algebraic reasons, those reasons being rooted in such considerations as the choice of system for shifting the set of Cartesian axes from central to non-central (which means infinitely far away). There is however no denying that all of the ruled surfaces under consideration here (whether central or non-central) might be called, unequivocally, *hyperboloids*. Refer to figure 4.10. The parabolic hyperboloid

is a non-central quadric surface of one sheet which might be seen as existing at one end of the elliptical throat of an hyperboloid where the major axis of the throat is infinitely long. In which case, of course, the elliptical throat is parabolic. The parabolic hyperboloid may be *rectilinear* (my terminology), or non-rectilinear, according to whether its principal axes (see figure 4.10) are perpendicular with one another or not. Refer for a formal algebraic treatment to Salmon [45] [§87, §115], or to later writers. Figure 4.10 has been given here to provide some visual clues to the actual shape of this surface which needs to be appreciated clearly, as we shall see, during the practical, trial-by-error business of synthesizing involute skew gear teeth. A curious matter to grapple with is that, given its two intersecting principal axes (or given, for that matter, any skew quadrilateral upon its surface), the parabolic hyperboloid has not only a *shape,* but (given its shape) also a *size*. The shapes having been determined, one length-parameter is necessary and sufficient to tell the sizes, for examples, of a sphere or of a cylindroid; we have radius R and thus $2R$ for the diameter of a sphere, and we have half-length B and thus $2B$ for the length of a cylindroid [1] [§15.13]. In a similar way, one length-parameter is necessary and sufficient to give the size of a rectilinear parabolic hyperboloid. *A convenient parameter to take for this purpose is the distance, say 2D, measured along either one of the two principal axes, and measured to straddle equally the central point, between that unique pair of generators — I wish*

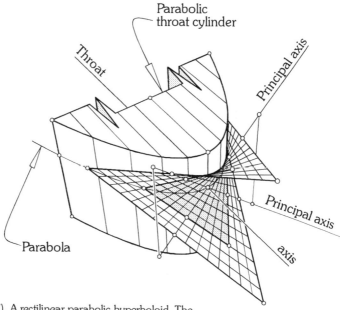

Figure 4.10 (§4.77). A rectilinear parabolic hyperboloid. The principal axes, shown, are here not simply symmetrically arranged at some acute angle to one another (as is the case in general), but are symmetrically arranged and mutualy perpendicular.

to refer to them as the two key generators — that are *perpendicular with one another*. We note that tan $(\pm\pi/4) = \pm 1$, and we note that, when D is zero, the parabolic hyperboloid takes of the special appearance shown for example at figure 7.13(b) in [1]. The reader will find related reading at §6.17 and §6.34. I have suggested here that for the rectilinear parabolic hyperboloid we might speak about its half-measure say D, and its *measure 2D*. If the parabolic hyperboloid were not rectilinear, its shape would be determined by the angle between its two principal generators, and its size might be measured (given the symmetry) by the same $2D$. Why are there *three* coefficients in the Cartesian equation given by Salmon at the foot of his page 111?.

Four different rectilinear parabolic hyperboloids

78. In the four paragraphs that follow, four different rectilinear parabolic hyperboloids (each of which is encountered at least once in this book) are qualitatively distinguished. They are distinguished from the three kinds of circular hyperboloids, summarized at §4.73 and displayed at figure 4.09, by use of the prefix RPH followed by an asterisk. They are distinguished from one another by the consecutive numbers 1 to 4. They are grouped here in this way simply for the sake of convenience; the nomenclature is to be used mainly for subsequence reference back.

79. *Surface RPH*1*. At §1.36 we met a first parabolic hyperboloid. Marked (d) at figure 1.07, its principal axes are the CDL and the pitch line. Its central point is at P. At each extremity of the centre distance C there are two generators of the cylindroid intersecting. Notice that the surface does not contain the shaft axes (which are the pair of conjugate generators of zero pitch of the cylindroid) but the other pair of conjugate generators which are intersecting there. Note that the surface, as shown, is truncated by a circular cylinder coaxial with the z-axis. The figure moreover shows only one family of its generators; the surface accordingly takes on the appearance of being a helix; but it is not a helix; there is of course the other family of generators whose members are perpendicular to pitch line.

80. *Surface RPH*2*. At §4.13 (in the absence of a diagram) we met another parabolic hyperboloid. We found there that, at all points Γ at the pitch line, the transversal at Γ cuts the pitch line perpendicularly, and that the whole population of these transversals make up the surface. This one has for its principal axes the CDL and the pitch line, its central point is at P, and it does contain the two shaft axes.

81. *Surface RPH*3*. A third RPH that might be mentioned is the rectilinear array of velocity vectors that we met at §4.26 and will meet again in chapter 6; its principle axes are (a) the velocity vector at [Qf], and (b) the equilateral transversal JEHYK; refer to figure 6.01. Another in this category of course is the RPH mentioned in the middle of §4.64; it was there the linear array of velocity vectors spread out along the path of Q.

KEY ASPECTS OF THE GEOMETRY   §4.82

82. *Surface RPH*4*. At §4.26, we mooted another. Like the one of §4.13, it contains the two shaft axes, but it has for its principal axes the CDL and the FAXL (§1.41). Its central point, accordingly, is not at P but at O. It appears as a major feature of figure 5B.01, and in chapter 6 this most important RPH will be further considered. *This one is the F-surface.*

83. All four of these parabolic hyperboloids are different in size (see the half-measure D at §4.76), and they are certainly different in location and mechanical significance, but they have one characteristic in common. They are all *rectilinear* parabolic hyperboloids, rectilinear in the sense that their principal axes, which intersect at the central (saddle) point of the surface, intersect there perpendicularly. *The four should not be confused*. Recapitulating, I say this: see figure 1.07 for RPH*1; see figure 22.13 in [1] for an impression of RPH*2, see figure 6.01 for RPH*3; and see figure 5B.01 for RPH*4, this latter being the F-surface. One could make a special, separate figure simultaneously illustrating all four of these; but I leave this project as an instructive exercise for the reader.

Appendix A at chapter 4

84. As promised at §3.66, I return now to Olivier [40]. He expressed very early (1839) an important idea. It was to swivel (in the imagination) one of the wheels of a planar involute spur gear set, about one of the straight-line paths of the points of contact Q, through some acute angle with respect to the other wheel of the set. Having explained this idea (and I explain it again in the figures that follow), he went on with extensive ruminations about its possible usefulness in the solution of the unsolved problem of finding a workable kinematics for offset skew involute gearing. To begin, I wish to say that this appendix, while largely historical, contains also my contribution to the literature constructively critical of Olivier's idea. I will be referring more carefully soon to the works of Olivier himself [40] (1839, 1842), Willis [12] (1841, 1870), Mac Cord [18] (1883), and Disteli [77] (1898-1911). After making a brief survey of these, I will write my own contribution to this old discussion. I do this for two simple reasons: spatial involute action comprises almost exclusively the scope of my present argument outlined in this book; and Olivier's work will be seen by many to have been at the root of it.

Refer to figure 4.11(a). This is my picture of Olivier's spur gear set. While he chose for his example 24 teeth, 16 teeth, and 20° for the two equal angles of obliquity, I have chosen 24, 18, and 15° for these; but these specific numbers and physical magnitudes are irrelevant to the main gist of the argument. The main thing is that the teeth are involute teeth (after Euler at that time). Please recognize the *central reference plane* in figure 4.11(a) and the two symmetrically placed paths of the two Q (one for drive and the other for coast, or *vice versa*) set within it, and see that, having chosen one of these paths (any one), I have called that chosen one *Olivier's line*. Looking ahead, see also how I give that same name to that same line, but under different

§4.84   GENERAL SPATIAL INVOLUTE GEARING

circumstances, at the figures 10.05(a) and 10.09(a) both explained within the supporting text at chapter 10. That the non-chosen path appears parallel with the edge of the paper is not significant; it is an accident of the drawing. Nowhere else in this book does a picture appear of a simple planar spur gear set which shows as clearly as this one does that, given the necessary accuracy and/or flexibility of the links, one can imagine *straight-line contact* between the teeth, and that one must, for the sake of geometric analysis, settle upon a

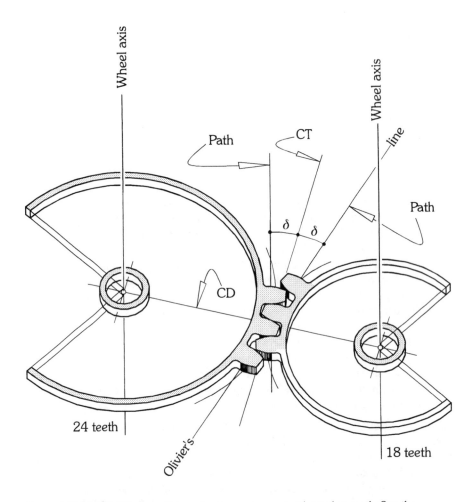

Figure 4.11(a) (§4.84). An ordinary planar spur gear set with involute teeth. See the common tangent (CT) to its two key circles perpendicular to the centre distance (CD) and the two alternative paths for the points of contact Q ranged at equal angles delta on either side of it. One of these (any one) has been chosen by me and called by the name, Olivier's line. This figure is designed to recreate the starting point for Olivier's purposeful, imaginary maneuvres, which maneuvres are further depicted in the following figures.

150

KEY ASPECTS OF THE GEOMETRY §4.84

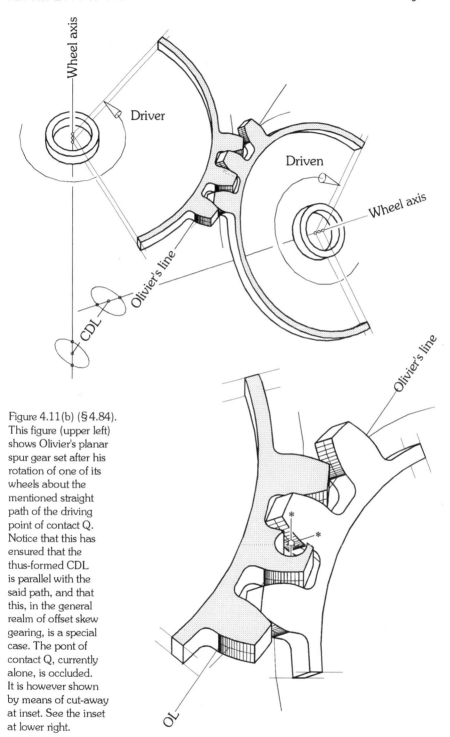

Figure 4.11(b) (§4.84). This figure (upper left) shows Olivier's planar spur gear set after his rotation of one of its wheels about the mentioned straight path of the driving point of contact Q. Notice that this has ensured that the thus-formed CDL is parallel with the said path, and that this, in the general realm of offset skew gearing, is a special case. The pont of contact Q, currently alone, is occluded. It is however shown by means of cut-away at inset. See the inset at lower right.

reference plane which is normal to the said straight line of contact and which contains both of the straight-line paths of the two Q. Notice in figure 4.11(a) that the point of contact Q on Olivier's line is positioned, just now, exactly at the pitch point P on the CDL where the coplanar key circles meet.

What Olivier does, effectively, is this: keeping the left hand wheel (24 teeth) fixed, he rotates in the imagination the right hand wheel (mine has 18 teeth) about his nominated line through an arbitrarily chosen, acute, finite angle (40°) to relocate the right hand wheel anew. The maneuvre is shown completed in both parts of my figure 4.11(b). For the worked example in his book, [40] [page 118, chapter 2, and plate IV], Olivier did indeed use 40° for this *folding* of the set into its new configuration; and I have used the same angle. But notice now what has happened. The shafts have become, as might be expected, skew with one another, and the unique common perpendicular between the two, the CDL, of length $C \sin \delta$ incidentally, has sprung into being. *More importantly, when seen in the present context, the CDL turns out to be parallel with Olivier's line.* The angle of the folding, moreover, translates into the angle between the two shaft-axes, namely my angle Σ.

Next came Olivier's seminal observation. It was as follows. Refer again to figure 4.11(b). He saw that, if the *fronts* of the teeth were to be seen as those flanks in contact at Q, as shown by my nomination of the *driver* and the *driven* in the upper part of figure 4.11(b) and by the crossed pair of rulings (*) engaged in single-point contact in the lower part, *and if the backs of the teeth were ignored (because there is some complicated interference evident there which is best looked away from)*, then the gear set would operate quite satisfactorily in its folded configuration. He next went on to explain (a) that, if the teeth on the driver were permitted to generate directly the teeth on the driven, by means of (1) his one-to-one *gears-for-strength* approach (§3.66), continuous straight-line contact between the teeth would be achieved, and (b) that that would be a new idea for further contemplation. In other words, he was not sufficiently satisfied with mere single-point contact at this juncture, which would give a satisfying overall mobility of unity for the mechanism as the teeth stood, but determinedly held out for line contact, seeking strength at the expense of overconstraint. Having said that by way of private reservation here, I can agree with Olivier's line of thinking thus far.

He went on next however to assert that the flanks of the generated teeth will be *involute helicoids*, that their straight-line rulings will be tangential with a helix drawn coaxial with their shaft axis, that the radius of that helix will be that of the base circle for the original planar involutes of that wheel, and that the helix angle of the helix will be precisely Σ [40] [page 26]. I stand ready to be corrected here, but I cannot agree with this. It seems to me, after carefully constructing the drawing at figure 4.11(c), that the newly generated surface of the driven tooth (whose shape is shown in this figure by means of its rulings) cannot be an involute helicoid or even be mistaken for one. The successive rulings are traces upon the generated surface imprinted by the successive

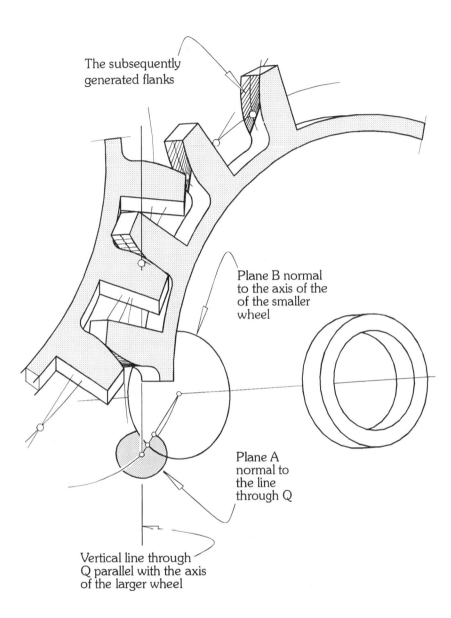

Figure 4.11(c) (§ 4.84). This is my drawing of Olivier's scenerio after (a) the folding about Olivier's line has taken place, (b) the consequently interfering backs of the teeth have been removed, and (c) the original flanks of the undisturbed wheel (on the LHS)) have generated by direct envelopment their counterparts for conjugality (with straight-line contact) on the teeth of the folded wheel. The generated flanks are not involute helicoids. They are of some other shape not known at this address.

rulings of the generating surface. I have taken small steps. For each 1.5° of angular displacement of the bigger wheel, I have matched that with 2.0° of the smaller wheel, the inverse ratio being 24/18. The continuous contacting of the crossed planar involutes has been observed by me to remain proper as the relative motion proceeded, and it was clear me that, at every step of the process, the straight-line ruling of the generated surface must be contained within the plane of adjacency pertaining at that step. Each ruling of the surface is tangential to the original surface at the shown planar involute there, and the small shortest distances drawn between the successively generated generators show (as a matter of interest) that the original planar involute constitutes an orthogonal trajectory across the newly generated, ruled surface. There is no cuspidal edge to be found, the surface does not present itself as being everywhere satisfactorily convex-outwards; the surface is clearly non-developable; Olivier's argument that "by thinking about this a little bit" [40] [page 26 line 2] the contention would be proven is, it seems to me, thus shown by these facts to be false.

For the sake of the record, just here, I give my account of what I have found to be the more likely shape of Olivier's newly generated flanks. The sentence just written is couched conditionally because the following comments come, not from a rigourous algebraic analysis, but (with the usual uncertainty about the generality) from the computer-driven graphics. Refer to figure 4.11(d). This is a parallel orthographic view of the smaller, namely the newly generated wheel, looking directly along its axis. The planar curve labeled *planar involute curve* is in the reference plane, which is in the plane of the paper. Although *not* in the plane of the paper, each one of the shown rulings of the newly generated surface actually cuts the said planar involute curve at the appropriate point but appears (in this view) to be tangential with it; see this more clearly at figure 4.11(c). The surface is twisted, non-developable, and lies tangential (along the said planar involute) with the cylindrical involute surface of the original spur-tooth flank. The shown intercept (A–B) in the figure is the shortest distance between the so-called vertical line through Q in figure 4.11(c) and the axis of the wheel. This (A–B) is parallel with the CDL, is of length $[C \sin \delta] [k/(k+1)]$, and is situated, at a distance $[r \cos \delta \cot \Sigma]$ along the wheel-axis, below the reference plane. The quadrilateral ABTQ is a rectangle. In figure 4.11(c), this is vertically arranged, but here in figure 4.11(d) it is inclined at angle Σ to the vertical. The intercept (A–B) is contained within a plane at B normal to the axis of the wheel and thus below, but parallel here with the plane of the paper; and, within this thus-defined *lower parallel plane*, the labeled *planar Archimedean spiral* lies. This spiral is, remarkably, yet not wholly without surprise, the locus of the points of intersection of the rulings with said lower parallel plane.

There is much more that could be said about Olivier, about his subsequent suggestions pursuant upon (a) there being — as there were, in his view — involute helicoids involved in this above-mentioned kind of gear set, (b) there

KEY ASPECTS OF THE GEOMETRY §4.84

being the obviously needed, but absent provision for coast as well as for drive, and (c) there being entirely other forms of gear teeth that might be possible, exhibiting both single-point and/or curved-line contact. I should also mention that a well made, wooden, working model of the gear set at Plate IV in his book, complete with front flanks but without the troublesome backs, and with (or without?) its erroneous aspects, remains safe behind glass at *Musée des arts et métiers*, in Paris.

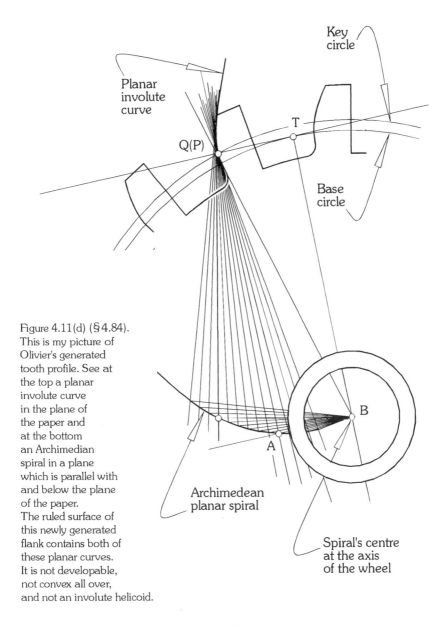

Figure 4.11(d) (§4.84). This is my picture of Olivier's generated tooth profile. See at the top a planar involute curve in the plane of the paper and at the bottom an Archimedian spiral in a plane which is parallel with and below the plane of the paper. The ruled surface of this newly generated flank contains both of these planar curves. It is not developable, not convex all over, and not an involute helicoid.

But I turn now to Willis [12]. Willis, in his famous book, [2nd ed., 1870, §201], had occasion to mention Olivier. "Suppose now", he wrote, succinctly as usual, "the plane of one wheel be inclined to the other by turning on the line [the line indicated], in the manner of a hinge, so that this line shall be the intersection of the two planes, but that the position of each wheel in its own plane with respect to this line shall not be altered." Willis cited, here, only Olivier's earlier paper of 1839 [102]. He spoke not about the false generation of the new profile discussed at length in Olivier's book [40], but he did mention that the common perpendicular which springs into being (the CDL) was parallel with the path. He contended himself by writing simply as follows. "Involute wheels, therefore, may be employed to communicate a constant velocity ratio between axes that are inclined at any angle to each other, but which do not meet." This however was followed straight away by the following. "But the demonstration supposes the wheels to be very thin, since they coincide with the planes that meet in the line [the line indicated], and the invariable points of contact are situated in this line."

Next I wish to write about Mac Cord [18]. The earlier cited [§368-§400] in Mac Cord constitutes Part 2 of his Chapter XII which begins (2nd ed.) on page 232. The said Part 2 is sub-headed, *The teeth of skew wheels*; and, taken as a whole, it deserves careful reading. Mac Cord's figure 236, however, along with his text nearby at §382, is particularly relevant to the general observations I am making here. It draws graphic attention to the above-mentioned folding maneuvre also attributed by him to Olivier. My figure 4.11(b) is a reconstruction of Mac Cord's figure 236 which is, in turn, a figure after Olivier. Mac Cord explains of behalf of Olivier that his folding maneuvre produces (on the drive sides of the teeth) circumferentially arranged sets of unsupported planar involutes originating in the two base circles that might be seen to drive one another satisfactorily with the shaft axes skew, while still maintaining a constant k. The crossed, unsupported curves of Mac Cord's figure explaining Olivier's idea are of course the original planar involutes of the teeth but, as I shall later show (appendix D at 6), these are, actually, in the special circumstances of that particular figure which comes indeed from Olivier, *slip tracks* (§3.16). Slip tracks, in general, are not planar; they are twisted curves. But in the said figure the tangent TT, namely *Olivier's line* as I have called it, which contains the path of the point of contact Q, is specially arranged. *Firstly it is put to be parallel with the CDL, and secondly it is arranged to cut the pitch line perpendicularly.* The path is, accordingly, legitimate by the first law (§2.06); and it is, therefore, not surprising that it functions properly. The circumstances are however very special. I wish for obvious reasons to mention this; but any clear explanation will firstly require an understanding of what there is to follow, namely a first explanation of how (a) the required architecture of a proposed gear set, and (b) an intelligent application of the fundamental law [14], can result in the satisfactory synthesis of genuine, spatial, single-point-of-contact, involute motion, gear-transmission. Go to the chapters 5 and 6.

Separately from this, Mac Cord exposed in this said Part 2 of his chapter XII a great deal of his exploratory thinking about straight-line-contact in the field of 'hyperboloidal' skew gearing, taking his cue, in many respects, directly from the 'involute helicoidal' teeth of Olivier. He naturally became critical of the geometric consequences of Olivier's futuristic proposals and finished, in the long run, by dismissing his own tentative hypotheses as being less than hopeful. I have made various other remarks in review of Mac Cord at §2.37, §5B.12 and §5B.39.

Finally I wish to speak about Disteli [77]. Disteli wrote a connected series of four lengthy papers extending over the period 1898-1911. The series appeared in *Zeitschrift für Mathematik und Physik* — see [77] — and the titles of the papers (put somewhat freely into contemporary English) might, acceptably, run as follows: (1) *About centrodes and axodes* (in two parts 1998-01), the second part of which is titled, *C. The axodes for crossed shaft axes*; (2) *About velocity screw axes and the meshing of hyperboloidal wheels* (1904); (3) *Several theorems in kinematic geometry which relate to the meshing in cylindrical and conical (bevel) gears* (1908); and (4) *About the meshing of hyperboloidal wheels with straight-line contact* (1911). Clearly part II of (1) and the last of these namely (4) are the most interesting in the present context, and I write down for consideration here my impression of their contents.

In part II of paper (1) Disteli goes into the initial business of explaining, in respect of two relatively moving rigid bodies, what the instantaneous screw axis is, and how, over time, it will sweep out its two axodes, one in each of the bodies, much as I have done in [1] [§5.58, §22.61] where I spoke about open ended axodes, and in this book at §1.26 *et seq.*, §5B.02 *et seq.* and elsewhere where I speak about closed, circular, hyperboloidal axodes. He draws attention to the application of these general ideas in the special realm of gearing between skew axes, where spherical (bevel) gearing and planar (spur) gearing are special cases. See my figure 1.06 for the relative rotational motion of mating gear wheels about skew shaft axes fixed in space, and be aware that there (as Disteli explains) there is not only pure rolling about the ISA at the ISA, but also pure sliding along the ISA at the ISA. See my remarks about the imaginary spline-line teeth of axodes at §1.33. Disteli goes on to infer however that, although the relative motion between the ruled profiles of gear teeth in straight-line contact allows crosswise sliding also to occur at the straight line of contact, there is nevertheless a kind of similarity between this relative motion between teeth and that at the axodes, and that there may, accordingly, be some possibility that involute skew gear sets exhibiting straight-line contact might lend themselves to synthesis.

In paper (4) Disteli goes on to expand even further the idea that the rolling-with-crosswise-sliding action at the as yet undiscovered teeth of the looked for, line-contact kind of gearing can somehow be related to the non-crosswise-sliding, rolling motion of the axodes which are, for any one pair of gears, unique. Here I wish to mention Reuleaux [95] (*circa* 1865), and in particular his then expressed, and for long time widely taught, concepts involving *primary and secondary centrodes (and axodes)*. I have written briefly about these at §9B.12 *et seq*. Disteli freely uses them in his paper (4). He expresses his belief that, if the generation by envelopment of the ruled flanks of teeth by other ruled surfaces (and in particular planes for example), there might be the possibility that these kinds of gears could be successfully synthesized geometrically and thus be capable of being built. He speaks about the spherical indicatrix and other such devices for handling the classical ruled surfaces, and he makes mention of the cylindroid and the linear line complex that both play their parts as we know in the chorus; but like Olivier and those that came between Olivier and himself Disteli speaks almost exclusively and with no small enthusiasm about *line-contact*; he speaks *not* about the possible usefulness of *crossing* the rulings upon the mating flanks of ruled teeth, and thus about the quite straightforward possibility that *point-contact* might indeed be an answer.

We should recall here, it seems to me, Olivier's alternative, *engranage de précision*, and not be afraid that, given low curvatures at the mating surfaces and satisfactorily consistent crosswise slidings at the points of contact, point-contact will lead to disaster. Given elasticity, lubrication, and the real world of *imprecision*, it will not. Refer to §5B.37 for my angles χ. One might say with a measure of resignation now that this long controversy is water under the bridge, that, although the various workers worked sincerely to solve the many problems, one doesn't need to know exactly what they did. But it does matter to me to have explained here, not that Olivier investigated spatial involute gearing and showed that it was hopeless (as many continue to argue), but that his gearing was not involute (as Disteli agrees) and that it was hopeless.

Appendix B at chapter 4

85. Originating at §4.32, this appendix relates to Dr Sticher's general algebraic analysis of intersecting circular hyperboloids sharing a common generator. I have already mentioned at §4.32 and elsewhere those parts of his conclusions that are directly relevant to the main matters at issue here, namely those within the direct mechanical context of gearing. I reiterate that the mentioned algebraic work was undertaken at my suggestion. The proffered material however goes well beyond the scope of this book, and should now find its proper place elsewhere. Its elegance, moreover, would be diminished by any short account I might try to make of it here. Dr Sticher currently intends to write, in due course, a paper sufficiently rigorous for journal publication; and he will, by so doing, be able to give his argument the thorough exposition it deserves. I write here with his approval; and I, like him, retain for safe-keeping his first rough draft of the forthcoming paper.

CHAPTER FIVE (A)
THE SIMPLICITY
OF EQUIANGULARITY

Transition comment

01. Early in chapter 4, I said that that chapter might be omitted upon a first reading. I stand by that. From here on, however, reference back to specific items in chapter 4 will occur from time to time. Having noted the characterizing parameters C, Σ and k for a gear set of whatever kind in chapter 1, having established the fundamental law of gearing (a law for all gearing) in chapter 2, having studied the kinematics of spatial involute action in chapter 3, and having investigated for scientific reasons some key aspects of the geometry of pairs of hyperboloids sharing a common generator in chapter 4, we begin to study now, in this and the following chapters, the synthesis of offset skew involute teeth.

02. This chapter 5A is in many ways a watershed. It stands between the analyses which have preceded it and the syntheses which follow. There is however no clear dividing line. While most of the working material for this chapter comes from earlier chapters (as one might hope), much of it comes from those that follow; and what arises from this is that the general thread of argument may seem here to be illogical. I try to extract for isolated examination the central idea of equiangularity nevertheless, and I hope that the reader will bear with the general uncertainty for a little longer yet. Please read this chapter knowing that its contents are important, but knowing also that it is being written now for backward reference later.

Prognosis

03. As in all kinds of gearing, we will soon find, here among involute gearing, the need for two travel paths, there being the twin needs for drive and for coast (§2.14, §4.72). Discussing these, we will encounter questions about the general directions in which we need to go, and these will soon reduce to the primary question of *how to begin*. We will discover however in the present chapter a particular, simple, geometrical layout of the important lines in a gear set that may (if we wish) be established as a basis for design. We will next accept this as a convenient benchmark, a firm plateau as it were,

from which to view the more mountainous problems associated with other, less simple layouts. I will call the plateau, *equiangular architecture*, and, in connection with it, *we will locate an important straight-line locus called the E-line*. This locus, parallel with the pitch line, intersecting with and perpendicular to the CDL, will be seen to contain an infinity of points E, exclusively associated with *equiangularity*. We will learn how to locate two convenient, legitimate paths for the two moving points of contact Q, and we will, in the event of equiangularity, choose these to intersect at one of the said points E. Their point of intersection with one another, called [Qx], will thereby fall, in the event of equiangularity, directly upon the said E-line. A worked example (a synthesis) of an equiangular, offset, involute skew gear set with $\Sigma = 50°$ and $k = 0.6$ will be pursued first in chapter 5B, this to be followed by a second, more difficult worked example involving a similar equiangular set with $\Sigma = 90°$ (square) and $k = 9/41$. More difficult other architectures, called *polyangular* [44], will be looked at by means of other worked examples later.

Synopsis of the chapters 5

04. In chapter 5A we deal with (a) some new aspects of the pitch line and its helices, (b) the special points (or *stations*) [Qt], [Qm], [Qp], [Qc] and [Qx] along the two paths of any Q, (c) the *measure of offset* in offset skew involute gears, and (d) the *circumstances of, and the criteria for, equiangularity*. In chapter 5B, in connection with the worked examples already mentioned, we deal, within equiangularity, with (a) the cone-hugging *naked wheels* of the chosen equiangular set, (b) the angles of obliquity δ; (c) the said paths of Q which, intersecting, must be in the same polar plane, (d) the changing orientation of the triad of planes at Q, (e) the curvatures of the flanks of the teeth at contact; (f) tooth interferences and the related contact ratio; (g) changing directions of the sliding velocity, and (h) the somewhat bewildering scope for overall optimization.

The common objectives of the chapters 5B and 6

05. Also in chapter 5B, and as previously explained, I will call the straight line that may be chosen for the path of the point of contact Q, which path will be seen as traversing the fixed space 1 of any involute gear box, *the travel path of Q in 1* or, more simply, *the travel path* (§3.29). Correspondingly I will call the curved paths of the point of contact Q drawn upon the mating surfaces of the teeth of the wheels, which wheels are the links 2 and 3 of the 3-link loop, *the slip tracks of Q on 2 and 3*, or, more simply, *the slip tracks* (§3.65, §4.70). The travel path and the two slip tracks continuously intersect, of course, in the moving Q. This important central theme (independently of architecture) pervades both of the chapters 5B and 6.

06. In chapter 6 we begin to deal with polyangularity but we confirm in both of the chapters 5B and 6 that the travel path in all skew involute gearing is always a straight line traveled at constant speed, and that the sliding veloc-

ity varies across the flanks of the teeth but never, in general, reverses direction. These facts augur well for the absence of vibration and a smooth continuation of film lubrication independently of architecture. All offset skew involute gearing will be found to be, within practical limits, insensitive to errors at assembly given rigid wheels correctly cut, and thus, also, to relative movements of the shaft axes in the event of flexible or changeable mountings. We will also find that involute teeth of whatever architecture may need to be provided with unequal angles of obliquity at the backs and the fronts of teeth according to circumstances. We will find nevertheless that (not barring undercut) the teeth may always be cut by NC milling machine or, indeed, by CNC hobbing. Throughout the worked examples of the chapters 5B and 6 — of which there are four, there being two in each chapter — the known mechanics of and the various methods for designing and cutting ordinary planar involute teeth are taken for granted.

Two important theorems that need to be studied

07. There are two important theorems that need to be studied. The first, to be given at §5A.08 below, relates in general to all kinds of gears, and in particular here to our choice of legitimate travel path. The second, formulated and listed with its various corollaries at §5A.19-26, relates to involute gears alone and to equiangular architecture exclusively.

Aspects of the pitch helices not well studied as yet

08. I refer to the system of helices of the same pitch as that of the pitch line which coaxially surrounds that line in any gear set. See [1] [§9.06, §11.30] and [5] [§3.5]. There is an ∞^2 of these helices. Through each point in space only one of the helices passes, and, collectively, they fill the whole of space. In gears, the direction of the rubbing velocity at any point of contact Q between teeth will be tangential to the unique helix passing through that Q, The rubbing velocity occurs, in other words, in the direction of the helitangent (§1.38, §3.51) [1] [§21.15]; see figure 1.01, 1.08 and the figures 3.13. Unless this were so, the surfaces of the mating teeth at Q would be separating from one another or digging in (§2.22). The tangents to the helices (the helitangents) at all of the points form a *quadratic complex of rubbing velocity vectors* [1] [§11.63]; the normals to the helices at all of the points form a *linear complex of right lines* [1] [§9.17, §11.34]; and it is a known consequence of this geometry that the following theorem may be written down.

> *Theorem concerning the pitch helices.* Any line in a polar plane of the linear complex of lines among the pitch helices of a gear set at a point say Q in the fixed space (namely in the box of the gear set) will not only cut its own helix at Q perpendicularly but cut all of the helices it meets perpendicularly. Any other line through Q, that is any line not within the polar plane at Q, will not have this special property.

This observation thus stated is important. It means that, if, in gears, we wish to choose our path for Q to be not only straight but also continuously collinear with the contact normal, we may do so.

09. We may do this, however, only if we pay attention to the above theorem, that is, only if we ensure that not only the contact normal at Q but also the whole path of Q is chosen according to the fundamental law (§2.02). *Only out of such a step will come the circumstance of pure involute action.* If we do thus aim for involute action, we will find (a) that the path has its own unique point [Qp] upon it which occurs at the shortest distance q from the pitch line, (b) that, at this point, the path is binormal to the helix which passes through there, and (c) that the sliding velocity vector, v_{2Q3} or v_{3Q2} [39], which occurs always along the helitangent, and which is always perpendicular to the path, is least when Q is at [Qp]. Refer to figure 2.01. Put this way, these facts boil down to being, simply, a restatement of the fundamental law.

Problems concerning our choice for the travel paths

10. Unless for some special reason in design the mating teeth of the wheels are to have only one working flank each, they will need not only fronts (for drive), but also backs (for coast). To avoid collisions between teeth, the travel paths for drive and coast will need moreover to be different. This is a fact for all gears whether spatial or planar. It is however true in planar involute gearing that the polar plane is in the plane of the paper (the reference plane), and, with equal obliquity angles δ, the two travel paths are thereby also in the reference plane and symmetrically arranged about the CDL. They reside as we know along the two common tangents to the two base circles, which latter are also within the reference plane. Accordingly, a special symmetry can obtain in the plane. But in spatial gearing, except for the special cases of the crossed helicals where the path of Q actually cuts the CDL, a similar symmetry cannot obtain, and this has the general result that the backs of the teeth can never be the same as the fronts.

11. Selection of the travel paths will be a necessary early step in most successful syntheses, but we are not yet ready to take this step. Both of the travel paths for Q (one for drive and the other for coast) must be legitimate according to the fundamental law, but, within that restriction, the said paths might be chosen almost anywhere, and they need not intersect. Given however that a point on one of the paths path might be coincident with another point on the other path, the double point might be called [Qx], the x denoting intersection. Having this in mind, it might be said in general that, legitimate paths having been chosen to be just anywhere, points [Qx] on the paths might not exist at all. For tidiness in design, however, *the paths may be chosen, not to be just anywhere and thus skew, but to intersect*, in which case a double point [Qx] will exist (§9.02). If the paths *are* chosen to intersect, which, unless denied, will be held to apply henceforth, the paths will together define and occupy the same polar plane and, at the centre of this po-

lar plane, the coalesced, double [Qx] will reside. This double [Qx] will thus become another important point fixed in both of the paths. It is incidentally not possible in general for two chosen paths to have their respective points [Qp] coincident with one another at [Qx], for at each point [Qp] the unique binormal there is already occupied by the one path, and the other cannot be collinear. If however both paths are chosen to cut the CDL (as must occur in planar, and may occur in spatial gears), it is possible that the two points [Qp] will be coincident at [Qx].

12. Were we to fix a pair of paths (either by guesswork or otherwise), each would be found to have its own intercept of length Z_l located upon it (§2.12, §3.27). We could find these Z_l by erecting the four shortest distances (the four radii a) between the two paths and the two shaft axes. Also, as we know (§3.57), we get equal principal curvatures of the surfaces in contact when the point of contact Q is at the mid point of its Z_l. These mid-points of the intercepts Z_l are thus also important fixed points upon the paths. At §3.57 I have called them [Qc]. I wish to make the remark that, in the planar case, these Z_l are simply the two common tangents to the two base circles which are coplanar, and, given the same base circles for both drive and coast (which is ordinary practice), we get the same Z_l for both paths, the two [Qc] falling at equal distances from the pitch point P. A similar symmetry can occur in the special case of the crossed helicals (§5B.60). *Generally in space, however, each path will have its own Z_l (Z_l being the segment of path between the throat circles of the relevant base hyperboloids), and these Z_l, along with the locations of their mid-points [Qc], will differ.*

13. So, summarising, we might say that, in ordinary *non-exotic* offset spatial cases (§5A.19), where the paths are chosen to intersect, there are four identifiable fixed points upon each path: I have called them here as follows: [Qx] where the paths have been chosen to intersect; [Qm] the mid points according to §3.56; [Qp] where the relevant path comes closest to the pitch line; and [Qc] where the said radii of curvatures of the surfaces at the point of contact (which sum to Z_l) are equal. See chapter 4 for more intimate details of some of these and other such points. In the highly special case of planar spur gears with k = unity and where the paths are symmetrical about the common tangent to the pitch circles at P, [Qm], [Qp] and [Qc] on both paths coalesce at [Qx], and there are no special points upon the paths except [Qx] itself, which is at the pitch point.

14. These ruminations about the travel paths and the special points thereon are of not much use in design, however, unless we better understand the central problem of *where to begin*. There are various facts of the matter that preclude the possibility of our proclaiming in a vague way that we might start with a starting point [Qx] *just anywhere*. Such an uninformed approach will result, most often, in our becoming lost. Refer for example to Merritt [58] [§10.11], and to my analysis of his suggestions at §8.06 *et seq*.

Radial offset R and the FAXL

15. Throughout the analyses of chapter 3 we were trying, not to *design* gears, but to establish scientifically the basic principles. The distance from the CDL to the actual scene of the meshing was, accordingly, of secondary importance there. See for example the CDL belatedly discovered at figure 3.08. Naturally, now, in design, we might reasonably wish to describe this distance accurately somehow, to define it, and to set it down as one of the fixed variables of a proposed, new, skew gear set. And this, of course, turns the matter on its head. It presents us with the kind of problem characteristic of the well known and previously mentioned analysis-synthesis dichotomy in the kinematics of mechanism (§1.13). *The sweeping of water uphill is not the mere reversal of sweeping it down.*

16. I wish to remark therefore — and I do this without either justification or proof just yet — that a unique line through any proposed, double point [Qx] will always be drawable to cut both of the two shaft axes somewhere. Such a line may be called the *transversal* through [Qx] [1] [§15.36]. If the proposed double [Qx] is chosen cunningly, however, the relevant transversal will cut the two shaft axes at equal angles ξ. See for example the line JK and the equal angles ξ at figure 5A.01. This equality of these angles ξ is in no way related to *equiangularity* incidentally; the whole idea of equiangularity is something else, yet to be introduced. Returning to the said proposition, it means, in reverse, that a proposed, double [Qx] might always be chosen to lie upon the doubly ruled surface illustrated at figure 5B.01. The surface is called by me the F-surface; it has been discussed already in chapter 4; go to the discussions about the point [Qf] at §4.27 and §4.63-67; go also to check the four different rectilinear parabolic hyperboloids at §4.78; the F-surface at figure 5B.01 is not the same parabolic hyperboloid as the one illustrated at figure 1.07; the two are different. The generators of the F-surface that cut the two shaft axes clearly cut those axes in such ways, not only that the angles ξ at the intersections are equal, but also that the distances between the CDL and the intersections namely (A–J) and (B–K) are equal. It is for this reason that I refer to the intercepts cut by the axes from these transversals, shown as (J–K) in the various figures, as *equilateral transversals.*

17. Refer now collectively to the figures from 5A.01 to 5A.03 inclusive where the shapes (but not the orientations) of the figures are the same. Perceive in the figures 5A.01 and 5A.02 especially, and in the other figures of this and the following chapters that between the equilateral transversal (J–K) and the CDL there is a common perpendicular, and that this lies precisely along the algebraic axis O-y of figure 1.01. I wish to call the intercept (O–Y) upon this axis, which is the length of the said common perpendicular, *the radial offset.* The term applies to any gear set whose [Qx] is anywhere upon the prevailing (J–K). The length of this radial offset is designated R.

THE SIMPLICITY OF EQUILANGULARITY §5A.18

18. Before leaving this matter of R (we return to it later), I say again that the axis O-y has already been called by the acronym FAXL (§1.43). The acronym derives from the conglomerated word 'F-axis-line'. This line called the FAXL, which can here be seen to have, not only its being, but also its unambiguous meaning, well delineated within the practical context of all gear sets of whatever kind, will, henceforth in this book, be used freely.

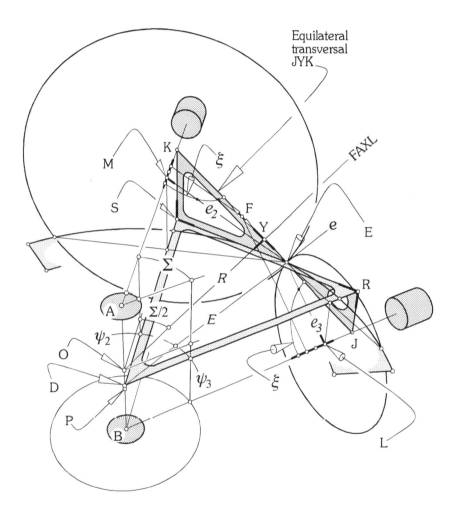

Figure 5A.01 (§5A.17). Genaral view of the geometry of equiangularity. The offset R extends in the radial direction from O to Y, O being the mid-point of the CDL (A-B), and Y being the mid-point of the equilateral transversal (J-K). The point E appears at the intersection of the E-line (marked D-e) and the said transversal. The two E-radii, e_2 and e_3, appear also. The pitch line is not shown but, emanating from P (which is shown), it lies just below and parallel with D-e. Numerically speaking, $\Sigma = 50°$, and $k = +0.6$.

165

The fortuitous aspects of an equiangular architecture

19. I have previously referred to the architecture of a gear set (§2.32, §2.34). I mean by this the general layout of important lines such as the shaft axes, the CDL, the FAXL, the transversal at [Qx], and the pitch line, and important planes such as the polar plane at [Qx] and the transnormal there. The architecture is, as it were, the skeletal shape of a gear set, excluding the actual shapes of the teeth. I wish to speak here however about a special, simple arrangement of the architecture which I will call, for reasons to be clarified, *equiangular architecture*. Other architectures, to be collected together under the descriptive terms *plain polyangular* and *exotic polyangular*, will be dealt with later.

20. Refer again to figure 5A.01. This is a skeletal diagram drawn for the geometry of this chapter. It illustrates the most elementary aspects of equiangularity. Refer in advance to figure 5B.01; this is a re-creation of figure 1.07 with certain important additions and a few modifications. Notice also the twin figures 5A.02 and 5A.03. These show the same equiangular architecture (the same as the architecture of figure 5A.01 and the same as those of one another), but they are viewed from different aspects, the second of which corresponds to the aspect of figure 5A.01. Let there be a point D on the segment of line (A–B) otherwise known as the centre distance. The length of (A–B) is C. Let D be so positioned that it divides C into two segments whose lengths are as follows: $C/(1+k)$ and $Ck/(1+k)$. This D is the same D that was mentioned at §1.41-42, and the two mentioned lengths are the lengths of the radii d_2 and d_3 which were also mentioned there. I repeat that D is not at the pitch point P; the points P and D are distinctly separated points (§1.14, §1.41); they are both indicated at figure 5B.01.

21. See figures 5A.01, 5A.02 and 5A.03 and let us draw an endless line D-d through D parallel with the pitch line. Go to equations (10) and (11) at §1.14 for the direction, namely the two angles ψ_2 and ψ_3, of this line. The equations are redundant; either one of them will do.

> *Theorem concerning the E-line.* At any point E along this special line D-d the chosen point E will have, among other special properties to be listed soon, the following special property: if we drop perpendiculars from E onto the two wheel-axes, we locate a special pair of G-circles which intersect at that E and whose radii are in the ratio k.

Any pair of the said G-circles are such special circles that I wish to call them E-circles of the equiangular set; and I refer to their radii, not as g_2 and g_3 (§1.07, §3.41), but as e_2 and e_3. See §5A.23 for a repetition of this statement and a list of its ramifications, and see figure 5A.01 and the text at §5A.27 for some sample proofs. The whole central gist of equiangularity is not elucidated yet, but these remarks are leading up to it.

THE SIMPLICITY OF EQUILANGULARITY §5A.22

22. I wish to call the line D-d, the E-line of the gear set, and any point upon that line, an E-point. The unique D thus becomes, not only a special point on the CDL, but a special E-point. Please see (at each of the figures 5A.01, 5A.02 and 5A.03) the E-line namely D-d of unlimited length. See also

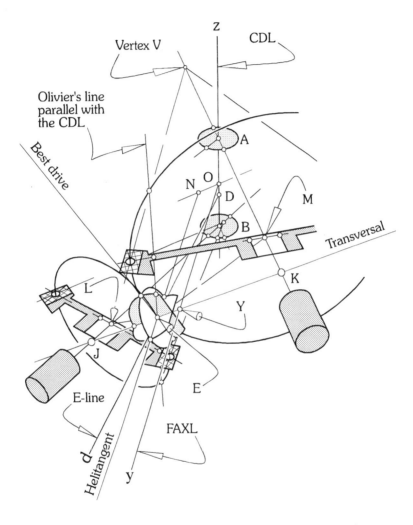

Figure 5A.02 (§5A.17). This figure, like figure 5A.01, is illustrating equiangularity, but we are looking here from Q at JK towards the CDL. All points upon the E-line are called E, and a point of contact Q is here one of them. Notice the points L and M at the feet of perpendiculars dropped from E onto the shaft axes. These points are the centres of the E-circles; the E-circles intersect at E. Notice the line ON. ON is parallel with JK, N is upon the helitangent. the helitangent is parallel with the FAXL, and the figure EYON is a rectangle. All three planes including the polar plane and the two planes of the E-circles intersect in the same line, and that line is Olivier's line.

the segment (D–E) marked upon it. Be aware next that, if we choose in design to put both paths for Q to pass through the same chosen E-point (that is, if we choose to employ equiangularity), the mid points [Qm] of the paths will be, as may be shown, coincident at that E (§3.56). It is under circumstances such as these that I wish to speak about equiangular architecture. By using the above, inset, simple, but spectacular statement — I call it a theorem — and its numerous corollaries to be listed next, we can avoid blundering about in the architectural space and quickly find a convenient, regular starting point for ordinary, equiangular syntheses. We need simply go far enough along D-d, from D to E, to satisfy our wish to put the zone of the meshing (as yet a vague region) somewhere suitably remote from the pitch line.

23. *The theorem concerning the E-line and its corollaries.* In the event of equiangular architecture we have along D-d (namely along the E-line) a continuum of points E. At each E:

(a) the transversal at E cuts the shaft axes at points equidistant from the CDL (see J and K); see figures 5A.01 to 5A.04 incl.; the transversal at E thus cuts the FAXL perpendicularly at Y;

(b) the best facet, which is that plane defined by the said transversal and the helitangent at E (§3.48), cuts the CDL at its mid-point O; see figure 5B.02; and see that, parallel with the transversal through E, a line through O cuts the helitangent in N;

(c) as already said (§5A.21), the E-radii of the two E-circles at E are proportioned according to k; neither of the two E-radii lie in the plane of the best facet incidentally; the plane of the two E-radii (another plane) will be called (see later) the *radiplane*;

(d) the helitangent at E lies, not only within the best facet (as it always does), but within the radiplane as well; the radiplane, in other words, cuts the best facet in the helitangent;

(e) the plane of the transnormal at E cuts respectively the axes of the shafts at two points V whose distances from the centres of the respective E-circles are in the same proportion as are the radii of those circles namely k; this means

(f) not only that the half angles κ of the cones of Wildhaber are equal (for this is the case in all involute gear sets unless the set is exotic), but also that the swivel angles λ are equal; see figures 5B.02(a)(b), and get some more information at §6.39 and §10.37;

(g) the heliradius h at E, which is the segment (N–Q) upon the helinormal there — see figure 3.11(c) — is parallel with the CDL; accordingly the helitangent at E lies in a plane normal to the CDL;

(h) accordingly also, the said helitangent is equally inclined to the two shaft axes; the angles are each $\Sigma/2$; and this means that

THE SIMPLICITY OF EQUILANGULARITY

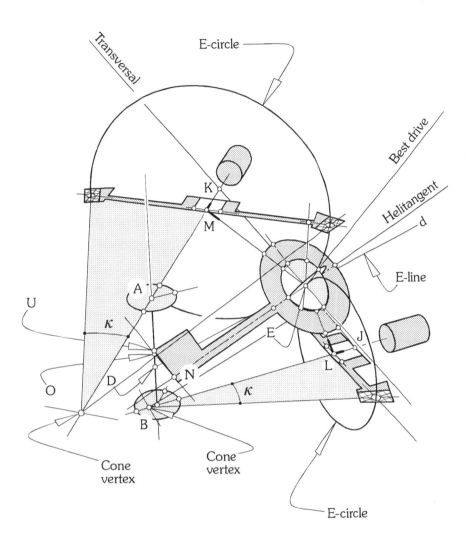

Figure 5A.03 (§5A.17). Another layout of the same equiangular architecture showing more clearly this time the plane of the E-radii. The plane of the E-radii is indicated here by the key-shaped lamina, which, as will be seen, is not in the plane of the best facet. The plane of the E-radii and the plane of the best facet do intersect along the line of the helitangent however, and, although this is not quite clear in the figure, a close examination will reveal that it is clear. It is important to perceive — but this is a feature unique to equiangularity — that the two E-radii and the helitangent are coplanar. The half angles κ of the Wildhaber cones are moreover equal. This is a feature not only of equiangularity but of polyangularity as well. This phenomenon (the equality of the two κ) thereby distinguishes all of the involute gears of this book from most of the hypoid gears of current engineering practice.

§5A.24   GENERAL SPATIAL INVOLUTE GEARING

(j) the twist angles τ of the two naked wheels are equal also, each being the said $\Sigma/2$; go for further clarification of this to §5B.04;

(k) see figure 5B.04; the throat radii a_2 and a_3 of an interacting pair of base hyperboloids, obtained by swinging any one legitimate path about the shaft axes in turn (§3.67), are in the proportion k, and the corresponding angles α are equal; thus

(m) the pitches of the core helices of the two mating helicoidal involutes, given by $a \cot \alpha$ (§3.12), are in the same proportion k;

(n) when a travelling point of contact Q arrives at its point E, the sum of the radii of the two G-circles at Q, namely $(g_2 + g_3)$, arrives at its local minimum, namely $(e_2 + e_3)$; this refers to my remarks about the point [Qm] on a path already made at §3.56.

The items listed (a) to (n) above are collectively spectacular because, in general, when [Qx] is not upon the E-line, no one of them obtains. Together the items characterize the concept of equiangular architecture. Not all of them are proven here however. For further details of items (a) to (d), go to appendices A at 5A (§5A.34) and B at 5A (§5A.35). For items (e) to (j) go to appendix A at 5B (§5B.66), and for items (k) to (m), see later. With respect to the mentioned point N, please read note [57].

24. *Comment.* Items (a) and (b) in the list above mean that we can locate the best facet at any E very easily. The best facet at any E is clearly defined by the line JYK of the relevant transversal and the mid point O of the centre distance; following this, and by virtue of such remarks as those that are made at §3.44 and §3.52, we see that items (g) and (h) hold true; and item (h) means that, because the helitangents at all points E along a given E-line are parallel, the polar planes at all points E along that line are parallel. Accordingly the legitimate paths which cut the E-line for a given k, namely the legitimate paths for all solutions given that k, constitute a linear congruence of the fifth kind [1] [§11.29]. Speaking more comprehensively we can now say (for equiangular architecture) that, *whatever the value of k, and because the sliding velocity at any E is parallel with the FAXL, the polar planes at all ∞^2 of E (for a given C and Σ) are parallel with one another and normal to the FAXL.* This convenient fact will find a relevance later, at §5B.22. It will also be mentioned again (in retrospect) at §6.27 and §6.55.

25 *Further comment.* Item (c) in the list above means that, if we cut both wheels (the real toothed wheels) by the plane containing the two E-radii, the resulting sections through the wheels will be diametral sections, and these will be seen to contain the helitangent namely the line of the sliding velocity at Q when Q is at [Qx]; see in advance the naked wheels at figure 5B.02(a). Item (n) similarly relates to the same phenomenon; when we project orthogonally across the plane of the two E-radii, we see [Qm] at E with the rubbing velocity contained within that plane; the sum of the E-radii is the least sum referred to, and we see that [Qm] in the path of Q is some kind of

special point. Whereas the best facet contains O (namely the mid-point of the centre distance), the radiplane does not. For proof that the sliding velocity resides in the event of equiangularity within the plane of the E-radii, namely the radiplane), refer to appendix A at 5A. The radiplane comes up again at §5B.37. Nor have we finished yet with the question of [Qm].

26. Equiangular architecture accordingly requires by its own definition that [Qx] be put at the E-line, and that we have under those circumstances five planes intersecting along the helitangent there. The planes are (1) the transnormal at [Qx], (2) the best facet at [Qx], (3) the plane of adjacency to the mating tooth profiles for drive when the relevant point of contact is at [Qx], (4) the plane of adjacency to the mating tooth profiles for coast when the relevant point of contact is at [Qx], and (5) the plane of the two E-radii (namely the radiplane), somewhat mysterious as yet, but to be discussed again (§5B.38).

A selected few of the algebraic formulae

27. Before we leave figure 5A.01 it should be mentioned that this figure and extensions of it can be seen as the vehicle for a great deal of algebraic argument. Proofs for all of the propositions listed (a) to (n) at §5A.23 may be seen to reside within it. It may be shown for example that, by elementary trigonometry, the radii e_2 and e_3 (namely the E-radii) may be written

$$e_2 = [1/(1+k)] \sqrt{[4R^2 \sin^2(\Sigma/2) + C^2]},$$
$$e_3 = [k/(1+k)] \sqrt{[4R^2 \sin^2(\Sigma/2) + C^2]};$$

and from these it may be seen incidentally that, for the same R and a varying k, namely for a range of equiangular possibilities with a given R, the two E-radii sum to a constant, namely

$$e_2 + e_3 = \sqrt{[4R^2 \sin^2(\Sigma/2) + C^2]}.$$

Go to §6.06. That paragraph deserves consideration here. Carefully studied, it will extend the ramifications of what has been written here. For a right (as distinct from a left) handed architecture, and for my conveniently adopted conventions for verticality and horizontality, go to appendix B at 5A.

28. Although an established E is clearly a point of importance along an E-line, we will find it convenient to nominate the location of an E, not by stepping the segment (D–E) directly along the E-line from the CDL to E, but by approaching E indirectly as follows. For the sake of symmetry and thus for the sake of a simpler algebra, we first step the segment (O–Y) along the FAXL from the CDL to point Y. Notice that the FAXL is collinear not only with the algebraic axis O-y of figure 1.07, but also with one of the two principal axes of the ubiquitous cylindroid [1] [§15.23]; see figure 5B.01; and see there that the FAXL bisects the shaft angle Σ. The FAXL and the said cylindroid, in other words, together lie at the root of the geometry displayed in both of the figures 1.07 and 5B.01. Having chosen R in the course of a syn-

thesis, and thus having fixed the point Y, we need only to step, next, via the transversal JYK existing perpendicularly at Y, to locate the point E at radius E along the E-line pertaining. Clearly

$$E = R \sec(\psi_2 - \Sigma/2),$$

where ψ_2, as we know, depends on k (§1.14). For further details refer again to appendix B at 5A.

29. Refer again to figure 5A.01. I wish to mention parenthetically the circle of radius E, drawn coaxial with the CDL at centre D. It resides in the plane of the isosceles triangle DRS and cuts each of the two E-circles in two points. The said circle, the said plane of the triangle, and the triangle itself, are clearly delineated. *Let the CDL be said to be vertical.* The triangle is accordingly horizontal, and the planes of the two E-circles (parallel with the CDL) are thus, in turn, each vertical. They intersect in *Olivier's line*; see figure 5A.02 (§3.66, §5B.22). *The E-circles are not co-spherical.* Were they to be, the centre distance C would be zero, and the relevant gears would be bevel gears. This unnamed circle of radius E which has been mentioned parenthetically here is of passing interest. It should not be confused with the circular section of the so called R-cylinder, which is to be looked at later — and for a much clearer reason — at §6.18.

Two special values of R

30. *If on the one hand we chose R to be zero,* E would be at D, and we would be dealing with that special, equiangular set of crossed helicals based upon the two D-radii, d_2 and d_3 (§1.38). The naked wheels for these, which interact at D, are not the same as the axodes shown in the figures 1.07 and 5B.01. The axodes meet (and are tangential) at P (§1.26). It will be clear from figure 5A.02 that, if R were zero, [Qx] would be upon the CDL at D, the transversal would be collinear with the CDL, the transnormal at [Qx] would be normal to the CDL, the other two planes of the triad at [Qx] would be intersecting along the CDL, and (referring to figure 1.07) the helitangent would be occupying its relevant generator (not marked) upon the ruled surface there marked (d). These evidently simple gears would have equal angles τ ($\tau = \Sigma/2$) at the teeth — see angle τ at figure 3.10 and refer in advance to figure 5B.02(b) — and they might be called, in the present context, *the first of the two special equiangulars*. They and their more complicated polyangular derivatives employing the so called C-points along the CDL, and their notorious disadvantages with respect to wear and so on, are looked at again at §5B.60. *If on the other hand we chose R to be infinite,* E would be infinite also, and we would be dealing with the *second of the two special equiangulars,* namely the equiangular, *bevel* gears. But these are, quite clearly, and especially for values of k other than unity, not the so-called straight bevel gears of ordinary engineering practice (§5B.62). They are however known. They are variously called, as mentioned by Drago [43], *conical* bevel gears (a misnomer in my view) after Merritt [58], and Beveloid gears (§5B.61) [61].

The special case when C is at P

31. For all positions other than at D of the movable point C upon the CDL (§1.39), the wheels are clearly non-equiangular; they are, by my definition, *polyangular* (§6.02). If C were put at the other special point on the CDL, namely P, the helitangent would become collinear with the pitch line and the otherwise singly departed sheets, which when truncated form the naked wheels, would become, simply, *the non-departed axodes for the motion*, namely the mating surfaces marked (a) and (b) in figure 5B.01; and then, because P is not a point upon the E-line, the gears would not be equiangular; c_2/c_3 would not be equal to k (§1.26). This discussion of the movable C and its special locations at D and P (here and at §1.39) is reminiscent of chapter 22 in [1] where, it must be said in retrospect, these special equiangular crossed helicals at D and their polyangular derivatives based upon other points C along the CDL were alone the chief topic of study [1] [§22.34].

An interim mention of the polyangular

32. Ultimately, our choice for [Qx] is much less restricted than that suggested by the definition of equiangularity outlined above. *It is absolutely not necessary for [Qx] to be at an E. It is not even necessary, as has been said, that the travel paths intersect; and it is certainly not necessary that the G-circles at [Qx] have their radii proportioned according to the speed ratio.* What is being put forward under the guise of equiangularity is no more than a carefully chosen, restricted, tidy criterion, on the basis of which we may, if we wish, distinguish between the simple and the more complex. Most of the useful skew sets that we find in engineering practice are polyangular; and most of these are hypoids with curved line contact and undiscoverable paths (§2.25); but see for legitimate involute examples the various skew gears of chapter 6 and the worms of chapter 7. We have said that a beginning [Qx] in design might be chosen to be anywhere; but, as we shall see, there will be wisdom, not in setting off from the safe haven of the E-line in any old direction, but in setting off (step by step) along the equilateral transversal through E. There is such an equilateral transversal through every point E upon the E-line for any given k, and these transversals trace out the relevant F-surface which is, indeed, a parabolic hyperboloid. See the generally chosen point F upon the generally chosen equilateral transversal at figure 5A.01. The other set of generators of this same surface is in fact the locus of the E-lines for all possible k. So for *plain*, non exotic work, we might either go directly step by step along the equilateral transversal JEK, or wander anywhere upon the twisted checkerboard of the so called F-surface.

Last remarks and a transition comment

33. It is an isolated matter of interest that without an opposing set of mating flanks for reverse motion (or for coast) the single-sided gear-body set

at figure 3.10, although a specimen displaying pure involute action, cannot be classified according to the criteria under consideration here. The precise point in the fixed space at figure 3.10, where the known straight path of the single Q cuts the known F-surface, may be called by definition [Qf], but, without an opposing set of flanks and thus without another path, there is no knowledge of whether or not there is a [Qx]. This leaves entirely unanswered the subsidiary question of whether or not a yet to be contrived [Qx] might reside upon the F-surface there or be remote from it. Go to §8.01. Having said that, I return now to the central subject matter of this and the next chapter namely equiangularity, and thus begin the business of chapter 5B.

Appendix A at chapter 5A

34. This comes from §5A.25. Refer to figure 5A.04. Equiangularity prevailing, we see that the two shaft axes (each of which is perpendicular to the CDL) and the single line of the sliding velocity when Q is at E (which is the helitangent and which is also perpendicular to the CDL because it is parallel with the FAXL) form a threesome of lines that reside respectively in three planes that are parallel. The E-radii appearing in the figure intersect at E. The plane defined by these is called by me the radiplane. A parallel view taken against this plane is called by me the radiplan view (§5B.38).

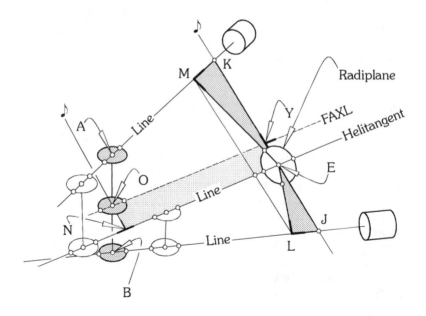

Figure 5A.04 (§ 5A.34). For appendix A at 5A. This figure is to be used for proving the proposition that, given equiangularity, the radiplane contains the helitangent. The difficulty here is that, although the triangles ELJ and EMK are similar, they are not coplanar.

THE SIMPLICITY OF EQUILANGULARITY §5A.34

The three mentioned lines are marked 'Line' in figure 5A.04. Note the shaded rectangle EYON, which was evident also at figure 5A.02). The side YO of this rectangle, namely the FAXL, is excluded from the mentioned set of three lines; it is not relevant in the argument here. Now we know that any set of three lines generally disposed defines a regulus that in turn defines an hyperboloid. But not so well known is the fact that, if the three lines reside respectively in three parallel planes, as the three selected lines here do, the hyperboloid they define is parabolic (§4.77) [1] [§7.68].

We are aiming to prove the hypothesis that, given equiangularity, the two E-radii and the helitangent, which three intersect one another at E, are coplanar. I propose to do this by showing that the line LM which is in the radiplane by definition, intersects the line of the helitangent. Although premature as yet, the mooted intersection is already illustrated at the figure.

It is a property of the parabolic hyperboloid that, if we isolate upon its surface any skew quadrilateral bounded by four existing straight-line generators of the surface (two from each family), we can proceed as follows: divide opposite sides of the said quadrilateral into the same unspecified number of segments, ensuring that the ratios of the lengths of opposite segments remain the same as the ratio of the whole lengths of the said opposite sides, join the corresponding points of division thus distributed along the opposite sides, and produce thereby fresh duplicate generators of the original surface. This theorem is well known among, and regularly applied by, those practising architects who pursue hyperboloidal roof construction.

In figure 5A.04 the angles ξ at J and K are equal (§5A.16), and the angles at L and M are both right angles. It follows that the right-angled triangles EMK and ELJ are similar, and, applying the above theorem among the known generators of the parabolic hyperboloid here, we can see that the line LM does indeed cut the helitangent. While the three defining lines are by definition generators of the defined parabolic hyperboloid, the lines LM and JK are generators of the opposite family. The starting hypothesis here, that the radiplane contains the helitangent in the event of equiangularity, has thus been proven.

It might be noted that the particular parabolic hyperboloid invoked here is quite other than the important F-surface of this book (§5A.16). I suspect it also of not being rectilinear, as the F-surface and others have been found to be (§4.26, §4.77, §4.78 *et seq.*). The reader might care to plot this new, somewhat strange parabolic hyperboloid *in situ*, and determine by whatever means, graphical or otherwise, the exact location of its central point and the angle between its intersecting generators there [75]. The reader might also speculate upon the general significance of this and other such accidentally inspected surfaces.

The here-tested hypothesis will certainly be provable also by regular Cartesian methods, but, characteristically of the field, such an algebraic effort might turn out to be, in the short run, if not difficult, then at least awkward. The author is aware of course that, for computer-programming in this particular epoch, that kind of algebra is preferable. I wait for the time, however, when computers can swim like fish among the direct concepts of screw theory as well as they swim like fish already among the matrices [27].

Appendix B at chapter 5A

35. This comes from §5A.27. The figure below illustrates the relative locations of some of the fixed variables of an architecture for equiangularity. The set here is *right handed*. I mean by this that its pitch p (whose sign for a given k is determined wholly by the layout of the shaft axes) is positive, not negative as in the previous figures of this chapter. Note the straight-line transversal JEYK. Note the CDL (A–B) whose length is C. Note the radial offset (O–Y) whose length is R. For Σ and the two ψ, see figure 5A.01. See the radiplane LEM whose sides (M–E) and (L–E) are the E-radii e_2 and e_3 respectively. See the formulae at §5A.27. That the otherwise irrelevant line LM cuts the FAXL is a phenomenon peculiar to equiangularity; in the event of *polyangularity* (chapter 6) this line does not cut the FAXL.

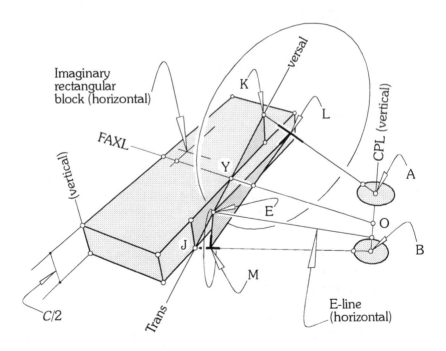

Figure 5A.05 (§5A.35). For appendix B at 5A. See text for explanatory comment.

CHAPTER FIVE (B)
SYNTHESISING AN
EQUIANGULAR SET

Introduction

01. This chapter deals with two worked examples in equiangular synthesis, the raw data for the first of which may be listed as follows: $C = 80$ mm, $\Sigma = 50°$, $k = +0.6$, $R = 140$ mm. The ongoing synthetical work (the designing of the gears) will dominate the argument, but I will, from time to time, interrupt the flow of this to make general comment as required. Relevant sub-headed sections will, I hope, clearly direct the reader in this respect. The raw data for the second example, which begins at §5B.45, are: $C = 40$ mm, $\Sigma = 90°$, $k = 9/41$, $R = 60$ mm. Almost without exception, the figures have been drawn to scale. The figures are mostly drawn, however, in perspective. Before beginning seriously with the design work at §5B.09, therefore, it will do no harm to take an exploratory look at some of the figures. By doing this we can deal at an early stage with certain matters of geometric principle that still require attention. Let me explain as well that correct visual perception will depend upon certain symbolic devices that I use.

Naked wheels

02. Please consult note [90]. Implicit among the aspects of equiangularity listed at §5A.23 are the circular hyperboloidal surfaces swept out by the helitangent at E as it is rotated in turn about the two shaft axes. These are the #3 hyperboloids of chapter 4, the cone hugging doubly departed sheets. A pair of these, with their indicated narrow circumferential bands (the naked wheels), is shown at a number of places: (a) at figure 5B.01 which is an omnibus picture intended for future reference; (b) at figure 5B.02(a) which shows some essential geometry of the naked wheels in contact; (c) at figure 5B.02(b) which is intended to clarify the difference between the twist angle τ of a naked wheel and the cone angle κ of its accompanying Wildhaber cone [6]; and (d) at figure 5B.03 which is a composite group of orthogonal projections showing how the naked wheels fit together, interact, and operate as an embryo gear set. The naked wheels in the mentioned figures straddle the

E-circles and they are of course ruled surfaces (§4.36). See them disassembled but still hugging their Wildhaber cones at figure 5B.02(b). They display by means of the lines inscribed upon them the rotated directions of the helitangent at the point of contact Q when Q is at [Qx] (which is at E) and thus the direction of the sliding or the rubbing velocity there; see §5A.23, item (g). The naked wheels are not the axodes, but, as has been said, the relevant pair of doubly departed sheets (§4.20). The four truncated parts of figure 5B.01, in other words, belong to two entirely different pairs of hyperboloids; firstly we have in the background the axodes for the motion; these remain from figure 1.07 (§1.31); and secondly we have, in the foreground, the naked wheels for the motion, whose shapes are determined by the length of the intercept (O–Y) namely the offset radius R.

03. The rulings shown at figure 5B.02(a) do not mesh with one another as the rulings at the axodes do. There is circumferential slipping at the helitangent at [Qf]. See more about this at §5B.08. The naked wheels are tangential with one another at [Qf] (which is here at E), but only at [Qf] (§4.35). This isolated point of tangency is an example of the special points [Γt] discovered in the discussion beginning at §4.03. Interaction between the naked wheels cannot occur without interference (intersection of the mating surfaces) unless the said wheels (whose teeth straddle the isolated point of tangency) are of infinitesimal width. Refer to figure 4.04(c) and the text at §4.07 et seq. These together deal in general with the criterion for tangency. Observe that, because the helitangent here (which is the common generator there) is by definition perpendicular to the transversal, the criterion for tangency is fulfilled at [Qx]. See too that [Qx] is here upon the E-line, not because of some inescapable law, but because, indeed, we put it there. I look again at the question of naked wheels at §6.29.

The angles τ, κ and λ at the naked wheel

04. The whole of this section may be omitted upon a first reading. It relates (a) to certain mathematical links between current analyses and the quite different analyses pursued by the non-involute (the hypoid) theorists, (b) to the practical business of actually cutting involute teeth, and (c) to some of the as yet obscure questions regarding friction and lubrication. Speaking in my terms and those of Wildhaber [6], my transversal is his pitch vertical, and my transnormal is his pitch plane; the corresponding terms are exactly synonymous. In general, among all gears of whatever kind, it is true to say (a) that the transnormal at [Qx] is tangent to both of the doubly departed sheets at the point E, but not elsewhere as can be seen with the help of figure 5B.03, (b) that the transnormal cuts the two shaft axes in the two apices, V_2 and V_3, of the well known cones of Wildhaber [6], (c) that the said cones thus enjoy line contact with the transnormal (on either side of that plane), (d) that, whereas the said cones enjoy point contact with one another at E, the doubly departed sheets enjoy tangency with one another at E while intersecting one

SYNTHESISING AN EQUIANGULAR SET §5B.04

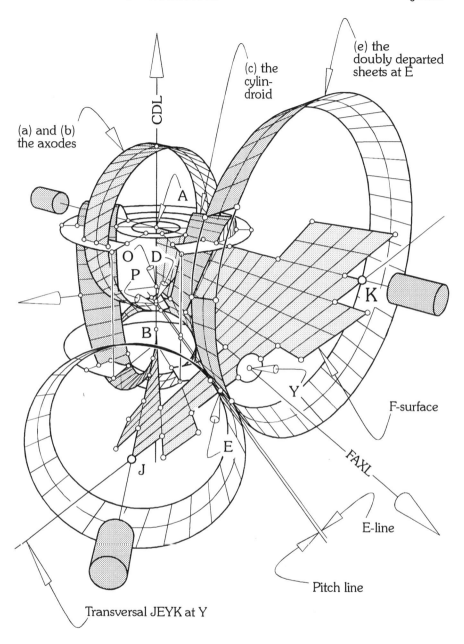

Figure 5B.01(§5B.02). Essentially, this is a repeat of figure 1.07, but with extra information added. See the point D upon the CDL, the E-line, and scattered, filled-in parts of the F-surface. The F-surface contains the E-line, the FAXL, the CDL and the two shaft axes; in general (unless k = unity) it excludes the pitch line. See also the doubly departed sheets determined by the sliding velocity along the helitangent at E, narrow circumferential bands of which have been selected to become the naked wheels.

§5B.05 GENERAL SPATIAL INVOLUTE GEARING

another along the helitangent at E, (e) that the pitch cones are tangential around the peripheries of the E-circles with the said doubly departed sheets (namely the naked wheels), but (f) that the said sheets (which come from me) are not the same surfaces as the pitch cones (which come from Wildhaber). The generators of the cones of Wildhaber originate in the intersecting intercepts $(V_2-[Qx])$ and $(V_3-[Qx])$ and intersect at the respective apices V of the said cones which have their special significance (§10.29). The generators of the naked wheels, on the other hand, originate in the direction of the unique helitangent at E and do not intersect; they form the shown hyperboloids.

05. Let the half angle κ at the apex of a Wildhaber cone be called the *conicity* of a wheel. This angle κ will later be called the *tilt angle* of the wheel, but the logic for this will not be made clear until chapter 10. Using geometry implicit in the figures 5A.02 and 5A.03, it may be shown that, with equiangularity obtaining, the conicities κ of the two wheels are equal and that the value of this κ is independent not only of which wheel is which but also of the speed ratio k. We also see that the radii t of the throat circles of the mentioned departed sheets are exactly in the proportion k under these special circumstances, and that the angles α of inclination of the generators of these surfaces at the throat circles are, upon the wheels, equal, and that this α also is independent of k; see figure 5B.02(a) and refer back to §5A.23. Let the twist angle τ of a wheel (namely that of the naked wheel) be measured as shown at figure 3.01. Please read note [90]. Refer to figure 5B.02(b). It will be clear that the twist angle τ of a wheel is not only (a) the angle (in space) between the helitangent at E and the axis of the wheel, but also (b) the half-angle at the apex of the asymptotic cone of the swung hyperboloid. We might say in rough summary that, given equiangularity, both the twist angles τ of the wheels and the conicities κ (which angles are not the same angles) are, when taken like upon like, the same for both wheels and (for given C, Σ and R) the same for all k. Given equiangularity, it may be shown from the geometry that, whereas

$\tau = \Sigma/2$,
$\kappa = \arctan [\{2R \sin (\Sigma/2) \tan (\Sigma/2)\}/\{\sqrt{[4R^2 \sin^2 (\Sigma/2) + C^2]}\}]$.

Recalling that we are, here, within the arena of equiangularity, notice that these expressions are, indeed, both independent of k. See appendix A at 5B. The said fact becomes important later. Given the current data, $C = 80$ mm, $\Sigma = 50°$, and $R = 140$ mm, the numerical results are:

[$\tau = 25°$ exactly]
[$\kappa = 21.12199°$ approximately].

For the relevant algebraic details (and more) refer to appendix A at 5B (§5B.66). Figure 5B.02(b) also shows more clearly that, whereas τ is the half angle at the apex of the asymptotic cone, κ is the half angle at the apex of the Wildhaber cone. It might be mentioned parenthetically that, with polyangularity on the other hand (chapter 6), while the angles κ will remain equal with one another, the twist angles τ will differ; see in advance figure 6.07. Both of the figures 5.02 show another angle, λ. This will later be called the *swivel angle* of the wheel. Its practical significance need not concern us here, but it is important, and will be dealt with later (§10.07).

06. In the special case of planar spur gears both κ, and both τ, are zero; but in the quasi-special case of cylindrical helicals, while both κ are zero, both τ are non-zero. Bevels are another special case to be separately dealt with later (§5B.61), but let us note in the meantime (and with satisfaction) that, when C is put to zero in the above expression for κ, κ breaks down to being, simply, $\Sigma/2$ namely τ; this must occur because, in the case of bevel gears (when C becomes zero and/or R becomes infinite), the two otherwise differing cones of a given wheel become identical.

Rolling, rocking and boring

07. This paragraph might also be omitted upon a first reading. The terms *rolling, rocking and boring* were first introduced by me in the context of points [Qx] occurring at positions along the CDL [1] [§22.41]. The three terms are generally intelligible, of course, only if seen against a background of some nominated or inferred triad of axes say Q-uvw set mutually perpendicular at Q. The axes inferred in [1] are, clearly, Q-u along the helitangent for rolling, Q-v along the best drive for rocking, and Q-w along the CDL for boring. It should accordingly be clear that the said terms were used in [1] in relation to the mating surfaces of pairs of naked wheels whose [Qx] was exclusively at the CDL. Referring to [1] [§5.47-49], where the motion duel or the six vector (ω, v) is explained, we see that, between the naked wheels of the *special equiangulars* (namely the crossed helicals), boring does not occur [1] [§22.41]. But boring at the meshing of naked wheels does occur when the said wheels are offset. At all points E along D-d and at all points G in space, boring between the naked wheels in mesh is always there. At all such points, all three axes Q-uvw as defined above are inclined to the pitch line. All three axes are thus inclined to the relative angular velocity ω_{23} which occurs at the pitch line. All three, accordingly have a component of ω_{23} occurring along them. In this imaginary case of the naked wheels, however, the length Z_A of the actual zone of contact, which is a portion only of the length Z_I of the ideal zone of contact (§3.56), is of course zero, so to talk about rolling, rocking and boring at such an imaginary, instantaneous point of contact Q is, in a way, stupid [1] [§22.71]. It will be much more useful in the long run to talk about rolling, rocking and boring, not thus instantaneously, but as it takes place at the real profiles of real teeth in mesh, as we move along a finite path (which may be straight or otherwise) of a point (or points) of contact Q (§4.64). Under these circumstances, we will see (a) that the terms rolling, rocking and boring will have significant meanings (in respect for example of lubrication); (b) that the phenomena will be of practical importance (in respect for example of kinematic interference in the case of non-involute teeth), and (c) that,

because no such interference can occur at the single point of contact between the convex surfaces in involute gearing, we can pay attention there to the problems of lubrication without kinematic complications. It might be reasonable, for future studies, to nominate the three axes Q-uvw as follows: (a) Q-u for rolling along the helitangent namely along the direction of the rubbing velocity at the point of contact, (b) Q-v for rocking along the best drive, and (c) Q-w for boring along the transversal. All of this might well be a subject for further study in the areas perhaps of friction and lubrication. However I wish pursue such matters no further here.

Watershed

08. We are just now at the change-over point in these chapters 5 between the problems of analysis on the one hand and the problems of synthesis on the other; and these are, as has been said, not mere reversals of one another (§5A.15) [1] [§7.61]. Accordingly we need to proceed with caution. At any beginning in design, however, the following two steps are unavoidable: (a) choose numerical values for C, Σ, R and k (§5B.01), (b) knowing the required relative directions of rotation (§1.23), locate the relevant pitch line and calculate its pitch (mm/rad). After this, the group of interrelated parameters thus determined, C, Σ, R, k, p, r_2 and ψ_2, characterize the gear set (§1.14). It is relevant to remark just here that I have not yet mentioned the question of helix angles. *The helix angle in general gear terminology is a property not of a tooth but of the profile of a tooth (either front or back), and there are, as yet, no teeth with profiles existing here.* The naked wheels have bald surfaces; and we need to see in the present context that the naked wheels, while free of course to slide upon one another in the longitudinal direction namely in the direction of the helitangent at E, *are at the same time free, by virtue of the mating generators being smooth (§4.36, §6.29), to slip circumferentially.* To be more exact I mean by circumferentially here, that direction in the plane of the transnormal which is perpendicular to the said mating generators. But remember that there are, as yet, no real teeth existing, and that there is, accordingly, no point in trying to discuss just yet the question of helix angles.

Beginning of worked example #1 (part 1)

09. *Data.* I will use, in text, the abbreviation WkEx#1(1). We deal here with the earlier stages of an exploratory synthesis of a simply constructed general equiangular gear set. The data to be used may be seen to be already used in the graphical constructions of figures 5A.01 and 5B.01. The radii r_2 and r_3 (measured along the CDL) are found, using equations (4) and (5) at §1.14, to be approximately 27.999 and 52.001 mm respectively (they add to 80); the angles ψ_2 and ψ_3 (measured around the CDL) are found, using equations (10) and (11) at the same place, to be approximately 18.35° and 31.65° respectively (they add to 50°); and the pitch p, using equation (12) at the

SYNTHESISING AN EQUIANGULAR SET §5B.09

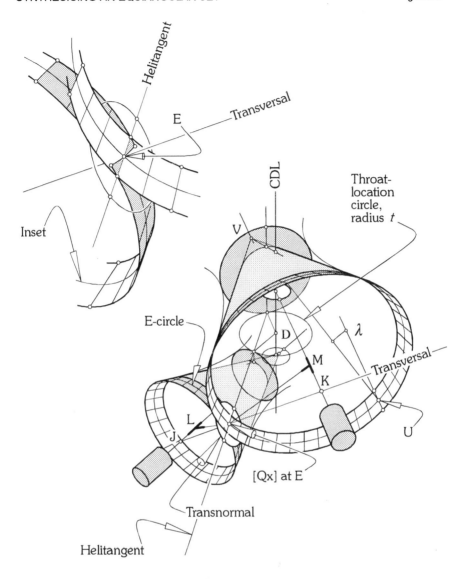

Figure 5B.02(a) (§5B.10). These are the naked wheels of the current exercise, shown hugging the cones of Wildhaber. The throat circles of the corresponding doubly departed sheets are seen to be located in agreement with the throat location circle. The throast location circle is at centre D and normal to the CDL. On opposite sides of the transnormal, the Wildhaber cones separately touch the transnormal along a generator, each cone touching the other at the isolated point [Qx]. Shown enlarged at inset is the double intersection of the naked wheels with one another along (a) the straight line of the helitangent, and (b) the curved line of a second intersection nearby. The two lines of intersection intersect one another at [Qx]. The naked wheels are tangential with one another and with the intervening transnormal only at the isolated point [Qx]. The subtler aspects of these ubiquitous, elusive, but important phenomena are discussed already. See figure 4.04(c) and §4.28 et seq.

same place, is found to be approximately, $p = -17.252$ mm/rad. These derived data are offered with more precision among the numbers at the *table of numerical results* (TNR) at the end of the book. Given the equiangular architecture, we have next located D, the distances $C/(1+k)$ and $Ck/(1+k)$ being 50 mm and 30 mm respectively. We next find the length (D–E), namely the distance R_E, to be $R_E = R \sec(\psi - \Sigma/2) = 140.948$ mm approximately (§5A.28). Next, by knowing items (a) and (b) listed at §5A.23 above, we can construct the transversal, the transnormal, the best facet and the polar plane at [Qx] at the discovered E. Otherwise we can, by knowing items (g) and (h) at §5A.23, erect the helitangent first, and then, by knowing that the transversal is perpendicular to the said helitangent at [Qx], we can locate the transversal. Thus, or otherwise, we can set up to begin.

10. *The naked wheels.* Please look at figure 5B.02(a), and see in advance figure 5B.03. The naked wheels of figure 5B.02(a) relate directly, as we know, to the departed sheets of figure 5B.03, the former being, quite simply, severely truncated versions of the latter (§5B.02). Having located the helitangent at [Qx], which gives the direction of the sliding velocity there whatever the actual shapes of the real teeth turn out to be, we swing a short segment of the helitangent, which evenly straddles [Qx], about the two shaft axes in turn. We find that the two naked wheels (the two hyperboloidal surfaces thus produced) share a common generator namely the said helitangent itself. Because of the prevailing equiangularity, this common generator lies in the plane of the two E-radii incidentally; see §5A.23, item (c). The two naked wheels generally intersect along the said common generator, but they are tangential with one another at [Qx]. This apparent paradox is explained by the following italicized passage which, although important, might be omitted upon a first reading. *The apparent paradox is explained by what was said, first at §4.20 in the form of a general theorem about the points [Γt] in the general scenario of Axis, Axis and Line, and again at §4.38 where information was offered about the points [Ht] among the naked wheels of real gears, that the two points of tangency along the common generator of the relevant intersecting hyperboloids occur at those two points where the said common generator and the local transversal are perpendicular. In the case of that particular point where the naked wheels meet at [Qx] (which is the one relevant point of the two), the local transversal is the unique, equilateral transversal.*

11. *The doubly departed sheets.* Whereas figure 5B.02(a) (which is in perspective) shows the naked wheels in contact and their corresponding Wildhaber cones soon to be mentioned (§5B.13), and whereas figure 5B.02(b) shows the same naked wheels (again with their Wildhaber cones) taken apart and lying flat upon the table, figure 5B.03 (which is strictly orthogonal) shows a series of interrelated views of the doubly departed sheets in contact. Note that these interact with one another in such a way that, while they are tangential at [Qx], their throats are not coincident at the CDL. Nor does the common generator, namely the helitangent at [Qx], cut the CDL. The throats of the departed sheets are spread equidistant in different direc-

SYNTHESISING AN EQUIANGULAR SET §5B.11

tions radially on either side of the CDL; the common generator is in that plane through D which is normal to the CDL; and, originating at A and B, directed distances S_t are measured along the axes to the centres of the two throat circles. Here, with [Qx] at E, these S_t are equal but, if [Qx] were not at E, they would not be equal [91]. Given items (g) and (h) at §5A.23, we can find by easy algebra that the equal distances S_t may be written,

$$S_t = R\,[(1-k)/(1+k)]\sec(\Sigma/2)$$

This shows that, whereas the angles τ and κ (and λ as we shall see) are constants within a given equiangular set independently of the speed ratio k, these distances S_t depend on k. Due also to the prevailing equiangularity, both departed sheets have their throat radii t_2 and t_3 (which are the same as

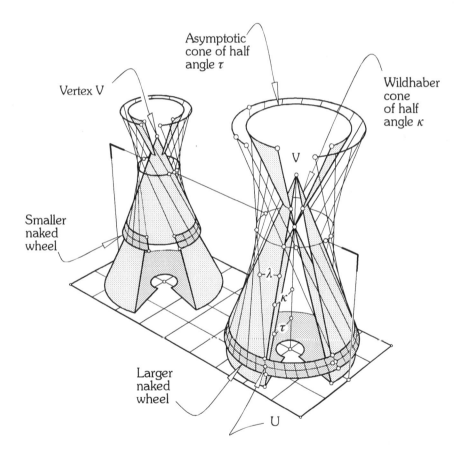

Figure 5B.02(b) (§5B.13). Having separated the wheels, we are illustrating here (a) the asymptotic cones with their vertices at the centres of the throat circles whose half angles τ are equal due to equiangulatity, (b) the Wildhaber cones with their labeled vertices V whose half angles κ are equal anyway, due fundamentally to the fact that the gears here are involute gears not hypoid gears, and (c) the swivel angles λ.

§5B.11　　　　　　　　　　　　　　　　　　GENERAL SPATIAL INVOLUTE GEARING

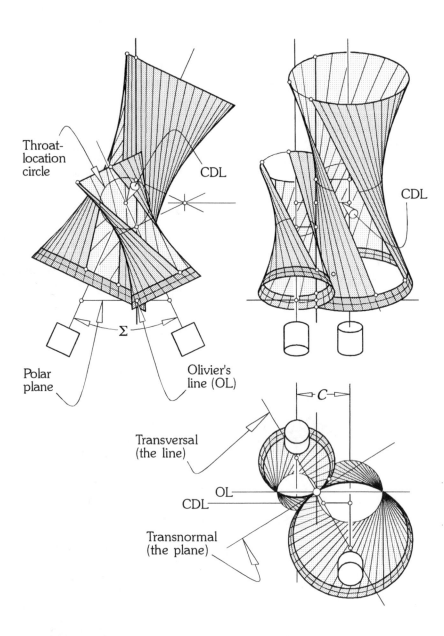

Figure 5B.03 (§5B.10). Orthogonal views of the doubly departed sheets containing the naked wheels of the current exercise. At top left, where we are looking along the CDL, we see flat-on the plane of the throat location circle, that circle appearing there accordingly. Notice also that, in the views at top left and bottom right, Olivier's line appears (as the CDL also does) respectively as a point, and as a line paralle with the plane of the paper.

the radii d_2 and d_3 of §1.41 incidentally) exactly in the ratio k. Notice in figure 5B.02(a) that the intersection of the naked wheels on either side of their isolated point of tangency at [Qx] has been illustrated at inset. Notice also in the mentioned figures that, whereas the vertices of the asymptotic cones occur at the planes of the throat circles exactly, the vertices of the cones of Wildhaber occur at positions along the shaft axes which are neither at those planes nor at the CDL. Because there is equiangularity here, the two angles λ are equal. Please remain aware also that naked wheels are not expected to mesh in the ordinary sense of that word; they intersect, and they slip with respect to one another circumferentially (§5B.08, §6.29).

12. *Charles William MacCord.* Go to MacCord [18] to find his earlier but forgivable misconceptions about these asymmetrical aspects of the assembly. Discussing generalized skew gears with offset (§5A.17), and with a tooth ratio k other than unity, he shows in his figure 108 two hyperboloids mating with different throat diameters, *but he shows them with their throats coincident, presumably at the CDL.* I mention here, as Mac Cord did, the Schröder models made in Darmstadt [41] which were, apparently, already existing at that time. By dealing with the crossed helicals only, and by working with radii located at point D (§5A.20), these models gave the unfortunate impression that the hyperboloids of such models (the departed sheets of this discussion) would always mate throat to throat. So we mention again the important but confusing fact we have just again discovered, *namely that the departed sheets of the skew naked wheels which are offset at a finite radius R are departed, not only in respect of their throat radii as was discussed at §1.34, but also in respect of their radial displacements from the CDL.* Along the CDL and away from the pitch point, the sheets are departed equally in that their throat radii continue to sum to the fixed length C, but the throats are also equally departed radially away from the CDL in that they occur at equal distances along the common generator on opposite sides of the CDL. The geometry of these phenomena can be studied with the help of the throat-location circle already mentioned and illustrated at figure 5B.02(a).

13. *The Wildhaber cones.* We thus come again to the pitch cones of Wildhaber [6], which for obvious reasons I wish to call, simply, the cones of Wildhaber or the Wildhaber cones. In figure 5B.02(a) they are 'ghosted in'; and one of them is partly hidden behind the other. Let us recall that these cones touch one another at [Qx] namely at [Qf], and that they are separately tangential (along straight lines) to the plane of the transnormal on opposite sides of that plane (§5A.23, §5B.04). *It needs to be noted here that in hypoid theory these cones originate at a freely picked point of contact M; see Baxter [15]; this so called mid-point M, which is also called by both Baxter [16] and Litwin [86] the pitch point P, does not in general reside upon the F-surface; it resides, within vague limits, anywhere in space; the Wildhaber cones in hypoid theory are, in general, accordingly non-equiangular; and this might lead to confusion when I say, as I do in the context here (§5B.04), that the said cones are equiangular.* We soon find in the arena of involute gearing that the

said cones are tangential circumferentially at the E-circles with the naked wheels (as might be expected), but that they are —see item (e) at §5A.23 — indeed equiangular. Figure 5B.02(b) shows the said cones being 'hugged' at the E-circles by the naked wheels which, like frilly garters, surround the cones (§4.36). Notice also in figure 5B.02(a) that the cones are shown to be piercing two annular discs which are equidistant on either side of the CDL. The outside peripheries of these discs are the circular throats of the departed sheets, the horizonal (*sic*) edges of which are sketched in. Figure 5B.02(b) shows firstly that the two kinds of cones (the asymptotic cones of §5B.04 and the Wildhaber cones of this paragraph) are different, and secondly that in each wheel the kinds intersect one another in two circles the second of which

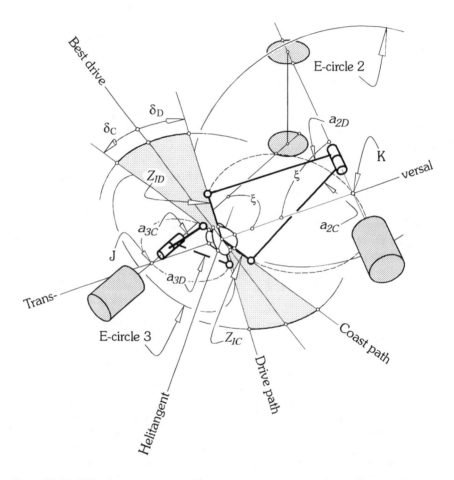

Figure 5B.04 (§5B.15). Inscription upon the polar plane of the two chosen paths for Q, showing the two angles of obliquity δ, both set at 20°. Constructed here are the four common perpendiculars to determine the four throat radii a, the four angles α, and the two distances Z_I. Notice the transversal (J—K) and the two equal angles ζ at J and K.

(being off the page) is not shown. Both of these kinds of cone are relevant to an understanding of gears in general, and the asymptotic kind is just as important algebraically as is the Wildhaber kind. Indeed we shall see in chapter 10 (which deals with the machining of the teeth) that not only κ but also another important angle λ can be written simply in terms of τ (§10.08). It might be observed that, in the limit as $R \to \infty$, the apices of the cones of Wildhaber move towards, and arrive simultaneously at, the CDL, the centre distance C becomes insignificant, the departed sheets become mere cones, and the relevant gears become bevel gears with intersecting axes. Please keep this special case in mind as the forthcoming argument develops. It will lead to another view of the correct involute shapes of the teeth in the case of straight bevels; it will give us cause for thought in the case of Bevelloids; and it will involve conceptual and pure mathematical difficulties yet to be spoken of. But look again at §1.50, §2.06 and §4.31, and go ahead if you wish to §5B.62.

14. *Selecting the paths.* This is the first, crucial step in design. Let us choose the two intersecting travel paths at [Qx] which is at [Qf] which is at the E-line. Refer to figure 5B.04. Both paths must be contained within the polar plane at E (§2.17, §3.50, §5A.11), and they must, therein, be at opposing obliquity angles δ to the best drive (§2.14); see Steeds [4] [§147] for his use of the term obliquity angle. The angles δ need not be the same, but certainly they should both be less than the very large angle δ_{REMOTE} that appears by chance in the illustrative example pictured at figure 3.11(b); in that figure the Q is not at its [Qx] and the variable δ_{REMOTE} might be greater than its δ; please recall §3.54 and §4.59. I have set the angles δ here both at 20°, this for the want, at this stage, of a better-informed choice. There are questions about these angles δ yet to be clarified, and among these is the matter of their precise definition. More remarks about these δ appear at §10.02. Please be aware, also, that we have not yet distinguished between the path for drive and the path for coast.

15. These next steps are now already determined. Look again at figure 5B.04. Having tentatively decided the single pair of intersecting travel paths, we may next examine the two resulting pairs of base hyperboloids (§3.67). One pair of these will be for drive and the other for coast. Although these two pairs of base hyperboloids are mounted upon the same pair of shaft axes, they are quite distinct from the single pair of doubly departed sheets (the naked wheels) mounted upon the same shaft axes. It is important to have this clear (§4.74-76), and please read note [90]. Allowing the subscripts D (for drive) and C (for coast) to convey their obvious meanings, we can determine the respective radii $(a_2)_{DC}$ and $(a_3)_{DC}$ of the throat circles of the base hyperboloids, and their angles of inclination $(\alpha_2)_{DC}$ and $(\alpha_3)_{DC}$ corresponding. This may be done by constructing the four shortest distances between the two travel paths and the two shaft axes and/or by calculating these. The bold spheres in the figure drawn at the ends of the resulting intercepts Z_j upon the paths correspond exactly with the S-joints at figure 2.02. We have here, however, two linkages RSSR, as discussed and predicted at §2.14. Please

construct in the imagination working models of the two different linkages RSSR in figure 5.04 and wonder whether the linkage with the longer Z_f is the one which predicates the path for drive. Notice that, whereas here the rods SS are intersecting (§5A.11), the corresponding rods discussed in §8.17 and shown at figure 8.05, although legitimate (§2.15, §2.21), are not. There is more to all of this than meets the eye. See also dotted-in in figure 5B.04 the ellipses of obliquity. They reside in the polar plane. In this figure they are tangential with one another at [Qx] due to the equiangularity, but in general they intersect. Due for next mention at §5B.40, and for further discussion at chapter 6, please let them remain here illustrated, but unexplained as yet.

16. *Some of the relevant algebra and a few numerical results.* The algebra becomes too awkward here for comfort within the text. Let it be said however that Killeen [54], followed by Sticher [52] and to a minor extent the author, have either written and/or sorted out the following results for immediate presentation here. The equations apply for equiangularity only. Not surprisingly they turn out to be, in the main, simplified versions of the more general equations later to be found for the wider circumstances of plain polyangularity (§6.38). See also chapter 10. If the distances (A–M) and (B–L) from the CDL to the centres of the E-circles be called S_{e2} and S_{e3} respectively, these distances may be written [52]

$$S_e = R\,[(k-1)/(k+1)]\,\tan(\Sigma/2)\,\sin(\Sigma/2) \pm R\,\cos(\Sigma/2).$$

Next the radii *a* of the throat circles of the base hyperboloids are the shortest distances between the two paths of the two points Q and the two shaft axes. After some awkward algebra the four of them, a_{2D}, a_{3D}, a_{2C} and a_{2C}, may be written [54] [52]

$$a = \frac{[w]\cos\delta\,[C^2 + 4R^2\tan^2(\Sigma/2)]}{\sqrt{\{[C^2 + 4R^2\tan^2(\Sigma/2)] + [\tan^2(\Sigma/2)][C\sin\delta - 2R\cos\delta\tan(\Sigma/2)]^2\}}}$$

where [w] is either $1/(k+1)$ or $k/(k+1)$ according to which wheel it is, and where δ is either positive for drive, or negative for coast. Formulae on the other hand for the angles of inclination α at the throats of the base hyperboloids may (after equally awkward algebra) be written [54]

$$\sin\alpha = \frac{[\sin(\Sigma/2)]\,[C\cos\delta + 2R\sin\delta\tan(\Sigma/2)]}{\sqrt{[C^2 + 4R^2\tan^2(\Sigma/2)]}}$$

where δ is either positive for drive, or negative for coast. By employing the found relationship $[2R/C]\tan(\Sigma/2) = \tan\Phi_B$ [52], however, where $\tan\Phi_B$ is the slope of the best drive, see figure 10.05(a) at §10.08, this reduces to

$$\sin\alpha = \sin(\Sigma/2)\cos(\delta - \Phi_B),$$
$$\cos\alpha = \sqrt{1 - \sin^2(\Sigma/2)\cos^2(\delta - \Phi_B)};$$

and following these maneuvres we can write (for each of the two δ which may not be the same) that

SYNTHESISING AN EQUIANGULAR SET §5B.16

$$\frac{a_3 \cos \alpha_3}{a_2 \cos \alpha_2} = k = \frac{a_3 \sqrt{[1 - \sin^2 (\Sigma/2) \cos^2 (\alpha_3 - \Phi_B)]}}{a_2 \sqrt{[1 - \sin^2 (\Sigma/2) \cos^2 (\alpha_2 - \Phi_B)]}}$$

which gives the speed ratio k in terms of the other chosen parameters, C, R, Σ, the two δ, the four a, and the four α, the latter eight of which are inextricably interrelated [54]. There is an important element of simplicity involved in the evident circularity of this somewhat enigmatic formula that has yet to be recognized incidentally; but remember that we are dealing here with equiangularity only. Refer in advance to the quite general, *law of the speed ratio*, revealed and explained at §6.40. It is a feature of equiangularity in any event, however, that the above-given extrinsic expressions for the four radii a and the four angles α are independent of k. Formulae are also available [54] for the distances from A and B at the CDL to the centres of the four base hyperboloids; these relate to yet other distances S (I call them S_a), two upon each of the two shaft axes. They are as follows:

$$S_a = \frac{[R \tan (\Sigma/2)][(1-k)/(1+k)][*_1 \sin (\Sigma/2)] + [*_2 C \cos \delta] + [2*_2 R \sin \delta \tan (\Sigma/2)] + [*_1 R \cos (\Sigma/2)] + [*_2 C][k/(1+k)][2R \tan (\Sigma/2) \cos \delta - C \sin \delta]}{[*_1 - (*_2)^2]}$$

where $*_1 = [C^2 + 4R^2 \tan^2(\Sigma/2)]$
and $*_2 = \pm [\sin (\Sigma/2)][C \cos \delta + 2R \sin \delta \tan (\Sigma/2)]$

Angle δ must be inserted according to sign and according to drive or coast, and the function $*_2$ will take the positive or negative sign according to shaft. Now it must of course be said that despite our best intentions here, these algebraic expressions remain grotesque. This fact leads naturally and obviously to the formulation of an exercise for the reader. It must also be said however that we need in the long run not the algebra, but the *numbers* for design; and a well-written computer-driven spread sheet will give lists of these almost instantaneously. By comparing the numbers sufficiently accurate for practical purposes but generated by my laborious, but visual, CAD with even more accurate numbers from the very much faster, but blind, Killeen spreadsheet we got, for example, in the case of this WkEx#1, the following numerical results that were found to check:

$[S_{e2} = (A-M) = 119.98564]$
$[S_{e3} = (B-L) = 133.78054]$
$[S_{a2D} = 116.962403]$
$[S_{a2C} = 127.905218]$
$[S_{a3D} = 131.966605]$
$[S_{a3C} = 138.532294]$ (mm)

At each wheel the centres of the base hyperboloids straddle the centre of the key circle, namely the E-circle in equiangular cases such as this, but they do so at unequal distances from the said centre.

17. *Kinetostatic phenomena.* See, among the above the expressions, those worked out for the *a* and the α. I use the plural here because there are of course four different *a* and four corresponding α. The various algebraic devices to deal with the multiplicity of values are clear to see. Notice in the following table of approximately quoted numerical results for this WkEx#1 however that, whereas the values *a* are all different, and whereas the values α appear in two sets of two equal values, the values *a* cos α appear in two other sets of two equal values. This can be explained by arguing that, with equiangularity, the polar plane is always vertical; this means that the BDL (the best drive line) and the two paths remain, for fixed δ and for all k, in the same directions. These phenomena are a feature of equiangularity, but different phenomena in the event of polyangularity lead to the same result, namely that the ratio of the two *a* cos α always works out to be exactly k (§6.40). The following are the relevant numerical results from WkEx#1.

		radius *a* ↓	angle α ↓	*a* cos α ↓
Drive side, wheel	2	81.80	04.83	81.51
Coast side,	2	49.08	04.83	48.90
Drive side,	3	51.82	19.31	48.90
Coast side	3	86.37	19.31	81.51

Let us go to the kinetostatics. If, in the absence of friction, **F** is the force acting at the point of contact and along the line of action, then **F** times the relevant *a* cos α is the torque at the shaft. Notice that the ratio of the two different *a* cos α in the table above is 0.6 exactly. Refer to the spreadsheet of numerical results (TNR following chapter 12) for numerical details of other worked examples with other k. In the absence of friction it simply must be so under all circumstances (whether the set be equiangular or otherwise) that the ratio of the torques at the input and output shafts is the inverse of that for the corresponding angular velocities. Otherwise the established principle in machinery of there being no lost power in the absence of friction will be vitiated, which is absurd.

18. *Some other phenomena.* While the common generators of each pair of base hyperboloids lie collinear in their respective travel paths, the relevant throat circles intersect with those paths at four different points. These points of intersection (see the two points SS at figure 3.08 for the matter in general and the four spheres at figure 5B.04 for the present example) are at the extremities of the relevant Z_l. The segments of length Z_l — see figures 3.08, 3.11(a) and 5B.04 — are those segments of the common tangents to the two throat cylinders that stretch between their points S of intersection with the throat circles. Let me mention here, and again (a) the single rod SS of figure 2.02, (b) the two intersecting rods of figure 5B.04 (§5A.11), and (c) the two non-intersecting rods at figure 8.05. Be reminded that in the present equiangular case the rods do intersect and the angles α at the extremities of each Z_l are equal; see §5A.23, item (k). Please remain aware also that in the

special case of planar gears Z_l is simply the length of the common tangent in the reference plane between the coplanar pair of the base circles. This planar Z_l is illustrated at figure 3.12(a) by the segment (T_2-T_3). I wish to mention, incidentally, while here at figure 5.04, the two ellipses drawn there with broken lines. They are coplanar with the polar plane which in turn contains the transversal JK; they intersect at [Qx]; their centres are upon the line of the transversal; they separately contain the points J and K at the extremities of the transversal; they separately contain two each of the four points S of the two RSSR (§2.12); and they are, in general, important. They are called by me the *ellipses of obliquity*. I will begin to deal with them soon, at §5B.40.

More on the lengths Z_l and related questions

19. Having now chosen without proper guidance in figure 5B.04 the two angles δ, and having constructed and examined there the four different lines of action whose lengths Z_l have depended upon our choices for the said angles, it might be wise, at this juncture, to look more carefully at this matter of Z_l in general. From previous work (§3.24, §3.30, §3.55) we know already that, at all stages of any meshing, the two principal radii of curvature of the mating profiles at a moving Q, H_2 and H_3, are by definition collinear and add continuously to the constant length Z_l of the prevailing line of action. So the greater a Z_l the better the chance that the mating teeth (which are yet to be designed) will turn out to enjoy benign curvatures prevailing. *Thus we should say, perhaps, that the longest Z_l should predicate the drive, and the shortest Z_l the coast; but is this the only criterion?* I wish to report in any event that, in figure 5B.04, the two lengths Z_l are measured to be as follows: Z_l(long) = 57.21 mm, and Z_l(short) = 38.14 mm. *There follows next a short exercise for the reader. The answer is given.* Construct by computer graphics in figure 5B.04 the two identical lengths of the symmetrically arranged lines of action in the special case that occurs when R becomes zero; construct, in other words, the two lengths Z_l in the special case when the set becomes in the limit as $R \to 0$ an ordinary pair of cylindrical helicals with crossed axes (§5B.60). Find that the sought value of the two Z_l is only 32.48 mm. This would appear to suggest that, as R is made larger, both of the lengths Z_l become larger, and the problems in connection with curvature and lubrication become easier; but let us not forget that, as R becomes larger (other parameters remaining equal), the wheels themselves become larger.

20. With [Qx] open to choice on the E-line, and the polar plane at the said [Qx] lying always normal to the helitangent which remains always horizontal (§5A.24, §5A.29), the best way to see the true lengths Z_l of all possible ideal zones of action (given equiangularity, a fixed δ, and a variable R) is to project orthogonally normal to the implied array of parallel polar planes namely in the direction not of the E-line but of the FAXL (§1.43). Refer to figure 5B.05 where the view is the same (but turned through 90° in the plane of the paper) as the view at the bottom RHS of figure 5B.03. The graphics of

§5B.20   GENERAL SPATIAL INVOLUTE GEARING

figure 5B.04 (but drawn for Z_{ID} only) has been repeated here not only for R = +140 (mm), but also for a series of other values of R. The series ranges from +280 to –280 in eight steps of 70. The shorter Z_I of figure 5B.04, which was called there Z_{IC}, has not been thus re-plotted here, but the results for these would be (as they are for those of Z_{ID}) line-symmetric about the CDL. The Cartesian plot at inset, however, shows that the minima of the two Z_I do not occur together at $R = 0$. While the minimum Z_{ID} occurs on one side of the CDL, the minimum Z_{IC} occurs at its mirror-image location on the other. Unless R is small, it may accordingly be said that the larger the R (whether positive or negative) the larger the Z_I and that, as $R \to \infty$ in either direction, the two Z_I do also. These phenomena do not speak well for the crossed

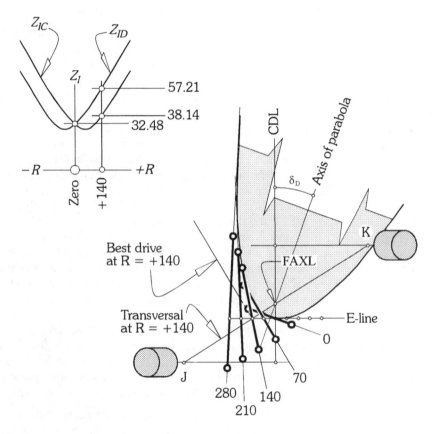

Figure 5B.05 (§5B.20). Variation in the location of the ideal zone of action (the line SS of length Z_I) as the offset R is allowed to vary, other things are remaining equal. We show an orthogonal view along the FAXL which exposes, strangely enough, what appears to be a parabolic cylinder to which all of the lines of action (each of them parallel with the plane of the paper) are tangential. The Cartesian plot at inset, relating to two symmetrically disposed minimum values of Z_I, is explained in the text nearby.

194

helicals where Z_l is short, and they do confirm the generally held belief that the curvatures there are somewhat too severe for adequate lubrication. But, as R increases (for a given C), the sizes of the wheels themselves increase; the improved curvatures come with this penalty. So what can we say incisively in support of the offset design? What we can say, as we shall see, is that, as R increases, it is not only the improving curvatures of the mating surfaces and the improving capacities of the geometrically determined Z_l to accommodate the practical Z_A that become important, *but also the growing angles (yet to be studied) at which the rubbing velocities at the various Q can be caused to cross the rulings.* These angles — I call them χ — are clearly of great importance in the areas of wear and lubrication (§5B.25).

21. Following my remarks about Olivier at §3.66, incidentally, see in figure 5B.05 that there is no restriction upon a line of action becoming parallel with the CDL. It must also be said that the picture here was made exploratorily, and that it exposed by accident the parabolic shape that may be seen. The author thus saw in retrospect that, because the series of equilateral transversals form a parabolic hyperboloid, not only the best drives at the points E, but also the unbounded lines of action (for constant δ), do also. We are looking here axially along the parabolic throat cylinder of the parabolic hyperboloid illustrated in general at figure 4.10. The said parabolic throat cylinder (for drive) is tilted away from the CDL at angle δ_D. A similar throat cylinder (for coast) is tilted away in the opposite direction at angle δ_C. The whole geometry of the various possibilities for the actual bounded lengths of the ideal zones of action Z_l is, in general, a much more complicated matter, for the angles δ_D and δ_C need not be equal (see in advance the *ellipses of obliquity* at §5B.40), and there is, of course, the matter of polyangularity yet to be examined.

22. But what we have found here brings into focus the clear question of which line of action might be best for drive and which for coast, and the less clear question of whether we have a satisfactory line of action at all. First we can see from the Cartesian R-v-Z_l plot at inset in figure 5B.05 that, due the obvious line-symmetry about the CDL, what we might choose to see as the drive line on one side of the CDL will be seen as the coast line on the other. But that answer begs the question. The real answer will turn, as we shall see, not upon the mere length of Z_l, but upon whether or not the intercept (S–S) of length Z_l can straddle the point [Qx] which is at E. By using the italicized result at the end of §5A.24, we will be able to show that the intercept Z_l along any one of the legitimate travel paths in any one of the said ∞^2 of parallel polar planes will straddle the relevant E; refer to §3.56, §4.18 and §5A.23 item (n). This will mean that, anywhere within the realm of equiangularity, [Qx] will always find itself located safely within the end-limits of the ideal zone of action; and this in turn will mean that both profiles of the mating teeth will always find themselves to be convex. *I am hinting here that in the wider realm of polyangular architecture we might not always be so lucky.*

§5B.23 GENERAL SPATIAL INVOLUTE GEARING

On the angles α and the shapes of the slip tracks

23. Refer to figure 5B.06. This is a composite figure dealing with various matters. It is primarily drawn however to show important aspects of the slip tracks soon to be mentioned. It shows also, not only the wheels of the current exercise with $R = 140$, but also the wheels of the corresponding crossed helicals where, with [Qx] at D (§5A.20), R is zero. Taken along the CDL, the view is orthogonal. It is the same view as the one at top-left in figure 5B.03, except that there the direction of projection is opposite and the view is angled in such a way that the common generator of the doubly departed sheets of the naked wheels is aligned with the edge of the paper. Two sets of four coaxial hyperboloids were mentioned at §4.76 and sketched at figure 4.09. Somewhat similar sets of six are shown here, not in general as they were there, but in particular, having been constructed with the numerical data of the present WkEx#1. Each set of six is coaxial with one of the two shaft axes. Each hyperboloid has been constructed by computer graphics, by sweeping the relevant generator in such a way as to leave such multiple copies of that line that the horizonal (*sic*) edges of the surfaces have been rendered visible. The unwanted multiplicity of lines on the screen has been

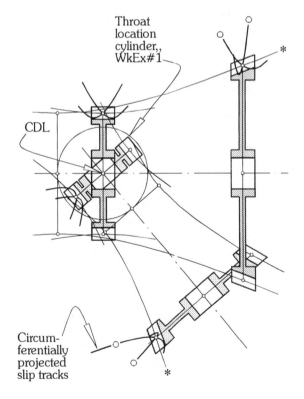

Figure 5B.06 (§5B.23). Ortogonal view along the CDL showing all three pairs of the hyperboloids intersecting. They intersect with one another (three-by-three) at the two E-circles. They are (a) the one pair of departed sheets illustrated with the symbol (*), and (b) the two pairs of base hyperboloids illustrated with the symbol (°). Notice the shapes of the slip tracks; they are projected circumferentially onto a radial plane in the plane of the paper. See also the crossed helicals for the same CDL superimposed.

196

removed. The shaft axes and thus the axes of the hyperboloids in figure 5B.06 are parallel with the plane of the paper and the horizonal edges of the hyperboloids take on the appearance of being their own planar sections which are of course planar hyperbolae. See the two doubly departed sheets for the naked wheels of the current exercise marked with the label star (*). Their throats of radius t, 50 mm and 30 mm, are equally displaced from the CDL (§4.37, §5B.05, §5B.11). Notice also the two singly departed sheets of the corresponding crossed helicals which are not marked; for these E is at D, R is zero, the radius u of the throat location circle is zero, and the two throat radii t are as before. Looking next at the horizonal edges of the base hyperboloids of the current example marked with the label ring (°), we can see that for each wheel there are three coaxial hyperboloids mutually intersecting at the periphery of the relevant E-circle; we see their horizontal shapes and thus their diametral sections clearly. They are, referring to §4.74-76, (a) the doubly departed sheet of kind #3 marked here (*), namely the cone-hugging naked wheel obtained by swinging the helitangent about the shaft axis, (b) the base hyperboloid (°) for the drive-sides of the teeth of kind #2 obtained by swinging the path of the relevant Q namely line Z_{ID} about the shaft axis, and (c) a similar hyperboloid (°) for the coast-sides of the teeth, also of kind #2, obtained by swinging the path of the relevant Q namely the line Z_{IC} similarly. The vertices of the hyperbolae (°), which are at the throat circles of the actual base hyperboloids, are symmetrically arranged on either side of the shaft axes as we might expect, but at each wheel the two planes of those circular throats and the single plane of the relevant E-circle do not coincide. Now we know from §3.19 that the actual base hyperboloids cut the actual flanks of the relevant teeth (namely the involute helicoids of §3.19) in the slip tracks, and we know that these latter are not flat but twisted curves. So the curves we are seeing in figure 5B.06 are not the actual shapes of the slip tracks (which are in reality twisted curves upon the convex ruled surfaces of the flanks of the teeth) but projected versions of them, projected circumferentially onto a radial plane containing the shaft axis. It is useful to study these projections nevertheless. The slip tracks appear projected as planar hyperbolae in figure 5B.06, and these planar figures are arranged symmetrically about the shaft axes in the orthogonal view which is taken along the CDL. The slip tracks accordingly appear not only to dip into the trough between the teeth, but to make a U-turn, and then to come out again; and this requires some comment now. Refer to the figures 3.13(a)(b). We see there that the U-turn occurs when the slip track (in space) crosses the throat circle of the relevant base hyperboloid. In circumferential projection this appears as the U-turn (in the plane) at the vertex of the hyperbola. It is precisely this location that should be avoided because, as soon as the angle ζ changes sign from positive to negative at value zero, the slip track from there on becomes unreal; it retreats away from the real material at the flank of the tooth (§3.59-64). We should always aim to employ the straighter parts only of the relevant branch of the slip track, avoiding any part of it too close to the U-turn. We must not let ζ go to zero in other words. As the relevant part of the slip track 'retreats away' at the U-turn, contact between the real surfaces (if they there both exist) of the flanks of the mating teeth is lost. The actual kinematic geometry of this might bear investigation; see figure 3.08; what exactly happens when the point of contact Q is taken to one or other of the end-points S?

Danger-locations for excessive wear

24. Having seen these things, we may look more closely now at some of the other phenomena. When R is zero, the gears are ordinary crossed helicals. The separate shapes of the two slip tracks, one on each of the opposite flanks of a single tooth, are then (a) symmetrical about a plane normal to the shaft axis, and (b) axially reversed mirror images of one another. But when R is non-zero the slip tracks on the opposite flanks of a tooth are markedly different, not only collectively from those of the crossed helicals (due to smaller α), but separately from one another. These phenomena are clear to see at figure 5B.06. The differing shapes of the slip tracks on the opposite flanks of a single tooth when the wheels are offset relates no doubt to the question of which flank is for drive and which for coast, but, having skirted around this matter already at §5B.19, I choose again not to deal with it. What I wish to draw attention to is this. *At any moving point of contact Q, there are two separate distinctions that need to be drawn: (a) the distinction between the direction of the rubbing velocity along the helitangent at that Q and the directions of the intersecting slip tracks there, there being one slip track on each of the contacting profiles of the mating teeth; and (b) the distinction between the direction of the rubbing velocity there and the directions of the intersecting rulings there, there being one of the said rulings on each of the contacting profiles of the mating teeth.* It is important to understand the implications of this. It means that, within the plane of adjacency and at any Q there are five separated lines intersecting. They are (a) the helitangent along which the rubbing velocity is occurring at the instant, (b) the tangents to the two slip tracks intersecting at Q, and (c) the two intersecting rulings of the touching helicoidal surfaces of the mating profiles of the teeth. *Just here I wish to define the angle χ as that angle existing between the helitangent and a ruling.* Now it might be expected that, whenever any one of the above five lines becomes near to being collinear with any other of them, either some kind of harmless special case is about to become extant, or some kind of dangerous practical trouble is incipient. If a slip track becomes near to being tangential with one of the rulings, that is if angle ζ is near to becoming zero, we are surely in trouble with looming unreality and undercut; but such an occurrence only occurs at the gorge of a base hyperboloid, only at one end or the other of a line of action Z_l in other words, and we should (by setting limits to our Z_A) be able to cope with that (§3.16, §3.22, §3.58-64, §4.49). If a slip track is found to be near to being tangential with its mating slip track, or if any pair of intersecting rulings become close to being collinear, it simply means that the gear set is close to being planar, which is special but no catastrophe. If the rubbing velocity is near to becoming tangential with either

one of the slip tracks, it simply means that the ratio k is either very large or very small, which, when understood, provokes no anxiety (§7.07). *If on the other hand the rubbing velocity becomes anywhere near to being parallel with either one of the rulings (in which case the angle χ becomes small), we are in deep trouble; and the trouble will be doubly deep if the said ruling and its intersecting mate are near to being collinear.* This last remark is made for those already worried about the likelihood of unsatisfactory lubrication, reduced efficiency and severe wear in the ordinary crossed helicals. *The mere occurrence of point contact (which is innate) is not the issue. The issue is deeper than that. It relates to the recommendation implicit here, that χ should never be small* (§5B.37).

Adverse circumstances at the crossed helicals

25. Having thus clarified some of the above matters, being aware of the received wisdom that the ordinary crossed helicals are in many respects a recipe for disaster, and working still within the ongoing WkEx#1 that began at §5B.09, we can now ponder further this question of what's amiss with the crossed helicals. A first obvious disadvantage of the ordinary crossed helicals is that, because the two Z_I (although continuously changing) are always relatively short (§5B.19), the two principal radii of curvature H_2 and H_3 (which add to Z_I) remain correspondingly small (§3.30). This means that the working flanks of the teeth are, relatively speaking, severely convex. This is surely not good for lubrication. A second disadvantage of the crossed helicals relates to the angle ζ. This is the angle between the direction of the slip track and the relevant ruling upon the flank of a tooth (§3.22, §5B.24). Refer to the sketched Cartesian plot at inset at figure 3.07 and see there that, as $\Omega \to 0$, that is as we go along a line of action towards the throat circle of the relevant base hyperboloid, or, in other words, as we go further towards one end or the other of an intercept Z_I, the angle ζ goes more and more rapidly towards zero. Now the crossed helicals clearly need to carry a real Q further towards the inside ends of the ideal zones of action than the corresponding offset arrangements need to do. This is because, as R diminishes, the length Z_A (§3.58) needs to extend itself further in this tricky direction. Accordingly we do run into the trouble of having to deal with a small ζ. In this connection it might be said with accuracy that, for a given contact ratio, Z_A (which is the NP times the contact ratio) will need to vary directly with the NP, which is $(2\pi\ a \cos \alpha)/N$, and thus directly with $a \cos \alpha$, but both a and α vary with R, so this needs to be examined. A third disadvantage of the ordinary crossed helicals almost certainly relates to the angle χ (§5B.24, §5B.37). This is the angle between the direction of the rubbing velocity at a point of contact and a ruling there. In the case of the crossed helicals the radius of the throat location circle is zero (§5B.11). But so what? *It becomes evident, due to geometry unique among the crossed helicals, especially among those where the tooth ratio k is near unity, that not only will (a) the diameters of the core helices of the mating involute helicoids be close to being equal to those of the key*

circles, causing the mating rulings to be almost parallel in the neighborhood of [Qx], but also (b) the direction of the rubbing velocity in the same neighborhood will be, substantially, in the same direction as the rulings. It is not the mere matter of single point contact that is the cause of the trouble here. It is the fact that, near the roots of the teeth especially, the sliding velocity continuously occurs in substantially the same direction as the rulings upon the flanks of the teeth. The reader may go in advance to figure 5B.13(a) where a satisfactory angle χ at [Qx] is shown, to §5B.37 where a graphical construction for χ is explained, and to the same paragraph where it is shown that, as R goes to zero, χ goes to be small.

On the sets of wheels combined at figure 5B.06

26. Looking more deeply at figure 5B.06, which was drawn for the main purpose of exposing how the slip tracks of a set might vary as R varies, we can see (if we wish) two sets of wheels in mesh simultaneously. The sets may be seen as being both fixed upon the same two shafts and running, with their respective pairs of coaxial wheels thus fixed to one another, in unison. Provided the teeth, which are, as yet, not drawn, were properly synchronized with one another, the imagined mechanism would be capable of continuous motion but, unlike the case where only one or other of the sets were present, *the assembly would be overconstrained.* Gone would be the possibility of 'lane changing' (or 'skating') across the slip tracks as a flexible frame was flexing (§1.01, §1.07), and gone would be the allowable inaccuracies of a rigid set upon assembly. The interesting question that next arises is the following: what kind of gear set with continuous line contact between teeth might result from the combination over a finite distance of an infinite series of different sets fixed infinitesimally close to one another? Finding an answer to this question (which is a false question in my view) has been a self-set problem for many writers, all of whom in seeking a solution have failed [40] [18] [77].

Beginning of part 2 of worked example #1

27. *Choosing the numbers N of the teeth.* Risking that the circumferential widths of the teeth (for strength) on the one hand, and the contact ratio (for general continuity of action and load sharing) on the other, will be mutually satisfactory, let us preliminarily fix the tooth numbers on the wheels at 24 and 40, the ratio 24/40 giving $k = 0.6$. These give the radial angles traversed in stepping from tooth to tooth as follows: 360/24 namely 15° for the smaller wheel, and 364/40 namely 9° for the larger. Knowing next the E-radii (look for these at the TNR), the circular pitch, namely

$$CP = (2\pi e)/N$$

works out to be, for each one of the two wheels, CP = 14.023099 mm. More importantly, however, the *normal* pitch, the NP, needs to be looked at now. *I hereby define the NP as that distance measured along a travel path and thus* normal to the relevant profiles which extends from any one profile to the next identical profile. Going back to §3.13, it will be seen that this may be written

$$NP = (2\pi a \cos \alpha)/N.$$

In the current exercise this works out to be, for all four profiles, NP = 12.802945 mm. Would there be different sets of these NP if the angles δ had been chosen to be, not the same (as they are here), but different? Yes, there would be different sets; see for example WkEx#2 beginning at §5B.45.

28. *Going to [Qx]*. Let us go now to [Qx], which, given R which is 140 mm measured from O to Y upon the FAXL, is at E where E, measured from D to E along the E-line, is 140.948 mm (§5B.07). I now take this fixed point, *this junction station at the crossing of the now established paths*, this [Qx], as a convenient stepping-off point for the forthcoming drawing work. I say drawing work, because we are, just here, about to find out how to construct the shapes of the teeth by drawing, *not to find out how to cut the teeth by machine tool*. The actual machine-cutting of the teeth is a question that needs to be looked at later; and that question will depend upon the discovered shapes of the teeth, which are, as yet, unknown.

29. *Waiting at [Qx]*. We are waiting, just now, at [Qx] but, independently of where we might be waiting, the whole unbounded shapes of the mating profiles of the involute teeth, although not discovered as yet, are already determined now. For each pair of mating profiles (given the need for both drive and coast) there is a pair of involute helicoids (infinitely long and infinitely wide diametrically) capable of driving one another continuously in the manner of figure 3.08 (§3.56). It is important to remember however that, whereas the involute helicoids of figure 3.08 belong at the upper end of the spectrum illustrated at figure 3.06 where the core helix is axially compressed, the involute helicoids of our teeth here at WkEx#1 are at the lower end of the same spectrum where the core helix is elongated. Whereas the two angles α arbitrarily chosen and used for geometric demonstration at figure 3.08 were 60° exactly and 50.48° approximately, the corresponding angles in our working example here (WkEx#1), which hopefully relates to simple practical teeth with no difficult problems, are only 4.83° and 19.31° approximately. See these already listed in the table at §5B.17.

30. *Generating the profiles*. Refer to the twin figures 5B.07(a)(b), which are different panoramic views of the same scene. Refer at the same time retrospectively to figure 3.03 and in advance to figure 5B.09. [Numerical notes: the twin figures 5B.07 are mid-spectrum at figure 3.06; drawn for $\alpha = 40°$, they relate to no specific flank quantitatively; figure 5B.09, on the other hand, is drawn precisely with the data from WkEx#1, where the two values of α are much smaller, $\alpha_{2D} = 4.83181°$ and $\alpha_{2C} = 19.31299°$; in figure 5B.09, accordingly, the pitches p_{FLANK} of the involute helicoids are rather large, –967.6486 mm/rad and –246.4442 mm/rad respectively; these latter are acceptable and manageable to scale in the CAD, but too large for easy illustration on the printed page.] To generate the four tooth profiles, we need, for

§5B.30   GENERAL SPATIAL INVOLUTE GEARING

each of them, (a) to locate the relevant base hyperboloid, (b) to locate the relevant core helix, (c) to construct the relevant involute helicoid by drawing the tangents to the said helix, (d) to locate the relevant slip track by tracing the intersection of the thus-found involute helicoid with the said base hyperboloid, and (d) to truncate the found profile so that in some way it begins to resemble the profile of a real tooth. In this and in subsequent worked examples we first recall §3.19 and construct as a basis for the drawing work the essentials of figure 5B.04. *We must in other words locate [Qx], the polar plane at [Qx], and the two paths for the two Q.* For illustrative convenience, the twin figures 5B.07(a)(b) are drawn for a general value α [$\alpha = 40°$]; this is much larger than either of the values inherent in WkEx#1; see figure 5B.09 for a construction taken directly from the 40-tooth wheel at WkEx#1. Starting

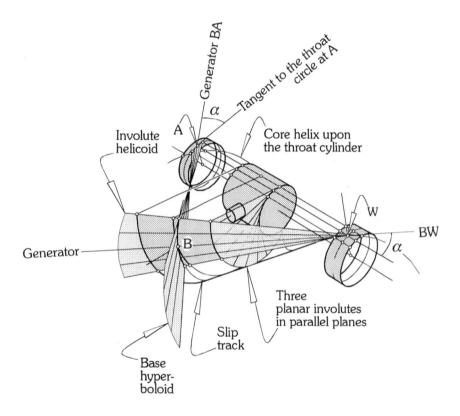

Figure 5B.07(a) (§5B.30). In this figure the points [Qx] and E (which are not shown) are coincident at B (which is shown); and the A and the B here are the A and the B of figure 3.03. The figure is illustrating the first stepping out (in the process of synthesis) from the starting point [Qx] (at B) to locate (a) the involute helicoid of one representative profile, (b) the slip track upon that profile, and thus, ultimately, (c) one of the two multiplicities of identical profiles illustrated at figure 5B.09.

SYNTHESISING AN EQUIANGULAR SET

from B in the figures 5B.07 (§3.26), and swinging in both directions away from [Qx], we rotate the chosen path, fixed at its constant shortest distance a, and in angular steps of say one tenth of the angle subtended by the CP (§5B.27), through an adequate range of Ω around the chosen shaft axis. We thus sweep out with its generator AB the relevant base hyperboloid. Using §3.19 or otherwise we may next locate the rulings upon the tooth profile. Noting A, B and W at 3.03, we see that all rulings BW must be perpendicular to the lines AB and tangential with the relevant helix at the throat cylinder. We may begin construction by locating the right angled triangle ABW with B at the key circle. See the importance of this triangle at figure 5.08. The right angle is at B; so erect the plane normal to AB at B and find the point where the line AW (parallel with the shaft axis) cuts this plane at W. Alternatively, by using the algebra at §3.17, we may go A→W, then go W→B, not forgetting the right angle at B. In any event repetition of these processes for successive

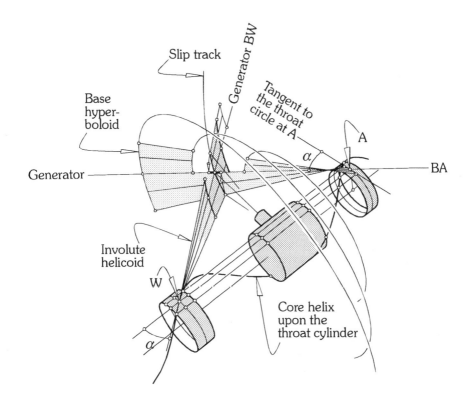

Figure 5B.07(b) (§5B.30). This figure is simply figure 5B.07(a) redrawn. For the sake of clarifying certain features, it looks from a different direction. For the sake of clarity also, B has no label here, but the slip track has been more clearly delineated. Notice the core helix with its same angle α. Numerically speaking, α is drawn at 40° in this pair of figures.

points B will construct the shape of the surface at the chosen flank and the slip track at the same time. The flank may be truncated at the side faces of a suitable cylindrical blank; see this again at figure 5B.07(b) and, of course, at the proper scale for WkEx#1, at figure 5B.09.

Unwrapping the roll of paper

31. This paragraph is of general importance. Although here interrupting again the flow of WkEx#1, it applies to all involute gears of whatever kind. It is to my knowledge new to gear theory. It helps to explain failed previous attempts to make sense of the impossible idea that a plane sheet of paper might be unrolled from an hyperboloidal bobbin [40] [18] [77]. Refer to figure 5B.08. [Here α is set to be 20°; there is no special significance in this; the figures 3.03, 5B.07(a)(b) and 5.08 have been drawn with α set to be 60°, 40° and 20° respectively.] The figures 5B.07(a)(b) naturally remind us of the well known argument (which is not wrong) involving the unrolling of paper from the base cylinders of the cylindrical spur and helical gears of current involute practice. The end-edge (either square-cut for spur gears or oblique-cut for the cylindrical helicals) of a roll of paper whose diameter is that of the base cylinder is unwound to generate with its relevant straight-line edge the helicoidal shape of the profiles of the teeth which in the special case of planar teeth are helicoids of infinite pitch where the rulings are parallel [92]. Separately I discuss the question of the so called base cones of so called straight bevels of ordinary engineering practice (which are not involute incidentally) at §5B.62; that discussion, involving the idea of a conical bobbin (which is at least plausible), will be of parallel interest here. What needs to be seen in general, however, is that the relevant bobbin must always be cylindrical; it must always be the core cylinder of the base hyperboloid (§3.10); *the requisite bobbin is not the hyperboloid itself.* Refer to figure 5B.08. An obliquely set square sheet, rolled up upon the core cylinder, and with its square corner protruding to match the triangle ABW of figure 3.03, is being wound both upwards and downwards from the reference position where B is at the known [Qx] (§3.20). The paper generates (a) with its end-edge BW the involute helicoid of the flank, (b) with its point B the particular planar involute curve on the surface of the helicoid which contains [Qx], and (c), with a steadily moving point in its end-edge BW, the unique slip track for the given [Qx]. We can by simple inversion, moreover, generate the path of Q in 1 which is of course the known straight line at [Qx] normal to the surface of the helicoid at [Qx]. The moving point in the end-edge moves from B in the direction towards W such that the incremental distance $[dJ] = (a \sin \alpha) [d\Omega]$, thus generating the slip track (§3.17). The algebraic symbolism comes directly from the figures 3.02 and 3.03, which is as might be expected, and the coefficient ($a \sin \alpha$) is a constant for the flank of course. This formula naturally gives further insight into the nature of the of slip track, which can be used, quite obviously, to advantage in design. The cases of the planar spur wheels

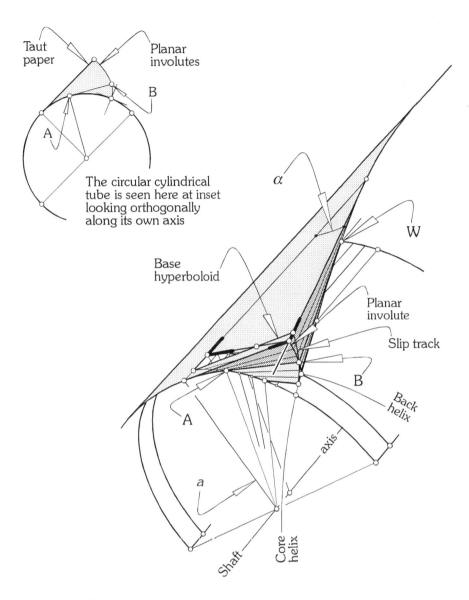

Figure 5B.08 (§5B.31). The unwrapping of the rolled-up oblique-wound square-cornered paper. This figure helps to make it clear that, while (a) the point at the front corner of the paper and all points along its leading edge (namely that edge adjacent to the angle α shown) describe parallel planar involutes emerging from their identical parallel base circles of radii a as shown, and while (b) the leading edge itself of the paper sweeps out the involute helicoid of the flank of a tooth, (c) a steadily moving point on the leading edge of the paper sweeps out a slip track. See figure 3.03 for the simple formula for the speed of the steadily moving point: $[dJ] = (a \sin \alpha) [d\Omega]$ where, because both a and α are fixed variables for the flank, $(a \sin \alpha)$ is a constant. See the circularity of the cylindrical roller and the taut paper at inset.

§5B.32   GENERAL SPATIAL INVOLUTE GEARING

and the parallel cylindrical helicals with their straight-line contact and the simple crossed helicals with their single-point contact fall out here as simple special cases.

WkEx#1(2) continued

32. *Achieving tight mesh at the mating profiles.* Next we need to transfer the discovered shape of one of the profiles, which is of course in the form of a series of ghostly rulings as yet only roughly bounded (and with no definable centre despite our talk of [Qx]), to all teeth of the relevant wheel. Refer to figure 5B.09. Selecting at first the relevant travel path, and knowing the NP, 12.802945 mm (§5B.27), we can, starting from [Qx], now step off successively along the travel path a few of the NP, each divided into steps of say ten. In all involute gearing, of course, whether equiangular or polyangular, this NP must, for a given path, be the same for both wheels. And the CP must correspond (§5B.27). Here are some handy hints: when constructing such profiles of the teeth against the background of an overall assembly drawing, keep the actual region of the meshing uncluttered with too many

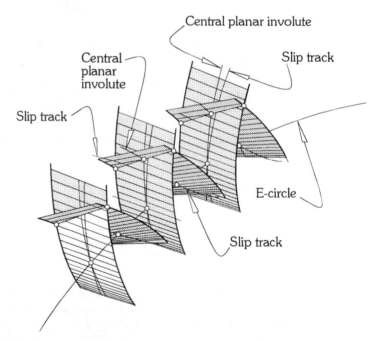

Figure 5B.09 (§5B.32). From WkEx#1, and drawn in wire-frame only, a multiplicity of both profiles of some of the teeth on the larger wheel (40 teeth) are shown here in correct relationship with one another; notice the equal spacings of the points [Qx] around the periphery of the E-circle. See also the sharp-nosed, untruncated teeth, with the twisted curve of the intersection of their involute helicoidal profiles at their crests.

SYNTHESISING AN EQUIANGULAR SET §5B.33

lines; one can do this by cloning the profiles in small groups elsewhere upon the circumferences of the wheels, swinging them into mesh for observation only whenever necessary. Notice also in the figures presented here that the profiles (the constructed involute helicoids) have been truncated for convenience by a suitable pair of parallel planes that might be imagined to be the parallel physical faces of real wheels. These planes have been located here equidistant axially, at 10 mm on either side of the planes of the E-circles, which planes are of course known. Having thus dealt with one of the sets of profiles of the teeth, say the fronts on both of the wheels, we can use the CP (§5B.27) — or, better, the angles subtended by the CP, namely 9° and 15° in this case — to deal with the other sets of profiles of the teeth namely the backs on both of the wheels. The shapes of the backs will be different from the shapes of the fronts, the travel path at its different distance a and its different angle α will be different, and the length Z_f will also be different for the backs. Don't forget to displace the sets of profiles circumferentially with respect to one another by half the CP, and check that the meshing is correct. This latter can be done by making the correct relative rotations of the completed profiles step by step (in tenths of the CP), while seeing at the same time that the point of contact is proceeding step by step along its proper straight-line path. The teeth of course are in tight mesh, and the planes of adjacency (both extant) must remain normal to their respective, but different paths of the two Q. The important kinematical aspects of the work are now completed. Refer to figure 5B.04 for general confirmation of the architecture.

33. *Truncating the teeth*. Revisit figure 5B.09. We need now to truncate radially the constructed profiles of the teeth to avoid collisions among the outer boundaries of the real material of the real teeth while preserving a contact ratio greater than unity. This naturally introduces the next question of how to shape the wheel blanks. The wheel blanks are the actual pieces of real material from which the real toothed wheels are to be cut. Some general remarks about the overall geometry of the radial truncations that may need to be made appear at §5B.54 *et seq.*; and suitable truncations to the widths of the teeth — the widths remain here set at 20 mm, 10 mm on either side of the planes of the E-circles — are discussed in greater detail later; see in advance figures 10.15 and 10.16, described and discussed at §10.66 *et seq.*

On the results of this synthesis

34. Refer to figure 5B.10(a). This is a general perspective view of the two finished wheels of the current exercise. See also figure 5B.10(b) which is a simplified orthogonal projection drawn to scale outlining the main dimensions. The numbers of the teeth are, as has been said, 24 and 40. Note the E-circles which are indicated in both of the figures, and remember that they intersect at the point [Qx], which is a fixed point in the fixed space. I draw attention to the four different shapes of the four tooth profiles; and (given color) I suspect that the brown and the yellow ones (which mate for coast) are

are somewhat unsatisfactory. See also the gashes that may be seen; these are the straight sided truncations (or slots) cut to give clearances at the troughs between the teeth; and I wonder whether the angles δ of obliquity might have been better chosen not to be equal. The radial truncations here have been crudely done simply to draw a rough compromise between the otherwise pointed heights of the top lands of the teeth and sufficiently long actual zones of contact Z_A within the ideal zones of contact Z_I (§4.57). See figures 5B.04 and 5B.11. See some measured numerical values (these are for WkEx#1 with lengths in mm) set out for comparison here:

[For drive, Z_I = 57.22, Z_A = 28.39, NP = 12.80, contact ratio = 2.2]
[For coast, Z_I = 38.14, Z_A = 16.78, NP = 12.80, contact ratio = 1.3].

See the slip tracks drawn at figure 5B.10(a). Recall (a) that these are twisted curves drawn upon the surfaces of the mating teeth, (b) that mating pairs of them are continuously intersecting at the moving points of contact Q, each such intersecting pair touching on either side its respective plane of adjacency there, (c) that each of the said planes remains always normal to the relevant

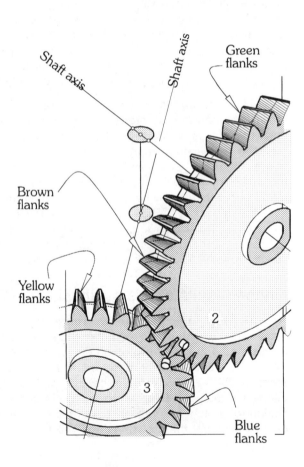

Figure 5B.10(a) (§ 5B.34). General perspective view of the meshing of the two synthesised wheels of WkEx#1. The numbers of teeth are 24 and 40. Notice the two shaft axes and the CDL. Other numerical details appear at Figure 5B.10(b).

travel path, and (d) that the slip tracks (like the teeth) differ in shape from one another. It is easy to plot at the same intervals (every tenth part of the NP) the changing locations of the slip tracks at their moving crossing point Q as Q travels its path, thus checking (a) that all is well with the construction, and (b) that, when a truncated edge of a tooth is met, the truncation there has not reduced too much the *contact ratio*. To obtain the contact ratio, we measure the length Z_A of the actual zone of contact along the travel path of Q, and divide this by the NP. In this example (WkEx#1), the truncations have been conservative, and the contacts ratio are not unreasonable.

The phantom rack

35. Refer to figure 5B.11. This is a close-up, perspective view taken directly along the helitangent. It is thus taken normal to the polar plane, looking directly at the fixed point [Qx]. The Q for drive is at [Qx]. The two travel paths (shown although hidden) are in the plane of the paper and we can see thereby the differing lengths of the two zones of contact Z_A. Notice that, because the line of sight is exactly along the helitangent (namely along the vector of the sliding velocity), and due to the severity of the perspective, we can clearly see the actual point of contact Q between the teeth. Please see the two cylindrical marker studs (5 mm × 5 mm) which are, believe it or not, set parallel with their respective shaft axes. The more comprehensive figure 5B.12, on the other hand, is an *orthogonal view* taken in the same direction of the same two wheels. In this view, looking from infinitely far away, the

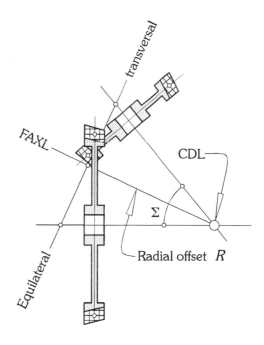

Figure 5B.10(b) (§5B.34). Plan view (along the CDL) of the data for, and the solution of, WkEx#1. The CDL is 80 mm, the shaft angle is 50 degrees, the speed ratio is 0.6, and the radial offset is 140 mm. The wheels appear to overlap, but be aware that the horizontal cross sections shown (which are normal to the CDL) are at different levels; they are indeed 80 mm apart. The actual meshing occurs in the space between. This is the most obvious way to draw, schematically, any skew gear set.

following important phenomenon is revealed. *Edge-on, the phantom rack may be seen*. This is a rigid, rectilinearly folded, ghostly surface (having no thickness) which, moving rectilinearly in the plane of the paper with uniform speed as the wheels steadily rotate, *mates with the teeth of both of the wheels* [88]. The view at figure 5B.12, incidentally, is exactly the same as that that is taken at the lower RHS of figure 5B.03. Because we are looking here (as we are there) perpendicularly at the CDL, the shafts appear in the view to be parallel. The phantom rack in the present case is equiangular in the sense that the sides of the flutes of the rack are sloped at the equal angles δ. The geometry of the phantom rack clearly relates with possible methods of manufacture, and this geometry is discussed more fully at chapter 10. See for another example figure 10.16(b), which relates to the next equiangular exercise (WkEx#2), where the angles of obliquity δ are *not* equal. Please note in passing that any sloping side of the rack is in fact the plane of adjacency

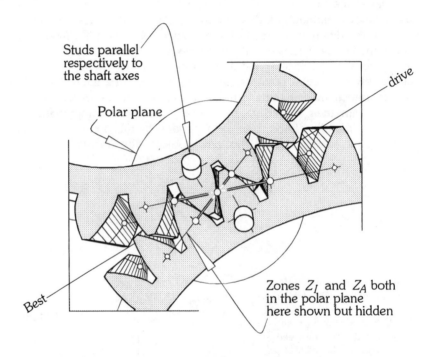

Figure 5B.11 (§5B.35). Looking inwards from the outside, along the helitangent and thus directly against the polar plane, this is a close-up perspective view of the meshing of the two wheels being synthesised here in this WkEx#1. The polar plane and the best best drive (which are of course coplanar) accordingly appear here in the plane of the paper. So also do the two lines of action with their respectively restricted zones of action, whose lengths are Z_I (ideal) and Z_A (actual). Althoughh hidden within the material of the gear teeth, these latter are here shown as if they were visible. Due to the convexity of the mating profiles, one of the moving points of contact Q, which is, just now, exactly at [Qx], is visible however.

SYNTHESISING AN EQUIANGULAR SET §5B.35

when the said side is at [Qx], and that the NP of the rack is the NP of the teeth. At §5B.26 I have calculated the said NP to be 12.802945 mm; but, as will be seen at chapter 10, *the CP of the rack is not the same as the CP of the wheel* (§10.05). I should also mention here that, while the two angles at the sloping sides of the phantom rack will in all cases of skew involute gearing be equal respectively to the two angles of obliquity δ, the slopes γ at the flanks of the teeth that are cut, except in the special case of spur gears, are a distinctly different matter (§10.14).

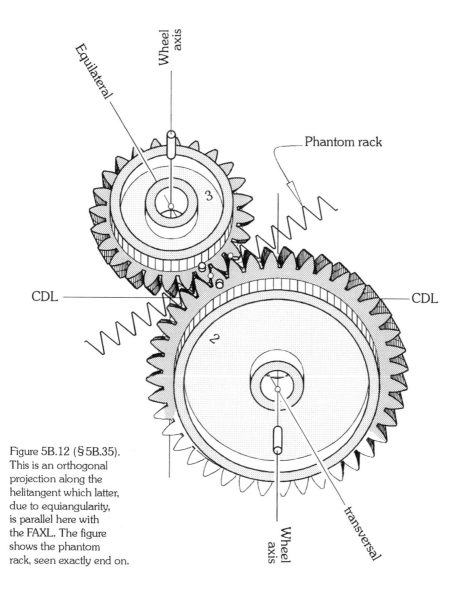

Figure 5B.12 (§5B.35). This is an orthogonal projection along the helitangent which latter, due to equiangularity, is parallel here with the FAXL. The figure shows the phantom rack, seen exactly end on.

The anatomy of a tooth

36. Go now to figure 5B.13 which shows in greater detail the shapes of the teeth of the bigger wheel. See first the shown hyperboloidal surface of the naked wheel. There are also the two base hyperboloids, only one branch of one of which is shown. All three of these intersect in the E-circle (see figure 5B.06), and the E-circle is clearly visible. The two base hyperboloids, by intersecting with the flanks of the teeth, define the slip tracks. The marked sets of planar involutes upon the flanks of the teeth are arranged helix-wise in parallel planes, all of which planes are normal to the axis of the wheel (§3.18). Notice, however, that the teeth appear to be tapered from heel to toe both radially and axially, and that, thereby, there is more to these teeth than meets the eye. Notice the bulbous nature of the flanks due to the non-parallel distribution of the rulings and the consequent shapes of the sharp edges at the crest of a truncated tooth; the top land of a tooth when truncated conical does not have straight edges; although for the sake of simplicity I often draw such edges straight, they are the twisted curves of intersection of the truncating cone and the flanks which are involute helicoids. Notice that the sets of opposite profiles of the teeth of this one wheel have been so located circumferentially that they equally split the circular pitch; the portions of the E-circle that straddle the gap between the teeth are equal in circumferential length to the portions that straddle the material body of the teeth. There are, as yet, the *helix angles* that need to be nominated, but these are beyond the scope of this chapter; go to §10.26. Read about the overall screw symmetry of a tooth profile at chapter 9A, and link this with the possibilities for milling or hobbing or other cutting methods discussed at chapter 10.

The very small angles χ at the crossed helicals

37. Look again at the angles χ. Go back to §5B.24 where the five lines within the plane of adjacency at a general Q were drawn in the imagination and where the angles χ were defined. Recall also §5B.25 where the chief disadvantage of the crossed helicals was alluded to. Look again at figure 5B.13 and notice the angle χ indicated there. This is the χ at [Qx] for that flank of that tooth, and by direct measurement its magnitude is 24.56°. This I have accepted as being satisfactory. But this or any other angle χ could have been earlier constructed and checked for size as follows: knowing the pitch line and its pitch, erect the helitangent at [Qx] by using the basic equation of the screw (§3.53) or by knowing that, with equiangularity prevailing, the helitangent at [Qx] is always parallel with the FAXL; (b) swing the relevant path of Q around the relevant shaft axis and determine by this means or otherwise the relevant *a* and α; (c) erect around the said shaft axis the core cylinder of radius *a*, and draw from [Qx] a line which is tangent (on the correct side) to that cylinder; (d) slide the line upon the one generator of the cylinder thus discovered until the angle α is correct; (e) note the angle χ between the helitangent and the ruling thus found, and measure it. Now in the particular case

SYNTHESISING AN EQUIANGULAR SET §5B.37

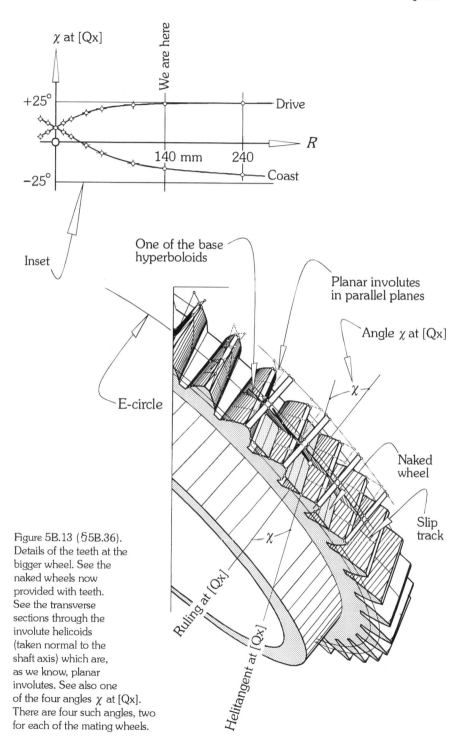

Figure 5B.13 (§5B.36). Details of the teeth at the bigger wheel. See the naked wheels now provided with teeth. See the transverse sections through the involute helicoids (taken normal to the shaft axis) which are, as we know, planar involutes. See also one of the four angles χ at [Qx]. There are four such angles, two for each of the mating wheels.

of this WkEx#1 the author has (while keeping the other parameters fixed) repeated this construction for a series of values of R. At inset in figure 5B.13 the values range from zero (which results of course in equiangular crossed helicals) upwards and step by step to the current value of 140 mm, then onwards to 240 mm (which latter is 3 times the centre distance C). I did the construction for both sides of a representative tooth of the bigger wheel, and the results are displayed by means of the plotted curves. It should not be impossible incidentally to write appropriate algebraic expressions for what might now be called χ_D and χ_C at [Qx]; but, just now, I wish to leave this as an exercise for the reader. The curves show the following: (a) that the nature of the symmetry is such that the zero values of the two χ at [Qx] occur, not at the origin where R is zero, but at equal values of $|R|$ on either side of it; compare the similarly occurring minima for Z_t in the curves at figure 5B.05 (§5B.20); (b) that when $|R|$ is small, say less than roughly $C/2$, we are in serious trouble with values of χ hovering near zero; (c) that a sure-fire solution to this difficulty is to introduce some amount of offset, an amount greater, say, than roughly C (and I say this despite the fact that, as Q moves away from [Qx], values of χ vary somewhat); (d) that, while χ_D at [Qx] rapidly approaches its limit at 25° as R increases, χ_C at [Qx] approaches the same limit similarly but more slowly; (e) that the said limit is indeed $\Sigma/2$, as may be shown; and thus (f) that our choice of numerical values at this particular example (WkEx#1) has justified itself; the resulting gear set is safely removed from the dangers associated with poor lubrication and wear inherent in the ordinary crossed helicals. *One might say indeed that this paragraph, standing alone, constitutes one clear justification for the employment of adequate offset. Future expansion of this uncovered material might lead to even deeper insights into (a) acceptable design for minimum wear and adequate lubrication in general spatial involute gearing, and (b) the unavoidable difficulties met in respect of both of these crucial matters in the narrow area close to the crossed helicals.*

On the plane of the E-radii and the radiplan view

38. In the context of equiangularity, the plane of the E-radii (the radiplane) was first mentioned at §5A.23, items (c) and (d), and again at the end of §5A.25, whence we went for details to appendix A at the end of that chapter (§5A.34). Having achieved the beginnings of a kinematic synthesis here in at least one sample equiangular case, it is time now to look again at this particular plane. It will be found to be useful in many ways (§5B.38-44). Refer now to figure 5B.14 and refer back to figure 5A.02. Notice that the tips of the two E-radii meet at E, and that their two separated feet are residing at the points L and M which are, in turn, the centres of the E-circles upon the two shaft axes as shown. Refer next to figure 5A.03 which is a different view of the same contents. See there the plane of the triangle LEM, namely the plane of the E-radii, the radiplane. It is indicated by means of a key-shaped

cut-out, the annular portion of which is concentric with and surrounding the triad of planes at [Qx]. It will clearly be seen in this figure that the plane LEM of the E-radii (the radiplane) is not coplanar with the plane of the best facet. Given equiangularity (but not without it) the mentioned planes both contain the sliding velocity which occurs along their line of intersection namely the helitangent (§5A.23). Look next at the point O, which is the centre-point of the centre-distance, and note that it resides, not in the plane LEM of the E-radii (the radiplane) as we might at first rashly suppose, but in the plane JOK of the best facet; see §5A.23 item (b). *The plane LEM of the E-radii intersects the plane JOK of the best facet in the helitangent at E; accordingly*

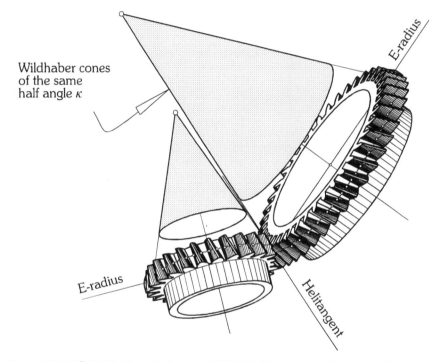

Figure 5B.14 (§5B.38). The radiplan view, WkEx#1. Here, equiangularity prevails, and the helitangent is coplanar with the plane of the E-radii; see §5A.23(e); otherwise it is not. In this figure a parallel view is taken orthogonally against the plane of the E-radii, so the two E-radii and the helitangent are together in the plane of the paper. Features of the view are (a) that the shaft axes and the planes of the E-circles are equally inclined to the plane of the paper so the major and minor axes of all ellipses in both wheels are equally proportioned, (b) that the cones of Wildhaber (which are truncated here to reveal all visible teeth) appear with their truly equal half-angles κ appearing also equal, and (c) that, while the Wildhaber cones (seen here separated) are each separately tangential along one of their generators with the plane of the intervening transnormal and thus not tangential with one another along a line, they are seen to be tangential at the one point which is [Qx]. That the half angles κ are equal here is not a feature of equiangularity in particular, but of involute gearing in general. It is only in hypoid gearing where we can find these angles unequal.

§5B.38 GENERAL SPATIAL INVOLUTE GEARING

the plane LEM is cut by the CDL at some point other than O. The actual point other than O is marked in the figure U. Refer to figure 5B.14. One useful aspect of the radiplane is that, by projecting drawings of the finished gears (or of the gear blanks) orthogonally upon this plane (thus employing the radiplan view), each wheel will appear with its diameter at [Qx] together with the other diameter at [Qx] in the plane of the paper. The shaft axes are equally inclined to the plane of the paper, so the wheels themselves and their cones of Wildhaber become directly comparable. Because the set here is an

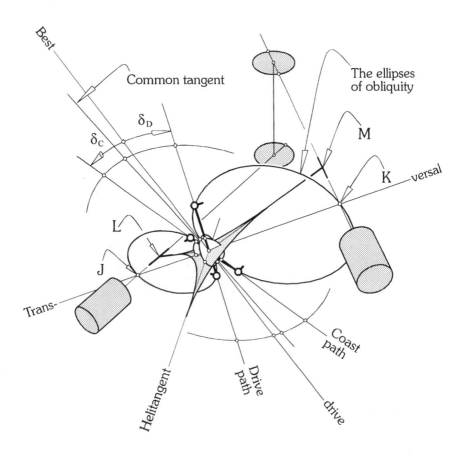

Figure 5B.15(a) (§5B.40). Panoramic view of the plane of the E-radii, WkEx#1. As already mentioned, and due to the equiangularity, the two E-radii and the helitangent are coplanar. The relevant plane, the so called radiplane, is represented here by the three-pointed star-like figure. It is important to understand that this plane (the radiplane) and the best facet are not the same plane. Nor are the two parallel. They intersect, indeed, in the helitangent. This figure's main purpose, however, is to illustrate for discussion the newly introduced, ellipses of obliquity. The said ellipses are are shown here, coplanar within the polar plane, intersecting with one another (indeed tangential because of the equiangularity) at [Qx], having their centres upon the transversal, and containg the points J and K.

equiangular set, the sliding velocity at [Qx] also appears in the plane of the paper; if the set were non-equiangular, this would not occur; see §5A.23 item (d). Remarkably, however, the shaft axes remain thus equally inclined to the radiplane in the more general event of plain polyangularity (chapter 6), and the Wildhaber cones continue to be, as they are at figure 5B.14, equiangular and separately tangential to the plane of the transnormal at [Qx]. Refer in advance to §6.46, and refer back to my remark #3 at §4.70.

39. Charles Mac Cord in his book [18] [§174-175, §382-399] drew a similar picture in 1883, but, as already mentioned (§2.36, §3.61, §5B.12), he made some understandable mistakes at that time. His picture of the two

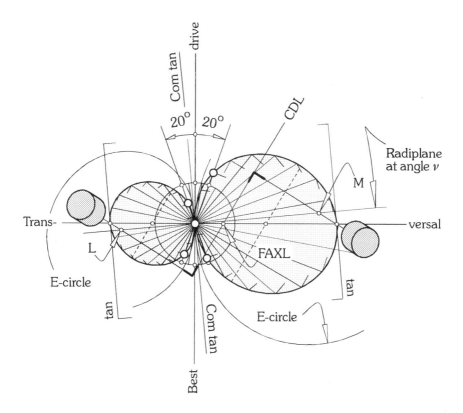

Figure 5B.15(b) (§5B.41). The ellipses of obliquity, WkEx#1. They appear here flat-on in the polar plane. The polar plane is represented by a circle. The overall orthogonal projection is parallel with the FAXL. The FAXL appears as a point, and the E-circles appear as ellipses. The paths of the two Q (both in the plane of the paper) cut the ellipses of obliquity (also both in the plane of the paper) at the four points S of the two equivalent linkages RSSR. Please note I have adopted the convention here (this to become standard) of putting the transversal horizontal across the page, whereupon the best drive becomes vertical. The radiplane is thereby tilted, and the common tangent to the ellipses of obliquity at E (which is normal to the radiplane) is tilted accordingly. Equiangularity obtains.

hyperboloids — his naked wheels? — mating throat to throat at his figure 109 can now be seen to be faulty. The gear set as drawn by him is not, moreover, nor is it claimed by him to be, drawn in the radiplan view.

The ellipses of obliquity

40. The plane of the E-radii is germane to another matter, the ellipses of obliquity. An sample pair of these, from WkEx#1, was ghosted in at figure 5B.04, and the general phenomenon of these ellipses was preliminarily discussed at the end of §5B.18. See the same pair of ellipses (from the same WkEx#1) now at figure 5B.15(a). See also at figure 5B.15(a) the plane of the E-radii (the radiplane) illustrated by the planar three-pointed star-shaped lamina containing the two E-radii and the helitangent. As the angles of obliquity δ are altered to cover the whole positive and negative ranges from zero to $\pm 180°$, the centres SS of the spherical sockets of the equivalent linkages RSSR, namely the extremities SS of the ideal zones of contact (whose lengths are Z_t), migrate around the said ellipses. The ellipses are coplanar within the polar plane and contain the line of the transversal. In general, across the whole wide arena of plain polyangularity (§6.10), the ellipses intersect; this occurs once at [Qx] and once again elsewhere in the polar plane. *Given the special circumstance of equiangularity however (which here prevails), we find (a) that the ellipses are tangential at [Qx], and (b) that their common tangent there is normal to the radiplane.* The said common tangent resides as a ray within the polar plane, as does the best drive, but be careful to notice that these two lines are not collinear.

41. See also figure 5B.15(b) which is an orthogonal view taken against the polar plane of the same scene. Given the prevailing equiangularity, this view is the same as that that might be taken along either the helitangent or the FAXL. These lines, parallel with one another, are both normal to the polar plane (§5A.23); they accordingly appear as points in this figure. Whereas the radiplane is designated by the triangle LEM at figure 5A.01, the said triangle appears in figure 5B.15(b) as the apparent straight line LEM. Still in figure 5B.15(b), the true shapes of the ellipses can be seen directly. The best drive can be seen normal to the best facet, and the common tangent to the ellipses can be seen normal to the radiplane. Looking along the helitangent, which is their line of intersection, the said planes JOK and LEM are inclined to one another at angle ν. This angle clearly increases with C, increases with Σ, but decreases with R. Indeed it can be shown (for equiangularity) that

$$\nu = \text{[work out the formula]}.$$

It might be observed as well that, given k, the ellipses of obliquity and angle ν are integral parts of the architecture. Also I wish to reiterate that, although we are here in a chapter dealing with equiangular synthesis only, the ellipses of obliquity are a general feature of involute gearing (§6.55).

SYNTHESISING AN EQUIANGULAR SET §5B.42

42. There are at least two methods available for constructing the ellipses of obliquity upon the polar plane. We may (a) simply follow the steps of this chapter and plot a series of points S (each one for a different δ measured positively or negatively from the best drive and within the polar plane), erecting on each occasion the shortest distance *a* (oblique in space) between the relevant ray of the polar plane and the relevant shaft axis; or we may (b) construct for each ellipse the particular circle upon the plane of the E-circle whose diameter is the actual operating radius *e* of that E-circle, then project this circle orthogonally in the direction of the relevant shaft axis upon the polar plane. A simple proof that this latter maneuvre is legitimate involves the well known elliptical plane sections of the cylindroid [1] [§15.26]. *It will be found to be a fact that the whole population of the said shortest distances between the rays of the polar plane and the two shaft axes comprises two cylindroids whose respective central axes are parallel, not with one another, but respectively with the said shaft axes.* These cylindroids figure prominently in the background geometry here. A theorem due to Cayley, newly explained and proved in a simpler way by Sticher [71], figures in the proof. Refer to appendix B at 5B (§5B.67). Looking orthogonally and parallel with the CDL, it is easy to see that, given equiangularity (which is here the case), the lengths 2B of the cylindroids [1] [§15.13] are in proportion with the E-radii namely with the speed ratio *k*.

43. The intercepts (S–S), which are of course the ideal zones of contact Z_I, both appear at true length in figure 5B.15(b). It becomes clear from the figure that the angles δ, chosen to be equal in the worked example just completed (equal at 20°), should perhaps have been chosen to differ. Examine the striking effect, for example, while keeping δ_D the same at 20°, of changing the value of δ_C from 20° to 25°. This 25% increase in the angle of obliquity δ_C would give an increase in Z_{IC} of roughly 50%, thus bringing Z_{IC} more into parity with Z_{ID}. Keeping in mind of course that in each case the length of the actual zone of contact Z_A will be less than the length of the ideal Z_I (§3.56), there is, quite clearly, scope here for adjusting the contact ratio at each of the paths. Knowing that the angles of obliquity may be altered separately, and that this is not important from the production point of view (chapter 10), we should be able to achieve the most comfortable circumstances not only for drive and coast but for production also. Look in advance at the next, the second exploratory exercise WkEx#2, beginning at §5B.45.

44. Please refer to figure 5B.15(c). This is another orthogonal projection of the same scene. It is taken, not as above along the helitangent onto the polar plane, but along one of the shaft axes onto the plane of the relevant wheel. Following the various explanations at appendix B at 5B (§5B.67), the reader might confirm at figure 5B.15(c) *that the relevant ellipse of obliquity appears, in the said orthogonal view, to be exactly circular.* Go for some further reading to [1] [§15.26] *et seq.* The same exactly circular appearance of the other ellipse occurs when we take the orthogonal projection not along the one but along the other of the two shaft axes.

Beginning of worked example #2

45. *The data.* It will be instructive to go now to the more difficult synthesis of a low-ratio ($k = 9/41$) square ($\Sigma = 90°$) equiangular set with unequal angles of obliquity. This is a new, equiangular example. I have chosen it for its exiting prospect, knowing well, however, the likelihood of a wholly unsatisfactory result. More demandingly than in the previous WkEx#1 (§5B.09), we take $\Sigma = 90°$, namely the maximum possible shaft angle short of that angle becoming obtuse (§1.22). We take moreover $C = 40$ mm, and $R = 60$ mm, which means that the slope of the transversal (Φ_T at §10.11) is steep. Given the chosen data, we take a relatively small speed ratio k with numbers of teeth on the wheels 9 and 41. Thus we employ $k = +0.2195$ approximately. After general observation, this small number of teeth on the pinion appears to be near to the challenging lower limit for any kind of gearing in the circumstances, below which inescapable difficulties probably occur. The chosen

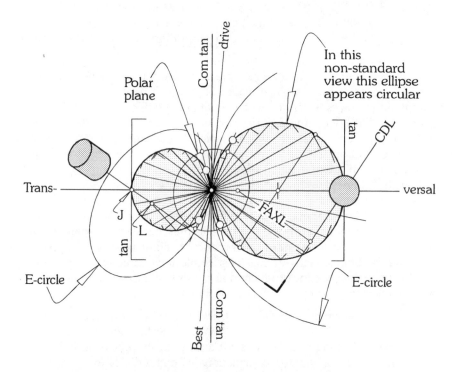

Figure 5B.15(c) (§5B.44). In the sense that figure 5B.15(b) was set to be a standard figure (see the caption there), this figure is non-standard. The parallel view here is not against the polar plane but along one of the shaft axes. It has been drawn exclusively for showing that the relevant ellipse of obliquity appears as a circle in this particular view. The matter at issue concerns the geometry of the cylindroid. It may be ignored upon a first reading.

data coincide very roughly with those of an old fashioned, but apparently successful, non-involute hypoid set taken from past automotive practice and known to the author. It is thus a challenge.

46. *The numbers of teeth.* Notice the prime number 41 and the non-prime number 9. This is not some magic minimum ratio. It is simply the case that in practice such pairs of numbers, where one of them is indivisible by the other (or whole-number factors of it), are often chosen with a view to distributing wear more evenly.

47. *The fear of failure.* Now it is openly acknowledged that the above data might lead to involute solutions that will become impossible due to severe undercut. The data might take us, indeed, into unresolved confusion, where all kinds of trouble might be found. Without prognoses, however, we can only explore (§5A.15). I therefore intend, by early judicious use of the ellipses of obliquity, to set out in such a way that the worst of the likely difficulties might be avoided.

48. *The architecture.* Peruse the series of figures 5B.16, all figures of which are derived from working drawings made in connection with the current example. Using already known methods worked out already at WkEx#1 (§5B.09), we can take the present example a long way. At 5B.16(a) the data of §5B.45 are drawn to scale. The figure, however, is in perspective. Determined by the data, the equilateral transversal, the helitangent at E, and the

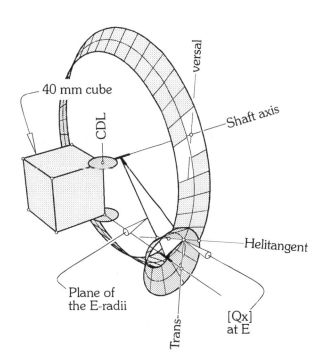

Figure 5B.16(a) (§5B.48). Panoramic view of the data for WkEx#2. The shaft angle is a right angle, the set is thereby a square set, and equiangularity obtains. Notice that the two E-radii and the helitangent at [Qx] are coplanar. The cubical box, whose one edge contains the CDL and whose other two edges contain the two shaft axes is drawn here simply to aid visualization.

§5B.49   GENERAL SPATIAL INVOLUTE GEARING

pair of naked wheels in mesh have been located. Equiangularity has required that [Qx] be put at E. Accordingly, and as seen in the figure, the plane of the E-radii (the radiplane) contains the helitangent at E. See §5A.23 item (d). Numerical results for this WkEx#2 appear at the TNR.

49. *The ellipses of obliquity.* Go next to figure 5B.16(b). Looking here orthogonally along the helitangent and thus directly against the polar plane in this WkEx#2, the ellipses of obliquity have been constructed (§5B.42). By starting from the shown common tangent to the ellipses, which (on account of the equiangularity) can in fact be drawn at [Qx], we next swing two separable rays of the polar plane outwardly away from it. These rays (as they go) cut the ellipses in continuously changing pairs of the points SS. These SS were previously mentioned and discussed at §5B.18 and §5B.40. As we thus separate the rays in the polar plane, we can judge (as we go) the lengths of the rods SS of the equivalent linkages that obtain. Thus arbitrarily, but not without guidance, I have settled upon $\delta_D = 5°$ and $\delta_C = 30°$. These chosen

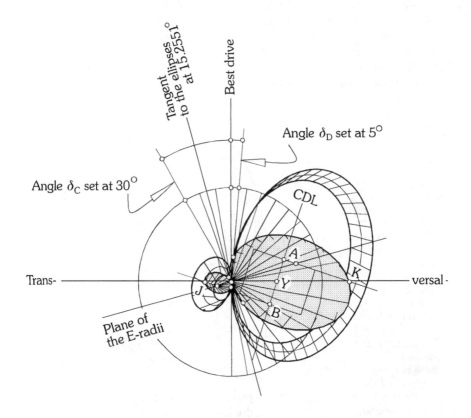

Figure 5B.16(b) (§5B.49). Standard view of the ellipses of obliquity for WkEx#2. Equiangularity obtains. The scale here is copied at figure 6.11, but see there for this.

SYNTHESISING AN EQUIANGULAR SET   §5B.50

values have led to the two Z_l, which, consistent with the two δ being not too large, have each been judged to be sufficiently long, roughly equal in length with the other, and each of them long enough to allow both of the contacts ratio to be greater than unity (§5B.43). The two paths of Q (collinear with the two Z_l) have thus also, by this choice, been determined.

50. *The concept of the tambourines.* Refer to figure 5B.16(c). The view here is a different view of the same set. It is not only taken from a new direction but also tightly focused in at the meshing. It is accordingly not easy to grasp upon a first looking. For guidance, therefore, a sketch of the whole

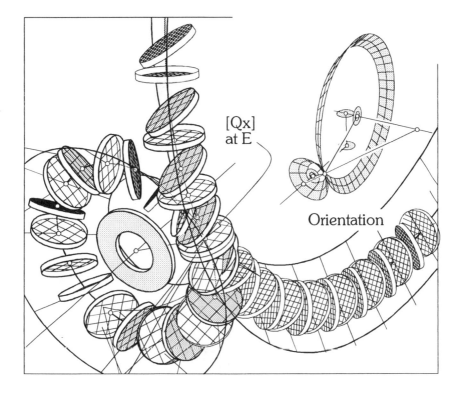

Figure 5B.16(c) (§5B.50). The interleaving circular facets of WkEx#2. These imaginary facets are attached to both convex flanks of all teeth at those points on the flanks which come into contact with their counterparts when Q is at [Qx]. Alternatively the said points upon the flanks are those points where the revevant E-circle pierces the flanks. Within the general region of the meshing the steadily moving facets come into coplanar contact with one another both suddenly and instantaneously. This occurs at that precise place where the facets come together at one or other of the two planes of adjacency at [Qx]. When not in contact the mating planar facets (shown shaded) intersect.

§5B.50   GENERAL SPATIAL INVOLUTE GEARING

scene is shown at inset. Except that it shows most clearly what I want it to show, the view is in no way special. Choosing the fixed [Qx] that we put at E as the meeting point for two reference flanks on the drive side, one may next draw the contact normal not only there but at all such reference points around the peripheries. There are 9 and 41 of them respectively. Employing the necessary maneuvre of next rotating the naked wheels in mesh through one half tooth, one may draw also the similar contact normals for the coast sides at all of the teeth. The actual shapes of the flanks are of course unknown as yet, but, normal to these contact normals at those mentioned points and concentric with those normals, we can erect what I wish to call for convenience here the *tambourines* that may be seen. Facing one another in

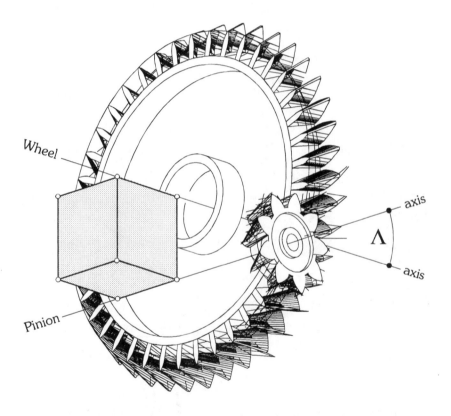

Figure 5B.16(d) (§5B.53). [Rough draft] This unfinished drawing relates to WkEx#2. It is a parellel view looking along the transversal, orthogonally against the transnormal. Let the transnormal be in the plane of the paper. The pinion is in front and the wheel behind. The true angle between the shaft axes (namely Σ) does not appear in this view; but the apparent angle in this view (which is Λ as shown) does. Angle Λ becomes important later, when we come to polyangularity.

SYNTHESISING AN EQUIANGULAR SET §5B.51

pairs like square brackets enclosing an algebraic expression, the pairs enclose (one by one) the teeth of the wheels. I repeat that the actual shapes of the teeth are unknown as yet, but these tambourines are tangential with the convex flanks at the already mentioned, evenly spaced points around the E-circles as may be seen. Watching the tambourines, now fixed upon the wheels as the wheels rotate at compatible speeds (9/41), interleave and merge, we can learn qualitatively a great deal about the ultimate interaction of the real teeth. We can make the rough judgment, moreover, that all appears to be going all right so far.

51. *Calibrating the paths.* For calibrating the paths of Q in terms of input angle we need to determine the one CP for the set and the two different NP (one each for the separate wheels). Graphically speaking, convenient finite steps might be taken of one tenth of the CP. The relevant fact is that, because the two angles δ at the phantom rack are different, the speeds at

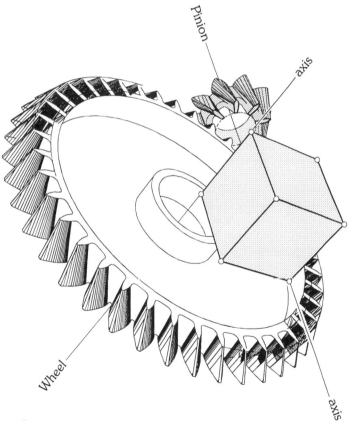

Figure 5B.16(e) (§5B.53). Again the gear set of WkEx#2. This is a non-specific parallel view revealing the inside toe-end of the pinion, and a possible method illustrated there for certain difficult cases where interference may be a special problem.

which the two Q travel their respective paths are different. See the formulae for CP and NP at §5B.27; and go to the TNR for details.

52. *The shapes of the flanks and the slip tracks.* Knowing next all four of the (a, α) — these are thrown up graphically at the ellipses of obliquity and may be checked quantitatively against the available algebra (§5B.16) — one may next construct an adequate range of the generators of all four of the involute helicoids. Having also constructed corresponding generators of the four base hyperboloids by swinging both paths about both shaft axes, one may discover next the shapes and locations of the slip tracks. These are spatial twisted curves as we know. They occur at the intersections of the base hyperboloids and the involute helicoids; and there are, of course, four sets of them. See both branches of a typical slip track illustrated at the twin figures 3.13(a)(b) and be reminded that only one of these (the one that finds itself drawn upon the real flank of the relevant tooth) is the useful one. The other branch departs from the real flank of the tooth, curling away (at the cuspidal edge of the helicoid) to become 'unreal'.

53. *The shapes of the truncated extremities of the teeth.* One might be tempted to say now that the job is almost complete, that all we need to do is truncate the teeth. But refer to the figures 5B.16(d)(e). These are two views of what appears to be the beginnings of a satisfactory result. Figure 5B.16(d) is taken orthogonally along the equilateral transversal; this reveals the angle Λ incidentally; angle Λ is a fixed angle of the architecture; it is the *apparent angle* (not the *true angle* which is Σ) between the shaft axes when the scene is viewed from this particular direction; angle Λ becomes important later (§6.46). Figure 5B.16(e) is another orthogonal view taken from a different, non-special direction; it is drawn to reveal what might be a way of dealing with the teeth of the pinion if the event of certain kinds of interference (§10.73), they are *cloven* in such a way that at the toe end the teeth have *prongs*. Externally the teeth have been truncated conical. The half angle at both of the conical cuts is somewhat greater than κ (§5B.04). Graphically, attention has been paid to the continuity of action: the contacts ratio both appear to be satisfactorily greater than unity, but the enclosing flanks of each one tooth are grossly decapitating one another at the toe ends. One soon begins to see from the shapes of the slip tracks, however, that this does not imply catastrophe. It is an expected phenomenon. But it is a new hazard. If the current example shows itself to have no credible solution, it will be merely instructive; but were it to be found that the absence of a solution is entirely inescapable, it would be disappointing. To avoid disappointment, one could (a) simply deny its likelihood and pass on, (b) try some different values of the angles of obliquity which are, of course, here open to choice, or (c) optimistically look more carefully at what might sensibly guide the truncations in this ongoing, but somewhat risky example. But go to chapter 10 where the details of this WkEx#2 are taken nearer to final completion (§10.73-74); and see in advance the associated figures 10.15(a)(b) and 10.16(a).

Mutual decapitation versus continuity of action

54. It is always likely that the radial truncation of pointed teeth, applied to avoid the mechanical mayhem otherwise threatened by interference with pointed teeth of the mating wheel, might be over-done at the expense of contact ratio. The boundaries of the envelope that enclose the geometric circumstances leading to this impasse require examination. Let us assume for the sake of argument that all truncations that might be made are made conical, that three external cuts to the wheel blank (at the tips of the teeth and at their ends) are made by lathe, and that an interrupted internal cut (at the roots of the teeth) is made, somehow, otherwise. Refer to figure 5B.17 and see (at top) an orthogonal view along the CDL of an unknown, arbitrarily truncated skew gear set. The said four conical cuts are shown to enclose at each wheel an annular ring of trapezoidal section that might be called, with justification, *the swept volume of the teeth of that wheel*. On first glance the teeth appear to be interfering; but the shaft axes are skew; the two swept volumes intersect in a way that is not clear to see in this conventional view.

55. In the same figure (below) is a parallel, but panoramic view. Firstly I wish to mention that the shapes of a pair of proposed swept volumes (as yet uncut) will in no way alter the fundamental kinematics of the interaction of

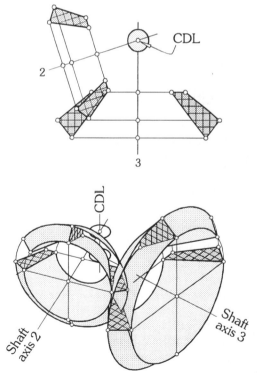

Figure 5B.17 (§5B.54). There are two unrelated views here of the same figure. They are of a general case of an arbitrarily but conically truncated skew gear set. Above there is a standard plan view of the set, in the style of figure 5B.10(b). Below there is a panoramic view. The cross-hatched areas are radial cross sections of the trapezoidal tori that may be seen. These tori are the so-called swept volumes of the teeth. The figures are dealing with the contradictory effects of (a) the likelyhood of mutual mechanical destruction on the one hand, and (b) the need for unhindered continuity of action on the other.

the mating pairs of already synthesized ghostly involute helicoids, suitable portions of which are destined to form the real flanks of the teeth. While the involute helicoids, each with its own (a, α), remain with their axes coaxial with the axes of the respective wheels, the said shapes of the swept volumes will simply determine whether or not the real teeth will avoid, on the one hand, clashing with one another, and continue, on the other, to mate for long enough to ensure that the contact ratio remains greater than unity. It might be said by way of metaphor that, provided the skull of a person is not cut by the barber, the mere style of the hair-cut does not alter the underlying shape of the person's head.

56. *Two rules for satisfactory mechanical action after truncation.* Upon reflection we can see (a) that, *in order for the teeth of the wheels to mesh at all, the swept volumes must interpenetrate one another, but, to avoid clashing of the teeth, neither of the inner truncated conical surfaces of the two swept volumes may be cut by the outer truncated conical surface of the other.* This might be said to be the rule for avoiding interference; and in this respect figure 5B.17 might be said to speak for itself. It needs to be seen however also (b) that, *for continuous mechanical action, necessary portions of the slip tracks (of which there are four) need to be enclosed within the swept volumes to ensure that both contact ratios (for drive and for coast) exceed unity.* This might be said to be the rule for ensuring continuity. But these two rules are inimical to one another. Careless, undivided attention to one will call into play the other, and we are obliged to look for the windows of opportunity. Within these windows both rules can safely apply. Without clear guidance, however, from geometric or algebraic clues, we cannot see (except by trial) where these windows are.

Transition comment

57. We could go on here to discuss the limiting shapes of the swept volumes explained with the help of figure 5B.17 above, go back to the computer-driven graphics and other arguments, and make some further progress. But the matters are complicated, they need to be considered in relation not only to equiangularity but also to polyangularity, and they are, in truth, beyond the scope of this chapter. So the above short discussion should be seen as being no more than an introduction to the various problems of interference occurring across the whole spectrum of general involute gearing. In connection with machining, and much more comprehensively, this particular aspect of WkEx#2, and similar aspects soon to be met at WkEx#3 and WkEx#4 in chapter 6, are taken up again and taken together at chapter 10. Discussion of WkEx#2 is now completed for the meantime. Before going on, however, I wish to mention just here some degenerate cases of the general gear set that are really or apparently occurring within the realm of equiangularity. All of them are quite well known, yet some of them (because they are, indeed, non-involute) are somewhat puzzling.

The special sets with Σ zero

58. *Cylindrical helicals with parallel axes.* Here we have that Σ is zero. We know from the general geometry of the involute helicoid outlined above that any planar section of its helicoidal surface taken normal to the axis o-o is a planar involute to its base circle which circle is simply the relevant normal section of the throat cylinder. See a few such sections drawn at figure 3.02 (and another one at figure 8.01). And we know from the well known theory of ordinary, cylindrical helical gears with parallel shaft axes that the surfaces of the flanks of the teeth (derived by 'twisting' the corresponding planar spur teeth) are indeed involute helicoids. Lines that were rulings on the original planar involutes become helices after the twisting, but other lines (originally curved) become rulings upon the new, twisted surfaces. Refer Baxter, section 1-10 of Dudley [16], to see there why straight line contact can be made among the cylindrical helicals. This straight line is of course parallel with the axes in the special case of the non-helical, planar spur gears. *The point here is that, in all gears with parallel axes, the pitch at the pitch line is zero, independently of the helix angles at the teeth.* This means that a straight travel path may be chosen to be any member of the special linear complex whose pitch is zero [1] [§9.41], and, accordingly, that a straight travel path must be chosen to cut the pitch line but may be chosen to cut it at any angle ϕ. If $\phi = \pi/2$ the resulting helical is an ordinary spur gear, but if ϕ is other than $\pi/2$ the resulting helical is a helical. In other words we can say that in ordinary cylindrical helical sets the fundamental law says that, while ϕ may be taken at any value, q must always be zero. Under these conditions, the rulings upon the involute helicoids mesh collinearly with one another; there is continuous line contact between the teeth. The straight travel path of the point of contact has bloomed [1] [§11.03] into a plane and this plane is called in the relevant literature the *surface of action*; see figure 1-15 in Dudley [47] first edition.

59. *Cylindrical helicals with parallel axes.* Let me ask the question again: why is it that, in the case of parallel axes, the so called helix angle of the teeth may take any value? To argue rigourously here, I should be speaking in the carefully defined terms of chapter 10, but in this special case of the parallel cylindrical helicals all I need to say for the moment is that the said helix angles are (a) equal, and (b) measured at the pitch line in a plane normal to the CDL namely at the touching point of a matching pair of representative coplanar E-circles there. In helical gears with parallel axes (where, as we know, the single helix angle may take any value), the linear complex of legitimate travel paths for a point of contact is, like the pitch line, of zero pitch. Under these special circumstances, all lines of the complex cut the central axis of the complex [1] [§9.41], and the geometry of the polar plane at any chosen Q becomes such that the said plane contains the said central axis namely the pitch line. So a legitimate travel path for that Q can exist along any straight line through Q that cuts the pitch line. The angle ϕ may take any value, in other words, provided q remains at zero (§2.02).

The special sets with R zero

60. *Cylindrical helicals with crossed axes*. Here we have that R is zero. It is not possible to transfer the simple concepts from the spur and the helical parallel cylindricals just discussed at §5B.59 directly into the realm of the crossed helicals suddenly blundered into here; for such an attempt would be to try, without guidance, to argue from the particular to the general (§5A.15). We need recourse to original general theory. Without going too far back, we could go to §5A.30 and §5B.37 where some definitive prognostic remarks have already been made. Despite their well known, poor reputation for load carrying, and their tendency to wear rapidly — even to seize up [89], the crossed helicals work in a proper way kinematically. They transmit with constant k provided each wheel is derived from the same basic rack, and they do so with contact occurring at a single point. There remain of course the two nagging questions: (a) the old one of how, simply speaking, can one explain this said-to-be-severe phenomenon, and (b) the new question of how does the present theory (the involute theory of these chapters) apply in the special case of the radial offset R being zero. The first question (a) has no answer because the matter is not simple, and the second question (b) has been addressed already at §5B.37. On the questions of wear, we have here an open field waiting for some controlled experimental investigation.

The special sets with C zero

61. *The straight Beveloids*. Take a retrospective look at §3.10 and read §5.31. See there the unwrapping of an oblique-cut sheet of paper from a cylindrical core. Note that the core there is not a conical core, but a cylindrical core (which is indeed the cylinder of the core helix) and aptly apply the exact gist of the matter there to the matter here. Here we have that $C = 0$ and (with the shaft angle Σ and the radial offset R both remaining finite) we get, in the ordinary course of events, gear sets with intersecting axes that are commonly called by the name *straight Beveloids*. Going to the general equiangular equations for a and α at §5B.16 and substituting there the special value $C =$ zero, we get

$a = [2Rw \cos \delta \, \tan (\Sigma/2)] / \sqrt{[1 + \cos^2 \delta \tan^2 (\Sigma/2)]}$
$\alpha = \arcsin [\sin \delta \sin (\Sigma/2)]$

These are the characteristic equations for the special, straight Beveloids, and they are exactly applicable at figure 5B.08. Taking the numerical data from WkEx#1 with the exception that $C =$ zero, we find that a_2 and a_3, the same both for drive and for coast, are 70.237 and 42.142 mm respectively, that the ratio of these is 0.6, namely k, and that angle α, the same both for drive and for coast (and for both wheels), is 8.3109°. Compare these values with those corresponding at WkEx#1 (§5B.17). The Beveloids are indeed a special case of the general theory. They are in this sense legitimate involute gears (§0.08). Unlike the straight bevel gears of current engineering practice (to be studied next), these Beveloid bevels are capable of being mounted with their nominal shaft angles Σ somewhat altered, and/or their nominally intersecting shaft axes somewhat askew, which means indeed that they may be regarded with justification as being proper involute gears. Refer to Beam [59], Drago [43], and Smith [61]. Contact by private communication with Smith [61] leads me to the view that I hold just now that, despite their dependence upon point contact, the straight Beveloids manufactured by Invincible Gear Company in Livonia Michigan USA do not suffer unduly from unwanted wear (§5B.64).

The so-called straight bevels of ordinary practice

62. *The so-called straight bevels.* Here, presumably, C and R are both zero; but these gears are much more of a puzzle than meets the uncritical eye. The first question to resolve is whether or not the gears are, or could be made to be, involute and thus conjugate. Much has been written about the possibility of 'spherical' involute geometry being applicable, and equally much about the probability that this will never be possible. See for examples Steeds [4] and Baxter [14]. A second question to ask is whether my particular treatment of involute geometry for gears in space is the only possible treatment; I say this in view of Bricard [82] and Fayet [98], who have separately shown (but only for the case of planar gearing) that, *if we want not only conjugality but also freedom to alter centre distance*, there is only one solution for the shapes of the mating profiles, the planar involute (§3.73). If we were to be dealing here (here in this §5B.62) with involute gearing (and this I dispute), we would be dealing with the highly degenerate case of my overall theory where, with Σ remaining finite, C and R have both become zero. Under these circumstances, we could, with logic, locate the equilateral transversal, the pitch point, the pitch line, the polar plane at [Qf], and a representative path of a moving Q, all of these intersecting at the same point. We could also see with logic that the base hyperboloids are both circular cones, having zero throat radius; and it follows from this that the core cylinders would also be of zero radius, and from the formulae at 5B.16 that each angle α (namely the inclination angle of the relevant base hyperboloid and the helix angle of the corresponding core helix) would be indeterminate. We would accordingly need to be envisaging a helix of zero diameter with an indeterminate pitch. There might then be available two alternate ways for thinking, (a) by considering folklore-wise the taper-cut paper unwrapping from a cylinder of zero radius (§3.10, §5B.31), and/or (b) by considering the relevant degenerate form of the involute helicoid as I have, so far, found it to be. That involute helicoid in this special case is reduced to being the single infinity of cones which are of half apex angle α and coaxial with the relevant shaft axis. This fits with the fact (which is intuitive) that the planar involute obtained by rolling a line around a base circle of zero radius comprises all the coaxial circles coplanar with the said circle. Is it possible for one circular involute to drive another? This last is not a valid question in space because it is the slip tracks that drive one another, not the involutes. But where are the slip tracks? Is

there in this case (where C and R are both zero) a single point of contact, or has it bloomed? In all gears with parallel or intersecting axes the pitch p at the pitch line is zero, which means that, while the path of Q must cut the pitch line, it may cut it at any angle ϕ (§5B.58)]. Many writers on the straight bevels (like Baxter for example) speak about octoid etc. [99], but few of them speak about the actual shapes of the cut teeth or claim that involute action is at all possible. Some of them (like Steeds for example) appear to contemplate the possibility that straight bevel gears can be made to be truly involute somehow, and thus conjugate as a consequence of that. I say on the other hand that, while Steeds proposed 'spherical' teeth are indeed conjugate (as many other non-involute teeth are), they are not involute. Does this mean that my theory is false? Does it mean that the currently sought (and eagerly expected) Bricard inverse proposition for space does not exist? No. Remember that no amount of inaccuracy of whatever kind is permissible with the Steeds spherical set; we cannot alter either C or Σ without catastrophe! Neither can we alter Σ in planar involute gears without catastrophe, but there is a valid geometrical explanation for that [1] [§22.38 *et seq.*]. I hold that the 'spherical' argument of Steeds is spurious, because (a) his twisted 'spherical involute' curves vary in size as they go along the straight line of contact, and (b) the normal plane sections taken through his spherical helicoid are not planar involutes (§3.18). The upshot to this discussion has got to be that the regular straight bevels of ordinary engineering practice never were, nor are they ever likely to become, however cunningly they may be cut, involute gears in this true sense. My theory here has already shown or will be showing that, if the flanks are true involute helicoids, properly mounted with their axes coaxial with the shaft axes and so on, then we will get conjugality right across the polyangulars and the worms etc., *independently of poor assembly*; and the straight bevels do not accord with this. Many other non-involute constructions (such as the hypoids) achieve conjugality but always at the expense of requiring accurate assembly, and none of these (including now the straight bevel gears) are proper involute gears. They may look like involute gears, and even be thought of as involute gears, but they are not involute gears. I predict that my prognosticated spatial analogue of the reverse proposition of Bricard will come to the rescue here; it will permit us unequivocally to define what we mean by *genuinely involute* in the context of gears; and it will, I hope, wrap the matter up definitively.

More transition comment

63. Having thus tentatively explored the equiangular arena, and having achieved thereby some limited objectives, we could, just here, go more deeply into directly related matters. We could for example look at the equiangular spiroids (which are offset) and the equiangular simple worms (which are not). But I intend to look first at general aspects in the wider polyangular arena of gears of the more ordinary kind in chapter 6, then to study in a more precise way the worms and spiroids in chapter 7. And now, before con-

cluding this chapter, I turn to mention the somewhat unrelated but practical matter of load bearing and lubrication that has been clamoring for attention.

Load bearing and lubrication

64. In plain spur gearing, where each point of contact Q has bloomed (by virtue of degeneracy) into a *line* of contact, and where each such line passes regularly through a condition of exact collinearity with the pitch line, there occurs, at the pitch line, a reversal through zero in the direction of the rubbing velocity. This is known to influence the hydrodynamic lubrication, either there exactly at the pitch line, or close nearby, where pitting may be found to occur. *It will be found however in general that the shapes of the profiles of the teeth in skew involute gearing turn out to be such that at Q the direction of the sliding velocity (along the helitangent) never varies suddenly, either in magnitude or direction.* It is true that in involute skew gearing contact occurs always at a single point (not along a straight line or a curve), but, because two already wetted surfaces are coming freshly and continuously into contact as the relative motion occurs without reversal of the sliding direction, the chances for continuous hydrodynamic lubrication at the roughly circular contact patches are enhanced. There is also the relevant fact that, due to the constant angular velocity ratio k across teeth and the straight-line path of the point of contact, the transmission force does not engage in cyclic fluctuations at tooth frequency. This benign circumstance, characteristic of all involute gears whether skew or planar, must surely lead to continuous, smooth lubrication. The exact kinetostatics of this continuous sliding in general and the abrupt reversal of direction in the planar case along with a constant force at all stages is quite difficult to explain, but it depends upon (a) the shape of any slip track as it crosses the throat circle of its hyperboloid, and (b) the fact that, in skew gears with radial offset R we can try to avoid those parts of the full slip track that bring us close to either end of Z_j. Please read again the passages §3.56-59 which are directly relevant here. We carefully select the working portions of the teeth, truncate them to suit, and operate, as it were, well within a safe portion of one branch only of the slip track. What has been said here applies with even greater force to the special cases of the crossed helicals.

Conclusion

65. Whatever the disadvantages of this new range of involute gears turn out to be in the long run, transmission error at tooth frequency due to wrong kinematic fundamentals will not be one of them. The great advantage of these gears, of course, will always be the flexibility, within limits, in the making of errors at assembly. Thus economical advantages in design, manufacture, and application in machinery accrue. We have however only studied involute gear sets with equiangular architecture so far, and there are (as we shall see) even more effective geometries to come.

Appendix A at chapter 5B

66. This appendix was first mentioned at §5B.05. It is to be mentioned again at §6.46. It deals with the algebra of κ and introduces the new angle Λ. The passages from §5B.04 to §5B.06 inclusive are themselves non-recommended for study upon a first reading, so this appendix might be seen to be doubly thus. While in some respects it goes prematurely beyond the current text, it belongs here nevertheless. While some parts of it are immediately relevant, other parts will be needed soon, at chapter 6.

Angle κ is the half angle at the vertex of the Wildhaber cones. Angle Λ is the apparent angle between the two shaft axes when the set is viewed orthogonally along the transversal. These two angles depend solely upon the basic architectural constants, C, Σ and R. They are independent of k, j and the angles δ. They are both recurring aspects of the overall algebra, and they are both of direct relevance to the practical business of the machining of teeth. Needless to say they are both expressible algebraically solely in terms C, Σ and R. Although at this juncture it is, perhaps, premature to say so, we already know (chapter 6) that, wherever F is taken to be, the plane of the transnormal there is oriented in the same way with respect to frame, namely normal to the transversal; we also know that both of the Wildhaber cones are always separately tangential (along their different generators) with this transnormal; and it follows from this that the angle κ (which is the angle at which the shaft axes pierce the said transnormal), for given C, Σ and R, and independently of whether equiangularity prevails, is always the same for both wheels. Whatever the different heights of the Wildhaber cones turn out to be, the half angles κ at their vertices, for given C, Σ and R, will always be the same. We might note, moreover, the following piece of simplicity: the same might be said about the equal angles ξ, both of which remain fixed for given C, Σ and R in any event, and it is clear to see that κ will always the complement of ξ. But refer to figure 5B.18.

On looking for κ. We find by trigonometry in figure 5B.18 (which is, admittedly, drawn for the special, symmetrical case where F is at Y) that the base radius of that particular Wildhaber cone, namely the radius of the F-circle, namely [(M –Y) true], may be written

Base radius = $\frac{1}{2} \sqrt{[4R^2 \sin^2 (\Sigma/2) + C^2]}$.

We can also show by judicious triangulation in the same figure that the axial height of the same Wildhaber cone, namely [(M–V) true] is

Axial height = $[4R^2 \sin^2 (\Sigma/2) + C^2]/[4R \sin (\Sigma/2) \tan (\Sigma/2)]$.

So we have (and this is in general) that

Tan κ = [base radius of the cone]/[axial height of the cone]
= $[2R \sin (\Sigma/2) \tan (\Sigma/2)]/\sqrt{[4R^2 \sin^2 (\Sigma/2) + C^2]}$

Which is the required result, a compact algebraic expression for κ. Notice the

SYNTHESISING AN EQUIANGULAR SET §5B.66

evident fact that this expression is independent of the speed ratio k and (if j is different from k) the poly-ratio j as well; this latter is a measure of polyangularity dealt with in chapter 6; and notice, above all, that angle κ is independent of which wheel is under consideration.

Angle κ is the same, in other words, for each of the two wheels. This fact is a feature, not of equiangularity in particular, but of all regular (both equiangular and plain polyangular) involute geometry. This excludes incidentally the exotic (chapter 8). In the event however of non-involute geometry — take the hypoid geometry, for example, or that of the straight bevels — the two angles κ of the two wheels will usually be different.

On looking for $\Lambda/2$. By virtue of the same maneuvre (by appealing to the same, symmetrical case), we find by trigonometry in the same figure that the base radius of the particular Wildhaber cone, namely the radius of the

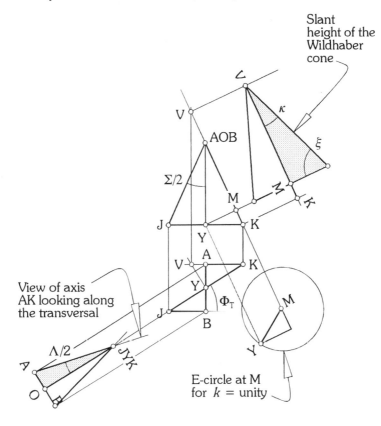

Figure 5B.18 (§5B.66). Refer to appendix A at 5B. Although using numerical data direct from WkEx#1, and although dealing here with the special case where F is at Y, this figure is the basis for algebraic formulations of a general nature. The results apply not simply for equiangularity incidentally, but for the whole wider arena of plain polyangularity as well.

F-circle when F is at Y, namely

[(Y–K) true] = $\sqrt{[4R^2 \tan^2 (\Sigma/2) + C^2]}$

We have by definition (§10.08) that,
where tan Φ_T is the slope of the transversal,

tan $\Phi_T = [C/2]/[R \tan (\Sigma/2)]$,
cos $\Phi_T = [2R \tan (\Sigma/2)]/\sqrt{[4R^2 \tan^2 (\Sigma/2) + C^2]}$, and
[OA projected] = $[(C/2) \cos \Phi_T]$.

So we can say that
tan $(\Lambda/2)$ = [OA projected]$/R$
= $[(C/2) \cos \Phi_T]/R$
= $[C \tan (\Sigma/2)]/\sqrt{[4R^2 \tan^2 (\Sigma/2) + C^2]}$.

which is the required result, a compact algebraic expression for $\Lambda/2$.

I wish to add a comment here about the already mentioned angle λ (§5A.23, §5B.04, §5B.11). Angle λ is such that $\lambda_2 + \lambda_3 = \Lambda$. In the special event of equiangularity (which has been the event so far), the two λ are equal and we have that $\lambda = \Lambda/2$.

Appendix B at chapter 5B

67. This appendix was first mentioned at $5B.42, then again at §5B.44. It deals with some aspects of the cylindroid and Sticher's treatment of Cayley's theorem concerning the planar pencil of lines and the cylindroid.

Cayley's theorem: Take a plane and a point A within it, draw all of the lines in the plane that meet at A, and take a line o-o (not in the plane) to cut the plane in B. Now draw all of the common perpendiculars between the line o-o and the lines in the plane. Find that the said perpendiculars are (a) arranged in the form of a cylindroid whose nodal line is o-o, and (b) intersecting the plane in points upon an ellipse. Find, furthermore, that the ellipse in the plane is located and proportioned as follows. The line segment (A–B) is a diameter of the ellipse, and the major axis of the ellipse is parallel with the projection of o-o upon the plane.

This theorem is more complicated than it might appear at first, and I give no proof of it here. But please go to [1] [§15.26-§15.31] [§16.32] for some explanations of it, its historical origins, and related matters. In [1] the theorem emerges as a simple corollary of the more general propositions being put and examined there. It emerges when the point Q in figure 15.12 in [1] is chosen to be, not just anywhere in space, but actually upon the surface of the cylindroid. Please note the explanation in [1] of the fact that, when viewed orthogonally along o-o the ellipse will always appear circular.

Sticher, at appendix A2 to his doctoral thesis [42], gives an independent treatment of Cayley's theorem, proving it definitively using matrix algebra. Reportedly [42] an earlier proof of another kind was given by Salmon [45].

CHAPTER SIX
THE PLAIN
POLYANGULAR OPTION

Recapitulation

01. On looking again at chapter 22 in [1] one sees that I was dealing there with *non-offset meshing*. I was dealing, in other words, exclusively with the crossed helicals. There is no mention of doubly departed sheets in [1]; neither is there of offset. Until the appearance of the fundamental law and its generally applicable ramifications became clear, however, it was not easy to make progress. But the said law has now appeared [14], and I have, thereby, been able to devise in this book (for comfort) a circumscribed, first approach to the otherwise chaotic scene of general spatial involute gearing. I described that first approach in chapter 5A, introducing there the term *equiangularity*. At WkEx#1 I took the case of a modest, non-extreme, somewhat special, but openly practical gear set exhibiting equiangularity. Find the data at §5B.01: Σ was only 50° (the set being thereby *oblique*); k was only 3/5; and the work began at §5B.09. At WkEx#2 a more confronting case was taken. Find the data at §5B.45. I found there some workable teeth for another equiangular set. Angle Σ was 90° (the set being thereby *square*), and k was a somewhat demanding 9/41. Recall the difficulties encountered in choosing the angles δ; these were overcome by using the newly discovered ellipses of obliquity.

Introduction to polyangularity

02. We now need to look at the scene more widely, and I wish to introduce for that purpose the term *polyangularity* [44]. The term applies to all gears. By this I mean to convey that the whole of the previously studied *equiangular architecture* is simply a sub-set of the *polyangular architecture* to be studied now. None of the new material of this chapter contradicts the material of the chapters 5. The new material is needed, however, to cope with various obvious requirements in gears, namely to be able to alter sometimes, while keeping the numbers of teeth the same (and thus the ratio k the same), *the relative sizes of the wheels*. We need this to deal with practical matters such as (a) the reduction in overall weight and the saving of space, (b) the

required non-reversibility of certain sets, for example worm sets, (c) the easing of certain difficulties in the realm of friction, for example an efficient drive might be more important than an efficient coast, and (d) the amelioration of certain other difficulties with respect to interference and undercut. This chapter further explores the rigid-body kinematics, and thus the practical application in real gears of the said wider possibilities. As before, the important but secondary matters of elasticity, wear and lubrication are mentioned only peripherally. The twist angles τ of the naked wheels, which, with [Qx] at E, are always equal, were first mentioned at §3.04; see figure 3.01 and refer to §5B.04. We are soon to become aware however that, for a given k, as we remove, as we may, [Qx] from E, and as the ratio of the sizes of the wheels is thus obliged to vary, the angles τ depart from equality with one another. *Overall, this inequality of the angles τ is the characterizing feature of polyangularity. It is only in the special case of equiangularity that the angles τ are equal.* Please recall the contents of note [90].

Timely exclusion of an as yet unwanted difficulty

03. Let us assert that, while willing to tolerate a change in the required ratio of the physical sizes of the real wheels of some equiangular gear set, and thus the likelihood of departure from equiangularity, we are unwilling to tolerate the complications that might be due to the pair of paths being chosen to be skew (§5A.11). Thus we reject as unacceptable, just now, the absence of an intersection point [Qx]. Let us correspondingly agree that, whatever else we may do in this chapter, *we will continue to choose a pair of paths that intersect.* We are choosing to do this, not because it is necessary, but because it is *comfortable* (§5A.11). For *discomfort*, see chapter 8. Having thus for the meantime removed the awkward difficulty of skew paths simply by outlawing it, the question next to be put is this: where can we put [Qx] if not at E? *The straightforward answer to this is that we may put it anywhere,* but in order not to become lost we need to be circumspect still.

On getting ready to begin

04. Refer to figure 4.07 and study the text nearby. For an early excursion into the polyangular, we might contemplate (with the tooth ratio k fixed) the removal of point [Qx], which is at [Qf], which is at E, from its shown position on the E-line to some other position not on that line. Given the satisfactory R and the satisfactory k, however, and thus a fixed E-line, and a fixed equilateral transversal JK through a fixed point Y namely the transversal JEYK, and given the contents of chapter 4, let us agree that the removal and repositioning of [Qx] might first and most conveniently be made by moving it to and fro along the said transversal JEYK. See figure 4.08 and then, for a change of view, figure 5A.01. Now in the same way as the E-line has been seen to be occupied by an infinity of points E, let the transversal JEYK be seen to be occupied by an infinity of points F.

05. See figures 5A.01 and 5B.01. With k fixed, that is, with the E-line remaining fixed in location, let us next see in the imagination the following. If, on moving to and fro among the points F along JEYK, we move [Qx] towards J, the smaller wheel will become smaller. If on the other hand we move [Qx] (among the points F) towards Y and thence beyond, the smaller wheel will become larger, eventually becoming, indeed, larger than the originally larger (namely the slower) wheel. As [Qx] moves in this way, in either direction away from the E-line, the originally equal twist angles τ at the teeth of the two naked wheels will become unequal; the wheels will thus become *other than equiangular*. Keeping k fixed, we deal in this chapter with this disciplined way of moving the path-intersection point [Qx].

06. Refer to figure 5A.01. See there that, corresponding with the unique radii e_2 and e_3, there is a smooth continuum of the sets of radii f_2 and f_3. See also that the equations at §5A.27 can be interpreted to show incidentally that,

> for a range of polyangular possibilities with a fixed R and a fixed k, [Qx] remaining upon the transversal JYK, the sum $(f_2 + f_3)$ of the F-circle radii at [Qx] remains a constant, and this constant is $\sqrt{[4R^2 \sin^2 (\Sigma/2) + C^2]}$.

See this again at §10.08; and we saw it before at appendix B at 5A. The F-circle radii are not properly defined as yet, only hinted at (§1.40, §4.66, §6.05); but study the contents of the inset here and those of the inset at §5A.27 and draw the obvious conclusion. This alone might help to show that adherence to a fixed R and a fixed k, as we roam about, is the way to go.

07. An important question to ask, here, is whether or not it is absolutely necessary to remain upon the straight-line transversal JEYK as we roam about looking for some new position for the point [Qx]. The answer is no, it is not absolutely necessary. We could stray from the transversal while remaining upon the F-surface — this important surface has already been mentioned extensively (§4.29(#3), §4.34, §5A.30), and see figure 5B.01 — or we could stray away from that surface altogether, directly into the G-space. The G-space is all of that space not occupied by the F-surface, which is, of course, all of space. *It might be said by way of definition now that any chosen removal of [Qx] in any direction away from the E-line will precipitate entry into the field of the general polyangular.*

On what happens in the G-space

08. Whereas (a) there is at the D-point, at the intersection of the E-line and the CDL, an ∞^0 of points, namely some finite number [1] [§4.01], which is only one point in this case given the impracticality of the alternative internal meshing; whereas (b) there is along the E-line an ∞^1 points; and whereas (c) there is upon the F-surface an ∞^2 of points; (d) *there is in the G-space an ∞^3 of points*. There is in the G-space a bewildering array of points; there is, in other words, an enormous freedom of choice for locating our new [Qx].

09. If we move our one starting point [Qx] out of the F-surface and into the G-space, the transversal through [Qx] becomes no longer equilateral; it no longer cuts the FAXL; and, whichever way we might go from the F-surface outwards, it increases in length. Looking again at §4.58-62, where we discussed this very mechanics *in vacuo,* it would seem that, if we do depart in this way from the F-surface, one set of teeth might become pointed, while the other (the mating) set might become, accordingly, undercut. Dudley [47] discusses the planar case of this phenomenon most convincingly, and its analogue in space is similar (see chapter 7). For what urgent reason should we always remain exactly upon the F-surface? The answer to this is that, if we do go even a small distance — taking our starting point [Qx] with us as we go — into the open G-space, we immediately go *exotic* (§6.10); we should do this only with well defined, clearly understood objectives in view.

The plain polyangular and the exotic

10. In deference to what has just been said, I wish to state now (by way of choosing a workable terminology) that, if we take our starting point [Qx] somewhere exactly upon the F-surface, we shall be dealing exclusively in the realm of *plain* polyangular architecture. If on the other hand we take our starting point [Qx] elsewhere in the G-space or if we choose the paths to be skew (in which latter case there will be no [Qx] at all), there will be *two* points [Qf] somewhere separated on the F-surface, and we will be dealing in the realm of *exotic* polyangular architecture. As will be shown in due course, various existing involute gear sets, appearing in their own light to be simple and thereby explicable, appear in the light of these definitions to be exotic and thereby, not surprisingly, somewhat inexplicable [59] [61]. Looked at from this point of view exclusively, hypoid gearing (always non-involute as we know) will almost always be exotic; whereas it is a fact that throughout the whole range of plain polyangular involute gearing the half angles κ at the vertices of the cones of Wildhaber are equal, in hypoid gearing these angles are almost always unequal.

Important aspects of the equilateral transversal

11. Refer to figure 6.01. So, at least for convenience while we wait for alternatives or for unseen troubles, let us stay with our chosen point [Qx] not only somewhere upon the F-surface (thus ensuring a non-exotic *plain* polyangularity), but also exactly upon the unique equilateral transversal determined by our choice of offset R. Look again at figures 4.07 and 4.08, look again at figure 22.08 in [1], and perceive the relationships (§4.66). As we relocate our point [Qx] — step by step by trial in synthesis — to and fro along an equilateral transversal, the direction of the relevant sliding velocity v_{2Q3}, and thus the direction of the relevant pair of mated rulings upon the naked wheels yet to be located, will be seen to vary according to a familiar tan-angle law. Please note incidentally that the CDL is merely a special one of these

THE PLAIN POLYANGULAR OPTION §6.12

transversals; it obtains when R is zero; see the equations at §1.37 written exclusively for the crossed helicals, and see equation (11) in the box at figure 22.06 in [1]. *The surface traced out by the range of possibilities for the location of the relevant velocity vector v_{2Q3} — this is the sliding (the rubbing) velocity between the flanks of the real teeth yet to be synthesized at the point of contact Q when Q is at [Qf] — along a given equilateral transversal at offset R is, as it was for the special case when R was zero, a parabolic hyperboloid.* See figure 6.01. The square prismatic rods that look like threaded beads upon a taut string in figure 6.01 — the equilateral transversal is the said taut string — are simply aids to visualization; the same may be said of the two labeled tubes that may be seen; correctly placed in relation to chapter 4, these pass longitudinally through a selected two of the said prismatic rods.

12. I have used the numerical data of WkEx#1 (§5B.09) for constructing figure 6.01. Figure 6.01 shows not only the parabolic hyperboloidal nature of the distribution of the linear velocity vectors along the equilateral

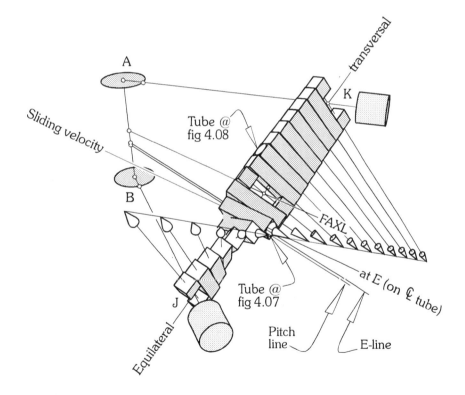

Figure 6.01 (§6.10). Panoramic view of the array of velocity vectors along an equilateral transversal. The numerical dimensions of this figure relate directly to WkEx#1 and WkEx#3. Both of them are catered for. The figure however is also of general significance. Much of the general geometric and algebraic discussion in this chapter is based upon it.

transversal in this particular case, but also that there might be much to be gained by traveling (in the various processes of design by trial) along this particular line. Notice first that the directions of the sliding velocities, which occur at the various possible [Qx], determine the shape of the swept-out surface (the parabolic hyperboloid); but notice next that the magnitudes of the vectors (with the basepoints upon the transversal) display themselves in a straight-truncated, rectangular array: the vectors are all perpendicular to the transversal and the tips of the vectors are in a straight line. See some examples of this known phenomenon appearing in chapters 3 and 9 in [1]; see the photo-frontispiece to chapter 3 in [1]; and see [1] [§21.08]. See also, in figure 6.01, not only (a) the square prismatic rods already mentioned, not only (b) the tubes already mentioned, but also (c) the shown velocity vectors collinear with the centre lines of the said rods and tubes, and (d), most importantly, that the rectilinear array of linear velocity vectors illustrated there is a *part only of the whole helicoidal array of linear velocity vectors surrounding the pitch line* (§1.09). Please note, incidentally, that the transversal JYK in figure 6.01 does not need to be (nor is it) perpendicular to the pitch line. The *whole helicoidal array* was first mentioned in this book at §1.09, mentioned again at §4.26, and described against its broader context within gears at §5A.08. It was also generally studied earlier [1], but in great enough detail there to justify our recalling here the photographic frontispieces and their captions at the title pages of the chapters 5 and 9 in that book. At the real risk of creating confusion (and this reference might be omitted upon a first reading), refer also [1] [§21.19].

13. Like the figures 4.07 and 4.08, figure 6.01 is in perspective. Its orientation does not exactly match that of the said figures, but its dimensions are the same. Note the locations indicated in figure 6.01 of the two particular tubes of the two said figures, and see those particular tubes now as part of a continuum. The continuum of tubes with its attendant array of velocity vectors is illustrating two things: (a) that the pairs of mated naked teeth may be put at a series of predetermined locations along the equilateral transversal, thus providing for the naked wheels and the gear sets that might be possible as alternative plain polyangular solutions in this case; and (b) that the sliding velocity increases as we depart further and further away from the pitch line, the fixed location of the pitch line being determined of course by the prevailing k ($k = +0.6$). The array of vectors swept out, as has been said, is a *rectilinear* parabolic hyperboloid [69]. The central point of this, however, is not at E. The central point, say H, is at the foot — at the transversal end — of the shortest distance between the pitch line and the transversal. This central point H is not at, but somewhat remote from, the central axis of the helicoidal array, which central axis is, of course, the pitch line, and the location of this depends, as has been said, upon the value of k. The two mutually perpendicular generators of the said linear array, this rectilinear, ruled surface, occur along (a) the said transversal JHEYK, and (b) the velocity vector at that point upon the transversal where that vector is least namely at H.

14. Thus we see, incidentally, that equiangular architecture, despite its spectacular display of special phenomena, does not display minimum sliding velocity at its otherwise important point E. By definition in equiangular architecture, we have our [Qx] at some point E, and at this E the sliding velocity is always in a plane normal to the CDL, but, unless the path itself is in this same plane, the point [Qp] on any given path (namely the point where the sliding velocity is least), is not at E (§3.53, §5.06).

An important observation

15. A fascinating aspect of figure 6.01 which might be useful may be noted as follows. *Independently of k, the sliding velocity vectors at J and K are respectively perpendicular, not only to the transversal JHEYK itself, but also to the shaft axes of the wheels 2 and 3 which obtain, conversely, at K and J. While the vector at K is perpendicular to the shaft axis at J, in other words, the vector at J is perpendicular to the shaft axis at K.* This somewhat spectacular theorem may be proved by observing that, because v_{3K2} at K equals v_{3K1} at K by virtue of the fact that the moving wheel 2 cannot slide upon its moving shaft 1 there, and because the segment (J_1-K_1) embedded in link 1 is fixed in length by virtue of rigidity, v_{3K2} must be perpendicular to the rotating shaft 1 at the fixed wheel 3. The corresponding argument for the point J on wheel 2 is of course symmetrical. Refer to [1] [§12.08] for the nomenclature. See also this book §12.06. Later, at §6.33, another important theorem is enunciated. It deals not only with this particular right line among the relatively moving bodies, but with all such lines [1] [§3.01]; and don't forget that for any given k all of the ∞^3 of right lines among the relatively moving bodies here are legitimate paths for the point of contact Q.

The figures 6.02

16. The material of this paragraph is somewhat dense. It will not be easy to understand immediately. Although necessary for a full understanding of the intricacies of polyangularity, it might be postponed until after the more straightforward material begins at §6.38. For a first visual impression of the matter, refer to the photo-frontispiece at chapter 3 in [1]. The figures 6.02, tagged in order (a), (b), (c) and (d), show a series of orthogonal projections of the rectilinear array of the vectors of figure 6.01. The projections are drawn by looking in turn in four different significant directions. We look firstly (a) parallel with the FAXL and thus parallel with the linear velocity vector at E and normal to the polar plane there; in this direction we see the vectors appearing parallel and we see the way in which a single traveling vector turns through a total angle of 180° on its way from one end of the infinitely long transversal to the other (§4.81). We look next (b) parallel with the pitch line and thus along the axis of the whole helicoidal array; in this direction we see the vectors appearing each perpendicular to its relevant radius h drawn from the pitch line outwards (§1.37); and we see the components ($\omega \times h$) of the

§6.16 GENERAL SPATIAL INVOLUTE GEARING

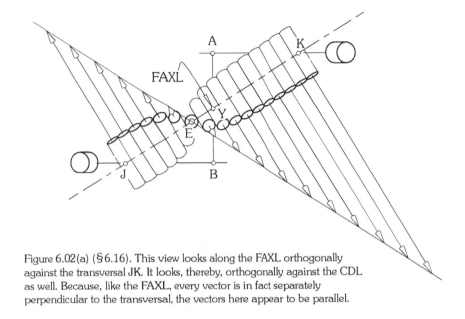

Figure 6.02(a) (§6.16). This view looks along the FAXL orthogonally against the transversal JK. It looks, thereby, orthogonally against the CDL as well. Because, like the FAXL, every vector is in fact separately perpendicular to the transversal, the vectors here appear to be parallel.

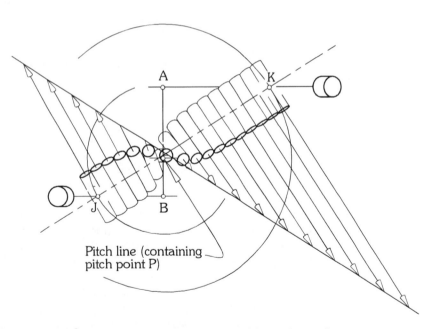

Figure 6.02(b) (§6.16). This view looks along the pitch line orthogonally against the CDL. It looks, thereby, non-orthogonally against the transversal JK. Because every vector is in fact arranged to siuit the helicoidal field surrounding the pitch line, they appear here tangential to concentric circles at centre P.

THE PLAIN POLYANGULAR OPTION   §6.16

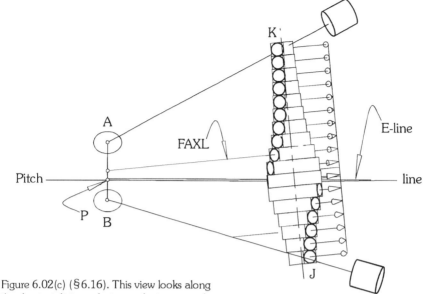

Figure 6.02(c) (§6.16). This view looks along the shortest distance between the transversal and the pitch line, namely along the intercept (H-[Qp]), and thus orthogonally against both of those two lines. It looks non-orthogonally against the CDL. Because the vectors are in fact each perpendicular to the transversal, and becaue the pitch line is in the plane of the paper, the vectors appear not only parallel but of equal length.

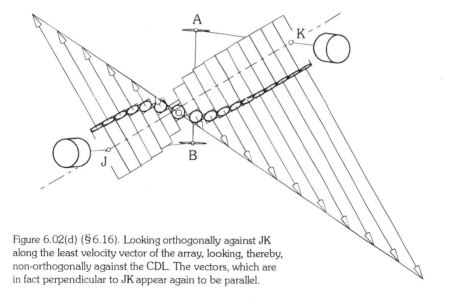

Figure 6.02(d) (§6.16). Looking orthogonally against JK along the least velocity vector of the array, looking, thereby, non-orthogonally against the CDL. The vectors, which are in fact perpendicular to JK appear again to be parallel.

vectors displaying themselves at their true magnitudes; the vectors do not appear parallel in this view; they appear tangential to a series of concentric circles drawn about the pitch line which itself appears as a point of course. We look next (c) along the shortest distance between the transversal and the pitch line; the length of this shortest distance will later be called q, and I wish to state now (as I did before) that it terminates at its transversal end in point H; *in this view we see the true magnitude of the angle ϕ between the transversal and the pitch line*; in this view also we see the velocity vectors again appearing parallel because, residing in a series of parallel planes all normal to the transversal, the said vectors are, indeed, all perpendicular to the transversal. That this is so is not surprising: *by virtue of definition the transversal is a right line in both of the relatively moving bodies*, namely the two gear wheels [1] [§3.01-70]. We look lastly (d) along the least velocity vector of the array at H; we see here why the centre point H of the array is likely to be more important algebraically than the earlier known point E of the array. *It is above all important to understand that H is not at E.* Generally speaking, however, please be reminded of what was said at §6.12 item (d), namely that this simple rectilinear array of an ∞^1 of velocity vectors distributed at points along this single equilateral transversal is only a miniscule part of the whole helicoidal array of an ∞^3 of velocity vectors tangential to the helices of pitch p coaxial with the pitch line (§5A.08). The whole complex of the velocity vectors is, incidentally, not a linear but a quadratic complex [1] [§21.19]. We have been reminded here that generally in skew gear sets the equilateral transversal JYK is neither perpendicular to, nor intersecting with, the pitch line. There is however the special case when k is unity. When k is unity the pitch line is at the FAXL, the E-line is at the FAXL also, and the transversal and the pitch line intersect perpendicularly. Finally let me say again that, while the ramifications of the above material are not easy to digest immediately, we need at least some of them for a clear understanding of polyangularity; and there is more to come (§6.33 *et seq.*). The simple, practical, worked example beginning at §6.42 (WkEx#3) will, I hope, be easier to understand, even if only superficially.

The figure 6.03

17. For further clarification of the four figures 6.02, figure 6.03 shows a second panoramic view. It shows, somewhat unconnectedly but similarly oriented, the generators only of the parabolic hyperboloid of figure 6.01. The changed overall appearance of the surface is due to the fact that the generators are now truncated by the coaxially arranged circular cylinder ghosted in. The axes coaxial are at H-z. The central point H (which is hidden) is at the intersection of the shown axes H-x and H-y. Note the special *key generators* of the surface at the quarter points marked $\pi/4$ (§4.77); and note the measure *2D* of the surface, which, seen as an intercept upon the transversal, evenly straddles the central point H (§4.77). Note that the generators located at the extreme limits of z, at $z = \pm \infty$, are parallel with one another.

Circular cylinder of radius R coaxial with the CDL

18. Some investigators may wish to explore the following alternative route to the general conclusion that the F-surface is of central importance in design. The route reveals other interesting phenomena. Erect in the mind's eye a circular cylinder of radius R, coaxial with the CDL. This might be called for convenience the R-cylinder at offset R. Understanding next that this cylinder might be mounted for a different family of gear sets having the same set of characterizing parameters C, Σ, k and R, we see that the unique location of a point E upon an R-cylinder depends upon the collection of exceptional phenomena already called the equiangular. Leaving aside internal meshing which is, as has been shown (§3.32), not possible with skew gears, the next question might reasonably be as follows: can every point other than the found E on the R-cylinder be a legitimate point for other architectures using the same defining data, or is there a more restricted locus of such points? Given the already established importance of the F-surface, the answer is that there is a restricted locus, and that it is a curved line.

19. Matters may be clarified as follows. Nominate upon the R-cylinder its mid point Y such that (a) a plane through O_1 and containing the CDL bisects the angle Σ, and (b) a plane through O_1 normal to the CDL bisects the centre distance C. The segment of line (O_1-Y) is thus of length R, and the axis O_1-y (see figure 1.07) continues to be seen as one of the several lines of

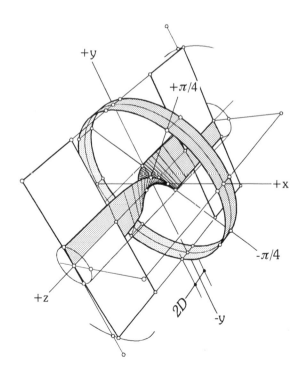

Figure 6.03 (§6.17). Panoramic view of the parabolic hyperboloid of the rectilinear array set at roughly the same orientation as the one at figure 6.01. See the central point H (occluded at the origin), the key generators at the 45° limits delineating the measure 2D of the surface, and the circular cylinder coaxial with the z-axiz which is evenly truncating the generators.

§6.20 GENERAL SPATIAL INVOLUTE GEARING

point-symmetry among the shaft axes. With a new origin O upon the surface of the chosen R-cylinder at Y, let us set up cylindrical axes in the circumferential and axial directions and ask the following question. Given some circumferential angle ε measured from O, at what distance T, measured axially along the relevant generator of the cylinder, will that unique point F occur where the sum of the lengths of the perpendiculars dropped from F onto the shaft axes be (for points upon the said generator) at a minimum? Recognizing that this will occur when the angles between the dropped perpendiculars and the said chosen generator are equal, and by using the two similar right-angled triangles thus displayed, we can quickly find that the minimum occurs at $T = (C/2) \cot (\Sigma/2) \tan \varepsilon$. Notice that, when $\Sigma = 90°$, $T = (C/2) \tan \varepsilon$. Notice also that T is independent of R; this means that there is a ruled surface in space (covering all R) upon which an ∞^2 of F-points can be found. From the above equation we can see that this surface is a rectilinear parabolic hyperboloid with its two principal axes set upon the CDL and the FAXL respectively (§4.82). It contains also the two shaft axes; and it is, precisely, the F-surface as was found before (§5B.16).

20. Refer to figure 6.04. Using the same numerical values as those used in WkEx#1 at chapter 5B and at figure 6.01, namely $C = 80$ mm, $\Sigma = 50°$, $k = 0.6$ and $R = 140$ mm, figure 6.04 shows a planar plot of T against ε

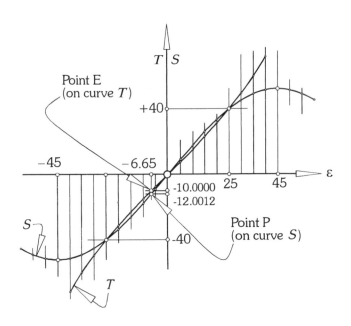

Figure 6.04 (§6.20). The tan-curve for T and the sine-curve for S drawn upon the R-cylinder, with numerical details from WkEx#1. The same curves (and the same numerical details) apply also for WkEx#3, beginning at §6.38.

THE PLAIN POLYANGULAR OPTION   §6.21

with origin at O. This line-locus of the F-points upon any R-cylinder is of course a tan-curve. It is symmetrical about its own centre at O, independent of k, and its equation is, as previously mentioned,

$$T = [(C/2) \cot (\Sigma/2)] \tan \varepsilon.$$

See the curve marked T. The value of k determines the location of that special F-point upon the T-curve which is the E-point for the gear set of that k. We can also plot in figure 6.04, the section cut upon the R-cylinder by the ever-present cylindroid which determines (for given k) the location of the pitch line. This intersection, using S for the ordinate, and using equation (3) in the box at figure 15.04 in [1] to calculate S, works out to be

$$S = [(C/2) \operatorname{cosec} \Sigma] \sin \varepsilon;$$

and this of course is a sine-curve, also symmetrical about its own centre O and also independent of k. See the curve marked S at figure 6.04. The two curves cross one another at their common origin O and at the shaft axes. Go now to the *square* offset skew gear sets mentioned at §1.14 and notice that, when $\Sigma = 90°$, $T = (C/2) \tan \varepsilon$, and $S = (C/2) \sin \varepsilon$.

21. Refer again to figure 6.04. I wish to discuss the locations of points upon the R-cylinder determined by the pitch line and the E-line. These lines appear represented as points P and E on the shown curves, and the points are of course related. Remembering that r_2 in the numerical example here (the same as at chapter 1) is 52.0012 mm (namely 40 plus 12.0012), and that ψ_2 is 31.65° (namely 25.00° plus 6.65°), please read the shown coordinates at the points P and E. While point P is on the sine-curve marked S (which curve is occasioned by the shape of the cylindroid), point E is on the tan-curve marked T (which curve is correspondingly occasioned by the shape of the relevant parabolic hyperboloid namely the F-surface). See the mentioned sine-curve standing out clearly at figure 1.07, and see the same curve and the mentioned tan-curve nestled together at an intermediate R at figure 5B.01. Please note that the tan-curve marked (d) at figure 1.07 is not the same tan-curve as the one under consideration here (§6.23, §4.29).

Some more-incisive remarks about the F-surface

22. So the discovered F-surface is, as we can see, of central significance in design. Given the two shaft axes (and of course their common perpendicular the CDL), the relevant F-surface may be ruled by two intersecting sets of generators as shown in figure 5B.01. The generators of one of the families cut the FAXL perpendicularly at all points along the FAXL and cut the shaft axes, and the generators of the other family cut the CDL perpendicularly at all points along the CDL and contain the shaft axes. The surface may be recognized geometrically, either as a rectilinear parabolic hyperboloid, or as a special regulus [1] [§11.15]. Its three principal axes intersect orthogonally at O and are coincident with the axes of symmetry already established by the two shaft axes and the CDL. It needs to be understood that, independently of

§6.23   GENERAL SPATIAL INVOLUTE GEARING

the fixed parameters of the gear set including even S, the shape of the F-surface remains always the same. It is not the shape of the surface that varies from figure to figure and gear-set to gear-set; it is, quite simply, *its size* (§4.77). Exercise: try to draw with pencil and paper — to some selected scale of course — either the whole extent (?), or a portion only of (!), a rectilinear parabolic hyperboloid.

23. As mooted already (§6.21), the F-surface of figure 5B.01 should not be confused with the surface marked (d) in figure 1.07. Both of these surfaces are doubly-ruled rectilinear parabolic hyperboloids and they have one of their principal axes in common (the CDL), but they have different central points and different sets of principal axes (§4.79, (§4.82). The central point of the surface marked (d) in figure 1.07 is at the pitch point P, and one of its principal axes is along the pitch line. The two surfaces relate, moreover, to different phenomena. The surface marked (d) in figure 1.07 appears with only one set of its rulings because these rulings are showing the directions of the rectangular array of sliding-velocity vectors at the CDL among a family of crossed helicals of a given k ($k = +0.6$). The surface marked (d) does not appear again in figure 5B.01. It has, for the sake of clarity, been eliminated there. The surface marked (d) in figure 1.07 may however be compared with the surface shown in figure 6.01. These two are both arrays of velocity vectors relating to points fixed in a straight line within a moving, rigid body. Refer to the many passages in [1] about motional right lines. See [1] [§3.20, §5.26, §10.49, §21.05] for examples, and be aware that such parabolic hyperboloidal arrays are characteristic of the helicoidal velocity field surrounding any motion screw — or pitch line, as is the case here.

Location of [Qf]

24. So we have now a well conditioned construction for finding the [Qf] upon any legitimate path for a Q, there being of course two Q, one on the path for drive and the other on the path for coast. We can write this: *for any chosen legitimate path for a proposed point of contact Q, the point [Qf] upon that path is that unique point where the path pierces the F-surface.* Find this already said at §4.27. We can always find the point [Qf]; it occurs where the transversal at the moving Q suddenly cuts, not only the two shaft axes, but also the FAXL, cutting this latter perpendicularly. We have implied here the proposition that, in the event of there being two [Qf], there is exoticism.

Finally some definitive remarks about the elusive [Qm]

25 But what about the elusive [Qm]? An enduring difficulty with [Qm] has been that I have always presented [Qm] as a point involving the local minimum of the algebraic sum of two G-radii (§3.56, §4.15-18, §5A.13 *etc.*), and that, hitherto, this definition has never thrown up a clearly delineated locus for all points [Qm]. We have found no delineated surface in the fixed space upon which all possible points [Qm] lie (§4.43 *et seq.*). Given the na-

THE PLAIN POLYANGULAR OPTION   §6.26

ture of the F-surface as now defined and subsequently discussed in connection with the R-cylinder, however, we can speak more accurately now about the local minima as follows. Choose any line parallel with the CDL; it will cut the F-surface somewhere. Such lines were being inferred of course at §6.19 above, where I sought such a local minimum as we dealt with a moving point in such a line. Such lines are not members of the complex, incidentally, unless they cut the pitch line. We were moving, there in §6.19, upon a representative generator of a particular R-cylinder, which line was nominated by the chosen values of the offset R and the angle ε. Following §6.19, we can say this: as we travel along any line parallel with the CDL, with a moving point say G, the algebraic sum of the distances from G to the shaft axes arrives at a local minimum when G arrives at the F-surface. This matter of the minimum algebraic sum of the two distances will be thoroughly obscured unless we pay attention to what I have just said. If we follow for example a point Q along its legitimate path (or take a point say G along some other path), the algebraic sum of the G-radii as we go will *not* arrive at a local minimum when the Q (or the G) arrives at [Qf], *unless the path is parallel with the CDL*. The minimum will occur, in general, somewhere else. This is due, quite simply, to the nature of [Qm].

> Points [Qm] are path dependent. This means in other words that a found [Qm] for one path will not be the [Qm] for another path that might be chosen to pass through that point.

Distributed evenly throughout the space on either side of the F-surface there is a multitude of points [Qm], and among these the iso-surfaces for equal values of $(g_2 + g_3)$ are distributed in such a complicated way that they warrant here no further kinematic investigation.

26. Recall, however, the various possibilities mooted previously. See the relevant passages listed under [Qm] at the Symbols Index. Recall in particular the discussion about the total strain energy at a point [Qm] in a given path. Somewhat laboriously, this extends all the way from §4.43 to §4.56. Whereas we now have on the one hand the undoubted fact that the F-surface, unambiguously made up of locatable points [Qf], is easy to envisage, easy to construct, and entirely substantiated as a kinematic device for designating the location of a mid point, we have on the other hand this well developed notion now that [Qm] belongs, not here among the immediate kinematics of the problem, but in the more distant realm of the meta-statics (§4.50).

Two special plain polyangulars

27. Although the helitangent at any Q, and thus at any [Qf], is always perpendicular to the transversal there (§3.41), and although, accordingly, the polar plane at any [Qf] always contains the transversal, the helitangent at any [Qf] does not (in general) lie as a generator of the F-surface. Before we can locate the triad of mutually perpendicular planes at any chosen point [Qf] upon the F-surface (§3.44), we need to know, not only that the polar plane

contains the transversal, but also the value of k. The value of k locates the pitch line (which is, as we know, generally not upon the F-surface); and the location and pitch of the pitch line determines the angular orientation of the polar plane at [Qf]. If the chosen point upon the F-surface is at the E-line determined by the prevailing k, the helitangent (without intersecting the CDL) is in a plane normal to the CDL and bisects the angle Σ (§5A.24); this is a special case, and we call it the *equiangular case*. If the ratio k is unity, and the chosen point [Qf] is upon the FAXL, however, we have a doubly special case; the helitangent will lie as a generator of the F-surface; it will lie indeed collinear with the FAXL; *and the corresponding gear-set will be an equiangular offset skew gear set with k unity and the two wheels identical.*

Again the plain and the exotic

28. *It might be said with conviction now that the points F upon the parabolic hyperboloid of the F-surface characterize the plain polyangular.* We have already said however that a [Qx] for an intersecting pair of paths might be chosen with legitimacy anywhere *off* the F-surface. It has moreover been said that the two paths of the two Q need not even intersect (§4.23, §6.10). We have been made aware of course that hopefully practical involute designs straying too far away from the plain polyangular, namely those employing points [Qx] ridiculously far away from the F-surface, will be subject to troubles such as hopelessly pointed teeth and the inability to follow through. Such designs will no doubt be investigated, however, but not, perhaps, in this book by me. See however §8.17 *et seq.* I have already called this arena the exotic polyangular or, more simply, the exotic. *So it might moreover be said that the remaining points G in the whole of space characterize the exotic.*

More about axodes and the naked wheels

29. Needing to deal more intimately in the forthcoming worked examples with the complex nature of the naked wheels, and in looking forward to chapter 7, where even more surprises come to light, I wish to write at this juncture a retrospective and more comprehensive view of the naked wheel in general. In chapter 22 of [1] I did little more than introduce the simplest possible concept of the naked wheel. Confining myself to the crossed helicals, where the single point [Qx] for the intersecting paths of Q is always upon the CDL, I spoke there (a) of the axodes for the relative motion, which axodes are unique for given C, Σ and k, and (b) of the *departed sheets*, which sheets are not unique for the given C, Σ and k. Speaking in the newer terms of this book, the shapes of the departed sheets in [1] depended upon the location chosen for the point C upon the CDL. The central portions of these departed sheets, which I would now call the naked wheels for the action there, were *singly* departed. They were not *doubly* departed as is first explained in connection with equiangular offset gears at §5B.11. Looking now in retrospect, I have earlier discussed the idea of the naked wheel in [1], and

THE PLAIN POLYANGULAR OPTION   §6.30

in this book at §1.31, §4.20, and again at §5B.02. Generally, in chapter 4, I took a more extensive view of the geometry of a mated pair of naked wheels, discussing there the single point of tangency between the intersecting surfaces which always occurs exactly at the intersection of the F-circles with the F-surface, namely at the point [Qf] (§4.20-23, §4.27). Even there, however, I failed to explain very clearly the nature of the complex interaction that takes place between mated pairs of naked wheels.

30. From the philosophical point of view, pairs of axodes may be imagined, with legitimacy, as being *finely corrugated* — being folded, that is, into a single infinity of straight, infinitesimally thin ridges and grooves that follow the generators of the axodes and which, thereby, *mesh* with one another without disturbing the underlying, kinematically correct, axial sliding and circumferential rolling characteristic of axodes. Speaking alternatively not of corrugations but of *spline-line teeth* — they are the same — I said exactly this at §1.33; and I say it again, despite the fact that the axodal surfaces associated with the *meshing* — as distinct from the *machining* (chapter 10) — in skew gears are non-developable. *The departed sheets, on the other hand, selected parts of which become the naked wheels, may not be so imagined.* Although they are similar surfaces — they are, like the axodes, circular hyperboloids — they need to be imagined firstly as being ruled but secondly as being *smooth*. They may not be seen as being able to drive one another satisfactorily. They must be seen to be independently driven. The teeth themselves (which are mere lines) are smooth. Let me go on to explain more clearly what I mean by this.

31. It is in the nature of the relative motion of any pair of meshing gear wheels (a) that a unique screw axis exists at any isolated moment, and (b) that, for the repeated cycle of the motion, a pair of corrugated axodes exist both of which are simple, circular hyperboloids. As we have seen, however, there is also the multiplicity of the naked wheels, each pair of which spring into being in response to a selected point [Qx]. By using the geometry of the system of coaxial helices surrounding the pitch line (§1.09), we can determine the direction of the sliding velocity at the chosen [Qx] and (by using this) sweep out the relevant naked wheels. In general, thereby, any mating pair of naked wheels intersect one another at all points along their common generator, each mated pair of naked teeth being collinear with and thus residing within the said common generator. The actual surfaces of the naked wheels are however also *tangential* with one another. *This may appear, at first, paradoxical and thus untrue; but the said tangency between the surfaces of the naked wheels occurs only at an isolated point; it occurs indeed at [Qf]; and, given a moving point of reference upon the straight line of the intersection, this is precisely where the angle between the intersecting surfaces, passing from positive to negative through zero as the said point goes along the line, changes sign* (§4.20, §4.38).

32. Having said that, I now say this. *No one of a pair of mating naked wheels, except in the special case where the pair is identical with the axodes, rolls on the other without circumferential slip.* The circumferential direction here is perpendicular to the rulings (§1.33); and this direction, incidentally, is not tangential with either one of the intersecting F-circles (§3.43). Any pair of naked wheels, being *departed,* exists in such a way that the line of their intersection (which contains the relevant sliding velocity) is not at but remote from the pitch line; this means that circumferential slipping at the said line of intersection must occur. The main thing that characterizes a properly interacting, interpenetrating pair of naked wheels is that the relative velocity at the point where the two F-circles meet is in the direction of the said collinear pair of rulings. This collinearity of the said vector with the pair of collinear rulings occurs only at [Qf]; elsewhere along the pair of collinear rulings the relative velocity is different (§10.41); and it follows from this, not only that the naked wheels must be seen to be free to interpenetrate, but also that they must be seen to be smooth; for otherwise they would not be free to slip circumferentially. Taking for example the departed sheets with their naked wheels in figure 5B.03, we see that, even though the circumferences of the E-circles there are in exact proportion with the speed ratio k due to the equiangularity prevailing, the naked wheels are not rolling without circumferential slip. It might accordingly be seen that, in the presence of any finite Σ, the naked wheels (in the absence, as yet, of real gear teeth) are naked in two respects. *They are naked firstly because, having become departed, they are no longer clothed with spline line teeth, as the axodes from which they spring are clothed, and they are naked secondly because they are not yet newly clothed with the real gear teeth yet to be synthesized.*

Geometric relations within the linear array

33. As we know already (§5A.08), the distribution of the possible sliding (or rubbing) velocities throughout a gear set is determined by the shape of the linear complex of lines surrounding the pitch line [39]. The said shape is determined solely by the pitch p of the complex which is the same as the pitch p of the relative screwing at the pitch line (§3.47, §5A.08). Due to the axi-symmetry of the complex, it is possible to encompass all possibilities for *right lines* within the complex by considering a series of bi-normals to the helices as we go step by step along any one chosen radial line extending from the central axis of the complex outwards (§5A.08) [1] [§9.20, §11.30]. I proceed now to think along those lines. In gears we may look at the equilateral transversal as being a special one of the right lines of the complex surrounding the pitch line, and study the distribution of the directions and magnitudes of the velocity vectors along it as being simply a geometric consequence of the complex. Figure 6.01 illustrates the rectilinear array of vectors along the equilateral transversal in our particular example here; the array shows the available rubbing velocities at [Qx] as, on going polyangular, we slide our chosen [Qx] outwards from E through the different available points

THE PLAIN POLYANGULAR OPTION　　　　　　　　　　　　　　　　　§6.34

F along the equilateral transversal. It shows in general, however, a twisted linear array characteristic of any right line of the complex, including those two, incidentally, that are the paths of the two points of contact Q. Except in the special case in gears where k = unity, the equilateral transversal at the chosen offset R never cuts the pitch line; the shortest distance q between the pitch line and the said transversal is always finite. I borrow q directly from figure 2.01 because, as will be seen, the fundamental relations here are the same as they were there. The *half measure D* of a linear array having already been explained at §4.77, it is easy to show now (and to say) the following: *the half measure D of the linear array on the particular right line that cuts the axis of the complex perpendicularly is of length p (mm) precisely.* This is interesting, but we find that D grows as we increase the shortest distance q from zero upwards. My held belief, based on sketchy algebra and some executed graphical work, is that the following equations obtain:

$q = p \cot \phi$, which is, in a broader context, the fundamental law, and
$D = p \csc^2 \phi$,

where q is the shortest distance between the axis and the right line, where ϕ is the angle between the right line and the axis of the complex (§2.02) [1] [§19.08], and where D is the half measure of the linear array. See appendix A at 6 (§6.58). Notice in these equations that, when q is zero, ϕ is $\pi/2$, and that as q increases from zero on its way towards infinity, ϕ decreases from $\pi/2$ on its way towards zero.

34. Along any motional right line drawn anywhere among the helices of an helicoidal velocity field, there will be its central point H; this is the point where the right line comes closest (at shortest distance q) to the central axis of the field, and where the velocity vector is least (§6.13); it is interesting also to note that, when the right line is a *path* (not a transversal as is here being considered), the point H is called [Qp] (§5A.09). Looking at the equations inset above we notice that these are saying that, independently of ϕ, the smaller the p the smaller the D, and thus the more severe the *twistedness* of the linear array in the vicinity of H. Twistedness might be seen as the reciprocal of D (§4.77). Recall figure 6.03.

The velocity vector at F

35. *The angle beta.* What we have worked out here might be used to determine directly the angle between the velocity vector at F and the velocity vector at H. I wish to call that angle β. As explained already in another context (§6.14), *this β is not the same as the angle between the velocity vector at F and the velocity vector at E*; H, as has been explained, is not at E. The shortest distance q from the pitch line to the transversal JK is not vertical (§5A.29), and the vector at H is not horizontal; but we have here (with H at the centre) a more satisfactory way of viewing the matter. The reason is that the happenings along the transversal JK in gears are in many ways more important than the happenings among the lines of the complex as a whole.

Nevertheless we have, now, two ways of viewing the matter of the vector at F: (a) in terms of H and the rectilinear array as explained above, and (b) in terms of the pitch line and the basic equation of the screw as explained at §1.51. Different methods of calculation based on these two equivalent views of the said matter will, of course, give the same result. See a worked example at §6.44, and go to appendix B at 6 for the geometric details. In any event and by whatever method we find it, the location of the chosen F and the direction of the velocity vector there will be used to find the two related angles τ we have been looking for. They may be found by dropping the shortest distances between the velocity vector at F and the two shaft axes. From the found τ (when found) we can go on to get the two needed λ (§10.08). We don't have as yet an algebraic relationship between the angle β at F and the angles τ, but for finding this relationship the construction of a unit sphere and the application of spherical trigonometry might be the way to go. Pending the unknown algebra, I intend to proceed both graphically and numerically with a chosen worked example. See the forthcoming WkEx#3.

The parabolic hyperboloid of the rectilinear array

36. *Local algebraic relations.* Now look again at figure 6.03, which is a free-standing picture of a parabolic hyperboloid truncated in such a way that all of its generators of the relevant family are cut to the same length and distributed evenly (§6.17). It is truncated in other words by a circular cylinder coaxial with one of its two identical principal generators. Another such truncated one was drawn at figure 22.08 in [1] [§22.50]. Figure 6.03 might be seen as the twisted parabolic hyperboloid of the linear array here whose central point is H. Looking at all of the figures of the linear array it is easy to see that, in any one array,

$b = D \tan \beta$, and
$v_F = v_H \sec \beta$,

where b is the length of the intercept (H–F), v_H and v_F are the velocities at H and F respectively, and the angle β (already defined above) is the angular displacement from an origin-zero, set at H. See appendix B at 6. In the special cases of planar and spherical (bevel) gearing p is zero and D is zero and the vectors of the array are, respectively, either wholly in the reference plane or wholly in the plane of the transnormal. Pictures showing the lines of the relevant regulus of some parabolic hyperboloids where D is zero may be seen at figures 7.12(d), 7.13(b) and 7.16(b) in [1]. Read also [1] [§11.14-17]. This matter is also mentioned briefly here (in this book) at §4.77. See also figure 7.12(d) in [1], the text for which appears at [1] [§7.68].

Need for a dedicated algebra

37. Not in this, but in the next paragraph I try to summarize the overall algebraic results obtained to date, partly by myself but mostly by others [42]

[54], to describe definitively the elements of the basic architecture, the naked wheels, the paths of the points of contact, the locations and magnitudes of the radii a and the angles α and the interacting base hyperboloids for the generalised gear set in the circumstances of plain polyangularity; and I wish to make beforehand the following comment. I trust that the overall structure of this complicated algebra will be made clear enough by me for its intended purpose in this book. Knowing that in the long run a dedicated algebra will be necessary for quick, effectively directed design by trial in this special field of involute gearing, its main purpose here in this book is not to do that, but to assist with the underlying logic. The logic here, I suppose, is (a) to show that such an algebra, although complicated, may indeed be written and effectively used, and (b) to suggest that in due course it may, somehow, be simplified. Perhaps the modern *screw-algebra* of Parkin and others [27], and/or the actual *screw-capable, electronic hardware* mooted Parkin [27], will have their parts to play, but all known alternative algebras and electronic devices do come down to matrices and numerics in the long run, so the matter remains a question. This book is however confining itself to the geometric structure, the general appearance and the methods of manufacture of the mechanism and machinery being discovered here, and for this purpose I am using computer-driven packages as investigative and design tools. Let us not forget in this connection that it is only by virtue of the underlying algebra built in to the software packages that the said tools are effective. To the extent that the said algebras *are* indeed within, we can — however slowly it may be — design and direct the machining of real gear sets without a dedicated algebra.

An algebra for the plain polyangular architecture

38. Here is the place to enlarge upon the equations that were listed for the various S, a and α, for equiangularity only at §5B.16. It is clear that the geometry of the polyangularity here is more complicated than that for the equiangularity of chapter 5B; and the algebra is also more complicated. It will be evident however from the already-described geometry of the paths of Q in polyangularity that the development of an algebra for plain polyangularity in gear sets might take the following steps: (a) from the pitch line [PLN] and its pitch p (§1.41), which together make a firm foundation for the algebraic work, we can proceed, using the basic equation of the screw (§1.51), to the relative linear velocity of sliding v_{2Q3} at [Qf]; this occurs as we know along the helitangent [HTG], and, as we know, this is one of the members of the fundamental triad of lines there, the others being the transversal [TVL] and the best drive (best drive line) [BDL]; thus (b) we can next get from the helitangent to the best drive by means of a fairly simple matrix transformation involving a single rotation through a right angle about the transversal; next (c) after introducing the angles of obliquity δ, both in the polar plane at [Qf], we can go from the best drive to the relevant path [PTH], of which of course there are two; and finally (d) we can write equations for the shortest distances a and the angles α between the two paths and the two shaft axes [SAX]. Us-

ing the acronyms suggested here, we can symbolically write the algebraic journey, jumping from located line to located line in the set, as follows:

$$[PLN] \to [HTG] \to [BDL] \to [PTH] \to [SAX].$$

Killeen [54] has written such an algebra. In recording his results I will use the above acronyms as subscripts to sets of unit vectors, for example ($\underline{l}_{HTG}$, $\underline{m}_{HTG}$ and $\underline{n}_{HTG}$), denoting the components in the directions x, y, and z of the directions of these located lines within an overall frame of reference O-xyz which is fixed in the body 1. This frame, used consistently by Killeen, is illustrated at my figure 1.01. The following summarised results, written by me, are a somewhat shortened account of his results, and they are paraphrased here to accord, not with his symbolism which comes in the main from Australian Standard 2075-1991 which closely follows ISO 701 International Gear Notation, but with the separately established symbolism of this book (§P.04). The complexity of the equations is once again, as it was at §5B.16, grotesque; and my remarks about that — Killeen concurs — remain as before (§5B.16).

39. To find the equations for equiangularity which (not surprisingly) work out to be special cases of those that follow here, go back to §5A.16. See also the expressions for the two radii e (easily obtained) at §5.27. See next the simple modification of the geometry to accommodate the two radii f for plain polyangularity at §6.06. The radii e and the radii f present no problem. Killeen has accordingly been able to show without difficulty in his work that equations for the two S_f (the distances along the axes from the CDL to the centers of the two F-circles of the set) may be written

$$\underline{S}_{f2} = \underline{l}_{SAX} [-R \tan (\Sigma/2)] [(1+j \cos \Sigma] / [j+1]$$
$$+ \underline{m}_{SAX} [R] [(1+j \cos \Sigma] / [j+1]$$
$$+ \underline{n}_{SAX} [C/2],$$

$$\underline{S}_{f3} = \underline{l}_{SAX} [+R \tan (\Sigma/2)] [(j+\cos \Sigma] / [j+1]$$
$$+ \underline{m}_{SAX} [R] [(j+\cos \Sigma] / [j+1]$$
$$+ \underline{n}_{SAX} [-C/2].$$

An effective algebra [54] can be initiated, on stepping from [PLN] to [HTG], by invoking the fundamental law (§2.02), which itself involves the basic equation of the screw (§1.51). My q at §2.02 is a special case of my h at §1.51 and, in the context here, either one of them is the shortest distance from the pitch line to [Qf], which point (in both of the paths) is coincident with the key point F, which key point (located upon the transversal according to j) is unique in the architecture. Killen thus finds the direction of the velocity vector v_{2Q3} at [Qf] and, by thus finding the components $\underline{l}_{HTG}$, $\underline{m}_{HTG}$ and $\underline{n}_{HTG}$ of the direction of the helitangent there, he next writes (by cross multiplication of the relevant vectors) the components $\underline{l}_{BDL}$, $\underline{m}_{BDL}$ and $\underline{n}_{BDL}$ of the direction of the BDL, which line belongs of course to the fundamental triad at F. Killeen goes next, using the two angles of obliquity δ_D and δ_C, to the paths [PTH]. This step is not any easy one. Having written the components $\underline{l}_{PTH}$, $\underline{m}_{PTH}$ and $\underline{n}_{PTH}$ of the direction of the path, and knowing the unique location of F, he can, with the known location and direction of the relevant shaft (of which of course there are two), he can next write expressions for the four radii a (the shortest distances between the shaft axes and the relevant paths), the four angles α (the inclination angles of the base hyperboloids), and the four distances S_a (the distances measured along the shafts from the CDL to the centres of the throat circles of radius a of the four different base hyperboloids upon which the four different involute helicoids of the four flanks of the teeth are mounted. Killeen's expressions follow.

$$a = 2[w]\frac{(C/2)[\underline{l}_{PTH}\cos(\Sigma/2) + \underline{m}_{PTH}\sin(\Sigma/2)] + R[\underline{n}_{PTH}\sin(\Sigma/2)]}{\sqrt{\{[\underline{l}_{PTH}\cos(\Sigma/2) + \underline{m}_{PTH}\sin(\Sigma/2)]^2 + [\underline{n}_{PTH}]^2\}}},$$

where $[w]$ is either $j/(j+1)$ or $-1/(j+1)$ according to which wheel it is and whether we are dealing with drive or coast. But which is which? Formulae on the other hand for the angles of inclination α at the throats of the base hyperboloids may be written [54]

$$\sin \alpha = \frac{\pm\, \underline{l}_{PTH}\sin(\Sigma/2) + \underline{m}_{PTH}\cos(\Sigma/2)}{\sqrt{\{[\underline{l}_{PTH}]^2 + [\underline{m}_{PTH}]^2 + [\underline{n}_{PTH}]^2\}}}$$

where the unit vector $(\underline{l}_{PTH}, \underline{m}_{PTH}, \underline{n}_{PTH})$ is a complicated function of the unit vector $(\underline{l}_{BDL}, \underline{m}_{BDL}, \underline{n}_{BDL})$ involving its cross multiplication with an immense 3×3 matrix including various functions of Killeen's angles ρ which relate in turn to my angles δ, and where $(\underline{l}_{BDL}, \underline{m}_{BDL}, \underline{n}_{BDL})$ is yet a further complicated function of $(\underline{l}_{HTG}, \underline{m}_{HTG}, \underline{n}_{HTG})$ which was mentioned earlier as having been derived from the pitch line. There must of course be four different α, and this is reflected no doubt in the algebra, whereby the above expression for α gives four different solutions. Next we may write, because we need it, and by means of Pythagoras directly, a similar four-valued expression for $\cos \alpha$.

$$\cos \alpha = \frac{\sqrt{\{[\underline{l}_{PTH}\cos(\Sigma/2) + \underline{m}_{PTH}\sin(\Sigma/2)]^2 + [\underline{n}_{PTH}]^2\}}}{\sqrt{\{[\underline{l}_{PTH}]^2 + [\underline{m}_{PTH}]^2 + [\underline{n}_{PTH}]^2\}}}$$

Killeen goes on to derive expressions for what I call the distances S_a of the centres of the base hyperboloids measured along the shaft axes from the CDL, but these (although evaluated below) are not included here. Killeen's methods could also be used, no doubt, to discover the shapes of the naked wheels in the form of the half-angles τ of their asymptotic cones (given the helitangent), but details are not currently available. I get the angles τ accurately enough at the CAD incidentally; but see also appendix C at 6.

These said distances S (both S_f and S_a), from the CDL to the centers of the various circles of radii f and radii a_D and a_C respectively of the wheels at their respective axes, are close in magnitude with one another. The differences between them are however highly significant, for without these differences the constructed teeth at the meshing would not be correct. The numerical values of these distances, it should be understood, are quoted here

§6.40   GENERAL SPATIAL INVOLUTE GEARING

(as follows) to demonstrate, not so much their relative importance or their actual magnitudes, but the fact that the algebraic expressions we have for them, while frightening me (and likely the reader too), do not frighten the spreadsheet and its computer. The following selected results for the various S in WkEx#3, taken directly from my computer-driven drawings, where the algebra (hidden within) is still unknown to me, accurately agree (within six significant figures) with the more accurate calculated results at the Killeen spreadsheet. See the TNR for a more complete evaluation of WkEx#3. The numbers there are taken more often from the Killeen spreadsheets for better accuracy, but the reader should be assured that with most acceptable CAD packages he or she can get (a) the visual picture, and (b) the numbers however low the chosen setting for accuracy may be.

$[S_{f2} = (A–M) = 122.0143]$
$[S_{f3} = (B–L) = 131.7519]$
$[S_{a2D} = 126.6976]$
$[S_{a2C} = 123.8787]$
$[S_{a3D} = 120.9384]$
$[S_{a3C} = 139.0682 \text{ (mm)}]$

It is instructive to notice that, while in the case of the smaller wheel the two centres of the two base circles on the shaft axis straddle the centre of the F-circle there, in the case of the larger wheel the said two centres *are together on the same side of* the centre of the F-circle there. This is related to the fact that the *beta shift* (§6.44) in going from WkEx#1 to WkEx#3 has been great enough to reverse the handedness (from left handed to right handed) of the hyperboloid of generators of the naked wheel of the larger wheel.

Law of the speed ratio $k = [a_3 \cos \alpha_3]/[a_2 \cos \alpha_2]$

40. Changing the subject now to that of the kinetostatics, go to my unguarded and as yet unproven statement at §5B.17 that for a given path in involute gearing the ratio $[a_3 \cos \alpha_3]/[a_2 \cos \alpha_2]$ will equal the speed ratio k whatever the circumstances. Although not surprising, and despite the ingenuous but compelling nature of the equivalent RSSR at figure 2.02, this statement — let me call it, for the sake of argument, *the law of the speed ratio* — is somehow amazing. Accordingly it and its ramifications are under their own sub-heading here. It is clear now (having come thus far in this chapter 6) that the appearance of this law and its explanation was not due at §5B.17 to some special property of the equiangularity prevailing, but to a general, geometrical property of the linear complex which applies in any event. Imagine this: take any point on the F-surface and erect the rays of the unique polar plane whose centre is at F; take any one of these rays and show that in general $[a_3 \cos \alpha_3]/[a_2 \cos \alpha_2] = k$, where k is the ratio (J–E)/(E–K), which latter can always be found by ordinary simple division at E of the length (J–K) of the equilateral transversal JEFK. Alternatively, one can now see, given C and Σ, and given the freedom to engage in polyangularity, that a

chosen k will locate the pitch line and its pitch (§1.14), that, given the chosen F, that data will determine the orientation of the polar plane at that F (§6.02), and that, having chosen any ray of that plane (having chosen in other words a single *legitimate path*), the said innate, geometrical property of the linear complex ensures that the law of the speed ratio holds.

41. Having looked at the kinematics, let us look now at the statics. In the absence of friction, the two equal and opposite forces **F** at the point of contact Q (wherever Q might be in its path) are engendered by the two opposing but unequal torques at the shaft axes whose values are, quite clearly, $[a_2 \cos \alpha_2]$ times **F**, and $[a_3 \cos \alpha_3]$ times **F**, respectively. We find thereby that the mechanical advantage (which is by definition the simple numerical reciprocal of the angular velocity ratio) is the reciprocal of the above expression for k. It would appear in other words that, given the truths of kinetostatics, *the law of the speed ratio is self apparent*. My argument here, however, although in a way convincing, seems curiously circular to me; and I for one will be watching in the worked examples that follow for convincing numerical evidence that might appear in support of the law (§6.53). In the special case of planar gearing, incidentally, the law breaks down to saying, simply, $[a_3]/[a_2] = [r_3]/[r_2] = k$ (§1.39-49). Finally also we can see here that, *in the presence of friction*, we enter a whole new area of study that is not only irrelevant to the matters at issue here but also beyond the scope of this book.

Beginning of worked example #3

42 *The data and the introduction of radius-ratio.* This is a somewhat timid modification of WkEx#1. It is a first excursion into the oblique, plain polyangular sets. Please be aware that, while the figures 6.01 and 6.02 have been used hitherto as vehicles for general argument, their physical dimensions are such that they relate exactly to the data of this exercise. They may now be seen, indeed, as directly pertaining. With this exercise we take a small, first step. We begin for convenience with the same initial data as was used for WkEx#1 (§5B.01, §5B.09); but we here modify the position of the path-intersection point [Qx]. Keeping the speed-ratio k fixed at the same value ($k - 0.6$), and thus keeping the E-line fixed, we remove our [Qx] from its old position at E to a new, chosen position at some F upon the same equilateral transversal. Please see, illustrated in figure 5A.01, the point F free to be slid about, and thus to be located at any position along the said transversal. It needs to be understood in the current figures that, as before, the points [Qx] and [Qf] are coincident; we put a newly chosen path-intersection point [Qx] at an F which is by definition upon the F-surface. The new circumstance is thus *plain polyangular*. It is not exotic. To be exotic the paths would need to intersect elsewhere or not intersect at all. Let us now say by way of definition that, whereas (J–E)/(E–K) = k, which is the tooth ratio (namely the speed ratio) for the set, (J–F)/(F–K) = j; and this j is something else. The ratio j might be called the *radius-ratio* (or the *poly-ratio*) for the set; it is the ratio, not of the numbers of teeth, but of the two key-circle radii f.

§6.42   GENERAL SPATIAL INVOLUTE GEARING

We might accordingly note that, while $e_3/e_2 = k$, the tooth ratio, which is always some vulgar fraction incidentally, $f_3/f_2 = j$, the radius ratio, which is often some rounded decimal. We can also state by way of definition (and I do) that the ratio j/k might be called i, where i is a measure of the *polyangularity* inherent.

Refer to figure 6.05 which incorporates, as explained above, the old data from WkEx#1. In this WkEx#3, however, we have chosen for j the value $j = 0.7$. This displaces F but puts it still on the E-side of Y. So the transversal

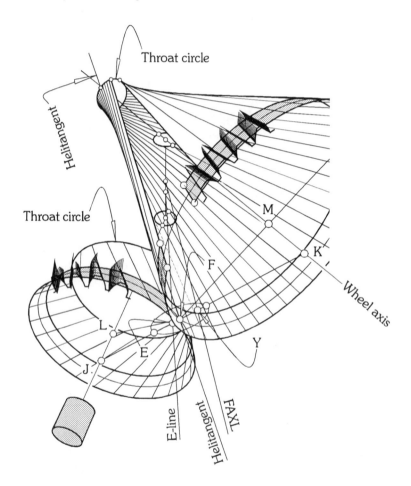

Figure 6.05 (§6.41). Standard perspective view of the departed sheets and the naked wheels drawn for WkEx#3. Polyangularity now obtains; [Qx] is now not at E but at an F; while $k = 0.6$, $j = 0.7$. While the twist angles τ of the naked wheels have become different, the half-angles κ of the Wildhaber cones have remained the same. These matters of τ and κ (not made clear here) are more clearly illustrated at figure 6.07. Notice that the helitangent at [Qx] is no longer parallal with the FAXL.

reads, from its lower end to its upper, JEFYK. See that, because $k = 24/40$ (namely 0.6) and $j = 0.7$, $i = (0.7)/(0.6) = 1.17$ approximately.

43. *The sloping velocity vector at F.* Unlike the equiangular situation at WkEx#1, introduced at §5B.09 and illustrated at figure 5B.02(a), where the sliding velocity at [Qx] was at E and in a plane normal to the CDL, namely 'horizontal' and thus parallel with the FAXL (§5A.23g), the sliding velocity now is no longer thus conveniently located and directed. Here at WkEx#3, its new location at F and its new direction must be found before we can go on. Peruse the figures 6.01, 6.03 and 6.05. Notice how quickly the direction of the velocity vector at [Qx] changes as j increases away from the value of k ($k = 0.6$), and become aware of why I have chosen the value of j, in this first instance, to be so little different from that of k. Notice also that, for no particular reason, I have chosen the value of j to be larger, not smaller, than that of k. This has meant — and this will be evident soon — that the smaller wheel, *the pinion,* becomes larger as the larger wheel, *the wheel,* becomes smaller. Go to appendix B at 6 (§6.67).

44. *The beta shift.* Now by a mixture of direct calculation and computer-driven graphical construction, working to four places of decimals, and appealing not only to the equations at §6.33 and §6.36 but also to the basic equation of the screw at §1.09, the following numbers have been established. For the picture go to appendix B at 6. Having measured ϕ to be 84.3338°, having erected the shortest distance q from the pitch line to the transversal, $q = 1.7117$ mm, and having thus checked, using $p = -17.2521$ mm/rad, that $q = p \cot \phi$, we locate H the central point of the array. The distances, measured downwards and upwards along the transversal from E to H and from E to F respectively, measure 1.0611 and 5.6296 mm. The total intercept (H–F) upon the transversal, which is of course the sum of these, measures 6.6907 mm. From the equation $D = p \operatorname{cosec}^2 \phi$ at §6.33, we calculate D; $D = 17.4219$ mm. Next, from the equation $d = D \tan \beta$ at §6.36, we calculate the relevant β, namely β_F; this $\beta_F = \operatorname{artan}(6.6907)/(17.4219) = 21.0088°$. This checks with the CAD. Having measured v_H (at ω = unity) to be 17.3368 mm/sec, we next calculate, from the equation $v_F = v_H \sec \beta$ at §6.36, that $v_F = 18.5713$ mm/sec. This checks with the CAD. Checking independently now by employing the contents of §6.35, we again erect the same vector v_F at F but by using this time the pitch line and the pitch p and the radius vector h which is of course not (F–E) but the perpendicular dropped from F to the pitch line. We find by measurement that h, namely (N–F), is 6.8745 mm; then, by the basic equation of the screw at §1.09, we find the angle of rotation from E to F about the pitch line to be artan (6.8745/17.2521) which works out to be 21.72595°. This is of course not the angle β, *but it gives within its own context exactly the same result,* namely that $v_F = 18.5713$ mm/sec. Overall we have now found that, in mm/sec (for ω_{23} = unity), while $v_E = 17.3689$, $v_H = 17.3368$, and $v_F = 18.5713$. It might be noted that, because we are working well within the half measure of the rectilinear array, these magnitudes of the various velocities are quite close to

one another. Finally here I wish for the sake of future reference (§6.45) to work out the angle turned by an imaginary moving vector of the linear array as we go along the transversal, not from the central point H to F, but from E to F. I call this angle *the angle involved in the shift* from E to F, and, while writing it symbolically β_{SHIFT}, I write it, colloquially, *beta shift*. Clearly now,

$$\beta_{SHIFT} = \beta_F - \beta_E.$$

Finding, in this WkEx#3, that β_E = artan (1.0611/17.4219) = 3.4854°, β_{SHIFT} works out to be 17.5234°. Please see in advance figure 6.08. Unexplained as yet, this figure shows the importance of the angle β_{SHIFT}.

45. *On figure 6.05.* This is the remains of a working drawing. It was previously well developed but, after determination of the unbounded involute helicoidal shapes of the flanks of the teeth, it was never completed; the ends of the teeth are truncated by the inside and outside faces of the flank, but the tips and the roots of the teeth are not truncated. Most of the confusing construction lines have been removed. Please re-perform in the mind's eye the following previously taken steps. Locate the sliding velocity at [Qx]; this is, of course, precisely the velocity vector v_F of the above paragraph; follow the contents of §1.14, §1.51, and/or §6.40 for this. Sweep out the naked wheels; follow the contents of §5B.02 and §5B.03 for this. Locate the polar plane at F which (no longer vertical) is normal to the sliding velocity at F, and locate within it the best drive which is perpendicular to the transversal; recall the triad at Q for this (§3.44). On either side of the best drive and within the polar plane, set out the two paths of the points of contact Q; the two angles of obliquity δ should be set, as they were set by me in this case to be, equal at 20°. Continue according to the principles of §5B.29 (figure 5B.09), §5B.30 (figure 5B.10), and §5B.31 (figure 5B.11). Ultimately, as will be seen in the figure, the unbounded shapes of the profiles of the teeth will have emerged, and these among other matters are now to be examined.

46. *Equal angles kappa.* In figure 6.05 look now at the filled-in portions of the doubly departed sheets (those of the naked wheels) extending from the F-circles to the throat circles of the sheets, and notice how, in relation to the equiangular wheels at figure 5B.02(a), these have become distorted. *But notice that the previously equal angles κ have (a) remained equal, and (b) kept their original value.* This was predicted at appendix A at 5B where it was, in effect, argued that, given C, Σ, R and k, κ would be independent j. In connection with this, and despite the changes in having gone from WkEx#1 to WkEx#3, a graphical check of the overall construction — including an examination of the contacts ratio — has already been made to show that, after obviously needed truncations, the mating sets of teeth of this particular plain polyangular set will be, most likely, mutually viable. Go back to figure 5B.02(b) and notice that the pairs of angles τ and λ, which are seen to be, within pairs, equal there, are no longer equal here; see in advance figure 6.07. Notice in other words that the two angular dispositions of the teeth — the term *angular disposition* has intentionally not been given a precise defini-

tion here — are such that, while drive from pinion to wheel is now made easier, with less trouble due to friction, drive from wheel to pinion is now more difficult, with more trouble due to friction. Thus we can also see that, as we might move even further with our point [Qx] in this direction away from E (and ignoring for the moment other troubles that might occur), the gear set might, at a certain critical distance (E–F) or, perhaps better, at a certain critical radius-ratio *j*, become *non-reversible due to friction*.

47. *Radiplan view*. Please imagine, there being no figure, the radiplan view of the same construction. The two F-radii, f_2 and f_3, will be in the plane of the paper and their lengths will be directly comparable (§5A.23c), but the helitangent at [Qf] will no longer be in the plane of the paper because the prevailing circumstance is no longer equiangular. The radiplan view might afford, among other things, a better comparative look at the magnitudes of the different angles λ but, as these do in fact appear in their true magnitudes and to better advantage elsewhere (at figure 6.08), I let the radiplan view of the matter lapse just here. Let me remark by way of interlude that what we are groping for here is an algebraic determination, firstly for analysis, of the two λ in terms of the two τ and of the two τ in terms of *j* and the two δ, and secondly, for synthesis, of these relationships taken in reverse order.

48. *The helitan view*. In figure 6.06 we see an orthogonal view taken directly along the helitangent. Here the phantom rack (as might be expected) reveals itself in true-shape. Its flutes remain, as they were at WkEx#1, parallel with the helitangent namely the direction of the sliding velocity at [Qx], and all rulings of all flanks of all teeth pass through the meshing on their proper allotted sides of the phantom rack. The slopes of the flanks of the rack are both δ, namely both 20° as set. Remember that, in general,

NP = $(2\pi$ a cos $\alpha)$/N

(§3.09, §3.13), and note the various formulae listed at §10.17. For each of the four sets of flanks of the teeth, there are different values of *a* and α, but the NP for each of the four sets of the flanks (and for the two flanks of the rack) work out to be the same. [NP = 12.375 mm]. Despite the polyangularity, this follows simply from the fact that the two angles δ of obliquity have here been set, although unnecessarily, to be equal. Whenever we choose to put [Qx] at E — take the WkEx#1 and the WkEx#2 for examples — the projections of the shaft axes appear parallel when seen in this particular orthogonal view; but notice that the projections of the shaft axes now, here in this figure 6.06, are clearly *not parallel*. This is a direct result of the newly introduced polyangularity. The helitangent (now at F) is no longer parallel with the FAXL and thus horizontal (§5A.29), but downwardly inclined. The view, thereby, is no longer a horizontal view. *The non parallelism of the projected shaft axes in the orthogonal helitan view, and the oblique slope of the projected CDL appearing at figure 6.06, are wholly characteristic of polyangularity*. Numerical details for the currently running WkEx#3 appear at the TNR.

§6.49 GENERAL SPATIAL INVOLUTE GEARING

49. *The separated panoramic view.* Let's take a panoramic view now of the separated wheels showing their respective angles κ, τ and λ. Recall figure 5B.02(b) and refer now to figure 6.07. It shows by contrast what has happened. *The angles λ are no longer equal.* I find by CAD that, *while the λ at the slower wheel has decreased by 17.5234° from its original value, the λ at the faster wheel has increased by exactly the same amount.* This phenomenon, newly recognized here, requires a rewrite of the criteria for equiangularity: *it is not that equiangularity is exclusively characterized by equal angles κ (for plain polyangularity is also characterized by equal angles κ), it is that equiangularity is exclusively characterized by equal angles λ.*

50. *The two λ changing with the beta shift.* Go back to §5B.53 for the first mention of Λ. Go back to the formulae for κ and $\Lambda/2$ already stated and proven at appendix A at 5B. Notice that both are independent of j and thus of τ also. In the special case when equiangularity prevails the two τ are both

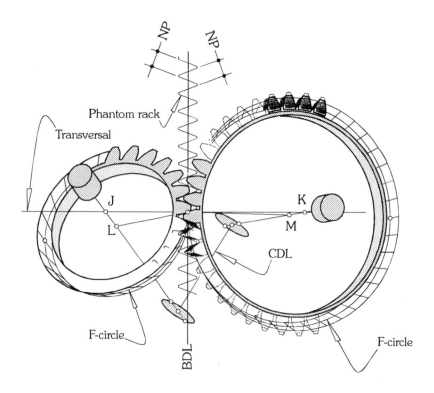

Figure 6.06 (§6.48). View along the helitangent in the case of WkEx#3. We have $k = 0.6$ and $j = 0.7$; the set is plain polyangular. This picture is showing among other things (a) that the helitangent is not parallel with the FAXL, (b) that the angles of obliquity (both 20°) are reflected in the shape of the rack, and (c) that the phantom rack is there as usual and, with the longitudinal direction of its flutes precisely parallel with the sliding velocity at [Qx], precisely fitting itself to the interstitial spaces between the teeth.

THE PLAIN POLYANGULAR OPTION §6.50

Σ/2, and when the two δ are equal also, the two λ are both Λ/2, *but otherwise this is not so*. Refer to figure 6.08 which compares the corresponding cases of the equiangular WkEx#1 beginning at §5B.01 and the polyangular WkEx3 beginning at §6.38. In each of these in figure 6.08 the same orthogonal view is taken; *but this is taken along the transversal, and Λ is revealed* (§5B.53). We have already, from appendix A at 5B, that, while

$\tan κ\quad = [2R\sin (Σ/2) \tan (Σ/2)] / [√(4R^2 \sin^2 (Σ/2) + C^2)]$,
$\tan Λ/2 = [C\tan (Σ/2)] / [√(4R^2 \tan^2 (Σ/2) + C^2)]$.

The latter of these reduces, in these examples WkEx#1 and WkEx#3, to the same numerical value, [Λ/2 = 13.691727°]. Refer now to figure 6.08. Note

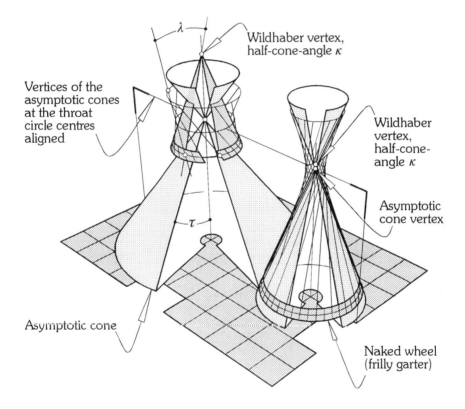

Figure 6.07 (§6.45). Separated wheels standing flat to show the fundamental conical surfaces for the plain polyangular wheels of WkEx#3. See the angles τ, κ, and λ. Compare this figure with figure 5B.02(b) for the equiangular WkEx#1. The naked wheels, each an integral part of its hyperboloidal doubly departed sheet, are the frilly garters upon, not the asymptotic cones whose vertices are at the centres of the throat circles shown and whose twist angles τ are different, but the Wildhaber cones whose vertices are elsewhere on the shafts and whose half angles κ are equal.

§6.50   GENERAL SPATIAL INVOLUTE GEARING

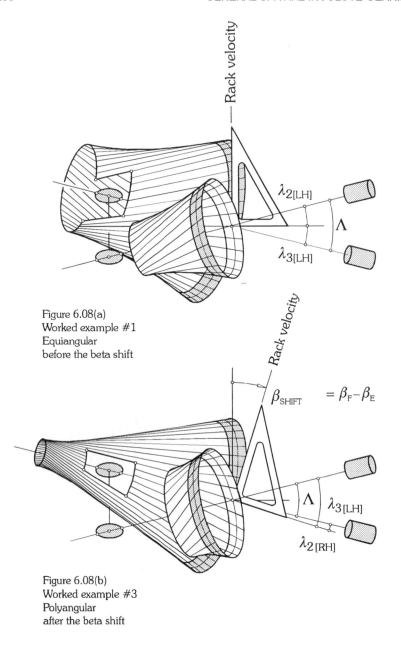

Figure 6.08(a)
Worked example #1
Equiangular
before the beta shift

Figure 6.08(b)
Worked example #3
Polyangular
after the beta shift

The figure 6.08 (§6.50). Matching views for comparison of (a) the equiangular WkEx#1 of chapter 5B, and (b) the polyangular WkEx#3 of chapter 6. Both projections look along the trasnsversal; in both of them the FAXL appears horizontal; and together they reveal what happens to the originally equal angles lambda when the set is 'twisted' from condition (a) to condition (b).

THE PLAIN POLYANGULAR OPTION §6.51

that, for given C, Σ, R and k (which is the case here in figure 6.08), angle β_E is a constant. Next we see in the same figure that, for polyangularity, which is the case at (b), while

$$\lambda_2 = \Lambda/2 - \beta_{SHIFT} = \Lambda/2 - (\beta_F - \beta_E) = (\Lambda/2 + \beta_E) - \beta_F,$$
$$\lambda_3 = \Lambda/2 + \beta_{SHIFT} = \Lambda/2 + (\beta_F - \beta_E) = (\Lambda/2 - \beta_E) + \beta_F.$$

These λ are of course different. They work out to be [− 3.8317° (negative)] and [+31.2151° (positive)]. These calculated numbers check exactly with the CAD. Notice in [1] [§22.37] that the terms *departure* and *swing* and the listed equations there, along with the neighboring figures 22.06 and 22.08, were, unwittingly at the time, foreshadowing the beta shift and the distance (E–F) newly under consideration here. In the simpler scenario there, the continuously changing set was certainly polyangular, but the offset was zero and the equilateral transversal was collinear with the CDL.

51. *Looking again at the angles* λ. Please regard figure 6.08 as showing two different stages — an initial one and an arbitrarily chosen, subsequent one — of a freely distortable geometrical model capable of encompassing and showing all possibilities throughout the nominated range of the plain polyangularity. The imaginary *set-square* (which can be seen) resides permanently in the plane of the transnormal which moves continuously parallel to the plane of the paper as the imagined moving F moves perpendicularly into the paper along the transversal. As suggested by the two matching parts of the composite figure, the set-square remains permanently pivoted as shown at the transversal, and its orientation continuously reveals the changing angle β_{SHIFT} (§6.40). This β_{SHIFT} has increased in the composite figure from zero (where the gear set is equiangular and the two λ are equal) to a value greater than $\Lambda/2$, where λ_3 has become greater than Λ and where λ_2 (diminishing) has become, indeed, negative.

52. Between the shown two stages of the apparatus, in other words, there is occurring a particular arrangement where $\lambda_3 = \Lambda$ and $\lambda_2 =$ zero. Notice in this regard that the handedness of the larger of the two naked wheels has changed from being negative namely left handed at the beginning of the change (namely at WkEx#1) to being positive namely right handed at the end of the change (namely at WkEx#3). *What this implies is that, throughout the range of the polyangular possibilities here, there is at least one special possibility where one of the wheels is straight conical.* An easy examination will reveal, indeed, that this will happen not only once but twice along the transversal. Consider the overall architecture of the set and the overall shape of the linear array and see quite clearly that, *whenever the velocity vector at F intersects one or other of the two shaft axes, this special condition will occur.* Let this somewhat unsurprising observation conclude, for the meantime, our work on this WkEx#3. The example has shown that no major difficulties occur in the realm of moderate k and moderate j. This and the above observations throw some retrospective light on the earlier works of Olivier [40] and Beam [59].

Numerical check on the law of the speed ratio

53. At §6.41 I promised to watch out, in WkEx#3, for numerical confirmation of the law of the speed ratio, which law was explained at §6.40. Keeping for comparison the contents of the table at §5B.17, which was for WkEx#1, I have inserted here (at the second row) the newly found numbers for the polyangular WkEx#3. The otherwise more accurately found numbers here have been everywhere rounded down to four significant figures.

	radius a ↓	angle α ↓	$a \cos \alpha$ ↓	
Drive side 2	81.80	4.83	81.51	Row 1, WkEx#1
	80.11	10.46	78.78	Row 2, WkEx#3
Coast side 2	49.08	4.83	48.90	
	78.84	3.72	78.78	
Drive side 3	51.82	19.31	48.90	
	50.10	19.32	47.27	
Coast side 3	86.37	19.31	81.51	
	57.90	35.26	47.27	[wrong boxes?]

For WkEx#1: [48.90]/[81.51] = [0.600], OK as before at §5B.17.
For WkEx#3: [47.27]/[78.78] = [0.600], OK again here.

See that the successive second row of numbers (these new ones for the polyangular WkEx#3) offer another striking confirmation of the law. For both drive and coast the ratio $[a_3 \cos \alpha_3]/[a_2 \cos \alpha_2]$ turns out to be, again in this polyangular WkEx#3, as it was before at §5B.17, $k = 0.600$.

Putting one of the paths upon the transversal

54. This paragraph might be omitted upon a first reading. Relevant here, it is a kind of puzzle. Be aware that although impractical, the transversal JEYK (like all transversals) is a legitimate travel path for Q. This is so because all transversals are by definition members of the linear complex surrounding the pitch line [1] [§11.36]. It is therefore not surprising to find that the said array of velocity vectors distributed along the said transversal is rectilinear. *An exercise*: go to figure 2.02, imagine the shown RSSR with input and output links of zero length, and investigate by means of the imagination the limiting case in practice that this unhappy circumstance represents. *Hint*: in the planar case, with ordinary spur involute teeth, the base circles are of zero radius, the involute profiles are mere circles (they are indeed the pitch circles, of indeterminate diameter, touching at the CDL), k is indeterminate, the angle of obliquity is 90°, and the two bald, toothless wheels are touching, but not functioning properly. *Another hint*: be aware of course that, for a given architecture, the legitimate paths, which are all *right lines* in the space, include the transversal which is, thereby, a right line also. Now consider the general, namely the *offset skew* case, and relate the investigation onwards to the highly special case where, with both the a at zero, $R \to$ zero. Refer §5B.62?

THE PLAIN POLYANGULAR OPTION §6.55

Beginning of worked example #4

55. *The data*. This is a moderate 2-step variation of WkEx#2 where first k is relaxed from 9/41 to 11/39, then a moderate j/k is introduced. We are looking here for a square, plain polyangular set where $k = 11/39$ and $j = 0.4$. The data for the worked example is the same as that for WkEx#2, $C = 40$mm and $R = 60$mm as before, but (a) we relax the severity of the ratio k (which becomes not 9/41 but 11/39 namely 0.2825), then (b) we nominate the point F to be not at the new E but at another point on the transversal where the poly-ratio j, differing from the new k, becomes 0.4. This j has already been explained at §6.38. After the two steps, the ratio j/k becomes, not unity as before, but (0.4) times (39/11), namely 1.42 roughly. By altering simultaneously both of the variables k and j, I have introduced a complication; this I know; but the 2-step maneuvre permits us to study not only the polyangularity of the case but incidentally also the contents of §6.15. Out of caution, a gentle polyangularity is introduced. On the basis of figure 6.04 (which is already explained), figure 6.09 has been prepared. It purports to illustrate pictorially the distribution among the variables S, T and ε upon the R-cylinder of the several worked examples of this book (§6.20). Although the figure is superficially clear, I offer its deeper aspects as an exercise for the reader.

56. Having thus seen this WkEx#4 as posing a new challenge to the theory (albeit a gentle one), and having thus set down the data, we are now obliged to carry on. It is in the general nature of the gear-set problem that, in the absence of experience and without a strong mathematics (and thus without a well devised spreadsheet for quickly achieving trial solutions by com-

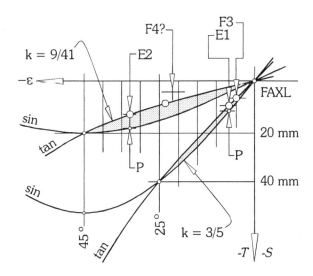

Figure 6.09 (§6.55). Charts for both the oblique ($\Sigma = 50°$) and the square ($\Sigma = 90°$) kinds of architecture drawn together on an R-cylinder of unit radius. The locations of the WkEx#1, #2, #3 and #4 are shown in comparison with one another here.

§6.57 GENERAL SPATIAL INVOLUTE GEARING

puter), we must, blindly, simply explore. The fundamental geometry of the interacting ghostly surfaces of the chapters 3 and 4 is sound, but what matters here is that the necessarily truncated edges of the said surfaces might in actual mechanical reality interfere with one another, either within the intersecting, inward-looking flanks of the convex shape of each tooth, or between the outward-looking, convex shapes of the flanks of either neighboring teeth of the same wheel or meshing teeth of the other wheel. This is a well known problem; it is endemic in gearing; and the general spatial involute gearing of this book is not immune.

57. Refer, to begin, to the series of figures 6.10. Using the data as set out, and using the geometrical principles already elucidated for the construction of the relevant sets of naked wheels, these superimposed, computer-drawn figures compare and contrast the architectures at the different stages

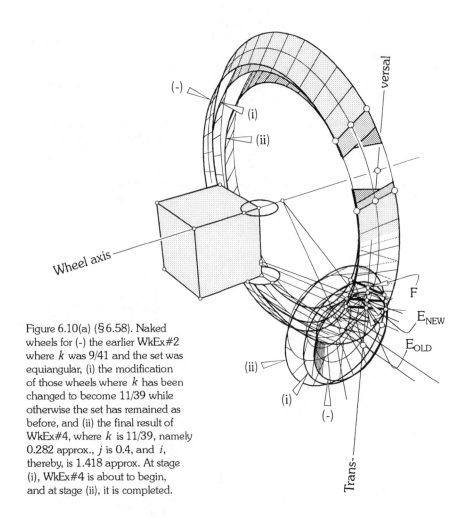

Figure 6.10(a) (§6.58). Naked wheels for (-) the earlier WkEx#2 where k was 9/41 and the set was equiangular, (i) the modification of those wheels where k has been changed to become 11/39 while otherwise the set has remained as before, and (ii) the final result of WkEx#4, where k is 11/39, namely 0.282 approx., j is 0.4, and i, thereby, is 1.418 approx. At stage (i), WkEx#4 is about to begin, and at stage (ii), it is completed.

of the mentioned 2-step process. Let us recall in these figures 6.10 that we are jumping — there are two steps — from an equiangular square set where $k = 9/41$, via an intermediate equiangular square set where $k = 11/39$, to a polyangular square set where j/k has been increased somewhat; we go from unity, via unity, to the final 1.42. These data have been chosen cautiously, for fear of geometrical trouble.

58. Figure 6.10(a) is a general perspective view of the various sets of the naked wheels. The view closely corresponds with the view at figure 5B.16(a). We see straight away that by moving from the old E at (-) where $k = 9/41$, through the new E at (i) where $k = 11/39$, to the F at (ii) where $j = 0.4$, we have (in each of those two steps) increased the size of the pinion and decreased the size of the wheel, thus progressively making the shape of the overall gear set (R remaining constant) more compact. It is not easy to see in the figure but it can be seen that the points E_{OLD}, E_{NEW} and F are upon the same equilateral transversal at $R = 40$ mm, and that the *angular dispositions*

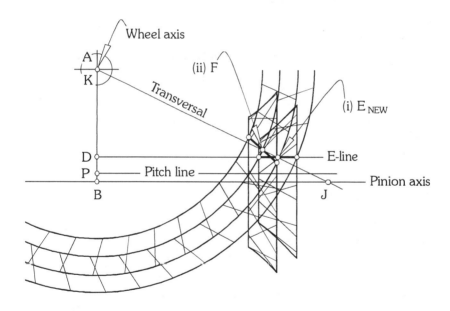

Figure 6.10(b) (§6.58). In respect of WkEx#4, we see in this parallel wire-frame view that is taken orthogonally against the key-circle plane of the bigger wheel (the wheel) how the naked wheels change as we move from the preliminarily chosen equiangularity at stage (i) to the ultimately required polyangularity at stage (ii). The black lines indicate the pairs of meshing generators of the pairs of naked wheels. Notice among other things that the slope of the helitangent at F has changed from being horizontal (at E) to sloping downwardly as it goes outwardly from F. This has been enough to change the spirality of the rulings upon the wheel from being left handed to being right handed. The slopes of the helitangents are more clearly elucidated at figure 6.10(c).

of the generators of the naked wheels — these remain only vaguely defined as before (§6.42) — have been considerably changed by the switch from (i) to (ii). See the spiralities of these more clearly compared at figure 6.10(b). Please see in advance figure 11.03(a). This shows another perspective view of two similar but other sets (where we go direct along a fixed transversal from an E to an F). It shows the transversal clearly and the two points E and F which are at the intersections with the transversal of the two different heli-tangents, respectively indicated by the two stars (*) (*), which may be seen. Figure 6.10(c) is special; it is the unique orthogonal view looking along the transversal where the angle Λ (the same for all three sets), the beta shift, which applies between (i) and (ii), and the two different angles λ, appear at their true magnitudes (§6.46). The equal angles κ (at the apices of the two Wildhaber cones) do not appear in any of the figures 6.10 but, using the general expressions for tan $(\Lambda/2)$ (and for tan κ incidentally) derived at appendix A at 5B, and by substituting into both of them the special value $\Sigma = \pi/4$ (to suit this WkEx#4), we get the following simplified formulae for Λ (and for κ incidentally) which are of course applicable to all square sets:

tan $(\Lambda/2) = [C]/[\sqrt{(4R^2 + C^2)}]$,
tan $\kappa \quad = [R\sqrt{2}]/[\sqrt{(2R^2 + C^2)}]$.

Recalling the material of §6.44 regarding β_E and β_F and the formula for β_{SHIFT} which is derived from them and explained there, recalling also the formulae for λ_2 and λ_3 explained at appendix A at 5B and explicitly set out at §6.50, and then by a combination of algebraic formulae and direct measurement from the CAD we discover for this current WkEx#4 the following numerical results to date:

[tan $(\Lambda/2)$ = 0.316228]
[tan κ = 0.904534]
[$\Lambda/2$ = 17.5484°]
[Λ = 35.10°]
[κ = 42.13°]
[β_E = °
[β_F = °
[β_{SHIFT} = 28.67°]
[λ_2 = –11.12°]
[λ_3 = +46.22°]

Notice that β_{SHIFT}, which is a clear measure of our departure from equiangularity in going from (i) to (ii), is of such magnitude that the handedness of the teeth on the wheel has changed from being moderately left handed (on the 41 teeth) to being moderately right handed (on the 39 teeth). The handedness of the teeth on the pinion on the other hand, on stepping initially from (i) to (ii), has remained left; this has become more severe however. One might fear that this severity of the handedness at the pinion might be a source of trouble; but — and look for this at chapter 7 — what if the pinion were a worm?

THE PLAIN POLYANGULAR OPTION   §6.59

59. Next in the processes of design we must allocate (at the polar plane) the two angles of obliquity. This is a tricky job. If we set them too small we run the risk of setting one or other or both of the Z_I too short (shorter than the NP); if we set them too great we run the risk of decapitating teeth by mutual interference of the flanks and thus of truncating the slip tracks; and, as we have seen already at WkEx#2, unless we choose their relative magnitudes sensibly, the two lengths Z_I will be markedly unequal (§5B.49). These hazards must be avoided. Go to figure 6.11.

60. Figure 6.11 is a parallel view looking along the helitangent at the naked wheels at stage (ii) of WkEx#4. We accordingly see in figure 6.11 the polar plane in the plane of the paper and the two paths of the two Q openly displayed within it. We need, once again, the ellipses of obliquity (§5B.40); and, at the risk of repetition, I wish to explain these ellipses again. First of all they wholly depend (exclusively of the paths) upon the established architecture of a set; they are independent of how we chose that architecture or of how we otherwise arrived at it. Now as we know all the legitimate straight-line paths for a moving point Q at a chosen F are confined within the radial array of lines that constitutes the polar plane at that F; the polar plane is

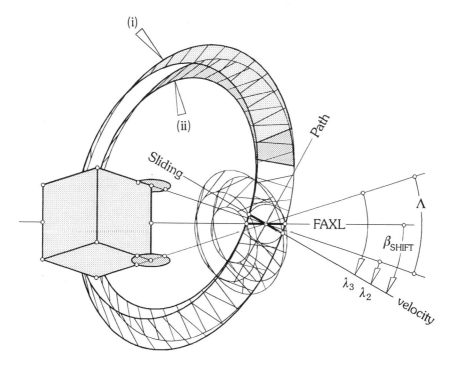

Figure 6.10(c) (§ 6.58). Orthogonal view along the transversal in the case of WkEx#4. Except for κ, we see all of the angles of §6.58 appearing here at their true magnitudes. This is directly comparable with figure 6.08(b) which showed the beta shift for WkEx#3.

275

§6.61   GENERAL SPATIAL INVOLUTE GEARING

normal to the velocity vector (the sliding velocity) at that F, and its central point is at the basepoint of that same vector. With polyangularity prevailing, here in this WkEx#4, the polar plane is no longer normal to the FAXL (§5A.24), so the cubical box at the figure is looked at obliquely. If, now, we construct the shortest distances between all of the legitimate paths in the polar plane at F and the two shaft axes, we construct in effect, all of the possibilities for the input and output links RS and SR of the fundamental RSSR of figure 2.02. This having been done, the two separate loci of the two points S become the ellipses of obliquity. *Notice that, independently of the two paths chosen, the ellipses of obliquity are a function of the architecture of the set; the ellipses are not a function of any one or other of the possible paths.*

61. Refer again to figure 6.11. It is, as has been said, an orthogonal view along the helitangent of the data of the current WkEx#4. The view is accordingly taken directly against the polar plane which appears in the figure as a circle; and, as usual now in this book, the picture has been arranged to

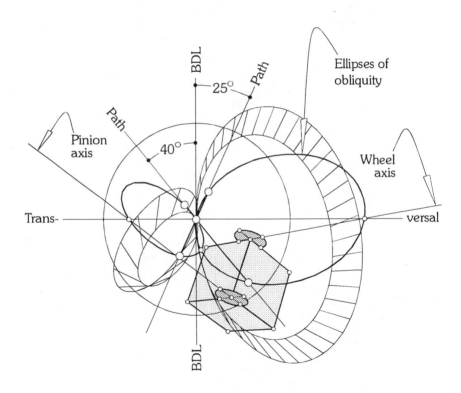

Figure 6.11 (§ 6.60). View along the helitangent (WkEx#4) showing the polar plane, the ellipses of obliquity (which are in the polar plane and intersecting at two points), and the chosen paths (also within the polar plane) set at 25 and 40 degrees to the BDL.

set the transversal (already in the plane of the paper) horizontal across the page. The ellipses of obliquity are quite clearly *intersecting* at the centre of the polar plane and, in comparison with figure 5B.16(b) where equiangularity obtained (§5B.40), they are no longer arranged with their minor axes parallel with the CDL. The CDL appears in the figure but, like the cubical box, it appears oblique and in the background. This oblique view of the cubical box, along with its CDL and the two shaft axes, is a natural result of the new set of conditions, namely the new polyangularity occasioned by the fact that the ratio j is no longer the same as the ratio k.

62. Trying judiciously now to set the angles of obliquity to optimize an outcoming result which is of course unforeseen as yet, I give in figure 6.11 my best considered choices for the locations of the rods SS. The angles of obliquity are set in the figure — whether they are for drive and coast respectively or for the opposite is unanswered as yet — at 40° and 25°. While setting these angles, it is both possible and desirable to envisage in space the shapes and dimensions of the two equivalent RSSR, with their less than hemispherical sockets at the joints SS (§2.14), and to gather from these the likelihood of higher than wanted transmission forces, of unhappy effects of friction due to high sliding velocities at low angles χ (§5B.37), of unwanted (or indeed wanted) absences of reversibility namely occurrences of non-back-driving, or of other unwanted troubles or special requirements. Without exceptional geometrical insight, or quickly available computer-calculated predictions based on a completed algebra (or long experience in practical fly-by design), this is not easy to do however.

63. Next in the processes of design we can follow the procedures already outlines at §5B.16 *et seq.* for a long way towards completion. Ignoring as yet the finer points of undercut, interference and contact ratio, we can come up with some preliminarily drawn pictures of the shapes of the flanks of the teeth and the actual meshing between the teeth. Refer to the series of figures 6.12 for a visual impression and to the TNR for numerical results.

64. One could argue that the figures 6.12(a)(b) which are unfinished drawings from WkEx#4 are self explanatory, and leave them at that. But here are some comments about the particular views that were chosen that might be useful. The figures are both looking (but in opposite directions) directly along the helitangent at [Qf], the one, namely (a), from the 'outside' looking inward, and the other, namely (b), from the 'inside' looking outward. They both reveal the true shape of the phantom rack, which is — due to the different angles δ which may be seen in retrospect at figure 6.11 and in advance at figure 10.03(b) — asymmetrical; but the projected shapes of the phantom rack (were it to be drawn) and the shapes and orientations of the wheels appear to be distorted. The reasons for this are two-fold: for the sake of clarity at the meshing, both of the figures are in quite severe perspective; the distances from the eye-point to the reference-point at [Qf] are in each case small; and they are moreover different. Refer to the captions.

Some ongoing comments in conclusion

65. By showing these working drawings incomplete I am indicating the kind of complexity involved in the drawings completed so far, but I wish to assure the reader that, as we shall see at chapter 10, *the practical business of setting the cutting machines and machining the teeth of actual wheels will (except for the usual ongoing problems of tool-wear, accuracy, quality control etc.) be a far simpler process.* I say it again: *based upon the shape of the phantom rack which carries directly the chosen angles of obliquity δ as we have seen, and fashioned to accord with the identically shaped imaginary rack that can be seen to generate by deformation, the actual cutting tools will be able properly to generate in one single pass both flanks of the teeth of spatial involute wheels even in the above-studied, general circumstance of plain polyangularity.* We can now move on to chapter 7.

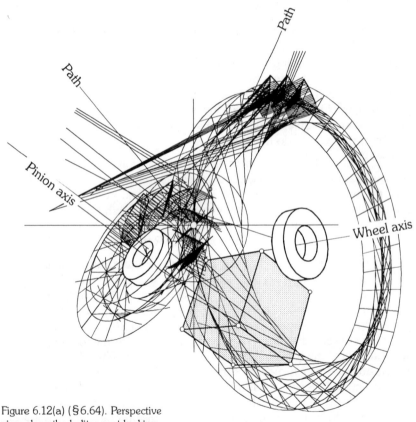

Figure 6.12(a) (§6.64). Perspective view along the helitangent looking inwards from the outside (WkEx#4). The reference circle at the polar plane is here 100 mm diameter, while the distance between its centre at [Qf] and the eye-point is 300 mm.

THE PLAIN POLYANGULAR OPTION   §6.66

Appendix A at chapter 6

66. The matter at issue here comes from §6.33. It comes up again at §6.44. It deals in general with the shape (the half-measure D) of the linear array if vectors attached at a generally chosen right line existing within a generally chosen, linear line complex. There is an ∞^4 of straight lines in the whole of Euclidean space [1] [§4.10]. There is an ∞^3 of these lines within any given linear complex [1] [§9.16 *et seq.*]; the straight lines of the complex however are called by this author *right* lines because each of them is characterized by having its attached vectors set perpendicular to it [1] [§9.20]. There is, in general, only an ∞^2 of the right lines that cut the central axis of the complex [1] [9.13]; and, while this ∞^2 cuts the central axis perpendicularly, the remaining ∞^3 of them are skew with the said axis, not

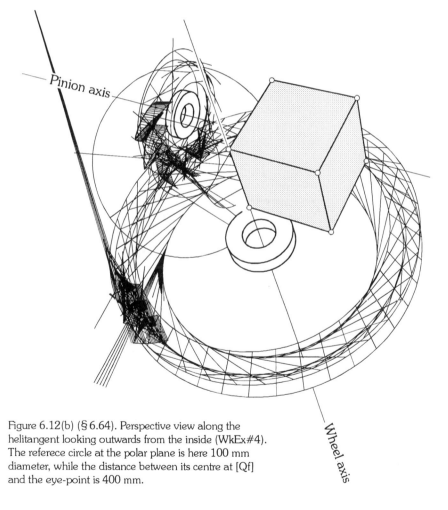

Figure 6.12(b) (§ 6.64). Perspective view along the helitangent looking outwards from the inside (WkEx#4). The referece circle at the polar plane is here 100 mm diameter, while the distance between its centre at [Qf] and the eye-point is 400 mm.

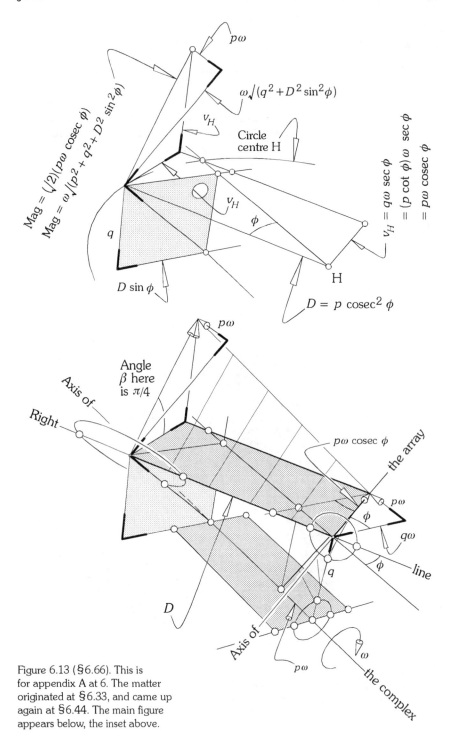

Figure 6.13 (§6.66). This is for appendix A at 6. The matter originated at §6.33, and came up again at §6.44. The main figure appears below, the inset above.

THE PLAIN POLYANGULAR OPTION §6.66

cutting it at all. Let each be set at some shortest distance q from the central axis and inclined at some angle ϕ to it. We may next say that the equation

$$q = p \cot \phi,$$

where p (mm/rad) is a constant, both defines and characterizes the complex. The reader will recognize this equation of course; the fundamental law of gearing has already been stated (§2.02) as follows,

$$q \tan \phi = p;$$

and, in the light of these two — namely (a) this definition of what a linear line complex is, and (b) this statement of the fundamental law of gearing — it would be tautologous to prove the truth of either one of them by invoking the truth of the other.

It is however possible within a complex of pitch p and independently of the said law, to show that, if D (mm) be the half-measure of the parabolic hyperboloid formed by the rectilinear array of vectors attached along any right line inclined at angle ϕ to the central axis, D may be written,

$$D = p \operatorname{cosec}^2 \phi,$$

which, although dependent on the shortest distance q, appears here not to be; but don't forget the known, inescapable relationship between q, p and ϕ, namely $q \tan \phi = p$, which remains extant.

For a proof of this relationship, the veracity of which is clearly important because the geometric correctness of the meshing and the proper accuracy of machining is at stake, go to figure 6.13. The main figure (below) and its inset (above) are intended to stand unaided, to be self explanatory. But the following comments need to be made. The interrelated variables p, q and ϕ having been established, the drawing has been firstly constructed by means of computer graphics to ensure that the distance D, measured from the central point H of the array to that point on the right line where the variable angle β is $\pi/4$ (§4.77), is correct as it should be, and be appropriate. Without loss of generality the drawings have been made for the special case where ω (rad/sec) is unity and, not forgetting that the scales for physical distances on the one hand and linear velocities on the other will naturally be different, an algebra of the magnitudes may be written (by means of judicious triangulation) as illustrated at inset. The gist of the formal proof resides in the fact that we have two different expressions, each arrived at independently of the other, for the magnitude of the indicated vector at the top left at inset. The vector in question is that unique one upon the shown wing of the array where the angle β is $\pi/4$. By equating the two expressions and solving the resulting equation for D, we can easily find, legitimately, that

$$D = p \operatorname{cosec}^2 \phi,$$

which is what was required. Notice that this D, this half-measure of the array at shortest distance q, q being $p \cot \phi$, is not surprisingly independent of ω.

Appendix B at chapter 6

67. This was mentioned at §6.35 and at §6.44-45 inclusive. We examine here the layout and the actual numerics of the two different ways of finding the velocity vector at F in the ongoing case of WkEx#3. We are dealing here with important bits of geometric detail that have been discussed in the text but not shown (for the sake of clarity) at the central portion of figure 6.05. All of the instantaneous motion illustrated here is happening at the relatively small but complicated region of the meshing. We are working in the tightly restricted space between the pitch line and the nearby points E, H and F. Figure 6.14 comes direct from the CAD, as did the figure 6.05 itself, and the direction taken for the perspective is the same, but here we look, uncluttered without the surrounding scenery, more closely at the detail. At appendix A at 6 (above) we have has already proven the algebraic relationships directly relevant, so there is no need to repeat that. I do wish to say however that the generally chosen magnitudes for the figure there, and the carefully chosen direction for the perspective view, is inherited here. At the following figure 6.14 we accordingly take pot luck with the view. Its exact placement within figure 6.05 I leave as an exercise for the interested reader.

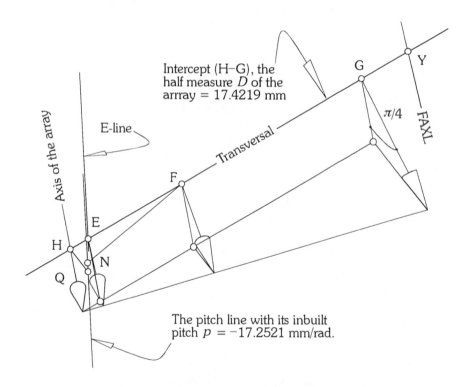

Figure 6.14 (§6.67). Layout and numerical detail at figure 6.05 central (WkEx#3).

Appendix C at chapter 6

68. Please read note [90]. At §6.39 (the origin of this appendix) it was noted by me that the algebra for working out definitively the (t, τ) for the shapes and sizes of the doubly departed sheets, slices normal to the axes of which surfaces are of course the naked wheels, was not available there in the context of Killeen's work. I suggested that we needed to do, yet, the algebraic job of swinging the helitangent at F about each of the two shaft axes, thus to find, like Killen did for the (a, α) of the base hyperboloids, definitive algebraic expressions for the distances S_t, the throat radii t, and

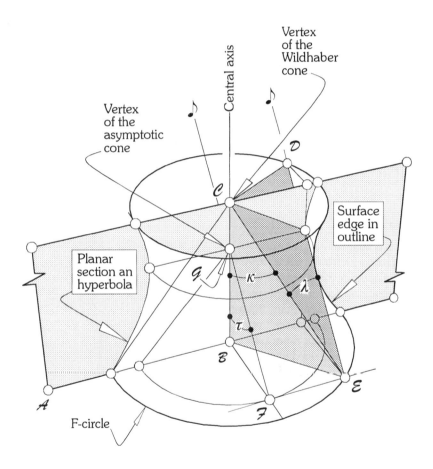

Figure 6.15 (§ 6.70). This figure is illustrating the mathematics of appendix C at chapter 6. It is helping to show that, across the whole field of plain polyangularity, $\cos \tau = \cos \kappa \cos \lambda$. Perceive that, while the Wildhaber cone is tangent to the hyperboloidal sheet of the naked wheel at the F-circle and has its vertex $\mathcal{C}$ somewhere generally upon the central axis, the astymtotic cone is tangent to the said hyperboloidal sheet at infinity, having its vertex $\mathcal{G}$ always at the centre of the throat circle precisely. Look back at figures 5B.02(b) and 6.07.

§6.68

the twist angles τ of the said departed sheets, thus the (t, τ) for the naked wheels; but there may be other ways to go.

First the radii t may, indeed, need to be determined by the same or a similar difficult method because the answer is not at all simple in the general case. In the special case of equiangularity, however, the radii t are precisely the same as the radii d that were first introduced at §1.41. We have for equiangularity that $d_3/d_2 = k$, and that $d_3 + d_2 = C$. So an expression for the throat radius t of the doubly departed sheets in the special case of equiangularity can be discovered easily. We can write for equiangularity that

$$t = [w]C,$$

where for t_2, $[w] = (1 - k)$, and for t_3, $[w] = k$.

With respect to the angles τ in general, however, please refer to figure 6.20. See there (a) the F-circle (the key circle) of a general naked wheel, (b) the circular throat of its hyperboloidal departed sheet, and (c) the vertex of the corresponding Wildhaber cone (at $\mathcal{C}$). In the skew quadrilateral $\mathcal{BCDE}$, $(\mathcal{B}-\mathcal{C})$ is the height of the Wildhaber cone, and $(\mathcal{D}-\mathcal{E})$ is the corresponding slant height of the straight-line generator of the departed sheet. Note that $(\mathcal{E}-\mathcal{F})$ is parallel with $(\mathcal{A}-\mathcal{B})$ and $(\mathcal{F}-\mathcal{G})$ is parallel with $(\mathcal{D}-\mathcal{E})$. It is clear that, while

$(\mathcal{D}-\mathcal{E}) \cos \lambda = (\mathcal{B}-\mathcal{C}) \sec \kappa$,
$(\mathcal{B}-\mathcal{C}) = (\mathcal{D}-\mathcal{E}) \cos \tau$;

so

$\cos \tau = \cos \kappa \cos \lambda$,

which is a spectacularly simple, general result. Should we wish to use the known expressions for $\tan \kappa$ (appendix A at 5B) and for λ (§6.50), we would be able (by knowing also $\Lambda/2$ and the beta shift) ultimately to write a general expression for τ in terms of the original variables C, Σ, R and k. So far as I can see, however, the angles τ, although of considerable academic interest, are not required for the machining. See chapter 10. In WkEx#4, incidentally, the half-cone-angles κ and τ_2 and τ_3 turn out to be, numerically, 42.13° (§6.58), and 43.35° and 59.13° (§6.68), respectively. Notice that, while 2κ (less than a right angle) is the same for both wheels, and while $2\tau_2$ is less than a right angle also, $2\tau_3$ is greater than a right angle.

CHAPTER SEVEN
WORM DRIVES
OFFSET AND NON-OFFSET

Introduction

01. The term *worm* appears to be applied to that kind of helical wheel whose number of teeth is small (which number may be, and often is, as low as one), and whose teeth have such helix angles and are so *wide* that, in going from cheek to cheek along the axis of the wheel, they make more than one whole circumferential circuit. Given one tooth, a worm is said to be a *one-start* worm, given two, a *two-start,* and so on; and one can actually see the start of a tooth by looking at the end of a real cut worm. What appears to be the *normal pitch* of a one-start worm is indeed a pitch in the ordinary sense of that term. Unlike the normal pitch of conventional multiple teeth, the said distance in a one-start worm is, directly speaking, *a known measure of the relevant involute helicoid.* The said measure is, indeed, the distance $2\pi a \cos \alpha$ mentioned unobtrusively at the end of §3.09 and illustrated without comment at figure 3.03. At §5B.27 we have shown in general, however, that NP = $[2\pi a \cos \alpha]/N$; and here, for a one-start worm, we have that N = 1; so the overall argument is consistent.

02. According to whether the point [Qx] for a general gear set be put at P, C, D, E, F or G among the architecture (§1.38, §5A.32, §8.03), we have hitherto called the relevant radii *r, c, d, e, f or g*. We have suitably subscripted these terms to distinguish the wheels, and have referred in the chapters 5 and 6, for example, to the E-radii and the F-radii of the wheels. Let us say now by way of definition — whether it be pinion, worm, or wheel under consideration (all of which terms are relative, qualitative terms in any event, there being no clear boundary lines between them) — that the relevant *key diameter* of a chosen wheel is twice the relevant radius, the P-, C-, D-, E- or the F-radius etc., of the wheel. The notion *key radius* has been mentioned and used already (§4.76 etc.); so this is not a new idea.

03. Generally in worms (as loosely defined above), the worm needs to be manufactured in such a way that its key diameter is greater than that of the shaft supporting it. This will mean, most often, that the point [Qx] in any

general worm reduction set (with radial offset) will not be at the E-line but somewhere remote from it. And this means in turn that such worm-reduction sets will be, most often, plain polyangular, the point [Qx] being put (for convenience) upon an equilateral transversal at F. In the simplest of worm drives, however, not only is Σ put to 90°, but also (and more importantly) R is put to zero. This means that in these special cases (a) the equilateral transversal becomes collinear with the CDL, (b) the point F becomes the point C (§1.38), and (c) the two paths of Q, which meet at [Qx] at C, meet at the CDL. It should accordingly be clear that, whereas the offset worm sets belong as a special case among the general offset skew gears, a simple, conventional worm set is a special case among the more restricted class we have already called the crossed helicals.

Recapitulation

04. It was at §3.13 that I first used the term *bed spring* to describe the shape of the twisted intersection of the circular hyperboloid of figure 3.03 and its involute helicoid there. The hyperboloid and the helicoid in the figure are of course intimately related. They share the same central axis and display the same (a, α). This means that the core helix of the helicoid finds itself wholly described upon the surface of the throat cylinder of the hyperboloid, and that each generator of the helicoid cuts its corresponding generator of the hyperboloid perpendicularly. Please study the bed spring. It goes continuously in figure 3.03, from its label (which says 'slip track') at its upper extremity at top left, all the way and via the throat (where it narrows), to its lower extremity at bottom left, where it has widened again. The involute helicoid itself, filling the whole of space, screw symmetric about its infinitely long axis, and having two branches incidentally (§3.62-64), remains identical at all locations along that axis. The slip track on the other hand does not remain identical at all locations along the axis. It is shaped otherwise. It is line-symmetric about a unique central point (which point occurs at the central point of the hyperboloid), but axially it is shaped in such a way that its radius (measured by dropping a perpendicular from a point on the twisted curve to the central axis), increases at an increasing rate (with the rate of increase of the rate slowly diminishing to zero) as we steadily move in either direction away from the said central point of symmetry.

05. Referring again to figure 3.03 it is important to understand that, as the angle Ω steadily diminishes (figure 3.02), the point W steadily travels downwards along the core helix. Carrying the tangent WB with it as W goes, the intercept (W−B) sweeps out that area of the involute helicoid shown which is bounded by the core helix and the slip track. For any given whole geometric construct, the intercept (W−B), of length is $J = a\,\Omega \cos \alpha \cot \alpha$ (§3.17), varies directly (as can be seen) with Ω. It steadily decreases in length as W steadily descends. As point B is obliged by the course of events to pass across the throat circle, however, it also passes across W, and the directed

length J, namely (W−B) which resides in the line WB of the tangent, changes sign. What this means in effect is that, at this crucial juncture (when B is at the throat circle), we *vault* (jump sideways over) the cuspidal edge of the helicoidal surface (which edge occurs at the core helix) and find ourselves upon the other branch. In conjunction with the two figures 3.13 this matter is discussed in close detail at §3.59 et seq., but please remain aware, while reading there, that the there-depicted angle α is not comparatively large but comparatively small. This recapitulation will, I trust, go to a more mature understanding of the worked example which is to follow.

Beginning of worked example #5

06. *Data.* We take an ordinary, single start, non back driving, right angular, worm reduction set with speed ratio 1/18. We take an ordinary cylindrical involute one-start worm engaging an ordinary cylindrical helical involute wheel. Let k be 1/18. The axes of the worm and the wheel are of course skew. Take that C = 63 mm, and that the set be *square*. Thus we have that Σ is a right angle, and that R is zero. To find the pitch point P and the pitch line, we go to the simplified set of equations (13) to (17) at §1.14. From these we find immediately that r_2 (for the worm) is very small, at 0.1938 mm only, and that ψ_2 (also for the worm) is very small also, at 3.1798° only. This puts P very close to A and the pitch line almost parallel with the shaft axis of the worm. Points A and B are shown at figure 7.01 and the pitch point P is sketched in. The segment (A−P) (which is the radius r_2) is not drawn to scale, however, and for the sake of clarity the pitch line (in its horizontal plane and intersecting the CDL at angle ψ_2 to the worm axis) is not drawn at all. Despite this, it must be understood that the exact location (r, ψ) and the pitch p of the pitch line are vital for subsequent calculations. The pitch p, found directly from equation (17) at §1.14, is −3.4842 mm/rad.

07. *Location of [Qx].* Let there be one tooth on the worm and 18 teeth on the wheel. The path of the point of contact cuts the CDL. There is of course single point contact (at a multiplicity of Q), and these Q will be travelling in tandem along the same straight path. Let the ratio of the C-radii be set to say 0.5, namely 1/2, meaning that the C-radius of the worm will be one half of the C-radius of the wheel. Thus we have that, whereas k = 1/18 = 0.0556 approx., j = 1/2 = 0.5 exactly. The point C on the CDL is thus so located that the radii c_2 and c_3 are 21 mm and 42 mm respectively. The set is non-equiangular. The radii d_2 and d_3, the E-radii for equiangularity, work out to be 3.1579 mm and 56.8421 mm approximately (§1.38); but for clear practical reasons (§7.08), we are not putting our [Qx] at E (which is at B); we are putting it at C.

08. *Comment.* These data show why it is that for worm drives in ordinary practice equiangularity is not a viable option. A bare shaft in a bored hole through the body of the worm could not be sufficiently robust in the case of this worked example, and the helix angles at the teeth of the worm

would be so small that non back driving, often a requirement of worm reduction sets, could not be ensured. I wish to mention also that, for the ordinary class of worm drive under consideration by way of worked example here, Σ need not be a right angle, the C-radius of the worm need not be smaller than the C-radius of the wheel, and the chosen worm might just as well be a multi-start worm. See the rough sketch at figure 22.07(d) in [1] which was inspired by a three-start worm cut integral with the robust cam shaft of a motor car engine. The speed-reduction set was driving, via flexible cable attached at the axis of the wheel, an engine speed indicator. See the three starts of the worm in the figure at top left.

09. *The naked wheels.* Refer to figure 7.01. Having determined, by means of the equations (13) to (17) at §1.14, the location of the pitch line and the value of the pitch *p* pertaining there, we can determine by means of equation (25) at §1.41 the direction of the sliding velocity at Q when Q is at [Qx]. [Qx], of course, is at C; and let us remain aware that the theorems expounded at §5A.08 about the field of coaxial helices surrounding the pitch line are applicable even here, even in this remote, 'distorted' field of the worms where the found geometry of the architecture might turn out to be surprising. By swinging the line of the sliding velocity vector about each of the two shaft axes, we can sweep out the naked wheels. See them drawn in the figure. Notice that, in accordance with previously established theory (§4.36-38, §5B.02, §5.10-11, and read in advance §10.32), they intersect along their common generator, are tangent to one another at the single isolated point [Qx], suffer ghostly interference elsewhere along the line of their intersection, and cannot roll without circumferential slip. The one and only

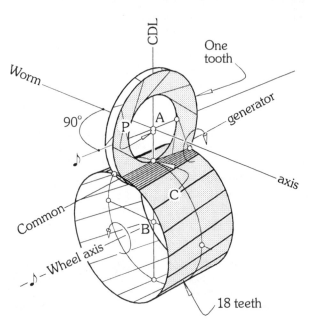

Figure 7.01(§7.06). Naked wheels for the worm set of WkEx#5. The set, having R = zero, is non-offset. It belongs thereby to the genus ordinary crossed helical.

WORM DRIVES OFFSET AND NON-OFFSET   §7.10

smooth tooth on the worm and only one of the 18 smooth teeth on the wheel are divided into 10 equal parts so that step by step the relative motion may be envisaged. Because the set belongs to what I have called the crossed-axes helicals (§P.01), the sliding velocity at C is perpendicular to the CDL, even though the set is also (by definition) polyangular. Understanding among the crossed helicals that the directed distance h from the pitch line to the transversal is simply the distance between the points P and C on the CDL, some numerical values to date are as follows:

[$C = 63$ mm]
[$r_3 = 0.1938$ mm]
[$\psi_3 = 3.1798°$]
[$p = -3.4892$ mm/rad]
[$c_2 = 42$ mm]
[$c_3 = 21$ mm]
[h $= 20.8062$ mm]

10. *Locating the fundamental triad.* Refer to figure 7.02, recall the convention adopted at §1.06, and look again at the contents of §3.49. Item #1 the equilateral transversal at [Qx] is the CDL itself, which is vertical; item #2 the helitangent, which is the line of the sliding velocity vector at [Qx], which is the common generator of figure 7.01, is horizontal; so item #3 the best drive, which is the third member of the triad at [Qx], is horizontal also. Notice that the triad as whole, although aligned with the CDL, is not aligned with either of the two shaft axes; looking downwards along the CDL, it is displaced angularly from the directions of both shaft axes.

11. *The throat circles of the base hyperboloids.* Refer to figure 7.02. Let us erect, in the polar plane which is normal to the helitangent at [Qx], one of the paths of the two Q inclined at its angle δ to the best drive. In view of the overall symmetry of the required action, I have taken the two δ to be equal; I have set them both to be 15°; but, as will be seen, only one of the paths — let to be the drive path — is shown. Next, I have erected the two common perpendiculars between (a) the said drive path, and (b) the two shaft axes. These intercepts, called a_{2D} and a_{3D} in previous work, and indicated each by an asterisk (*) here, are the throat radii of the relevant base hyperboloids and thus the radii also of the two core helices. The two outside circles of the shaded annuli are illustrating the said throats. We have that

[$a_{2D} = 8.0018$ mm]
[$a_{3D} = 40.5521$ mm]

The two RSSR (one of which is shown in figure 7.02) are line-symmetric with one another about the line of the CDL. In full accordance with the theory outlined in the middle chapters of this book, the two pairs of the four throats are pair-wise displaced axially and equally away from the key circles of their respective wheels. There is a right handed and a left handed RSSR. This kind of symmetry is obtaining here because (a) the offset R is here zero, and (b) two angles δ are here the same. We are dealing here with a special case.

§7.12 GENERAL SPATIAL INVOLUTE GEARING

12. *The base hyperboloids.* Two of these exist for each selected path. They are obtained by swinging the path itself about the two shaft axes in turn. None of the base hyperboloids is shown in figure 7.02, while only one of the them (one of the two on the worm shaft) is shown at figure 7.03. Refer to my earlier remarks at §3.12, and notice now the relative shapes of the hyperboloids. Because the angle a at the worm is relatively large, indeed very large, $\alpha_2 = 73.7435°$, the hyperboloid is slender, more slender, even, than the one at figure 3.03; and, because the angle a at the wheel is relatively small, indeed very small, $\alpha_3 = 6.1232°$, the relevant hyperboloid (which is not shown) is quite squat. Summarising the ongoing numerical results for this ongoing WkEx#5, we have that

[$\alpha_2 = 73.7435°$]
[$\alpha_3 = 6.3232°$]
[$a_2 \cos \alpha_2 = 2.2400$ (worm)]
[$a_3 \cos \alpha_3 = 40.3207$ (wheel)]
[speed ratio = $[a_2 \cos \alpha_2]/[a_3 \cos \alpha_3] = k$] [Refer §6.40]
[$k = [2.2400]/[40.3207] = [0.05555] = 1/18$, check]

13. *The slip tracks.* In general the slip tracks are at the curved-line intersections of the relevant base hyperboloids with the relevant involute helicoids. In the current example, the slip tracks on the *wheel,* as might be expected, are confined to a narrow, central portion of the teeth because the hyperboloids there are so squat. The slip tracks on the *worm,* however, are much more widely spread across the teeth (the one tooth), and thereby more

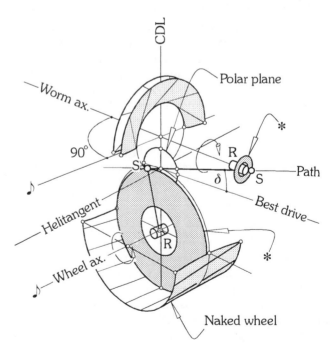

Figure 7.02 (§7.10). Given the angles of obliquity (both 20°), we find here the equivalent RSSR for drive. Note that instaneously a relatively large infinessimal displacement at the shorter crank, gives a relatively small infinitessimal displacenent at the longer crank. The ratio $k = 1/18$.

worthy of special attention here. Refer to figure 7.03. Angle α for the worm is very high (as has been said), so the base hyperboloid for the worm is very slender. It cuts numerous different turns of the involute helicoid within the annular space occupied by the real, single tooth of the worm, so the real slip track extends over several complete turns of its infinitely long bed-spring shape. See the equivalent linkage (the RSSR) marked in at figure 7.02. Examine there the relatively long length of the rod S—S. This is of course the abstract geometrical length Z_I (§5B.18), within which we will find (after truncation) the length of the actual zone of contact Z_A.

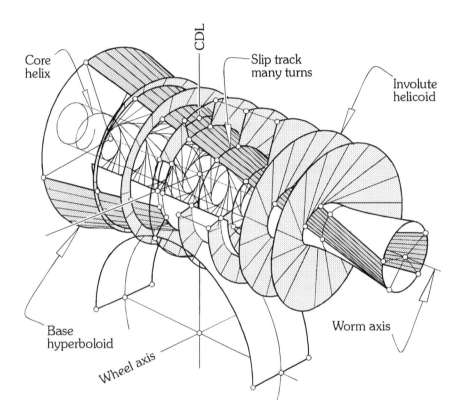

Figure 7.03 (§7.12). The base hyperboloid and the involute helicoid for the drive side of the single tooth of the worm at WkEx#5. These two surfaces intersect in the shown slip track. This slip track for drive, having a multiplicity of turns, diminishes in diameter in the figure as it winds (left handedly) from left to right. The slip track for coast and its associated surfaces are, for the sake of clarity, not shown here; but, with symmetry prevailing about the central plane of the wheel axis and the CDL, the slip track for coast diminishes in diameter similarly but from right to left. Notice the core helix for the shown involute helicoid which may be seen. Be aware that, while the ideal length of the path Z_I is determined as usual by the length of the rod SS, the actual length of the path Z_A will be determined as usual by truncations of the teeth.

14. *The two profiles of both sets of the teeth.* The single involute helicoid at figure 7.03, shown truncated cylindrically (coaxially with the shaft), might now be seen as the continuous driving flank of the single tooth of the synthesized worm. The coasting flank, not shown for the sake of clarity, will be a mirror image of this, not plane-reflected in the plane of the CDL and the wheel axis (which plane can be seen in the figure), but line-reflected in the line of the CDL itself. The opposing profiles might then be axially adjusted in location if necessary to suit the known half-CP at the key circle of the wheel. The shown slip track (with its many turns) belongs as an integral part of the shown profile, and reflects similarly. So we see that, while the shown slip track expands outwards from right to left along the working length (after axial truncation) of the real worm, the reflected slip track expands from left to right. The eighteen teeth of the wheel, similarly symmetrical, need no further mention here; their synthesis simply follows the rules of chapters 5B and 6. See a tolerably accurate sketch of the finished, conventional worm set of this WkEx#5 at figure 7.10, appearing later there for another purpose.

Looking more generally now at the plane of adjacency

15. Although prematurely, I wish to deal already with some highly relevant but as yet insufficiently developed matters. The *plane of adjacency* (§3.25) and the *conventional triad* (§10.30) are together of practical importance across the whole range of spatial involute gearing; and certain aspects of the combination of them are particularly relevant to worm gears. Refer in advance to figures 10.01 and 10.03, then go to figure 7.04 which relates directly to those two. Without loss of generality, all three figures relate to the teeth of the larger wheel at WkEx#1. In figures 10.01 and 10.03 the fundamental triad (§3.44) predominates, but in figure 7.04 (and in figure 10.07) the conventional triad (§10.30) does. I am awkwardly straddling the material of two different chapters here, but it will be clear I hope that the two tangents to the tooth's flank surface at any one [Qx], which have helped define the angles γ and θ in the relevant passages at chapter 10, together define (as other important lines also do) the plane of adjacency at [Qx]. Let the mentioned tangents be labeled (a) and (b), and refer to figure 7.04 which shows among other things the plane of adjacency defined by the angles γ and θ. One plane of the conventional triad, *the transverse plane*, gives us γ as can be seen, while another, the *peripheral plane*, gives us θ. There is a third plane of the conventional triad, namely the *radial plane* (§10.30), and it too cuts the profile of the tooth in some planar curve which will in turn be attended by its particular tangent in the plane of adjacency. This third, as yet mysterious tangent might be known, *pro tem*, as tangent (c). Its accompanying planar curve has yet to be studied (§7.16). There are of course two cases of this new curve and its tangents at the two [Qx], one for drive and one for coast. Refer to figure 10.06. As well as these three tangents derived from the conventional triad it will be clear from figure 10.03, where the sloping surface

WORM DRIVES OFFSET AND NON-OFFSET §7.15

of the phantom rack is in the plane of adjacency at [Qx], and where (by the way) the fundamental triad figures, that, also within the plane of adjacency at [Qx], there are (d) the helitangent which is in the direction of the sliding velocity, (e) the relevant ruling upon the surface of the flank, and (f) the hob slope line. These latter three derive from the fundamental triad at [Qx]. Although hitherto it has been ignored somewhat, due to pressing other considerations, the plane of adjacency is important. As it moves through the fixed space 1, it remains always normal to the relevant path which is fixed in the said fixed space. It has, however, not been seen as a triad-member since it belonged to the first of the triads, the *curvature triad*, at §3.21.

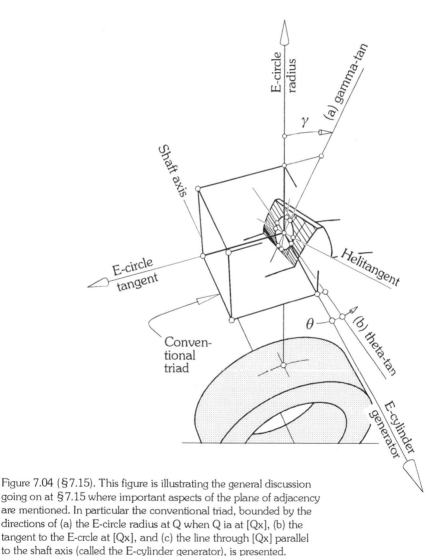

Figure 7.04 (§7.15). This figure is illustrating the general discussion going on at §7.15 where important aspects of the plane of adjacency are mentioned. In particular the conventional triad, bounded by the directions of (a) the E-circle radius at Q when Q ia at [Qx], (b) the tangent to the E-crcle at [Qx], and (c) the line through [Qx] parallel to the shaft axis (called the E-cylinder generator), is presented.

The radial planar section through the flank of a tooth

16. Going again to figure 10.06, please see the radial plane of the conventional triad drawn, for the sake of clarity, to be transparent. In the figure it bears the label, ρ-plane. Its two intersections with the sloping sides of the phantom rack are visible and marked each with an asterisk (*). I wish to call the relevant angles ρ (ρ for radial); see these angles more clearly marked at inset. By judicious triangulation, one can find that

$$\rho = \kappa \pm \text{artan}\,[\cot \delta \, \sin \gamma].$$

We have in figure 10.06 — which is drawn for not for the numerics of the worm or the wheel at WkEx#5, but for the numerics of the wheel at WkEx#1 — that, while $\rho_D = 11.914832°$, $\rho_C = 54.158817°$. The above general formula for ρ, although dissimilar to the similar two for γ and θ, which two are also associated with the conventional triad (see chapter 10), belongs with them as a group. Giving the group in advance — the formulae come from §10.23, §10.27, and §7.16 respectively — it runs as follows:

$$\tan \gamma = \tan \delta \, \sec \lambda \, \cos \kappa \pm \tan \lambda \, \sin \kappa;$$
$$\tan \theta = \tan \delta \, \sec \lambda \, \sin \kappa \pm \tan \lambda \, \cos \kappa,$$
$$\rho = \kappa \pm \text{artan}\,[\cot \delta \, \sin \gamma].$$

As we have now seen, each of these represents a tangent at [Qx] to a planar (or cylindric) section, namely a planar (or twisted) curve, taken through the involute helicoid of the flank of the relevant tooth; and each tangent lies, of course, in the plane of adjacency. The words within brackets in the foregoing sentence refer to the fact that in the case of θ we look not at the planar section, which is some unstudied planar curve, but at the *helix* upon the cylinder of the E-circle. This makes no difference, however, to the fact that we are dealing here with straight-line tangents at the points of tangency. They remain the same independently of the mentioned matters.

Beam's axvolute

17. The above material brings up now the following question: what is the shape of the planar curve cut by the third plane of the conventional triad, namely the radial sectioning plane? What shape would the curved edge of the cross-section be, in other words, if we chose physically to slice a wheel not transversely across the plane of the its E-circle but axially through one of its [Qx] by a sectioning plane containing the shaft axis? Refer to Beam [59] who asked this question also; he dubbed the curve 'axvolute' and wrote a convincing algebra for it; his approach, however, which was aimed at a better understanding of the phantom rack and its mechanics, differed from the one being taken here. Algebraically speaking, the question is asking: what is the nature of the continuum of planar curves upon the involute helicoid which is cut by the continuum of planes that contains the central axis? Refer

WORM DRIVES OFFSET AND NON-OFFSET

to figure 7.05 which is a computer-constructed picture of an involute helicoid that looks like the flank of the tooth of a worm where the core radius a is 100 mm and where the pitch of the core helix is 50 mm/rad (an example not previously used in this book) and where samples of the said continuum of planar curves upon the surface have been constructed and clearly shown. Look also at figure 7.06 which isolates the parameters and sets the scene for the establishment of the requisite equation. My first shot at finding the equation to the repeated curve of intersection produces

$y = a \cot \alpha \, [\Omega - \tan \Omega]$,
where
$\cot \Omega = a/\sqrt{(2ax + x^2)}$,

which has, as we shall see, the expected characteristics. The origin of coordinates is not at [Qx] but elsewhere upon the radial line through [Qx]; it is indeed at radius a from the central axis, which is at the periphery of the core helix. The x-axis points in the same, radially outwards direction; all real x are outside the range $\pm a$; those inside are non-existing; and the y-axis points with legitimacy in both directions parallel with the central axis. Seen in terms of figure 3.03, the origin might be seen to be at W, the x-axis in direction N→W, and the y-axis in direction W→A. Be reminded here that the involute helicoid has two branches (§3.15); so be not surprised to find that the completed curve for the above equation turns out to be as shown by the traces of

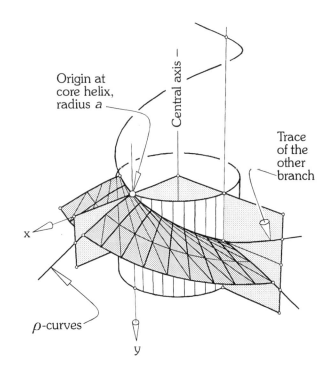

Figure 7.05 (§7.17). Generally chosen parameters are here showing the shape of an involute helicoid and its core helix. The angle α is relatively small however, so the surface has the characteristic shape of the flanks not of a wheel but of a worm. Previously unstudied, we see here the radial cross sections of the surface which are called by me in the circumstances here the ρ-curves upon the surface. They are not straight lines, but nearly so.

both branches of the involute helicoid in the shaded, radial plane of intersection at the right hand side of figure 7.05. As expected from figure 7.05, the relationship between x and y illustrated at figure 7.06, namely the ρ-curve sketched in there, is close to being a straight line through the said origin except for a gradual and then a tight turn into tangency with the x-axis at the said same origin. Sticher [52] has derived an expression for the straight-line asymptote to the mentioned curve, which is close to the curve itself when the tooth has (as it has here) a worm-like flank. Sticher's valued contribution is outlined at appendix A at 7 (§7.32), and the reader is referred to my comments on Beam at §10.31 for some related reading [59].

Visual recognition of properly cut involute worms

18. This radial plane of intersection and the shapes of the curves it cuts might be seen by some in the theory of gearing to be of academic interest only; and it might, in the case of ordinary spatial involute gearing, be just that; but refer as I have said to Beam [59]. In worms, and especially in those aspects of worms where we argue from time to time about whether a worm is an involute worm or some other kind of worm (and whether it matters or not), it will be more than that. Please refer to DIN 3975, as Litvin [5] has already done, and refer to Litvin himself who comments extensively about the various kinds of worm in his chapter 18. The main matter at issue just here however is this: if the surface at the flank of a worm tooth is truly an involute

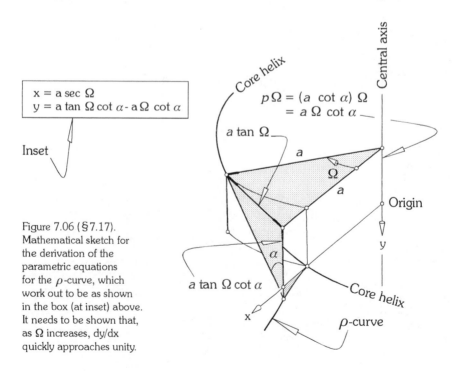

Figure 7.06 (§7.17). Mathematical sketch for the derivation of the parametric equations for the ρ-curve, which work out to be as shown in the box (at inset) above. It needs to be shown that, as Ω increases, dy/dx quickly approaches unity.

flank, its straight-line generators will be seen upon inspection to be arranged, not in such directions that they all cut the shaft axis at some angle, but in such directions that they all run tangential with the core helix. If moreover we actually take a machine-cut, axial cross-section through a real involute worm, which section must be cut, by definition (§7.16 *et seq.*), both through and along the shaft axis, that section will reveal, not straight lines at the flanks, but the already mentioned ρ-curves; see figure 7.05, and see appendix A at 7. If we wish to reveal the *planar involutes*, which are (as we know) always available upon the surfaces of the involute helicoids of all involute teeth of whatever kind, we must cut a *transverse* cross-section (§7.15); and this must be cut (by definition) *normal* to the shaft axis of the worm. Thus we will always see true planar involutes revealed at the plane-cut, transverse cross-sections that might be made at the two ends of a genuine involute worm. The collected remarks of this paragraph will help the reader to understand more easily, I hope, the unexpected intricacies of the next worked example. This, the forthcoming WkEx#6, will show that, despite gross departures from what we might call normality in the architectural layout of an offset skew gear set, and here we might take as an example of normality the set at figure 6.05 where all aspects of the layout (complete in the sense that polyangularity prevails and general in the sense that special cases do not obtrude) appear benign, the veracity of the theory of this book remains extant.

Beginning of worked example #6

19. *Preliminary remarks*. Well known in current engineering practice are (a) the offset skew gear sets generically known among theoretical workers as *the hypoids*, and (b) the offset worm reduction sets variously, but most often known among the same workers as *the spiroids*. Commercially registered trademark names such as *Hypoid* (Gleason Works of New York) for the former, and *Spiroid* and *Helicon* (Illinois Tool Works of Chicago) for the latter, are used by these famous makers of these gear sets to distinguish between the modes of operation, spheres of application, and the specific products competing with one another in the market. And speaking generically I wish to say that, whereas in chapters 5B and 6, at WkEx#2 and WkEx#4, I took examples of the hypoid kind to make my needed analyses, I take here a single example of the spiroid kind for the same purpose. Despite having made the implied distinction, I wish to explore here, not the evident but somewhat superficial differences between these various kinds of skew gears, but to show that, at least from the point of view of the involute theory exclusively outlined in this book, there is an essential sameness to be seen. Leaving aside the evident historical fact that most of the currently applied spatial gearing practices are non-involute and that these practices rest upon a wide variety of different theoretical bases, I wish to argue that all gear sets are, kinematically speaking, the same. Whether it will in the long run be for better or worse, whatever the effects of friction might be, and whether or not we encounter undercut and interference as we go, I wish to say, in other

words, that the involute theory developed in this book is applicable in all conceivable cases. In choosing this WkEx#6 to show just this, however, I do so with the uncomfortable premonition that some new phenomenon — new at least to me — is unexpectedly about to reveal itself.

20. This premonition comes from the seen evidence that a generally shaped, untruncated, involute tooth will have the appearance of those teeth collectively comprising the 'wire-frame' pinion illustrated at figure 11.01(a). The generally shaped tooth will have a sharp nose at *one* end (the nose-end) and a material body that tapers outwardly from that nose-end to the *other* end, which other-end must in the long run be truncated somehow, because otherwise it will become, in the long run, radially infinitely high and circumferentially infinitely thick. The reason that teeth are in general shaped like this is because they result from the intersection of two involute helicoids of differing core-cylinder radii and different pitches p_{FLANK}. See figure 5B.09 where such pairs of involute helicoids are shown at the inner intersection to be occluding the troughs, and at the outer intersection to be decapitating one another. Accordingly, and leaving aside the worm at WkEx#5 where the gross specialities of that problem there produced the single tooth on the worm that looked just like the continuous, symmetrical screw thread on a threaded bolt — see figure 7.10 — the following questions arise: *(a) what is the likely shape of the single tooth of the worm in this WkEx#6 where we deliberately set out to examine the single tooth on a one-start worm and where aspects of the essential asymmetry of the general worm-drive set are intentionally retained?, (b) how are the teeth on the wheel of the set shaped to cope with this?, and (c) what, apart from the shape of the blank is a 'face wheel' anyway?* With these questions unanswered as yet, let us proceed now to the selection of some beginning data.

21. *Data for this WkEx#6.* Let us take a right handed set with $C = 60$ mm, $\Sigma = 90°$, $R = 60$ mm. Take one tooth only for the worm and, in order to avoid a worm tooth that might be too narrow for a clearly seen graphical analysis, 20 teeth on the wheel; thus $k = 1/20$. This chosen k is much larger than it might otherwise be in the circumstances. For numerical convenience, take $j = 6/15$. Notice with respect to these k and j that, whereas the sum of the integers 1 and 20 is 21, the sum of the integers 6 and 15 is 21 also. From the chosen C, Σ and k alone, we can find by formula — equation (12) at §1.14 — that the pitch p for the set is $p = -2.992519$ mm/rad.

22. Refer to the interrelated figures 7.07(a), 7.07(b) and 7.07(c). The first of these shows a view of the set looking along the helitangent, the direction of which can be constructed or calculated from its location at F ($j = 6/15$) and the known value of the pitch p. Go for general guidance to equation (25) at §1.51. The view accordingly looks directly against the polar plane at F, upon which latter (in the plane of the paper) the ellipses of obliquity have been constructed. With the help of these, the angles of obliquity have been chosen, not necessarily but judiciously, to be equal. On opposite

WORM DRIVES OFFSET AND NON-OFFSET §7.22

sides of the BDL, they have each been set at 40°. In figure 7.07(b) the two chosen paths for the two Q have thus located in the polar plane the two ball-ended rods SS of the two equivalent linkages RSSR (§2.12). Each of these linkages is equivalent at the instant for the relevant relative motion, and each applies independently of where in its straight-line path the relevant Q may happen to be. Although the linkages RSSR are markedly different, this does not infer that the ratio k is other than 1/20 for both drive and coast. The difference is simply a reflection of the marked asymmetry of the set. Before we leave figure 7.07(a), see also there the CDL, the two shaft axes, the two key circles, and the shown naked wheel for the wheel; the naked wheel for the worm, too narrow to be seen clearly, is omitted.

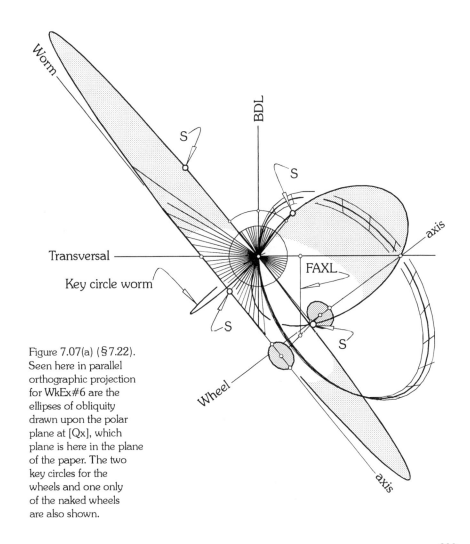

Figure 7.07(a) (§7.22). Seen here in parallel orthographic projection for WkEx#6 are the ellipses of obliquity drawn upon the polar plane at [Qx], which plane is here in the plane of the paper. The two key circles for the wheels and one only of the naked wheels are also shown.

§7.23 GENERAL SPATIAL INVOLUTE GEARING

23. Going next to figure 7.07(b), we see constructed there, as well as the two RSSR with their rods SS intersecting at [Qx], the four base hyperboloids. The drawn generators of each of these are shown extending from the throat circle to the key circle only, so the said hyperboloids appear in the figure to be not only truncated but also stuck together into two *stumpy elements*, one stumpy element for each of the two shaft axes; be reminded here of the two sets of two curves (o)(o) at figure 5B.06. Because each of the four

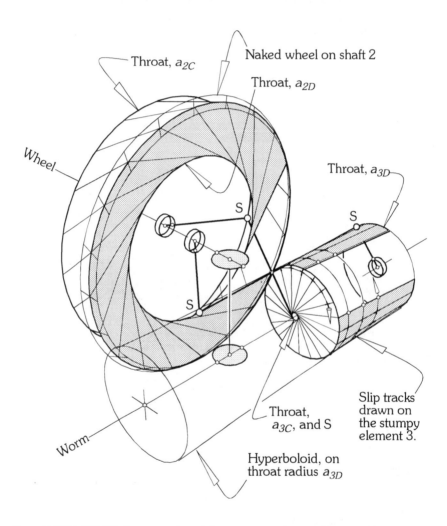

Figure 7.07(b) (§7.22). Panoramic (parallel) view of the linkages RSSR for WkEx#6. The four base hyperboloids are shown truncated separately at their throat circles and mutually at the relevant key circles of the wheels. The very short crank radius RS at right foreground is ommitted here for the sake of clarity. See this more clearly at figure 7.07(c). The two slip tracks drawn as a pair on the stumpy element 3 are both left handed.

WORM DRIVES OFFSET AND NON-OFFSET §7.23

slip tracks occurs at — or is, indeed, defined by — the intersection of a base hyperboloid and its relevant involute helicoid, the slip tracks may, upon discovery, be inscribed upon either one of the two said surfaces, the hyperboloid or the involute helicoid; be reminded here of the surfaces intersecting at figure 3.03. In figure 7.07(b) I have put the four slip tracks firstly upon the base hyperboloids; they may be seen there, each emanating in one direction only from [Qx]; in each case they are winding (bed-spring-wise upon the stumpy elements) from the key circle towards the relevant throat circle. Although they extend in actuality in both directions away from [Qx], they are not plotted in both of these directions; such a plotting, although satisfactorily

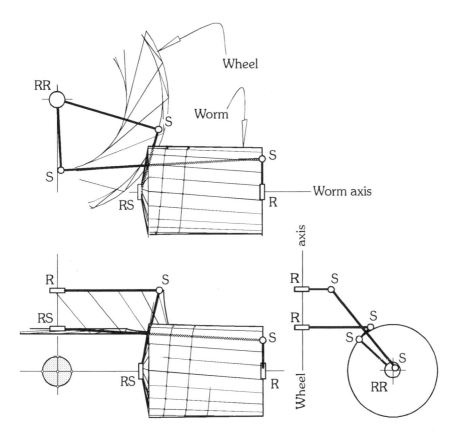

Figure 7.07(c) (§ 7.22). Three mutually perpendicular, othogonal views of the previous figure. This shows more clearly the slip tracks upon the stumpy element of the wormshaft and the layout in plan (looking along the CDL) of the equivalent revolutes R. Note that the joint S of the shorter input link RS (the one for coast) appears for the sake of clarity at one view only.

completing the figure here, would lead to confusion visually. Let us not forget that the involute helicoids, which constitute the flanks of the teeth (and these are not shown yet), unfold not from the base hyperboloids but from the core cylinders whose radii are the throat radii; be reminded here of the unfolding shown at figure 5B.08. This means that, in each of the four cases of there being a flank required, a flank is available for mechanical contact — but watch out for interference — anywhere from the core cylinder outwards. In figure 7.07(b) the slip tracks are grouped into pairs according to which stumpy element they are residing upon, and thus upon which of their respective shaft axes they reside, but they mechanically impinge upon one another according to the opposite system of pairing, namely that which asks to which of the two rods SS (namely to which of the two paths of Q) they belong. According to this other system of pairing, which is mechanically the important one, they are grouped into mutually impinging pairs where the members of the pairs, on opposite sides of the plane of adjacency, intersect at Q and are separately tangential there with the said plane, which plane (in the case of each of the two paths) remains, as Q moves, normal to the path.

24. Please look, before going on, at figure 7.07(c). This shows in simple plan and elevation, the state of the gear set so far. The two slip tracks plotted on the stumpy element 3 can be seen more convincingly: while the one is winding slowly down towards its wider throat at radius a_{3D} left handedly, the other is winding much more quickly down to its more narrow throat at radius a_{3D} also left handedly. Figure 7.07(c) also shows more clearly the layout of the four revolutes R of the two equivalent linkages RSSR.

25. *We can look already, now, for the shape of the tooth, the single tooth of the worm.* Somehow it will wrap itself, like a snake with its coils, about the two core cylinders coaxial with one another at shaft axis 3. It will have, as is usual for involute teeth, a *nose-end*, and it will somehow taper outwards from that to its *other-end* where, without truncation, troubles will occur (§7.20). Hopefully for the sake of adequate contact ratio (and this is of course a source of concern), the number of effective coils will exceed unity; but, as we know, it might accidentally not. Refer to figure 5B.09 and be aware that the geometry displayed there is directly applicable here. See also figure 11.01(a) and be aware that, whereas there are *nine* teeth on the pinion there, we have only *one* tooth on the 'pinion' here. We need to erect upon the stumpy element 3 the two relevant involute helicoids; these are at present intersecting one another at [Qx]; but they will need to be separated circumferentially by half the CP on the key circle namely 180° from one another before we can see, from the twisted line of intersection of the two flanking surfaces, the shape of the untruncated crest of the tooth. It should be mentioned in this connection that, because the CP extends, in this case of this 1-start worm, for the whole of the circumference of the key circle, the half-CP extends over 180° exactly, and, in the same way as it doesn't matter at figure 5B.09, it doesn't matter here about the direction of rotation in which the said circumferential separation of the slip tracks is effected.

26. Refer to figure 7.08(a). The eye-point for the perspective here is the same as that that was taken for the non-perspective, parallel view at figure 7.07(b). Here (and in the next figure) the shape of the worm has been constructed. The involute helicoids have been constructed for somewhat more than one turn, and their twisted curve of intersection, the crest of the tooth, has been located. The pictured worm (with its two outwardly sloping flanks) is seen to be mounted upon the larger of the two core cylinders, namely the one of radius a_{3D} = 29.0863 mm, while the smaller of the two namely the one of radius a_{3C} = 2.3400 mm is occluded within the larger. Notice that while one of the involute helicoids is of relatively large p_{FLANK}, and is thus only gently sloping downwards from the crest, the other is relatively small and thus more steeply sloping downwards from the crest. The formula is, quite simply, p_{FLANK} = $a \cot \alpha$ (§3.19); and the disparate numerical values are, respectively, 12.5144 mm/rad and 2.3073 mm/rad. The nose-end of the worm (which delineates itself quite quickly upon inspection) and the other-end (which is where I have halted the ongoing construction) are both occluded in figure 7.08(a). It is for this reason that figure 7.08(b) has been presented; it looks at the same scene, not from above, but from underneath. The slip tracks have already been shown in the figures 7.07; and, where not occluded, the same slip tracks appear here; but they appear here, not upon the base hyperboloids, but upon the involute helicoids of the flanks of the worm. The generators of these two pairs of surfaces intersect right-angularly in the slip tracks of course (§3.08), so there is no real problem with this; but the graphically active reader should not be surprised to find that — with the left and the right handed bed springs upon the right and the left handed hyperboloids respectively, and the opposite-handedness of the appropriate involute helicoids and the like — the known geometrical concepts are somewhat difficult to keep in mind. In due course, no doubt, an appropriate algebra based upon the here-found geometry will more amply clarify for the computer if not for many ordinary people the concrete essence of these phenomena; but there is no rest here for the geometrical research worker.

27. *Machining of slots at the trough.* It is often necessary, with involute teeth of whatever kind, to cut below the actual generated profiles (that is inside the relevant core helices), some kind of helical slot between the core helix itself and the curved fillet at the root of the tooth. This slot might be bounded by proper, ruled helices like those to be found within a square thread on a bolt or something similar. Such a cut slot is not shown here but needs to be imagined, to be mainly relevant at the steeply sloping flank of the worm where the core cylinder (the one that is shown) meets the steeply sloping flank. The two core cylinders in this WkEx#6, remember, are of disparate radii, 29.0863 mm and 2.3400 mm respectively, and the smaller one is deeply 'buried' within the larger. The cutting of such a helical slot of suitable shape will allow that slip track which is very close to the core helix on the steeply sloping flank of the worm to become more exposed and thus less hindered in its action when the worm is engaged with the wheel.

28. It would be a mistake to believe that the single worm tooth found just here will have a measurable, longitudinal pitch (measured mm/turn) in the same ways as has for example a screwed rod or a bolt. The worm tooth is, like any other involute tooth, bounded by two opposing, coaxial involute helicoids, each with its pitch p_{FLANK} (mm/rad); these intersect to form the crest of the tooth which is a twisted curve. Anthropomorphically speaking it could be said that the tooth behaves itself like the tapered branch of a creeping vine winding itself up a circular cylindrical pole, becoming thinner as it extends towards its nose, while paying attention (as it goes) more to maintaining the shape of its steadily tapering body than to the rate (mm/turn) at which it might be climbing the pole. Refer (looking upside down) to figure 11.01(a).

29. I wish to mention here the point marked 'Oops !' in figure 7.08(b). It is that point where the expanding slip track on the gently sloping flank of the worm oversteps the more slowly expanding crest of the worm itself and departs uselessly into vacant space. Because this happens before, not after, the completion of one whole turn of the worm from the nose-end, it is not good news. There is trouble with contact ratio. This does not invalidate the exercise, however, which is otherwise showing convincingly that the general principles of this book are applicable (except for accident) universally. It draws attention to the kinds of accident that can occur. In the same way as

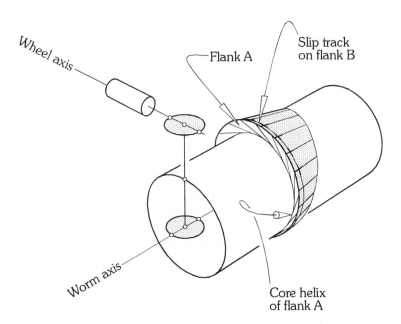

Figure 7.08(a) (§7.26). This shows the shapes of the two flanks A and B of the worm at WkEx#6. The worm is here shown as being supported upon the core cylinder for flank A.

there is within the realm of planar involute geometry, there is also here in space an as-yet ill-defined envelope of kinematic possibilities beyond which, unless we are looking for trouble, we cannot go. Refer to chapter 12.

30. To examine now the shapes of the twenty teeth of the wheel, we might next (a) examine the shape of the relevant naked wheel which is shown at figure 7.07(b), (b) look, in the CAD (as I did), along the helitangent again, and (c) look, one by one, along the two paths. On doing these things, and having noticed that the two p_{FLANK}, 797.1837 mm/rad and 60.8979 mm/rad, are somewhat disproportionate, which is however, given the worm, not surprising, one suspects that the finding of the slip tracks upon the stumpy element 2 (for the wheel) might present a problem also — a problem, as with the worm, not in principle, but in confusion among the different handed core helices and the bed-spring curves, and the disparate angles α, $\alpha_{2D} = 3.3080°$, and $\alpha_{2D} = 49.0574°$; and there was such a problem. Refer to figure 7.09. We can see there that the shapes of the teeth —ultimately found — are distinctly unorthodox; and I mean by *unorthodox* here that the shapes

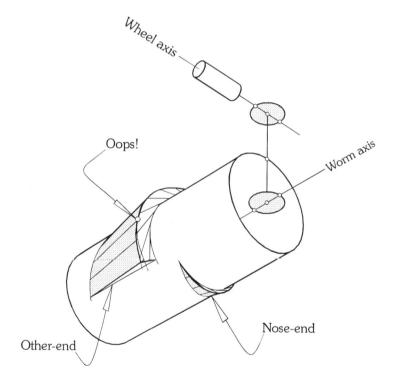

Figure 7.08(b) (§7.26). This is the same scene at that at figure 7.08(a), but seen here from underneath. The nose-end of the worm, the other-end, and the point of departure marked Oops! can be seen more clearly.

§7.30 GENERAL SPATIAL INVOLUTE GEARING

of the teeth at figure 6.05 for example, mounted as they are in the 'normal' way, 'externally' upon their blanks, constitute the orthodoxy. Notice that the teeth are best drawn for clarity, not upon the outside cylindrical surface of an orthodox blank, but on the annular face of a washer-shaped disc as shown in the figure. The flanks are lettered there A and B to conform with the lettered flanks of the worm; the flanks mesh A to A and B to B. Flank B on the wheel

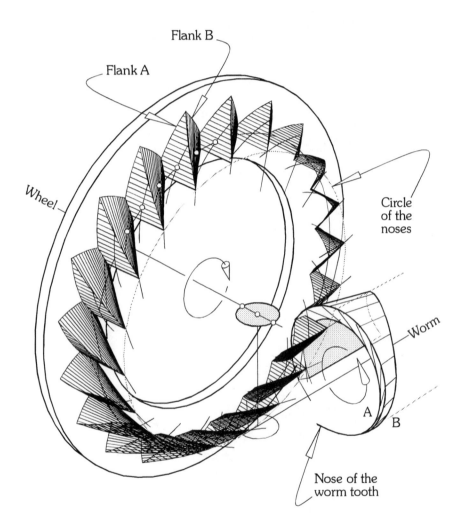

Figure 7.09 (§7.30). WkEx#6. This drawing is of a view taken from the same direction as that that was used for figure 7.07(b). It is however in perspective. We see here the discovered teeth of the wheel sketched in, and the single tooth of the worm in its proper engagement with those teeth. All teeth are shown by means of the active surfaces (working flanks) only. There is no supporting materail shown and no truncations.

306

goes all the way to its nose — the circle containing the twenty noses is ghosted in — and the core helices for the flanks B of the wheel reside in the same core cylinder which, although not shown, can here be imagined easily. The key circle for the wheel may be seen. It is intercepted by the successive flanks at equal intervals of 9°. Directions of rotation for a satisfactory *drive* (with the worm driving and the wheel resisting) are shown. The same directions of rotation for *coast* (with the wheel driving and the worm resisting) is not so comfortably free of friction; but recall in this regard the sometimes-essential non-back-driving of worm sets mentioned at §7.08.

31. *On the various kinds of truncation possible at the worm.* Having brushed over by means of passing mention the various *slottings* that may be necessary at the continuous trough between successive coils of the worm, and having mentioned (and shown) — see the 'oops' at figure 7.08(b) — the way in which premature mutual decapitation of the two opposing flanks of a tooth can suddenly terminate the mechanical reality of an ongoing slip track before it can achieve sufficient contact with its mate for proper contact ratio, we need to look at the kinds of truncation that might be necessity (or merely cosmetic) in connection with the meshing. Both Beam [59] and Brauer [103] indicate in their various explanatory diagrams of the straight Beveloid teeth a *conical* cut to make tidy the otherwise sharp (and thus vulnerable to damage) crest of the tooth. There is no reason to suppose that the cut must be conical however, there being as much reason to suppose that the cut might be *cylindrical*. The distinctions in respect of hardware that might be made between the given names *Hypoid* on the one hand (§7.19), and *Spiroid* and *Helicon* on the other §7.19), are sufficiently clear when seen across the wide range of the different non-involute geometries, but the generic equivalents of these are thoroughly blurred within the confines of the present analysis. *I wish to say again also (and this is a separate matter) that the idea that these generalised gear teeth, these general spatial involute teeth, are somehow essentially conical is a myth. They are indeed, and I draw attention again to the generally applicable figure 5B.08 at §5B.31, essentially cylindrical.* Refer to chapter 9A.

Load bearing and the small angles χ in the worm drives

32. The conventional, involute, non-offset, square worm set, as seen for example at WkEx#5, illustrated at figure 7.10, and as seen as a set of its genus, is simply one of the crossed helicals (5B.60); and the same remarks about high sliding velocities, small angles χ, rapid wear, and limited load carrying capacity apply here in the worms as well (§3.22 etc., §5B.24-25, 64 etc.). Detect the difference between ζ and χ. Because, however, we are here at a far-end of the spectrum of j among the crossed helicals, there is a multiplicity of Q in tandem along the paths of both Q. The contacts ratio are thus high, and each of the alternative forces **F** is well distributed among the multiplicity. Remember also that the mere presence of offset (see for example WkEx#6) greatly reduces the problem of the occurrence of small χ (§7.19).

Appendix A at chapter 7

33. This appendix comes from its second mention near the end of §7.17. It employs figure 7.10 (shown here) which, although coming from WkEx#5 at §7.14, belongs here as much as it does there. The appendix

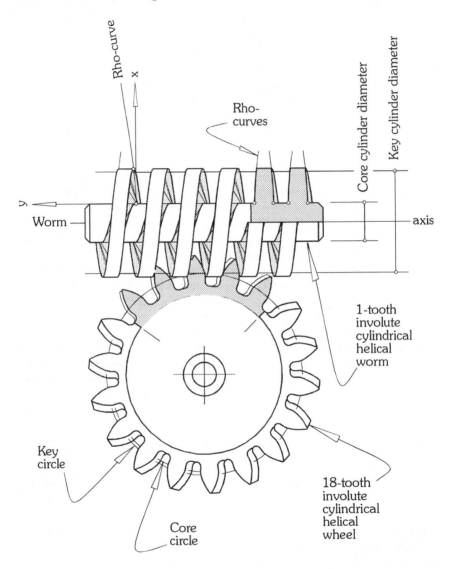

Figure 7.10 (§7.33). This is the conventional, square, worm set of WkEx#5. The figure is drawn to show, primarily, the shape and location of the representatve rho-curve plotted. Although the set is, essentially, a crossed helical set with the attendant disadvantages (§5B.24), the contact ratio here is high, so, due to the load sharing among the many Q in tandem along each of the two paths, trouble due to small angles χ is not so threatening. .

gives, almost exactly as written, the response from Sticher [57] to my request for an examination of the equations given by me at §7.17. These equations relate to the general shape of the ρ-curve, whatever the fixed variables; and the said curve for the special case of the conventional worm-reduction set of WkEx#5 is shown *in situ* here. The mentioned equations were implicit in Ω, and I naively requested from Sticher an explicit expression for y directly in terms of x. I asked also for some information about the asymptote. What follows is a modified version of Sticher's contribution, the modification consisting, mostly, in my having extended somewhat his explanatory text.

Given

$$y = a \cot \alpha \, [\Omega - \tan \Omega] \qquad \text{(i)}$$

and

$$\cot \Omega = a/\sqrt{(2ax + x^2)}, \qquad \text{(ii)}$$

we could, simply, substitute (ii) in (i) to achieve, independently of Ω, an equation for y explicit in terms of x. This could be used, as you say, to advantage in finding, should it be needed, the exact shape of the curve in question. It would of course read, for possible subsequent simplification,

$$y = a \cot \alpha \, \{ \text{arcot} \, [a/\sqrt{(2ax + x^2)}] \\ - \tan \, [\text{arcot} \, (a/\sqrt{(2ax + x^2)})] \}. \qquad \text{(iii)}$$

We could however, and more usefully perhaps, write from (ii) that

$$a \tan^2 \Omega = x^2 + 2ax \\ = (x + a)^2 - a^2$$

namely that

$$(x + a)^2 = a^2 (1 + \tan^2 \Omega) \\ = [a^2 \, (\cos^2 \Omega + \sin^2 \Omega)] / [\cos^2 \Omega] \\ = a^2/\cos^2 \Omega.$$

Provided x and a are both positive, it follows next that

$$x + a = a/\cos \Omega, \qquad \text{(iv)}$$

and accordingly that, when $x \to \infty$, $\Omega \to \pi/2$.

Let $\Omega = (\pi/2 - \zeta)$, where $\zeta \to 0$.

Whereupon

$$\cos (\pi/2 - \zeta) \to \sin \zeta \to \zeta,$$

and, from (iv),

$x + a = a/\zeta$, which is the same as
$$\zeta = a/(x + a). \qquad \text{(v)}$$

§7.33   GENERAL SPATIAL INVOLUTE GEARING

Then

$$\begin{aligned}\tan \Omega &= \tan (\pi/2 - \zeta) \\ &= \cot \zeta \\ &= 1 / \tan \zeta \end{aligned}$$

and, as ζ becomes small,

$\tan \Omega \to 1/\zeta$.

Therefore, for small ζ (that is for large x), equation (i) becomes

$y = a \cot \alpha \, (\pi/2 - \zeta - 1/\zeta)$,

which is (for small ζ only) the required equation.

Now, as $\zeta \to$ zero, this becomes

$$\begin{aligned} y &= a \cot \alpha \, (\pi/2 - 1/\zeta) \\ &= a \cot \alpha \, [\pi/2 - (x - a)/a], \text{ from (iv)}, \\ &= a \cot \alpha \, (\pi/2 - 1) - x \cot \alpha, \\ &= - x \cot \alpha + a \, (\pi/2 - 1) \cot \alpha. \end{aligned}$$

which is the required equation for the straight-line asymptote. Clearly its slope (in the xy-plane) is $[- \cot \alpha]$.

On the question of the slopes at points along the various curves, we can say in general that, from (i),

$$\begin{aligned} dy/d\Omega &= [a \cot \alpha] \, [1 - (1 / \cos^2 \Omega)] \\ &= - a \cot \alpha \, \tan^2 \Omega, \end{aligned}$$

from (iv),

$$\begin{aligned} dx/d\Omega &= - a \, [(- \sin \Omega)] / [\cos^2 \Omega] \\ &= [a \sin \Omega] / [\cos^2 \Omega], \end{aligned}$$

and, from these,

$$\begin{aligned} dy/dx &= [- a \cot \alpha \, \tan^2 \Omega] \, [\cos^2 \Omega] / [a \sin \Omega] \\ &= - \cot \alpha \, \sin \Omega. \end{aligned}$$

This last relates, of course, to equation (iii).

The significance of the above material relates of course to all kinds of involute gear teeth, but it relates in particular to the shape of an involute worm tooth. Any radial cross-section through an 'ordinary looking' tooth, such as the tooth depicted at figure 7.04, we see as not being important. We may call the shape, as I have done, *rho-curve* (ρ for radial), or, as Beam [59] called it, *axvolute*, his *ax-* having come from *axial*, the word he used to nominate the same cross-section. Taking a radial cross-section through a worm, however, clearly exposes the two rho-curves which clearly characterise the ordinarily perceived, cross-sectional shape of the tooth. To reveal the two opposing *planar involutes* at a worm tooth we must take the transverse plane (normal to the shaft axis) as usual, to do the sectioning, and this can be confusing .

CHAPTER EIGHT
EXOTICISM BY
ACCIDENT OR DESIGN

Preliminary remarks

01. In this chapter and henceforth I will for the sake of brevity refer to the linear complex of lines surrounding the pitch line, whose pitch is the pitch of the screwing there, and whose nature has already been described at §2.15, simply as the *pitch complex*. Next I wish to say that the essential thing about exoticism is not that the two paths of the two Q intersect (or do not) somewhere other than upon the F-surface but that the said paths cut the F-surface at two different points. This means that the question of whether or not the said paths intersect is irrelevant to the question of exoticism. Backhandedness is a somewhat related but separate matter (§8.12 *et seq.*).

Overview of exoticism

02. Because, in any one gear set, the angular velocity ratio k must be the same for drive and coast, there can be only one pitch line and thus only one pitch complex. If, in the realm of exotic design, we maintain our intention that the two paths will remain skew, the requirement is that the two paths must separately obey the fundamental law, which means, quite simply, that each of them must independently be a member of the one pitch complex. The two paths will inevitably pierce the F-surface at two separated points [Qf]. The common perpendicular between the paths will have no special significance, and it will be necessary to understand in the circumstances that the two different helitangents at the two different points [Qf] will not be parallel. These helitangents will necessarily and next be taken separately as the generators for two different pairs of naked wheels. The pairs of naked wheels (one for drive and the other for coast) will not be the same. The two polar planes of the complex whose centres are at the two points [Qf] upon the F-surface will not be parallel, and it is within these polar planes that the two paths must respectively reside. What this means is that we will need to envisage two imaginary racks, one for each flank of the sets of meshing teeth: it will not be possible for one two-sided rack to cut both sides of the teeth in a single operation. These mentioned matters are, of course, interdependent of

one another; and which of them is clarified first in design (to be followed in turn by the others) is a matter not of obligation but of strategy.

03. If, on the other hand (still in the realm of exotic design), we pursue a held intention (held for some particular reason) that the two paths will intersect, then (a) each of the paths, as before, and as always, must separately comply with the fundamental law, (b) the paths will be chosen to intersect at some point [Qx] which will in general not be upon the F-surface, (c) the paths will accordingly be coplanar, (d) the paths will accordingly reside within the unique polar plane whose centre is at [Qx], (e) the common perpendicular between the paths at [Qx] (which is of course the central axis of the single polar plane at that point) will be the helitangent there, but, as above, (f) the paths will pierce the F-surface at the two separated points [Qf], and (g) the helitangents at those points will be two different lines that will not be parallel. There will be, accordingly, and as there were before, two pairs of naked wheels. The items listed here are of course interdependent; and which of them is determined first in design (to be followed in turn by the others) is a matter not of obligation but of strategy. In the special case where (for some reason) [Qx] is chosen to be not in the open G-space but upon the F-surface, the resulting gear set will exhibit plain polyangularity. If (for more or other reasons) the skew paths are chosen not only to intersect but separately to cut the E-line, the polar planes at the two points [Qf] will be parallel (§5A.24). The two paths will then be members of the earlier mentioned special linear congruence of the fifth kind (§5A.24). If in the highly special case where the skew paths are chosen (for even more or other reasons) not only to intersect, but also to do so upon the E-line, the set will be equiangular.

04. If against a perceived background of effective machine design the above remarks are seen by the reader as being constructive, it might be worthwhile for him or her to pursue the mentioned matters further. There may well be powerful reasons, obscure to me as yet, for engaging in these various kinds of exotic design. Expressing no firm view at this time of writing, I content myself with these further few remarks.

05. An intended exotic design might be characterized firstly, perhaps, by the following three parameters: C, Σ and k. The designer might then be needing to choose the required R, of which there will be two (not only one). The given k will determine the unique E-line. The two different R will determine two different equilateral transversals; and these in turn will fix the two points E, one for drive and one for coast. Then the designer might wish to choose the two points F on the respective transversals where the two points [Qf] on the paths will be. The two best drives at these [Qf], each of which will be perpendicular to its own transversal, and neither of which will be in a vertical plane, will not be parallel and, inclined to these at the angles δ_D and δ_C, will be the two paths, each one of these in its own polar plane. It seems to me that the chosen locations of these paths will depend upon the designer's intent: it may be that the opposing flanks of the teeth are being relatively dis-

placed from one another to overcome some problem with undercut, or to help with non-back-driving due to friction, or for some other similar but complicated reason. In any event the four angles τ will be different, the asymptotic cones of all four of the naked wheels being different. The angles κ will be different too, but only between the pairs of naked wheels, for the two κ within any one pair will be, as always, the same (§10.08). So far in this book I have not needed the angles τ except to remark that, in the special case where equiangularity prevails, the two τ of a given set are the same; they are both then $\Sigma/2$. The formulae for the two angles τ in general, that is for the general case where the naked wheels are other than equiangular, have not been written yet; but take a look at appendix C at 6 where I show that $\cos \tau = \cos \kappa \cos \lambda$. As it was for the angles α in general (§6.38), the calculation for the angles τ in general will be somewhat complicated; each of them will depend upon the direction of the relevant helitangent at the relevant F and for this we will need the relevant beta shift and other such information. Refer to figures 6.05 and 6.07 for an illustration of the two different τ of a given set in the case of plain polyangularity, and see clearly thereby that here there will be, in the event of exoticism, *four* different τ as I have said. See appendix A at chapter 5B, see §5B.05, and see some relevant but interrelated general formulae (along with some special cases) at §10.08.

Analysis of the Merritt figure

06. Please go back to §P.02 and look more carefully now at figure 8.01; or, better, look again at Merritt's original figure from which my figure 8.01 was derived. The arbitrarily nominated dimensions (the angles φ in the plane of the paper) at figure 8.01 require some explanation: the values 25°, 35° and 45° were measured by me directly from the original figure, while the 55° is a slight adjustment from the original, made in the light of certain ambiguities there, and made for the sake of better clarity in my forthcoming figure 8.02. The diameters of the two circles are, again to copy Merritt, in the ratio 27/24. The four independently nominated angular dimensions and the ratio of the diameters of the circles are sufficient, in the absence, as yet, of the angular velocity ratio k, to nominate the essentials of the architecture. Read Merritt [58], chapter 10, and perceive that, although his details there were sketchy (as he no doubt meant them to be), his predictions were sound; and his use of the descriptive term *conical* in the circumstances was not unnatural.

07. Look next at my figure 8.02. This is simply another view, constructed by CAD, of my figure 8.01. I wish to use the given dimensions of the figures 8.01 and 8.02 because collectively they turn out to be an excellent vehicle with which to illustrate important aspects in the general arena of exotic architecture. As will been seen, exotic architecture may be consciously achieved with clear intent, or occur by accident,

08. I wish to mention again that Merritt used the symbol P to designate his initial point (presumably a point of contact between the flanks of mating

§8.09   GENERAL SPATIAL INVOLUTE GEARING

teeth) from which to make his start. I have assumed in any event that his P —
and I hereby call it P* to distinguish it from my P which I have used to desig-
nates the pitch point (§1.09) — is such a point of contact. Next it must be
said that, if P* is such a point, and if there are to be two paths chosen, one
for drive and one for coast (on both of which P* must presumably reside),
then Merritt's P* is my [Qx], and his paths must be intersecting paths (§8.03).

09. The following questions next arise. Is the point P* in Merritt's figure
located upon the F-surface of the set? Does the as yet invisible gear set of the
figure — invisible by virtue of having, as yet, no teeth — belong, in other
words, to the category plain polyangular? If not (and such a likely event
should not be surprising), what have been the geometrical consequences in
respect of teeth? The questions boil down to this: are correctly constructed
offset skew involute teeth at all possible in the circumstances of the figure?

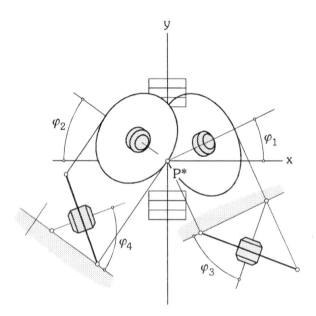

Figure 8.01 (§8.06). This figure is after Merritt [58] [§10.11]. It shows his way of explaining
in 1954 how a pair of offset skew involute gears might be seen in general to exist, and be
dimensioned in such a way that the layout of the set as a whole might be adequately
described. The rectilinear, ladder-shaped portion is his imaginary rack in the plane of the
paper. The point P* is at the intersection of two circles. Both circles are tangential to the
plane of the paper at P*, one circle being above the said plane, the other below it. For the
sake of generality the diameters of the circles are different, and the four angles φ for locating
the circles (two for each circle) are also different. Notice that the two shaft axes appear
to intersect in P*, but this illusion occurs only because the direction of the overall projection
(against the plane of the paper) is in the same direction as the unique line through P* which
may be drawn to cut both shaft axes. The angles φ_1 et seq., all in the plane of the paper,
have been estimated by me to be 25, 35, 45 and 55 degrees respectively. .

314

10. Without change of dimensions, as I have said, figure 8.02 is a reconstruction of figure 8.01. In the reconstruction I have first drawn the common perpendicular between the shaft axes, thus locating the centre distance C, namely the segment (A–B) of the CDL (see figure 1.01). I have next reoriented the figure in such a way that the CDL stands vertical, as has been the standard practice in this book (§1.06, §5A.29). Finally I have drawn the transversal jP*k, namely the unique straight line through P* to cut both of the two shaft axes (§4.40).

11. Now the fact that the two shaft axes appear at the Merritt original (and in figure 8.01) to intersect at P* illustrates that the orthogonal view in both of those figures is directed exactly along the said transversal. This puts one in mind of my figures 5B.16(d) and 6.08, for examples, where the same orthogonal projection (along the transversal) was employed. The next fact to notice in the said figures is that the planes of the circles (the circumferences of

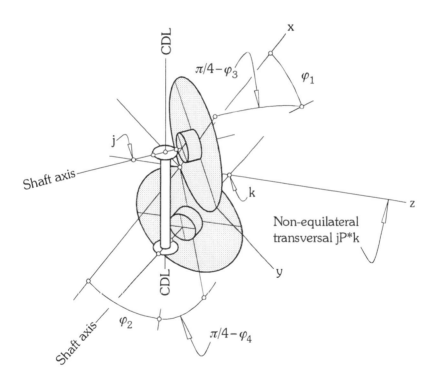

Figure 8.02 (§8.10). Merritt's original figure 10.11, see page 169 at reference [58], is already re-presented by me at my figure 8.01. This next figure, figure 8.02, is a re-drawn version of that figure 8.01. It is re-drawn in such a way that the CDL is now vertical with the shaft axes horizontal. This new figure thus conforms for the sake of the present argument with this book's adopted convention (§ 5A.29).

which intersect in space in P*) are tilted at different angles to the picture plane. This illustrates that the said transversal is not an *equiangular* transversal (§5A.16), but just an ordinary transversal. This in its turn shows that this P* (with the arbitrarily chosen dimensions of this example) does not reside in the F-surface of the set (§5A.16, §6.05). It shows in general, in other words, that an arbitrarily chosen P*, as Merritt's was, will not fall upon the F-surface.

12. Figure 8.02 has shown quite clearly that the prognosticated gears of Merritt's original figure were, in the sense of the word as defined at §S.01, §1.37, §1.53 etc., *offset*. If we look next at figure 8.03 which, in being another view of figure 8.01 is yet another view of Merritt's original figure, we see much more. Figure 8.03 is a plan view. It looks directly along the CDL, and it shows even more clearly that the point P* is remote from the CDL. It shows moreover that the transversal jP*k is not an equilateral transversal; the segments (A–j) and (B–k) are clearly not equal (§4.58); and this means that P* cannot be upon the F-surface. But it shows as well (and this newly discovered aspect is grist for the mill at §8.15) that the point P* is not within the range of the acute angle Σ; it is within the range of the corresponding *obtuse* angle Σ (§1.24). The simple concept of a single F-surface (a single parabolic hyperboloid) existing alone within the range of the acute angle Σ appears as a result of this analysis to have evaporated; but refer again (and now signifi-

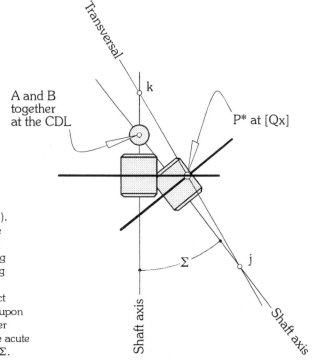

Book figure 8.03 (§8.12). This is a plan view of the same prognosticated set, looking downwards along the CDL. It shows among other things that the proposed point of contact P* at [Qx], which is not upon the F-surface, is moreover within the obtuse not the acute range of the shaft angle Σ.

cantly at last) to figure 1.02 at §1.11 where the two separated pitch lines *star and dagger* (*†) were carefully delineated. Refer also to the forthcoming §8.15 where a logically deduced but somewhat unexpected geometric concept, which I intend to call the *wholly-completed* F-surface for a given pair of shaft axes, is newly contemplated.

13. Given all of chapter 1 and all that was said there about the 'location in space' of the 'region of the meshing', we have found here a case where the second branch of the F-surface needs to be recognized. The F-surface always has two branches, one with its meaning within the acute angle Σ (as illustrated for example at figure 5B.01) and the other with its meaning within the obtuse angle Σ as found in this new figure 8.03. Refer for the idea in general to figure 8.04. As will be seen (and as already mooted at §1.10 *et seq.*, and clearly suggested at figure 1.05), there are in fact two FAXLs always intersecting perpendicularly at the origin O, and they, along with the CDL (counted twice), are the principal axes of the two interrelated parabolic hyperboloids which are (as can be shown) the said two branches of the F-surface. The branches intersect one another along the two shaft axes and along the CDL, and they have their relevance according to whether the region of the meshing is within the acute or the obtuse range of the angle Σ (§1.11). This matter was discussed in general terms in the five paragraphs beginning at [1] [§22.61] incidentally. Merritt's axodes are accordingly intersecting, and this bodes trouble in any event, with high rates of crosswise sliding (§1.11). But the double valued nature of the purely geometrical problem of the wholly-completed F-surface relates in gearing practice (a) to the separate possibilities of external and internal meshing, and (b) to the two alternative locations of the pitch lines (*) and (†). I have already mentioned that internal meshing is not possible with spatial involute action unless the shaft axes of the gears are parallel (§3.32), but we must understand that it is entirely possible for viable gears to exist in the more unusual realm where the pitch line resides within the obtuse angle Σ.

14. But where are the paths in Merritt's figure? And where do they cut the F-surface? For answers to these questions we need the polar plane at P^*, and for that we need the pitch line, and for that we need the speed ratio k, and none of these is available. Are the ladder-like lines of the single rack that appears to be shown in the plane of the paper showing us something? Yes they are showing that Merritt must have taken it for granted that only one rack would be enough. But we have shown that the set is exotic. So here is a contradiction. Returning now to the main matter at issue, namely the question of exoticism by accident, we don't find in Merritt's figure (or in his data) any information about his angles of obliquity δ; thus we have no information about the directions of his intended paths of Q; nor do we know his intended speed ratio k. So we are, in effect, snookered. But what I personally have found from this is that the whole area of exoticism is not easy to grasp, let alone to explain. That I have rambled around the subject matter of Merritt's figure for these many paragraphs without clearly extracting the main issues

§8.15   GENERAL SPATIAL INVOLUTE GEARING

for clarification indicates that in preparing his figure (a rough sketch of a dimly seen, new idea) Merritt was not wasting his time. He was drawing attention to the need for careful study in this newly emerging area of spatial involute action, and it would be churlish simply to reveal without respect for him the undiscovered aspects of his figure.

The wholly-completed F-surface

15. *An incomplete geometric concept.* Although the above discussion draws attention again to the fact that the point [Qx] at the intersection of intersecting paths can be put to occur not upon the F-surface and thus to produce exoticism, it draws attention also to the fact that such points of intersection (which are not, don't forget, obligatory) can be put within the obtuse not the acute range of the shaft angle Σ. Applications in design of the facts just mentioned are not mutually exclusive; they can be made with legitimacy either separately or together; but legitimacy (as we have seen) does not ensure practicality; and the overall questions regarding what will work and what won't — with or without mutual decapitation at the intersection of the opposite flanks of individual teeth, and with or without judicious truncation of the

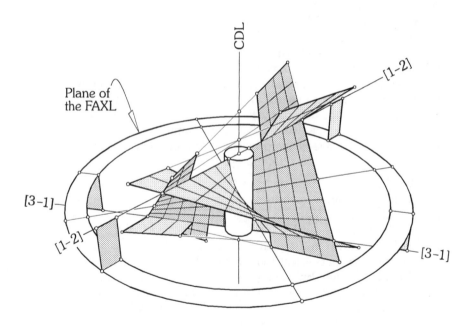

Figure 8.04 (§8.16). The complete F-surface consisting not of one but two parabolic hyperboloids intersecting one another (a) along the infinitely long CDL, and (b) upon both of the two shaft axes [1-2] and [3-1]. At the periphery, as the offset radius R tends to infinity, the surface tends towards becoming coplanar with the shown plane of the cruciform FAXL.

otherwise interfering but legitimate crests of mating teeth — are not at all simple. Despite my various warnings to the contrary — at §1.25, [1] [§22.61] and elsewhere — gear sets will in due course be intentionally designed with their points [Qx] inhabiting the difficult but legitimate, obtuse region of the shaft angle Σ. It is for this reason that I embark (although with some trepidation) upon the next paragraph.

Forehanded and backhanded design

16. *A tidy geometric repair.* If we define the F-surface as that surface generated by the single infinity of straight lines that can be drawn to cut a generally chosen but given pair of shaft axes (which axes will of course be skew) in such a way that each line cuts the two axes at equal angles, we will define a surface that has the appearance of figure 8.04. Clearly it includes not only that portion we have known hitherto as the F-surface existing within the acute range of the shaft angle Σ, but the other portion existing within the obtuse range as well. The FAXL now becomes not one but two lines mutually perpendicular, and the completed surface (wholly characterized by its C and Σ) clearly has two branches. All possibilities are now catered for. These include not only the alternative pitch lines star and dagger §1.11, but also the generally possible, external meshing we have been working with all along and the generally impossible, internal meshing mentioned for example at §1.10 and §3.32. The completed surface comprises two parabolic hyperboloids intersecting (a) along the infinitely long CDL, and (b) along both of the two shaft axes. As the radius $R \to \infty$, both surfaces become planar, approximating more and more closely to the shown plane of the cruciform FAXL. While the crossed helicals belong at the CDL, the straight bevels belong at $R = \infty$. As will be seen the surface is not *easy* to envisage, let alone easy to draw, but it does provide a vehicle for the assertion that, if overall design within the acute Σ is seen as conventional, or *forehanded* (in the sense that the driven wheel rotates in the expected direction), design within the obtuse Σ might be seen as unconventional, or *backhanded* accordingly. Refer to the photo-frontispiece at chapter 22 in [1].

Non-intersecting paths

17. It is clear from Merritt's hypothetical example, dredged by me in the present chapter for its hidden content, that he began with the reasonable proposition that the two paths of the points of contact Q might, at least to begin his general argument, intersect. In his figure, indeed, the paths are put to intersect at his point P*, and he goes from there. Hitherto I have done the same; but ever since chapter 4 of this book (§4.72 *et seq.*), I have said that the paths need not intersect, and under the heading of exoticism here, it behoves me now to explain what I mean by this.

18. We have seen already, geometrically speaking, that the mechanics of (a) *the meshing for the contacting fronts for drive*, and (b) *the meshing of*

§8.18   GENERAL SPATIAL INVOLUTE GEARING

the contacting backs for coast, are separable phenomena. Provided we avoid having two different sets of values for the numbers N of teeth on the wheels, and thus two different velocity ratios *k* (which would quickly lead to disaster), we can reckon as though the phenomena were *mechanically* separable also. Anyway I hypothesize thus, and I risk my neck, as Merritt did, as follows.

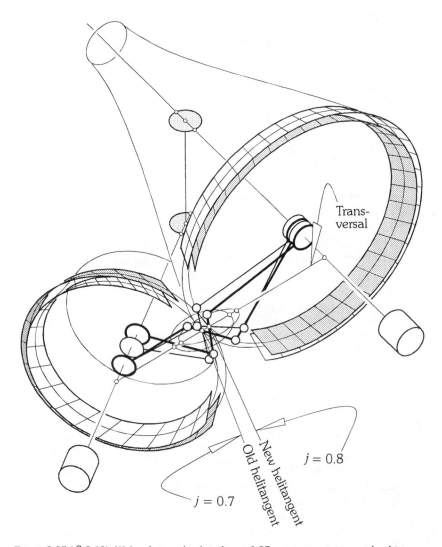

Figure 8.05 (§ 8.19). With reference back to figure 6.05 as a convenient examle, this figure is illustrating the effect of changing the the ratio *j* in a gear set while keeping the point [Qf] upon the same equilateral transversal. It is showing more importantly, however, that the two alternative sets of the linkages RSSR thrust up can be selectively used to achieve a limited exoticism.

320

19. Refer in retrospect to §4.72, §5A.11 and §6.28, and refer now to figure 8.05. To choose a convincing example that is moreover clear to illustrate, I begin with the data from WkEx#3. This is illustrated at figure 6.05: $k = 0.6$, $j = 0.7$, the two δ are each 20°, and, because I don't intend to change the overall architecture ($C = 80°$, $\Sigma = 50°$, $k = 0.6$), the pitch line and its pitch remain the same. Refer even further back if you wish to §5B.09, where WkEx#1 will be seen as an even earlier origin of the present adopted values; there the original point E appears, F is coincident with E there, and $j = k = 0.6$. What I have done in figure 8.05 is to maintain the value $j = 0.7$, as it was in chapter 6, but to introduce as well another value for j, namely 0.8. This has meant that I have now a new location for a second F where $j = 0.8$, and at this point a new, second, polar plane has been erected. See the labels for the 'new' and the 'old' helitangent which delineate the new and the old polar planes, and notice that their slopes are consistent with the range of slopes at figure 6.01. For convenience I have kept for this second polar plane the same two angles of obliquity δ, and constructed the two new paths accordingly. Next the common perpendiculars between each of the four paths and the two shaft axes have been constructed; thus the four equivalent linkages RSSR have been determined. The eight spherical joints, clearly marked, fall into four closely assembled pairs, where the members respectively of each pair relate to the old and the new configuration. Note next that a selected pair of the RSSR, a *drive* linkage from one of the polar planes, and a *coast* linkage from the other, has been drawn more boldly than the remaining pair. This non-matching pair, involving $j = 0.7$ for one of the sets of mating flanks, and $j = 0.8$ for the other, is next for consideration.

20. *Some remarks about exoticism as exemplified here so far.* If we choose (as we have already done in figure 8.05) a pair of the linkages RSSR which is a *non-matched pair*, in the sense that the pair comprises a new and an old RSSR, we will choose to be stuck with the following consequences: *two rods SS and thus two paths for the two Q that do not intersect; and (b) two different sets of naked wheels, one for one of the sets of meshing flanks, and the other for the other.* See figure 8.05. These consequences, neither of which is illegitimate according to the fundamental law, might be seen however as being either stupidly constrictive in design or wholly liberating according to one's point of view. They are constrictive in that they will render impossible the use of a single, straight-sided rack (usually a guided hob) for machine-cutting of the teeth (chapter 10); in which case some other method such as the face-milling of the separate flanks becomes necessary. The new circumstances are on the other hand openly liberating in that an opportunity for exploiting this exoticism will almost certainly provide another avenue for overcoming geometrical difficulties in design. In this last respect I quote for example the clear geometrical difficulties associated with WkEx#6. If special gears for special purposes are worth designing at all (and presumably they are), they are worth designing well.

The wider generality of exoticism

21. Please notice that the steps in the argument above have taken us only so far. I have earlier said that, provided the two chosen paths for the two Q are legitimate by the fundamental law, they may be placed, in design, *anywhere in the fixed space*; and this means, in the wider generality, that the two paths must be seen to cut the F-surface, which is fixed in the fixed space, *anywhere upon that surface*. It therefore behoves us lastly to discuss some further convenient, easily defined, graphical step from the second of the two F to a third, this third being elsewhere than upon the same equilateral transversal that contains the first two. I make this argument here, partly for the sake of completeness, and partly by way of recognizing that adjustments to avoid geometrical difficulties might need to be subtle, but mostly for the sake of the algebra. If the full possibilities of exoticism are to be taken into account in computer aided design, the algebraists will need some kind of describable geometric scaffolding. I propose therefore that we take one more short step to a third point F along the *local generator of the horizontal system of generators* of the F-surface (§6.22). Having done this, the helitangent and the polar plane at this third point F can be determined and two more new paths chosen. In this way we could arrive at pair of equivalent linkages RSSR where the two fundamental triads at the two Q, when the Q are at their respective [Qf], *will have no two of their respective directions parallel.*

On the difficulties surrounding truncation

22. This chapter has been showing by example (a) that, provided all working surfaces are ghostly surfaces, general spatial involute methods can used to deal with any arrangement of the axes of a gear set in producing geometric solutions that are kinematically (if not always statically) viable and relatively easy to visualize, but (b) that, when the physical reality of the single sidedness of the machined surfaces bounding the rigid material bodies of the teeth and their root-connections to the blanks, and the precipitate edges that abound, are taken into account, questions concerning the mutual mutilation of the surfaces at the crests of teeth and the likely collisions of the bodies in space obtrude, and matters becomes no longer easy to visualize. All questions of truncation are complicated, and it should be a reflection on none of us that cut-and-try methods in the development of the processes for mass production are openly pursued.

CHAPTER NINE (A)
HELICALITY TIGHT MESH
AND BACKLASH

Screw symmetry at the flanks of teeth

01. A source of confusion in the language of geometry is that, whereas the surface generated by a line moving screw-wise in space as it continues to intersect a fixed line perpendicularly is called a *helix*, the twisted curve of intersection between that surface and a coaxially drawn, circular cylinder is often called by the same name. I, as many others have done, will use the term *helix* indiscriminately. Any surface that is *screw-symmetric* might also be described as being *helical* incidentally. The same might be said for any screw-symmetric twisted curve. In any event a screw-symmetric surface or a screw-symmetric curve remains endlessly the same in the direction of its own axis.

02. At the flanks of teeth in involute gearing, the screw symmetry of any one involute helicoid (endlessly the same in the direction of its own axis) is such that its axis is, in all cases, *the axis of the relevant wheel* (§3.25). We say *helicoid* because, although the surface resembles a helix, it is not a helix. It resembles a helix in that all curves cut by intersecting coaxial circular cylinders are helices (twisted curves) upon it; but it doesn't resemble a helix in that its rulings do not cut the axis of symmetry perpendicularly. Its rulings are, as we know, tangent to a certain helix, the core helix (§3.12). *The ruled surface of the involute helicoid (a, α) comprises, quite simply, all of the tangents that may be drawn to the said helix. The pitch of the helix, and thus the pitch of the surface itself, is a cot α.*

03. See figure 9A.01. This is an omnibus figure intended for reference throughout the whole of this chapter. It shows without loss of generality the lower flanks of the five teeth of a 5-start worm. Look first at these flanks and see the various, clearly separated sets of curves upon the surfaces of the flanks, and be aware that the figure has been carefully proportioned to make these separations clear. See that the involute helicoid (which is, as we have seen, the surface defining the shapes of the flanks in all involute gearing) is not only (a) a ruled surface with a continuum of straight lines upon it (§3.12), and (b) a surface with a continuum of planar involutes, drawn in a contin-

uum of parallel planes normal to the shaft axis, upon it (§3.18), but also (c) *a surface with a continuum of coaxial helices of the same pitch, coaxial with the axis of symmetry, upon it.* In this last, important respect the involute helicoid resembles the shape of an ordinary screw thread.

04. I wish to remark in passing that, in the special case of the ordinary cylindrical helicals with parallel axes, the pitches of the involute helicoids on the opposite flanks of the teeth are usually the same. First comments on this confusing matter were made at §1.49 and §5B.60, and further comment will be made at §10.29. In the doubly degenerate case of planar spur gears, the pitch of the involute helicoid has become infinite and the helices have become straight lines; they have become moreover parallel with one another, and thus identical with the rulings.

05. Any screw-symmetric surface is such that we can effect a rotation of the surface about its own axis by a pure translation along that axis, or a pure rotation about it, or by a combination of these. It doesn't matter. The final result of any such combination of movements can always be seen as being the same, a pure rotation. This fact requires careful understanding whenever we try to consider not merely the replicated, ghostly, involute helicoids themselves as seen as pure geometric phenomena, but, more importantly, the replicated real flanks of the real teeth of actual gear wheels [64]. The bounded surfaces of such sets of real flanks are carefully selected identical portions taken from whole involute helicoids; they are physically connected via the material of the teeth to opposing, different sets of flanks; and the shapes of the selected portions will often be confusing. The portions will in practice not be truncated 'cylindrical'. They will, most often, be truncated in various 'conical' ways, at both their radial and axial edges. They may for example be truncated not only conical at their peripheries but also cut 'dish-ended'. We have seen already the importance of the edges as we travel radially across the teeth (at the crests of the teeth and at the troughs), but there are edges also as we travel axially along the teeth (at the two axial ends of the teeth where we meet the faces of the blanks), and the truncated shapes of these ends may also be important for the maintenance of contact ratio.

06. These remarks applying to the screw symmetry of the flanks of teeth apply whether the gears under consideration are (a) *ordinarily* skew, in which case we have 'cylindrical' skew involute helicals namely ordinary crossed helicals, as mentioned only incidentally in this book, or (b) *offset* skew, in which case we have what have been called, in the literature, 'conical' involute helicals, which are, as has been said, the main subject matter of this book. The word 'conical' seems to have come from Merritt [58]. He wrote predictively but somewhat hesitatively about the likely discovery of this kind of gearing in his [58] [§10.11] (§P.02). I have already used his perceptive work to advantage in large parts of my chapter 8; but I say again (and I do it just here), that the term *conical*, as I intend to show, was at that time, and continues to be, a dangerous misnomer.

On the misnomer

07. In the sense that all gear wheels in this general spatial involute area are always truncated somehow, there is, kinematically speaking, nothing special about a wheel being truncated 'cylindrical'. What this means in effect is that any cylindrical helical gear might well be truncated 'conical'. And what

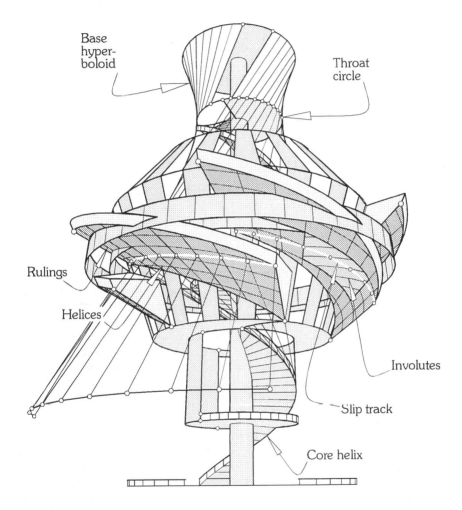

Figure 9A.01(§9A.03). An omnibus figure standing in for the greater part of this chapter. It shows the various identifiable sets of lines and curves that may be drawn upon the involute helicoidal surfaces of a set of identical flanks. As is the case with most others in this book, the fgure in drawn in three-point perspective. The helical staircase is there to show the overal left handedness of the figure which would be confusing otherwise.

this means in effect is that any cylindrical helical gear might with equal justification be said to be conical. *It's not the truncation that counts; it's the shapes (a, α) of the flanks*. Unless we understand the crux of this misnomer, we might never be able to put into modern perspective the idiosyncratic explanations of Olivier [40], or relate my ways of explaining this geometry to the current terminology of the Beveloid people [61].

The fronts and the backs

08. Each set of flanks on a given wheel — there being not only the fronts of the teeth but also the backs — consists in an evenly spaced multiplicity of identical involute helicoids coaxial at the axis of the wheel; see figure 5B.12 (§5B.35). But, as we know, the flanks at the fronts and backs of the teeth of a wheel are in general different. They have, among other differences, *different pitches of their respective core helices* (§3.17). This means that the pitches of the helical shapes of the two profiles are different. One might think that this might mean incidentally that, if the teeth are to be somehow cut, the cutting of the respective flanks must be executed separately; but see chapter 10. It also means of course that we can never, in general, speak about the pitch of the *teeth* of a wheel; we may only speak about the pitches of the *flanks*, (a) at the fronts, and (b) at the backs, of the teeth. There are no helices drawn on the flanks of the teeth at figure 5B.12; a smooth continuum of helices exists at every flank, of course, but, too easily confusable with the drawn rulings, all helices (even a few representative of them) are omitted in figure 5B.12.

09. In figure 9A.01, however, a few such representative helices, on a single set of flanks of an imaginary, half-constructed pinion-wheel with five teeth, are drawn; in this particular figure (the characteristic dimensions of which are lifted directly from figure 3.03 incidentally) the angle α (which is set to be 60°) is large enough for the helices to be distinguishable from the rulings; please notice also in figure 9A.01 that the size and location of the throat circle of the base hyperboloid is clearly indicated, that the helicoidal surfaces are truncated not according to the shape of the slip track shown but otherwise, and that figure 3.03 has been turned upside down.

10. I wish to repeat. The two sets of flanks of a given wheel are two sets of suitably truncated portions cut from two sets of *different but coaxial* involute helicoids. If N be the number of teeth, all 2N involute helicoids of the wheel are coaxial at the shaft, but each multiplicity of evenly spaced, identical, involute helicoids, one multiplicity for the fronts and one for the backs, has its own pair of defining parameters, namely its own (a, α).

Thick teeth

11. If we decide, during the design of a single wheel, to *thicken* the teeth (or to narrow them), we can displace the helicoids of the front flanks of the teeth away from the helicoids of the back flanks either by translation or

by rotation. As already said, it doesn't matter. Given that the pitches of the involute helicoids have a *sign* (either positive or negative), however, the translations or rotations must be made in the correct direction. In any event it will be clearly evident, after a such an imaginary movement or combination of movements of a set of flanks has thus been made, that the sets of flanks have been displaced from one another *through a certain angle purely circumferentially*. Refer for example to §5B.10 and §5B.30 for evidence that we have already seen the truth of these remarks and that, in the processes of synthesis, we have acted accordingly. It does remain of course true that, as with planar teeth, if with the axes fixed we thicken the teeth of one wheel, we must thicken the troughs (namely narrow the teeth) of the other. This might or might not be associated with changing the heights and depths of teeth (the radial edges) and, in offset cases, changing the locations of the faces of the blanks (namely the axial edges of the teeth). Overall, however, it doesn't matter in so far as the single initial position of the two coincident [Qx] is concerned. It is only the lengths Z_l or the locations of the of the zones of actual contact that may alter. Kinematically speaking, the overall motion will remain the same whatever we do.

Conical blanks

12. There is not only no difference in effect between rotation and translation in the context here, there is no 'conical' aspect of the kinematic argument at all. The only way in which these offset skew involute gears might be conical — and herein lies the misnomer — is that the external shapes of the truncated blanks [65] and the internal shapes of the gashes that need to be cut between the teeth will often be cut conical. Refer to figure 5B.13, and look again at figure 9A.01. This is simply due to the nature of the machine tools that might be used for the trimming. 'Conical' or 'multi-faceted pyramidal' are among the most convenient shapes for the trimming of excess material that might be obstructing the real, non-ghostly [64], interactions between the real surfaces of the real teeth which have, as we know, real flanks and, above all, as we shall see, real *axial edges* (§9.17, §9.23).

Drive and coast

13. Let us assume, just yet, that we are dealing exclusively here with gear sets that have been designed on the basis of two chosen legitimate paths intersecting at the F-surface. This means that we are dealing, just yet, with plain polyangular sets in general, and, as we go, with equiangular sets incidentally. Take, now, *a gear box and a pair of wheels in tight mesh*, namely a whole gear set with all three links extant, accurately constructed and accurately assembled according to initial design, and having zero clearances. Let us distinguish between (a) the kinematics of the *drive*, which involves the two front sets of the flanks of the teeth, and (b) the kinematics of the *coast*, which involves the two back sets of the flanks of the teeth. In general, and as already mentioned, *all four of the said sets of flanks (except for their being*

coaxial with one another in pairs upon their respective wheels) differ from one another. Next we need to see that the kinematics of the drive and the kinematics of the coast are not only thus different from one another, but, in many ways also, *independent of one another.*

14. This means for example that the flanks for coast may be displaced (in design) from those for drive through any angle we see fit. Without loss of generality, *displaced* in the context here means *circumferentially displaced* (§9.05). Such a relative displacement of the sets of flanks of wheels, however, will in no way alter the [Qx] and the two paths previously fixed (by virtue of design) in the frame. Such a set of displacements, moreover, will not interfere with the constancy of angular velocity ratio during transmission in either drive or coast. It also means for example that we can, with equanimity, alter the axial locations of the wheels upon their respective shafts without upsetting the fixed locations of the paths. We might in so doing alter the lengths and locations of the zones of action Z_h but overall the kinematics of the motion will remain the same.

15. None of these remarks preclude the actual separation of drive and coast that would clearly become apparent as soon as we adopted any design — I have called such designs *exotic* — involving either (a) a nominated point [Qx] placed either accidentally or intentionally remote from the F-surface, or (b) a nominated pair of paths that do not intersect at all. In either of these two cases the two paths will pierce the F-surface at two separated points [Qf]. I have looked already at such exotic designs (as thus defined) in Chapter 8.

The fixity of the paths

16. What I have said above is that, once the teeth have been designed and constructed for a pair of straight polyangular wheels on the basis of a chosen [Qx] and a chosen pair of legitimate paths, the [Qx] and the pair of paths remain fixed in the architectural space independently of how the two real wheels might be modified circumferentially or be disengaged and axially or circumferentially moved (tooth by tooth) and then re-engaged with one another. What follows next is in further explanation of what I mean by this.

Axial withdrawal from tight mesh

17. It needs to be explained before we begin here that, with offset skew sets (and with shaft axes fixed in location), it will, in general, always be possible to disengage the meshing teeth from tight mesh with one another by displacing one or other of the wheels *axially* along its shaft. Naturally this axial withdrawal from tight mesh can only be achieved by sliding the chosen wheel in one direction only. Should the wrong direction be chosen, the chosen wheel will simply remain immobile. That such axial withdrawal is possible in one direction only is due (as we shall see) to the fact that the pitches of the four participating flanks are different, and that the differences between these pitches, taken two by two, are different also. *It should be noticed incidentally*

that, in the degenerate planar spur, parallel helical and crossed helical arrangements, this kind of withdrawal from tight mesh is not possible. This is so for the simple reason that the necessary differences between the pitches of the relevant flanks are not available. *It is a severe disadvantage of these latter degenerate arrangements that, in order to adjust or allow for backlash, the centre distance must be altered.*

18. For the sake of argument, take a straight polyangular pair of wheels such as the pair illustrated at figure 6.04. By doing this, the argument will also apply to an equiangular set such as the set at the figures 5B.10. Both the 'ghostly' faces of the naked wheels and the 'real' faces of the wheel blanks in these cases are drawn to be equidistant on either side of the planes of the F-circles [65]; these chosen locations of the faces have merely been due to convenient conventions adopted so far; neither of them have any geometric validity; and the flanks of the teeth are circumferentially located in such a way in the figures that the widths of the teeth and the troughs are equal when the wheels (both kinds of wheels) are in their drawn locations. I use this latter to define what I mean by the *central reference locations*. These occur when the successive flanks of the teeth cut equal intercepts upon the F-circles when the real wheels, in mesh, are located at these central, reference locations. The wheels might, with justification, be then said to be *flush* (§9A.20).

19. By way of experiment now, let us erect two axially located shafts (both freely rotatable) within a fixed frame (link 1) to accommodate the said wheels (links 2 and 3). This will involve taking account of the C, Σ and R of the gear set. Having loosely mounted the wheels upon the said shafts, slide them now axially upon the shafts to find a suitable combination of the two axial locations to achieve tight mesh. *Within the limits determined by the wheel blanks continuing to straddle the fixed planes of the original F-circles (which latter remain embedded in the frame 1) there will be a smooth continuum of combinations of such locations.* Provided, in other words, each blank continues to straddle the plane of its original F-circle, which latter remains fixed within the architecture, the blanks may 'overlap'.

20. There is, however, a unique pair of central reference locations whereat both blanks are physically set at their original design relationship with the F-circles and with one another. In clarification of this I might argue that, imprinted into each wheel blank (which is of course a movable body), there is a copy of the relevant F-circle at its original location, and when that copy is actually at the F-circle, which has a fixed location in the architectural frame (link 1), the wheel is in its *central reference location*.

Shifting of the slip tracks

21. It would appear from these remarks incidentally that, by thus adjusting the relative axial locations of the wheels while keeping the teeth in tight mesh, we could arrange to traverse a smooth continuum of different pairs of slip tracks (§3.34). The slip tracks are the pairs of paths of Q in 2 and 3. We

could thus distribute across the flanks, from time to time as we saw fit, the wear due to sliding. One should recall here in passing that the direction of the rubbing velocity at any Q is not, in general, in the same direction as either of the two slip tracks which are of course intersecting at that Q (§5A.05). Indeed we might expect in best practice that the rubbing velocity at any Q might cut both of the two slip tracks intersecting there at some large angle.

22. Whenever the centre distance C or the shaft angle Σ is changing either erratically or periodically, due either to flexibility within a non-rigid gear box, or to deliberate changes being made, as might be made for example at the articulated joints of robots where rotary motion is being transmitted by means of suitable involute gears, the slip tracks will be changing too. Provided the necessary clearances are available at the 'make and break' points of contact within the gear set, these phenomenon are clearly benign; for (a) the freedoms are inherently available for this to occur (§1.05, §2.09), and (b) the wear is being distributed. I would like to call this activity, whatever its origin, *skating*, or *lane changing* at the slip tracks.

On mobility

23. Refer to figure 9A.02. This figure will be used in this chapter to illustrate the kinematic chain of the gear set as it may change (or be changed) from time to time as the argument here proceeds. Unlike the 3-link loops of mobility unity illustrated in the figures 1.03 (a)–(j) in [1] [§1.25–30], the 3-link loop here has mobility two [1] [1.32–34]. The freedoms f at the joints are such that $\Sigma f = 2 + 2 + 4 = 8$, and thus that the mobility $M = 6(n - g - 1) + \Sigma f = 2$, as stated. As shown, with the f2 cylindric joints at A and B at the shafts, and the f4-joint with its two points of contact between the ball and the V-slot at C, the chain is representing the gear set with (a) both rotatable wheels free to slide axially upon their shafts (which latter are fixed as parts of the frame), and (b) the teeth remaining in tight mesh with one another. Figure 9A.02 as it stands is telling that, with the wheels thus free, and independently of the intended transmission of the rotary motion, the wheels may move axially upon their shafts to new pairs of locations while maintaining tight mesh at the teeth. These two physically distinguishable activities (the rotary motions of the wheels and the slidings at the shafts) can occur either separately or together, this being the essence of the fact that the mobility M of the mechanism as a whole is 2. This important first observation points to the fact that the shown pairs of locations of the wheels in the figures 5B.10(a) and 5B.11, for example, is only one of a single infinity of such pairs of locations of the wheels for which tight mesh is possible.

24. If we suppress the freedoms for axial motion at the shafts by making the two cylindrics there into revolutes, the sum of the freedoms becomes $\Sigma f = 1 + 1 + 4 = 6$, and the mobility becomes $M = 6(n - g - 1) + \Sigma f =$ zero! This means that, unless there is some geometric speciality [1] (and there is), the mechanism becomes immobile. It is only by virtue of the cor-

HELICALITY TIGHT MESH AND BACKLASH §9A.25

rectly mating involute helicoidal shapes of both sets of the flanks of the teeth — and this is, indeed, the said speciality — that the rotary motions of the wheels are in ordinary practice possible. This continuous transmission of rotary motion from wheel to wheel, however, occurs (in tight mesh) in the presence of overconstraint; and that is at least one of the reasons why, in practice, where perfect accuracy can never be achieved, that some clearance allowing for backlash is always necessary.

25. If on the other hand (in figure 9A.02) we fuse the links 1 and 2, that is if we suppress the freedom of link 2 to move at all, the link 3, free to rotate about and slide along its shaft at B, must (when it moves) withdraw its ball from one or other of its points of contact at the V-slot, remaining free thereupon to remain in contact at the other. Having thus withdrawn from one of them, it could next slide with its ball in contact at the other, remaining point of contact and open up a *clearance* between the side of the ball remote from the said point and the opposite side of the slot. Under these circumstances the mobility M of the mechanism would be unity, and no internal stresses would be occasioned by the said (fixed-axis) screw movement. Let us now take up, in the following, the concrete argument from there.

Relative re-location at the axes of the wheels

26. The above mentioned continuum of pairs of axial locations for an existing pair of wheels in tight mesh is not very easy to understand. We could usefully think in a concrete way, however, like this. With the two wheels in

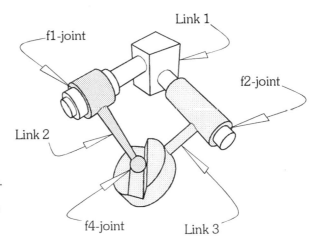

Figure 9A.02 (§9A.23). The three-link loop of the gear set. This figure is applicable for those aspects of this chapter where the freedoms at the joints and the overall mobility of the mechanism are at issue.

331

tight mesh and in their central reference locations (§9A.20), let us, with one of the wheels (say 2) remaining fixed upon its shaft with that shaft fixed, release the other. Manually applying torque say in the drive direction, we may now withdraw the loose wheel 3 axially from tight mesh. *Notice that, on being withdrawn, the loose wheel, (a) though sliding there, remains in contact at the fixed point of contact Q which remains fixed in both 1 and 2, (b) screws upon its shaft with the fixed pitch of that particular one of its flanks remaining in contact at that Q, and (c) inscribes the path of that fixed Q upon itself, which path is a helix.* Think of the moving wheel and its relevant set of flanks in motion in the circumstances, not as being some mysterious conical thing, but as being an ordinary multi-start, cylindrical, screw thread, for that, indeed, is essentially what it is. The fixed point of contact Q (fixed in 1 and fixed in 2) is acting as a mechanical guide for the pure screwing motion of 3 as 3 is being withdrawn. *This guiding action ceases however as soon as the relevant axial edge of the moving tooth arrives at Q.*

27. As the said withdrawing motion proceeds, of course, a clearance is being opened up between the opposite pair of flanks. Now, after withdrawing a certain limited distance (we must stop before the relevant axial edge of the sliding tooth arrives at Q), *rotate* the loose wheel 3 about its axis, exploiting the clearance now available between (a) the extant 'drive contact' with 2 on the one side of the gap, and (b) the incipient 'coast contact' on the other. Contact will thus be remade on the other side of the gap. I wish to remark in passing here that such an externally controlled, pure rotatory motion of the loose wheel might be seen to constitute exactly what we will be calling, in due course, the *backlash*.

28. Now, while manually applying a torque in the opposite (the coast) direction, slide the loose wheel 3 axially back into tight mesh again. This time, it will screw upon its shaft with the pitch of that other set of its flanks, its coast flanks. The two said screw movements (first the withdrawal and then the reinsertion) might be of different sign or the same sign, and the magnitudes of the respective pitches a cot α (mm/rad) will of course be different; but the axial distances traveled (on first withdrawing and then going in again) will be the same.

29. In summarizing the passages §§9A.26-28, and by way of further comment, we could say this. We (a) withdrew, by means of a pure screw movement about its own axis, one wheel from the other, sliding from tight mesh (a hard bottom datum) out along a freely chosen side of a wedge shaped opening, (b) rotated, at the wide end of the said opening, the withdrawn wheel to the other side of the said opening, then (c) returned the wheel, sliding in along the other side, whereby we found that, with a similar pure screw movement but with a different pitch this time, we came again, suddenly, to the same hard bottom datum. It must be said of course that we could equally well have withdrawn the loose wheel and returned it without paying attention to the sides of the opening at all, in which case, however,

The flush configuration

30. The term *flush*, tentatively used above (§9A.18), can have no useful physical meaning in relation to the wheel blanks unless the axial extents of those blanks are somehow related to the point [Qx] which is, as we know, fixed in the architecture [65]. It should be abundantly clear by now (and speaking figuratively) that, *while the extents of the wheel blanks might come and go, the point [Qx] will remain forever fixed in link 1*. Refer to the figures 5B.02(a) and 6.05 and notice that in both of them the smooth teeth of the naked wheels (§4.36) [65] are arranged at [Qx] in such a way (a) that they are collinear with one another in the direction of the sliding velocity at [Qx], (b) that they are of equal length, and (c) that, superimposed, they evenly straddle [Qx]. Seen against the theory of the naked wheel, this makes sense, but seen against drawing-office practice (in which arena the axial ends of the blanks might be put to be anywhere), it makes no sense. If however we imagine for the sake of the present argument that the extents of the wheel blanks coincide with the shown extents of the naked wheels in these figures, the term *flush* can have a meaning, and thus be a useful term. In any event we need to see quite clearly that, as we slide real wheels in mesh to and fro upon their axes, the circumferential widths of the teeth and those of the troughs at the F-circles, while continuing to sum to the fixed circumferential pitch of the teeth ($2\pi f/N$) on both wheels, will vary as we go. In the ordinary course of events, therefore, the flush configuration might also be seen to occur when the widths of the teeth and the troughs at the F-circles are equal.

Axial adjustment for location

31. The above discussion has shown that, if we wish to adjust the axial locations of real wheels while they remain in mesh, we may achieve the adjustments step by step. We could, as it were, withdraw one wheel to open up a small clearance, then advance the other (the previously fixed one) to take up the opened clearance, and repeat this (moving to and fro as necessary) until the pair of wheels are satisfactorily arranged. *We will of course find that each axial advance of the second wheel will not be equal to the axial withdrawal of the first.* It will become evident in other words that, when the wheels become so arranged that the central planes of the blanks are displaced away on either side of the relevant F-circles (which circles are fixed in the frame), the blanks will not be flush with one another. *It will also be evident that that particular set of two locations resulting in wheels that are flush with one another will be unique.* If we care to imagine the absence of friction (or sufficiently lubricate the teeth), we may simply push one or other of the wheels, the 'protruding' one, gently resisting the consequent motion of the other, until the wheels as a pair are satisfactorily arranged. Once again, of course, we will find the flush locations to be unique.

32. A matter that needs a mention here is that, whenever the wheels' locations in tight mesh are thus arranged to change, and while [Qx] and the paths remain the same, the points of contact Q are thrust into different, new positions upon their paths. Such changes in the tight-mesh locations, moreover, alter the ends of the zones of contact (§9A.11, §5B.34).

Axial adjustment for backlash

33. We see in any event that, even in the presence of clearance, and even, therefore, with the provision for backlash, and given equiangular and plain polyangular sets, the originally chosen point of intersection [Qx] upon the F-surface and the paths of the two Q will remain (whatever else may happen) fixed in the link 1. It follows next that the provision for backlash can be made by moving one or other of the wheels of the set (or both of them) axially upon the respective shafts until the required amount of clearance is achieved. It is possible also to see that a kind of centralized provision for backlash can be achieved by moving both of the wheels by an equal amount in the correct (opposite) directions on their respective sides and away from the mutual central location. In connection with backlash I draw the reader's attention to a paper by Innocenti [62].

34. *Flexibility of the box.* This and other dimensional irregularities can be accommodated whenever sufficient clearance is provided within which the necessary backlash can occur. There is no problem in this regard especially when Σ is sufficiently great, that is, when Σ is not too small. If Σ begins to approach zero we are, as we know already, in real trouble. We must then rely on the awkward business of adjusting the centre distance, which is in all respects a nuisance.

A general observation

35. This helical property of the teeth of the wheels in general spatial involute gearing is, of course, of profound significance. Without it we would, for example, not have the advantage of being able to adjust for backlash easily. Without it, also, we could not tolerate a flexible gear box, nor could we entertain the idea of cyclic (or non-cyclic) intentional relocation of the shafts of the wheels without an interruption to the steady mechanical transmission of the rotary motion. We can do almost everything with a fixed (even a wrong) centre distance. This property is not shared by the hypoids, of course, which rely for their action on *curved line* contact between the teeth, and thus on the occurrence of overconstraint and the consequent need for great accuracy of assembly within a rigid gear box.

CHAPTER NINE (B)
MECHANICS OF
THE PHANTOM RACK

Introduction

01. Until now in this book we have treated the involute helicoid as a surface generated by a moving line, the line remaining always tangential to a given helix. We have used for the radius of the helix the symbol a, and for the angle of the helix (the helix angle) the symbol α. See figure 3.06. Thus (a, α) has characterized the involute helicoid. We have argued incidentally, not only in general that the pitch p of the involute helicoid is $p = a \cot \alpha$, but also in the special context of gear teeth that the pitch p_{FLANK} of the flank of a tooth is $p_{FLANK} = a \cot \alpha$. Early in chapter 9A we studied the various sets of lines and curves to be found upon the surface of the helicoid. They were (a) the already mentioned straight generating lines, (b) the planar involutes occupying planes drawn normal to the axis of the helicoid, (c) the helices drawn upon circular cylinders coaxial with axis of the helicoid, and (d) the slip tracks. In this chapter 9B, we deal now with other geometrical aspects of the involute helicoid that are important not only in general but also in that they relate to the mechanical processes of the machine-cutting of teeth.

First look at a puzzling matter

02. Figure 9B.01 illustrates a well known theorem not found in most books about gearing. It may however be found in some books about kinematics. Look for example in Prudhomme and Lamasson [93]. The theorem is as follows:

> *Theorem 1.* If we screw, with pitch p, a plane about an axis fixed in space, which axis makes an angle α with the said plane, we describe in space a continuum of planes that successively intersect one another in a continuum of straight lines. These lines sweep out the involute helicoid (a, α) where $a = p \tan \alpha$. It might correspondingly be said that the screwing of the plane *sweeps in* the same involute helicoid. The involute helicoid is, in any event, the so called *envelope* of the continuum of planes.

Grant (1899) came close (but not convincingly) to stating this theorem exactly in connection with the teeth of Olivier (1842). See Grant [94] [§175].

03. One way to imagine the envelope here is to imagine that the plane is at the flat outer face of a smoothing tool (like a trowel) screwing about an axis fixed in the fixed space and fixed in the handle of the tool so as to sweep in (as a trowel might do) the surface upon a lump of wet plaster. The surface here must be seen as being convex outwards, and the lump of plaster must be seen as being smaller than the working face of the smoothing tool, *for the tool has edges*. Otherwise the working face of the tool must be seen to be of infinite extent. The presence of edges relates directly to the question of undercut in the machining of real gear teeth.

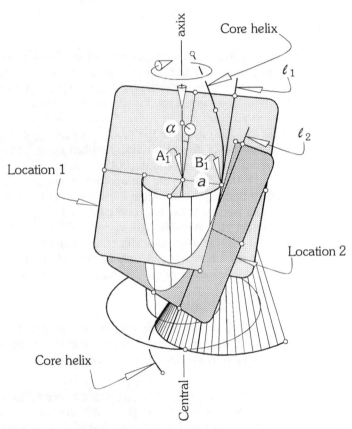

Figure 9B.01(§ 9B.02). A plane (seen here in locations 1 and 2) is screwed with a pitch p mm/rad about an axis inclined at angle α to the plane. The ruled envelope swept in by the plane is an involute helicoid. Its equation is $p = a \tan \alpha$, where a is the radius of the core helix.

04. It does need to be said however that the involute helicoid is more than the machined surface of some mechanical body. It is in general a ghostly surface with two branches (§1.29, §3.15). It is (as we know) a ruled, screw-symmetric surface, with a curved cuspidal edge (or curved line of regression) at a helix of radius a and angle α. There are no parts of the surface within the circular cylinder of radius a defined by the said helix. There exists within the surface in other words an axially arranged, circular cylindrical hole. I have called the cylinder the core cylinder, and the helix the *core helix*. The surface is of course infinitely long in the direction of its axis and infinitely wide in its radial directions. Each branch, ruled, is developable. See the inner portions of two branches opening out from a helical cuspidal edge illustrated at figure 3.13(a). See there the smooth curved trajectory of a slip track as it crosses over the sharp cuspidal edge; see too the *base ruling*, which is the unique generator tangent to the helix at the point of crossover. It might moreover be mentioned that different branches continuously intersect one another in helices of the same pitch as we go further and further radially outwards from the central axis. Infinitely far away in these directions, the surface consists of asymptotic cones (of half-angle α) intersecting one another in circles of infinite radius.

05. In figure 9B.01 the said generating plane is shown at an initial location 1. Note the same plane shown in a subsequent location 2. See the screw axis cutting the plane (at location 1) in point A_1, and see the angle α indicated there. Point A_2 with its same angle α is occluded. Now there is a unique line in the said generating plane that remains continuously collinear with one or other of the rulings of the swept-in surface. There is in other words a unique line in the generating plane along which the generating plane continuously touches the generated surface. If we draw in location 1 of the plane a line ℓ_1 parallel with the orthogonal projection upon the plane of the line of the screw, at a distance $(A_1–B_1) = a = p \tan \alpha$ from A_1 (and on that side of the core cylinder where the line is tangential with the helix of the screw), that line is the said unique line. Go back to figure 3.03: see there the generator through B of the involute helicoid and the so-called facet there which is tangential to the surface along that generator. By screwing the facet (left handedly) with pitch p about the central axis of the figure, we cause the facet to sweep in the envelope, the involute helicoid. At the same time we cause the unique line of the facet's contact with the surface simply to sweep out the same involute helicoid. This same circumstance (except for the opposite handedness) is illustrated again at figure 9B.01.

06. *Proof of the theorem.* Take an infinitesimal angular displacement $d\Omega$ about, and an infinitesimal linear translation $p\, d\Omega$ along, the axis of the screw and study the two infinitesimally displaced locations of the unique line which, fixed in the generating plane, has moved with it. Find that all measures of the movement from the original location of the line to the next are small quantities of the second order. This shows that the said pair of planes

§9B.07 GENERAL SPATIAL INVOLUTE GEARING

intersect in the said line and that the theorem is thereby proved. I should mention incidentally that, because the locations 1 and 2 of the plane in figure 9B.01 are not infinitesimally close to one another (being separated representatives of the continuum of planes), their shown line of intersection in the figure is not a generator of the helicoid.

Second look at the same matter

07. The statement and scope of the theorem above might first give the impression that it tells all that needs to be known about the mechanical generation of the flanks of a spatial involute tooth. However consider the following. When we go to the practical business of (a) the basic elements of the original design of, and (b) the day to day settings of those elements for the various jobs to be done within the various kinds of gear-cutting machine, we tend to think in terms of the *physical components of the machine* — the main frame, the hob, the cutter, the slide, the cradle etc. Each of these has its own established household name, and all of them are *bodies*. But what do we do with the phantom rack? It is evidently a thing, but, for reasons yet to be explained, it is not a body.

08. The above relatively simple Theorem 1 (§9B.02) came under consideration here because it obviously relates to the mechanics of the phantom rack and thus to the cutting of involute teeth. The theorem wholly relies, however, upon the contrived circumstance that there are *two bodies* involved, (a) the first body (the so called reference body) fixed in the fixed space within or upon which the involute helicoid is inscribed, and (b) the moving body which is the said plane, having not only the said straight line fixed within it but also the said line of the screw rigidly attached to it. Within the kinematic meaning of the word *body*, the plane by itself is not a body; nor is the generated involute helicoid by itself a body; and herein lies an unexpected difficulty. See [1] [§5.44] and the forthcoming §9B.09. Theorem 1 moreover assumes that the screw (the relevant ISA) continuously remains fixed in the fixed body. It assumes in other words that the fixed axode for the motion is the mere screw itself while the mating, moving axode is the *same* mere screw itself. All of this is a highly special case of the generality. A few elementary remarks about axodes in general appear at [1] [§22.61]. Other relevant material might be tackled at Rosenauer and Willis [29]; they discuss the generation of envelopes (although in the plane exclusively) by means of various planar curves moving under the guidance of sets of centrodes, there being always a fixed and a moving centrode to guide the curve.

09. Theorem 1 gives the mistaken impression that the extended flat surface at the flat flank of a phantom rack during the sweeping in of a tooth profile must be screwed about an ISA fixed at the axis of the wheel. But this is not the case; *for neither the extended plane itself nor the phantom rack that carries it are bodies*. Each of them are ghostly surfaces which may displace themselves in certain directions (the plane in two directions within itself

and the phantom rack in the single direction of its flutes) without changing the outward form of itself. Similarly the involute helicoid is a ghostly shape that may freely screw about its own axis with the same perplexing result. If a thing can move while appearing to remain at rest, the thing (by definition) cannot be a body. Certain solid things, like a wine glass (with no handle), or a smooth cylindrical bottle (with no label), are also not bodies in this special sense of the word, because they may, without our *seeing* it, freely rotate about their own axes. One might say in the context here that, before a thing can be a body, it must have a set of no less than three, generally disposed, recognizable points within it which move (when the body moves) integrally with it. See the idea *tick* at [1] [§4.04] and read note [95].

10. On this second look, when we actually handle the physical parts at the machine tool, it begins to appear that the phantom rack, in order to do its correct job namely the sweeping in of the profile of the tooth, might, indeed, be screwed about some ISA *other* than the above-assumed, *central* ISA. This other axis or these other axes all appear to be and to remain parallel with central axis of the wheel however. It moreover appears that any chosen, other ISA will need to be moving steadily around the surface of some cylinder concentric with the axis of the wheel; the fixed axode for the motion then being that cylinder while the moving axode is some plane in the rack rolling around that cylinder while sliding at the same time along the said ISA at some appropriate constant rate.

11. These impressions come from the well established geometry of the cutting of teeth in the case of planar involute wheels in which (a) the pitch circles are agreed to be acting as the centrodes for the relative motion of the two wheels, (b) the relevant pitch circle and the common tangent to the opposing pitch circles at the pitch point are agreed to be acting as the centrodes for the generation of the regular flanks of the teeth by means of envelopment by the flanks of the rack, and (c) the relevant base circle and the common tangent to the opposing base circles (the so called line of action) are agreed to be acting as the centrodes for the generation of the teeth by means of a single point describing its involute path in the reference plane as the rolling proceeds. See Steeds [1] [§141 *et seq.*] for a reasoned discussion of these planar but somewhat mysterious matters; see *Gear Handbook*, 1st ed. (not the 2nd), page 1-9, for a revealing pair of pictures by Baxter [16]; and see in advance my figure 10.08(c), reading its caption.

Axodes in general and the ideas of Reuleaux

12. In the regular planar kinematics of rigid bodies, we have a unique pair of planar centrodes for the relative motion of any two selected bodies (the bodies being represented by irregular laminae with three points fixed thereon) and, normal to the reference plane, a screw of zero pitch that may move as the motion proceeds. In the regular *spatial* kinematics of rigid bodies we have a unique pair of axodes for the relative motion of any two se-

lected bodies (the bodies being represented by irregular volumes of space with tetrahedra or ticks fixed therein) and a screw of finite pitch that may vary continuously in location magnitude and direction as the motion proceeds. In the well known and often invoked circumstance in regular spatial practice of there being a known *periodic* relative motion of two rigid asymmetrical bodies, *there is always only one discoverable screw at any one instant and one pair of axodes available repetitively across time to represent that relative motion* [1] [§5.43, §12.14].

13. Reading Reuleaux (1865) [8] [§9, §30-37], however, we find that he deals with matters beyond these simplicities, these firmly established, easily understood, modern certitudes. Although in two dimensions only mostly, he comes to grips with the kinds of ephemeral matters mooted at §9B.07 *et seq.* above. He explains (a) that for any coming together of pairs of profiles, namely for *any kind of planar cam action, sliding or non-sliding*, there is only one pair of *primary (non-sliding) centrodes* that will reproduce the history of the motion, but also (b) that the same motion may be represented by a multitude of sets of three or more *secondary (sliding or non-sliding) centrodes* that can work together to reproduce the history of the motion. He goes on in his §38 to discuss the axodes for the relative motion of spatial profiles, and predicts the kinds of difficulty I am having here. This early insightful work of Reuleaux does not completely elucidate the current issues — nor does [29] — but it does help us to understand that the general matters are far from being simple, and to explain that for a given spatial tooth profile (that is for a given involute helicoid) there will be an infinity of different sets of axodes available for reproducing the history of the working against one another in space of the two relevant profiles namely the curved flank of the actual tooth to be cut and the relevant plane flank of the phantom rack. It should accordingly be clear at outset here that the occurrence of a multiplicity of pairs of secondary axodes is, to follow the ideas of Reuleaux, a possibility. We must understand in other words that, if we deal with working (cam like) profiles in space in predictable, mechanical, *sliding and/or rolling*, contact with one another, we will walk directly into this unexpected, more difficult arena.

Another theorem

14. Go now to figure 9B.02(a) and see there two locations of a generating plane laid out in a way similar to that that was used in figure 9B.01. The two locations of the plane are not only bounded by similar blunt-cornered rectangles as before but also marked by means of three circular portions inlaid within them. A fixed, *primary screw axis* is set vertical at the left of the figure, and about this axis the plane is shown to have suffered a finite twist (D, Δ) from location 1 to location 2. The continuously instantaneous rate of twisting is held to have been D/Δ (mm/rad) where D is the finite linear displacement along the ISA and Δ is the finite angular displacement about it. At the top right hand of the figure a horizontal circle is drawn (coaxial with the central ISA) whose radius is a, where, as before, $a = p \tan \alpha$. This circle

MECHANICS OF THE PHANTOM RACK §9B.15

is accordingly called a circle of the core cylinder upon which latter is inscribed the core helix, and this helix is the cuspidal edge of the involute helicoid. The two labels l_1 and l_2 are showing the two locations of the unique line in the moving plane that, during the steady screwing of the plane from 1 to 2, sweeps out the involute helicoid; both l_1 and l_2 are tangent to the core helix. Without loss of generality in figure 9B.02(a) the following numerical values for the CAD have been adopted: [$\alpha = 60°$, $D = 50$ mm and $\Delta = 20°$; and from these by calculation, $p = 143.2390$ mm/rad and $a = 248.0979$ mm.]

15. Now the main object of the exercise here is to show that, if we take no account of the fixity of the circular portions within the infinite extent of the plane, that is if we assume there are no *fixed markings* on the plane, *there is an infinity of pairs of axodes with which we may screw the plane in order to sweep out the same involute helicoid*. But the pitches of the ISAs are not all the same; and all except one of the ISAs, the central ISA, *move as the motion occurs;* so a subsidiary object of the exercise is to explain the distribution of these different pairs of axodes and the pitches of their different screwings.

16. Let me call the three marked circular portions inlaid within the parent plane in its two locations the *buttons*. Let us suppose that the intermediate button marked in the plane is generally disposed, but with its centre upon the radial line of the centres of the buttons, which line is clear to see in both locations. Now if in location 1 we take the intermediate button and effect its finite screw displacement (D, Δ) about the fixed central ISA, we may see this as occurring in two distinct stages, (a) a pure rotation through angle Δ from centre-point (i) to centre-point (ii) in the horizontal plane of the circle marked star (*), and (b) a pure translation through distance D from centre-point (ii) to centre-point (iii) along the vertical line marked hash (#).

17. Let us next consider the same finite screw displacement taking place in two stages otherwise. Let there be in the horizontal plane of the circle marked star (*) a taut flexible cord attached to and wrapped around the said circle as shown. See this more clearly in the plan view at figure 9B.02(b); this might be studied in conjunction with the figure 9B.02(a) which remains relevant. The length of the straight unwrapped portion, extending from (ii) to (iv), is the same as the length of the arc of the circle extending from (i) to (ii). During rotation of the button through Δ, the central point of the button might be seen to have traveled along the shown planar involute from the cusp of the involute at (i) to its new position upon the same involute at (iv). What needs to be seen next is that the plane of the button at (iv) is now parallel with the parent plane in location 2, and that a pure vertical translation through an as yet specified distance will bring the button, *which was itself the parent plane in location 1*, into exact congruence with the parent plane in location 2. If I (for intermediate) be the length of the pure translation from (iv) to (v), which I is clearly less than D, the pitch of this particular finite screwing of the plane from location 1 to location 2, around a moving ISA (moving around the circle marked star in plan), is not D/Δ but only I/Δ mm/rad.

341

§9B.18   GENERAL SPATIAL INVOLUTE GEARING

18. Going now to the outer button of the three, starting with its centre-point at position (vi), let us perform the same series of maneuvres. We find that, having first performed the required rotation through Δ and the contemporaneous relocation of the central point of the button from (vi) to (vii) in the horizontal plane of the circles, *the subsequent pure vertical translation required to bring the button onto the plane in location 2 is precisely zero.* At

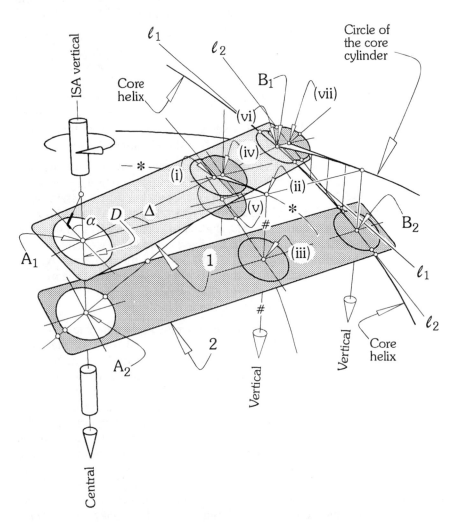

Figure 9B.02(a) (§9B.14). Two successive locations of a plane being screwed about a fixed ISA relative to frame. Superimposed there is an intermediate location drawn to allow for the finite rotation to be taken separately first, to be followed next by the finite translation. The figure is being used to argue that, provided the pitch of the screwing is adjusted, there are many more than only one ISA that can be used to reproduce (at any stage) the instantaneous motion.

342

the unique circle of the core cylinder, in other words, namely at radius a from the central ISA, the distance l is zero.

19. Next we notice that the geometry inherent in the figure clearly reveals that l (which is in the context a direct measure of the pitch) varies directly with the radius measured outwardly from the central axis. What we notice is that the three points A_1, (v) and (vii) have found themselves in a straight line upon the surface of the plane at location 2 and that the corresponding buttons are newly inlaid there. So without further ado we can write the following theorem.

Theorem 2. Whenever a plane is generating an involute helicoid by the process of envelopment as explained at Theorem 1 above, the reference location of an axis of a substitute screwing may be seen to be anywhere parallel with the central axis (where the pitch is p), provided (a) in choosing the reference location we remain

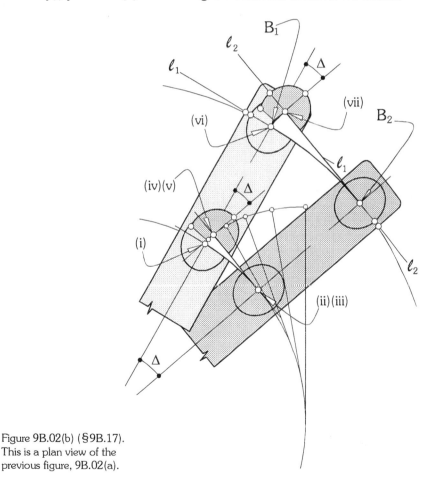

Figure 9B.02(b) (§9B.17). This is a plan view of the previous figure, 9B.02(a).

exclusively upon the common perpendicular (A–B) between the central axis itself and the above-mentioned unique generating line fixed in the generating plane, (b) the pitch at the substitute axis varies directly with its displacement from the central axis outwards, and (c) the pitch of the substitute axis at displacement *a* from the central axis is zero. At all displacements (including zero displacement) the fixed axode for the motion should be seen as the coaxially arranged cylinder of that radius, while the moving axode should be seen as the plane at that radius that slides there axially while rolling there circumferentially without slip.

An important aspect of this theorem is that, except at the central ISA itself where the movement of the substitute ISA is by definition zero, each subs-

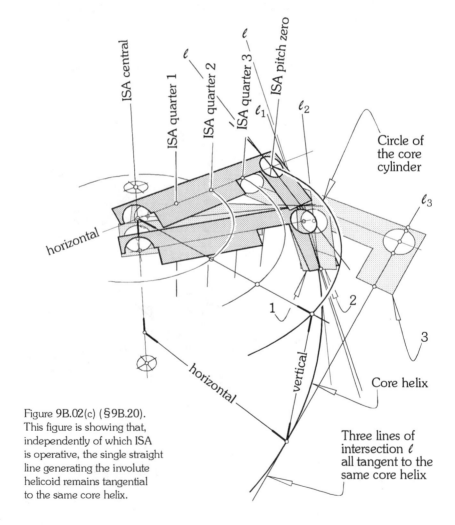

Figure 9B.02(c) (§9B.20). This figure is showing that, independently of which ISA is operative, the single straight line generating the involute helicoid remains tangential to the same core helix.

titute screw axis migrates steadily around the surface of its cylindrical fixed axode as the motion proceeds. Another aspect is that, with increasing displacement, and as we pass the unique displacement *a* at the core cylinder, the variable pitch changes sign. Another is that, at infinite displacement, the pitch is infinite.

20. The foregoing text, along with its three figures to date, together purport to contain sufficient proof of Theorem 2. Figure 9B.02(c), on the other hand, purports to further clarify the said theorem and to show how it harbours the possibility of spectacular extension. Very much simplified from the original CAD, it is a panoramic view that has been used by the author to confirm the following. Independently of where the ISA is taken — I have taken the three quarter-points along (A–B) — and provided the pitch of the

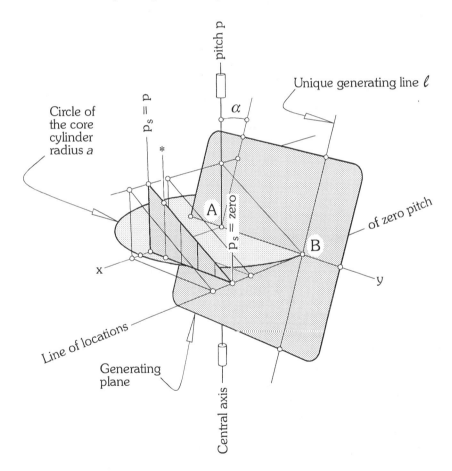

Book figure 9B.03 (§9B.21). The distribution over the plane A-xy of the the pitches of the substitute ISAs that may operate (independently with their different pitches) to sweep out by the generating line or sweep in by the generating plane the same involute helicoid.

screwing is chosen appropriately, the generating *line* of the involute helicoid (determined by the intersection of two infinitesimally close locations of the generating *plane* at those locations) is tangent to the same core helix. The author knows that figure 9B.02(c) is poorly presented and poorly explained, but the following technical information may be of help. The dimensions for the exercise are the same as they were for figure 9B.02(a); see the data at §9B.14. The finite screw displacements (D, Δ) executed by me at the three quarter-points have been 50 mm, in 26.66675°, in 40°, and in 80° respectively, the rates being 0.75 p, 0.50 p, and 0.25 p, namely 107.42925, 71.6195, and 35.8098 (mm/rad) respectively. The construction has meant that the centre lines of all seven of the displaced segments of plane, ignoring the small irregularities due to the infinitesimal (very small) screw movements I have made, *are coplanar with one another.* The lower, *final* centre-line (screwed about the original, ISA central) is 50 mm below its upper counterpart, the *initial* centre-line. The three pairs of infinitesimally separated planes (shown simply as three planes) have been used in the CAD to locate the lines of intersection ℓ between the said planes. These lines of intersection, ill conditioned as they are, have located themselves quite accurately as three tangents to the core helix. It has been interesting but not surprising to find that, upon effecting the infinitesimal movements at each of the three intermediate locations, the intersection of the pair of intersecting planes gave the generating line ℓ always passing through the point B of the lamina where, in the initial location, the ISA of zero pitch was passing. I make no claim that the construction of figure 9B.02(c) constitutes by itself a formal proof of the proposition. It is, however, a fairly convincing demonstration of its veracity.

On deeper consideration

21. Refer to figure 9B.03. I have so far in this chapter employed the term *reference location* in respect of a substitute ISA without a clear definition of what that term might mean, and I need to deal with that now. Any effective substitute ISA — and there are many more (as we shall see) than those along the line AB [37] — has the capacity to re-generate, with the given inclined plane and its angle of inclination, the original, given, involute helicoid. In figure 9B.03, the *core circle,* which may be seen, has centre A and radius (A–B). The plane A-xy, which contains this important circle, might be called the reference plane. I use the variable p_s for the pitch of the substitute ISA; but, as we shall see, any substitute ISA considered in terms of its pitch alone, without reference to its location (x, y), will be meaningless.

22. It will be found upon investigation that the original, given involute helicoid may also be reproduced by means of yet other substitute ISAs that, when seen as points in plan, are not exclusively upon the line (A–B). Their reference locations are scattered everywhere about the central ISA as evidenced for example by the reference location marked with a star (*), which is a generally chosen location. The pitch p_s at all values of x however is the pitch for that value of y; and, as we know, p_s varies as shown (from p at A

to zero at B) along the line AB. Thus the shown inclined plane in the figure illustrates the distribution of p_s. The slope of the plane is cot a. Such lines of argument lead to the following inset presentation of Theorem 3. An exercise for the reader might be to check the extent of the inclined plane. I take it to be infinite.

> *Theorem 3.* Refer to figure 9B.03. Whenever a plane is generating an involute helicoid by the process of envelopment as explained at Theorem 1 above, the axis of a substitute ISA may be found anywhere parallel with the central ISA. The pitch of the substitute ISA will be determined however by its reference location. See the plane containing (A–B) which is normal to the central axis, namely the plane A-xy in the figure. The reference location of a substitute ISA may be quoted (x, y), and the consequent magnitude p_s of its pitch by the ordinate z. Naturally the equation to the surface for p_s does not involve y. It reads, simply, $p_s = p(1 - x/a)$.

With care, this last theorem might be more comprehensively written to incorporate and thereby to supercede both of the previous two, but I leave all three of the theorems standing independently. Each one includes by default the substance of the preceding one. Although somewhat inelegant, this might better clarify the overall complicated situation.

Concluding remarks

23. First of all I wish to express my somewhat fanciful view just here that the buttons of the above argument, unseen by us, have been shifting from place to place upon the surfaces of the planes, or more precisely that the planes have been shifting with respect to the buttons, and that, unless we are careful, such ghostly phenomena will cause confusion. In my view, and doubtless in that of the reader, the above three theorems are not easy to explain or to be understood. They are neither. They are, moreover, quite complicated to apply in the processes of gear cutting. They are however crucial for clarifying some otherwise effective but obscure aspects of chapter 10. The matters at issue at chapter 10 concern the clear fact that a suitably driven single hob simply set up in a suitable machine can correctly cut in one machining operation both flanks of all teeth on a given wheel provided the set is not exotic. This means in other words (although it is somewhat abstract to say so in such words), that both flanks of the teeth can be cut with the suitable motion of a suitably shaped, *straight sided rack.* When, in the special case of planar gear cutting, we think in the reference plane in terms of the rack (and we habitually do), we see the rack rolling without slip on the pitch circle of the planar wheel as it moves continuously to 'generate' the teeth. The generation of planar envelopes by means of planar curves and sets of rolling centrodes is treated generally by Rosenauer and Willis [29], but here with general spatial involute gearing we are in open space, and we need

to think (and we do) in terms of the rack inclined but rolling with crosswise slip, *not upon one or other of the two core cylinders of the flanks of the teeth of the wheel, but upon a single fixed axode which, although cylindrical, is neither one of these two cylinders*. I have said in the Prologue (§P.05 *et seq.*) and elsewhere in this book that there is no place in general spatial involute theory for the concept *pitch surface*, as though it were somehow the spatial analogue of the simpler concept *pitch circle* in the plane; and I say it again. It is against the background of such matters as these briefly mentioned here that the above theorems are needed at chapter 10. The reader might refer in advance to figure 10.13, where a single ISA with a non zero pitch has been found to generate, with two different planes, both flanks of a gear tooth in one continuous sweep.

CHAPTER TEN
MATHEMATICS OF
THE MACHINING

Recapitulation

01. When speaking of the whole geometric construct in chapter 3, it was explained (a) at §3.12 that the pitch of an involute helicoid (mm/rad) is the pitch of its own core helix, namely $a \cot \alpha$, (b) at §3.04 that the said surface was accordingly screw symmetric about its own axis, (c) at §3.15 that the core helix is inscribed upon the throat cylinder of the relevant hyperboloid, and (d) at §3.18 that planar sections through an involute helicoid taken normal to the said axis reveal ordinary planar involutes whose base circles (of radius a) are the corresponding sections through the said throat cylinder. All of this will now be well understood but, looking again at figures 3.03 and 3.06, be reminded that, as the inclination angle α of the hyperboloid becomes smaller, the pitch $a \cot \alpha$ of the core helix becomes larger and the rulings upon the involute helicoid become more nearly parallel with the axis of the construct. Notice that, whereas for the sake of clarity and comparability the figures 3.03 and 9A.01 are drawn employing the same, comparatively large α ($\alpha = 60°$), the more practical angles α in the figures of WkEx#1 at chapter 5B are much smaller; the two of them are, there, only 4.83° and 19.31° respectively. See also $\alpha = 40°$ at figures 4.07(a)(b), and $\alpha = 20°$ at figure 5B.08. Be aware in this connection that, in the limit as $\alpha \to 0$, the said rulings become parallel with the said axis and thus with one another, and that the associated general spatial tooth profile becomes an ordinary, planar, spur-gear tooth profile.

02. Go back to WkEx#1 at chapter 5B. The steps of the exercise were explained in two main installments, (1) at §5B.09–18, and (2) at §5B.27–36. Recall that we took the case of an equiangular, general spatial gear set, whose k was mid-range. I mean by mid-range here that k was near unity (namely 0.6): it was neither near zero nor near infinity (§1.21): the wheels were so evenly proportioned, in other words, that it would have been inappropriate to call the smaller wheel the 'pinion'. The results of that exercise are illustrated pictorially in the collection of figures 5B.12–15. Figure 5B.14 is a drawing to scale of the shapes of the teeth of the bigger wheel; the picture however is in perspective. Figure 10.01 is another perspective view of the

§10.03   GENERAL SPATIAL INVOLUTE GEARING

same shapes of the same teeth, drawn in such a way here, however, that the angles δ, ξ and γ are made more clearly visible. Angles δ and ξ have been mentioned already (§5B.15, §4.34), angle γ is about to be defined (§10.15), and angle δ needs further consideration. See in all of the mentioned figures that the tooth profiles have been drawn, by choice, in such a way that the intercepts upon the key circle delineated by the profiles of the teeth are of equal arcuate length. At the key circle, in other words, the circumferential thicknesses of the teeth have been arranged to be the same as the gaps between the teeth.

Some early hints about milling and hobbing

03. In figure 10.01 a continuous taper has been turned at the tips of the teeth; it is turned about the shaft axis (at no particular taper angle) merely to remove protruding irrelevant material on the one wheel that might otherwise clash with similar, irrelevant material on the other. Located regularly between the teeth, as well, various odd-shaped slots have been cut to allow penetration of the mating teeth without diminishing unnecessarily the remaining root material. Such truncations and slottings are of course standard practice with gear teeth, and the details need to be considered, but, if we wish to consider exclusively the fundamental kinematics of the continuous motion, they are irrelevant and confusing. We see in the absence of distractions that, *because each profile is an ordinary involute helicoid with its axis collinear with the shaft axis* (§10.01), *there exists upon that surface a series of ordinary planar involutes occurring in the same way as such planar involutes occur upon the profiles of helical teeth cut for ordinary cylindrical helical sets with parallel axes* (§3.18, §8.03). See some of these planar involute curves in all of the figures 5B.14, 9A.01 and 10.01; see also parts of the same set of curves drawn at the various horizontal plane sections partly exposed at figure 9A.01. See in particular at figure 10.01 the sub-set of *coplanar* involutes circumferentially arranged and contained within the plane of the E-circle. See next the phantom rack revealed end-on at figure 5B.12; the gears there are equiangular. See also the different phantom rack revealed end-on at figure 6.06; the gears there are plain polyangular. These mentioned figures have been drawn to show that, at least for all plain polyangular sets, the direction of the flutes of the phantom rack coincides with the direction of the sliding velocity at [Qx]. Figure 10.01 should be read in conjunction with figure 5B.13. These figures look from different directions but deal with the same material, Figure 5B.13 helped to make clear the intersection, at the relevant E-circles of the wheels, of the two sets of coaxial hyperboloids, each set consisting, (a) and (b), of the two different base hyperboloids for the two different flanks of the teeth (§4.20), only one of which is shown incidentally, and (c) of the cone hugging naked wheel (§4.09). The intersection of these surfaces at the E-circle was also illustrated by means of a sketch at figure 4.07. The said hyperboloids in turn determine, (a) and (b), the different shapes of the slip tracks on the opposite sides of each one tooth, and (c) the same directions of

the sliding velocity at [Qf] on opposite sides of the same one tooth. Figure 10.01, on the other hand, helps to make clear that the planes of the E-circles cut the profiles of the teeth, not in the slip tracks as the base hyperboloids do, but in the previously mentioned, coplanar sub-set of planar involutes. Distinguish these involutes from the slip tracks. Recognizing the shown [Qx] as being the temporary point of contact on the drive side of the teeth (the other wheel being absent), we can see in figure 10.01 that, (a) while the best drive and the path of Q for drive, which are both in the polar plane at [Qx], are delineating the angle δ, (b) the line of the E-radius at [Qx] and the tangent to the planar involute at [Qx], which are both in the plane of the E-circle, are delineating the angle γ. Please see in advance that, whereas δ is the angle of obliquity set by choice for drive, γ is another angle. *Although related, the angles δ and γ are not the same.* The matter of γ is taken up again at §10.11.

The rack triad

04. Which straight lines on the rack are collinear, as they pass, with the rulings upon the flanks of the teeth? Two examples of the said lines may be seen at figure 10.02. Much of what follows is prognostication; but formal proofs for important propositions are implicit here. The rulings that may be seen in the figure (see the two sets of double lines) are parallel, and they both cut the shown path of that Q perpendicularly. They belong of course to a planar continuum of such lines, the whole continuum cutting the path of that Q perpendicularly. At any moving Q, (a) the relevant member of the said continuum, (b) the path of that Q, and (c) the tangent to the slip track there, complete a triad of lines mutually perpendicular. This is a newly mentioned triad; it needs a name; and I wish to call it the *rack triad*. The continuous presence of this triad at that Q results in a continuum of parallel lines existing not only in the fixed space as explained above but also upon the relevant flanks of the rack. It is somewhat hard to explain but it is, as we shall see, intimately related to the cutting mechanics at both flanks upon the flutes of the imaginary rack (§10.53). Due to there being both drive and coast, there are of course two rack triads, one sliding along and gradually rotating about the path of the said Q, and the other sliding along and gradually rotating about the path of the other Q. Radially with respect to the wheel the two rack triads move in opposite directions, one radially inward the other radially outward; their respective origins are continuously sliding upon the opposite flanks of some one tooth.

On the nagging question of the 'conicality' of teeth

05. Looking at the pictures distributed throughout this chapter, it is difficult not to ask again the same old question. With the phantom rack tilted with respect to the axis of the wheel through the two angles κ and λ as shown, how can the teeth of the wheels be other than 'conical'? One answer to this wrong headed question is to explain that the rack does not cut like a paper-punch or a cookie-cutter moving always parallel with its flutes with respect to

§10.05 GENERAL SPATIAL INVOLUTE GEARING

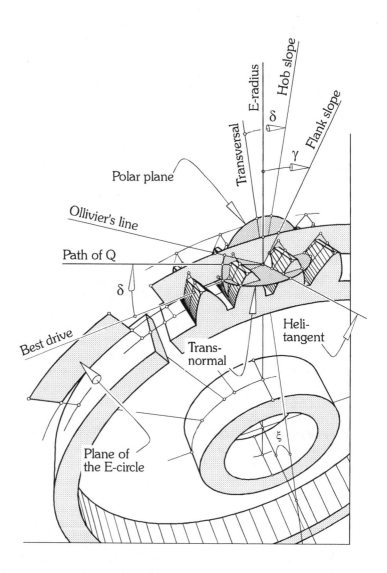

Figure 10.01(§10.02). This figure is a follow-on from figure 5B.12. It helps to relate (a) the angle of obliquity δ which is measured between the path of Q and the plane of the transnormal at [Qx], and (b) the hob angle γ which is measured between the sloping side of an imaginary rack cutter and a radial line of the wheel through the same [Qx]. Among the cylindrical helicals (among which are the planar spur gears) these angles are the same angle. They are jointly known there as the 'pressure angle'. Among the general spatial offset gears however (among which are the cylindrical helicals) the angles are, as can be seen, in general different. Clear to see in the figure are (a) the transversal at Q with its transnormal (§3.45), (b) the helitangent at Q with its polar plane, and (c) the path of Q with its tangent plane at the point of contact (§3.21). The angles δ and γ are shown as clearly as possible. Note the angle ξ.

MATHEMATICS OF THE MACHINING §10.05

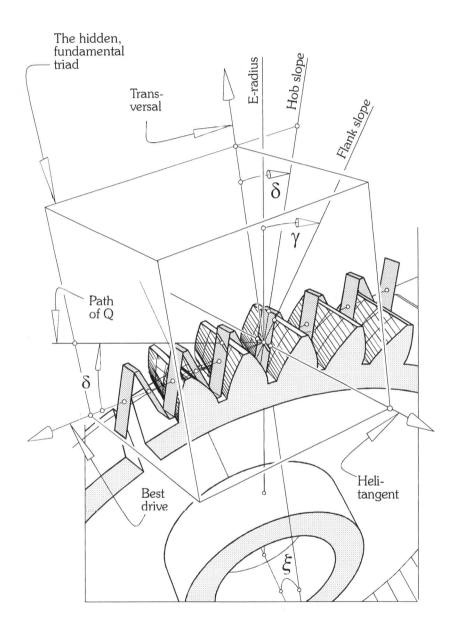

Figure 10.02 (§10.04). Detailed perspective view of a few teeth of the bigger wheel, WkEx#1. See the oblique disposition of the phantom rack. See also the fundamental triad. Along the path of Q, see at each position of the moving Q the straight line of contact (along a ruling) between the flat profile of the rack and the curved profile of the tooth.

§10.06   GENERAL SPATIAL INVOLUTE GEARING

the job, but otherwise. When I have explained in due course in this chapter (a) that the rack *screws* with respect to the wheel, and (b) *that it screws about an ISA which remains always parallel with the axis of the wheel*, the main question being asked might be the following more enlightened one. Given the evident *generating* action of the cutters (the cutters being of whatever kind), and leaving aside the *shapes of the edges* of the interacting surfaces, how can the cutting machines possibly produce *other than cylindrical teeth*, where the shapes of the flanks of the teeth might not be axi-symmetric about the axes of the wheels?

More on the shapes of teeth

06. In the case of ordinary helical gears with parallel axes, the helicoidal involute profiles on the two opposite flanks of a given tooth are of the same hand and have the same pitch p_{FLANK} (mm/rad); and this pitch may, as we know, be set to any value (§5B.15). The two profiles appear to be different, however, and this difference is due, not to the question of handedness or of pitch, but to the fact that the two flanks of the same tooth are cut from opposite branches of the same helicoid. See the figures 3.13(a)(b) and recall the text at §3.15 and §3.62. The said helicoid may itself be either left or right handed of course. But in the *general* spatial arena illustrated even by the equiangular figures 5B.13, 10.01 and 10.02, the pitches $a \cot \alpha$ of the profiles of the opposite flanks of the one tooth *are* different. We know this because, as we know, the radii a and the angles α are both different, and, given the different conditions for drive and coast, the products $a \cos \alpha$ are different too. This may better be seen in figure 5B.13 where both flanks of the teeth are more clearly visible. *Strangely, however, we can cut all of these and the plain polyangular teeth (all of which are, as has been shown, asymmetrical) using a single hob whose two rack angles δ and the relevant, face-to-face dimensions are cut by the toolmaker to suit.* The two different profiles of the teeth to be cut, with their two different pitches p_{FLANK} (mm/rad), and their different angles γ and θ (§10.26), may be cut at the same time during the one pass of a single, but asymmetrical tool (§10.17).

The swivel angle lambda

07. *Definition of lambda.* Various important angles have been identified and algebraically quantified in previous chapters, and many of these (for examples α, δ, τ and κ) are destined to figure in this chapter. Another important angle, briefly mentioned previously, but not yet properly defined, needs to be looked at now. Refer to the figures 5B.02(a)(b), 6.06 etc. Intersecting at any chosen point say U upon the key circle of a naked wheel, there are two easily identifiable lines: (a) the relevant ruling upon the naked wheel, and (b) the generator there of the Wildhaber cone (this joins U to the vertex V of the said cone). Notice that, while the mentioned ruling is contained within the hyperboloidal surface of the naked wheel, the mentioned generator of the Wildhaber cone is not. The angle between these lines is the angle in question. I

call it, for a reason to be made clear, the *swivel angle* at the teeth; and I use for it the symbol λ. Kindly distinguish between the swivel angle as here defined and the twist angle τ of the teeth; they are not the same; see §3.04, §5B.04, and refer to figure 5B.02(b). *Notice too that the axis of a pure rotation of magnitude λ, required to move from the relevant ruling of the naked wheel at U to the relevant generator of the cone at U, is not in the plane of the key circle, but perpendicular to the said generator and along a line such as BB as shown in figure 10.08(a).* This swivel angle λ at the teeth becomes important later (§10.33, §10.39), but take it for now, simply, as having been illustrated. In the case of equiangularity a tidy expression for λ in terms of τ can be found; it is simply that $\lambda = \Lambda/2$, where $\Lambda/2$ is given as shown at the second of the two insets at §10.08; this comes from appendix A at chapter 5B; details of the more complicated algebra for the two *different* angles λ when polyangularity prevails is worked through at chapter 6. Look for this at §6.40 or thereabouts.

The angles kappa and lambda taken together

08. For the sake of argument yet to be pursued, I wish to mention together here the algebraic expressions (as derived so far) for two important, interrelated angles. The angles are firstly κ, which comes from the chapters 5B and 6, and secondly λ, which has, at last, been accurately defined just now. We have already, from §5B.04, referring to figure 5B.02(b) and seeing appendix A at chapter 5B, that

$$\tan \kappa = [2R \sin (\Sigma/2) \tan (\Sigma/2)]/ \sqrt{[4R^2 \sin^2 (\Sigma/2) + C^2]},$$

and we have as well, also from appendix A at chapter 5B, that

$$\tan (\Lambda/2) = [C \tan (\pi/2)]/ \sqrt{[4R^2 \tan^2 \tau + C^2]}.$$

But notice, remembering β_{SHIFT} at figure 6.08 (§6.40), that, while κ is independent of τ, λ depends upon τ. In the special case of equiangularity, $\tau_2 = \tau_3 = \Sigma/2$, and $\lambda_2 = \lambda_3 = \Lambda/2$; but in general the two τ (and thus the two λ) vary with the location of [Qx]. When [Qx] is not at E but at some F, plain polyangularity prevails and the two τ take different values. See figure 6.07. Their values clearly depend upon the radius-ratio j. We have already seen (and I speak here colloquially) that, while $\lambda_2 = \Lambda/2$ minus the beta shift, $\lambda_3 = \Lambda/2$ plus the beta shift. Checking here for some of the special cases, we notice incidentally (a) that, for equiangularity when $C = 0$, $\kappa = \tau$, which checks, and (b) that, when $R = \infty$, $\lambda = 0$, which also checks. These last remarks relate, not to the so-called *ordinary straight bevels*, which are, as has been shown, not involute (§5B.62), but to the *Beveloid straight bevels*, which are (§5B.61).

Incidentally some interesting interrelated formulae

09. Let L be the length of the equilateral transversal, and let F be the sum of the F-radii. It is easy to show, for a given set where C, Σ and R are given (and for all k and j that might obtain), that not only is L a constant

(which is not surprising), but also that F is a constant. Expressions for these may be written, respectively, as follows:
$$L = \sqrt{[4R^2 \tan^2 (\Sigma/2) + C^2]},$$
$$F = \sqrt{[4R^2 \sin^2 (\Sigma/2) + C^2]}.$$
These formulae might or might not be useful. It might also be wise to notice, in advance, that the following simple relationships, independently of the status of the set (whether equiangular or plain polyangular), apply,
$$[C/F] = \sin \Phi_T,$$
$$[C/F] = \sin \Phi_N,$$
$$[C/2R] = \tan \Phi_Y,$$
where Φ_T, Φ_N and Φ_Y are fixed, angular variables of the gear set which (although already illustrated) have not been mentioned yet. They are illustrated variously at figures 10.05(a)(b), and actually used at §10.19 *et seq*. They might be defined as follows: they are the angular *slopes*, respectively, of the equilateral transversal, the key radii (of whatever kind), and the line segment (Y–A). I wish to include here a mention of Φ_B. We have that
$$\tan \Phi_B = [2R/C] \tan [\Sigma/2].$$
Cot Φ_B is the reciprocal of the slope of the equilateral transversal and, for equiangularity only, tan Φ_B is the slope of the best drive. This expression for tan Φ_B was used as a clever substitution by Sticher [52] at §5B.16.

Two other important angles, gamma and theta

10. Two more angles need to be introduced. First there are the *flank slope angles* γ which are to be dealt with next, §10.11 *et seq*.), and second there are the *helix angles* θ which will be looked at later (§10.26). Please see the single angle γ illustrated already at the figures 10.01 and 10.02, and be aware that there are different angles γ on the different sides of a given tooth. Although the angles γ are associated with the angles of obliquity δ, one is bigger and the other smaller than the equal angles δ in the mentioned figures, and the relationships are not yet made clear (§10.19, §10.23). Go in advance to figure 10.08(a). While the views in the figures 10.01 and 10.02 are essentially the same, the view in figure 10.08(a) is somewhat different. Whereas in figures 10.01 and 10.02 the wheel is seen to be upright with the rack aslant, in figure 10.08(a) the rack is seen to be upright with the wheel aslant. Observe in any event, preliminarily, the following: (a) that δ is shown as the *rack angle* of the phantom rack (the line of the *hob slope* is shown to be located centrally within the sloping plane surface of the rack); (b) that γ is shown as *the flank slope angle* of the tooth (the flank slope line is also contained within the sloping surface of the rack); (c) that there is straight-line contact between the rack and the profile of the tooth; (d) that this occurs along the straight-line ruling upon the surface of the tooth (§10.04); but (e) that the helix angle θ is, as yet, nowhere to be seen.

Matters to do with delta and gamma

11. *General matters.* Following earlier remarks at the end of §10.03, we need to look very carefully now at (a) the angles of obliquity δ which are the same as the angles δ at the sloping faces of the phantom rack or the cutting edges of the required hob cutter (§5B.15), and (b) the flank slope angles γ which are different angles (§10.10). In the special case of planar spur gearing, the angles δ and γ are the same, and the one of them most often mentioned, namely δ, is called, most often, *the pressure angle*. There exists a vast literature about the optimum size of this important angle. As we know, it can range in ordinary planar practice from say 10° to 25°. In *general spatial involute practice,* however, not only can δ be varied within very wide limits as we have discovered; see §5B.45 and subsequently; but *the two angles γ for any one tooth are not the same, and neither of them is the same as its corresponding δ.* It is time now to study these matters more carefully than hitherto.

12. *Definition of delta.* I define the angle δ as follows: *the angle of obliquity δ is that angle which obtains at [Qf] between the contact normal there and the plane of the transnormal there.* Given involute action, the line of the contact normal is the path of Q (§3.07). The path of Q is in the polar plane and, because the *best drive* (which is at the intersection of the polar plane and the transnormal) is also in the polar plane, we can define the angle of obliquity (for involute gearing) also as follows. *The angle of obliquity is the angle at [Qf] between the path of Q and the best drive.* Notice in figure 10.01 the plane of the transnormal at [Qx], which plane is normal to the transversal there (§3.44), and notice that it intersects the polar plane in the best drive (§3.44). *It may accordingly also be said that, when Q is at [Qf], δ is the angle between the plane of adjacency and the best facet.* In the special case of ordinary spur gears, each of the above alternative definitions breaks down to becoming a statement defining the well known, *pressure angle*. In these simpler gears (and in the cylindrical helicals), the angles δ_D for drive and δ_C for coast are most often set to be the same, but (as we have seen) this equality of the two angles δ is not necessary for correct involute action.

13. *The variable angle δ_{REMOTE}.* The phenomena associated with the variable angle δ_{REMOTE} as Q moves along its path and its special value δ which occurs when Q is at [Qf] have already been looked at. See §3.50 and the figures 3.11(a)(b). See also §4.64 and the figures 4.06(a)(b)(c)(d). The upshot of these passages may be encapsulated now. *The special value δ of δ_{REMOTE} occurs by definition at [Qf]. This angle (which is both the angle of obliquity at the tooth and the rack angle) needs to be known for machining.* The *maximum* value of δ_{REMOTE}, which is a somewhat irrelevant, other angle, occurs not by definition but by virtue of the mechanics at another point in the path namely [Qs]. The precise value of this maximum and the actual position of [Qs] in the path are of academic interest; they bring into play the otherwise obscure *line of striction of the girdling hyperboloid* (§4.60), and they might, in due course, be found to have some practical significance.

14. *The plane of adjacency, a matter to do with γ.* As mentioned earlier (§2.07, §3.25), I have called the tangent plane at the point of contact by a special name. I have called it, not simply the *plane of contact*, for that would imply surface-to-surface contact across a *planar contact patch* (which would be misleading), but the *plane of adjacency*. Please note, as matters of further interest in figure 10.01, (a) that the plane of adjacency is shown there unobtrusively but correctly, and (b) that the said plane intersects the plane of the E-circle in the line which is tangent to the relevant planar involute, which line I have called *the flank slope tangent*. In figure 10.02 the same flank slope tangent is shown (as it should be shown) contained within the relevant, shown plane portion of the phantom rack.

15. *Definition of gamma.* For the sake of clarity in the figures 10.01 and 10.02 the centrally appearing point [Qf] remains unlabeled. It is however at the origin of the well known, fundamental triad of lines, the transversal, the helitangent and the best drive, each one of which (mutually perpendicular with the other two) is labeled. Please refer to the various other lines emerging upwards and outwards from [Qf]. For the sake of being able to generalize our definitions here into the wider realm of plain polyangularity, draw also (in the mind's eye) the same lines in figure 6.05, which figure comes from the plain polyangular WkEx#3. *Whereas in all of the figures the angle of obliquity δ, shown with its apex at [Qf], resides in the polar plane at [Qf], the flank slope angle γ, shown with its apex also at [Qf], resides in the plane of the key circle.* It should further be seen that, *whereas δ is the angle between the transversal and the hob slope tangent of the hob, γ is the angle between the key radius and the flank slope tangent of the wheel.* It might otherwise be said that γ is the angle revealed, after cutting, at the outside planar surfaces (the flat cheeks) of a plain cylindrical wheel blank. Speaking more precisely we can say this: *the angle γ is that angle in the plane of the key circle which obtains (given tight mesh) between (a) the radial line of the gear wheel drawn to cut the tooth profile at the key circle, and (b) the flank slope tangent namely that line in the plane of the key circle which is tangent to the surface of the tooth profile at that same point.* The flank slope tangent of which I speak is also tangential, as we know by definition (§3.18), to the planar involute cut by the plane of the key circle in the helicoidal involute of the tooth profile.

The real rack

16. Having thus looked preliminarily at some of the algebra connecting some of the many angles requiring consideration, we might now begin to look at the actual shape, dimensions and cutting action of the rack. I have in mind here not the *phantom rack*, which is a severe geometric abstraction in the form of a thin, rigidly folded surface that continuously travels with and between the teeth of the wheels in tight mesh, but the *imaginary rack* of ordinary machining practice. The imaginary rack is more real than the phantom rack, in that it is either (a) a real rack in the sense that it actually generates teeth in a continuously indexed blank by reciprocating mechanical action, or

MATHEMATICS OF THE MACHINING   §10.16

(b) stands in for the real tools that may in fact be used, such, for example, as
a hob cutter or a combination of milling cutters variously driven in the presence of continuous rotation of the blank. Here, however the rack is both real
and rigid in the sense that it can by its continuous movement (yet to be studied) generate teeth in an imaginary, plastically deformable blank. Please see a
real rack introduced at figure 10.03(a); the figure is an orthogonal projection;
it is a simple rehash of figure 5B.12; the circumstance is equiangular; but (as
we shall see) it might as well be polyangular. I have removed the smaller
wheel (link 3) and solidified the relevant side of the phantom rack to produce
the real rack; and this real rack is in contact with the bigger wheel (link 2) exclusively (§5B.35). If we imagine the wheel to be free to rotate about its own
axis (shown inclined) and the rack to be free to translate crosswise in the
plane of the paper, we can see that the a combined movement may occur
without mutual interference at the teeth. *Don't be surprised to find soon,
however, that the speed of the translating rack in this horizontal direction
cannot be the same as the circumferential speed of the rotating wheel at
its E-circle* (§10.07, §10.39). We see in this figure (by virtue of the accurate
CAD) that no portion of the material of a tooth occupies at the same time the
space occupied by any portion of the material of the rack. Please see also in
figure 10.02 that, relative to the axes of a Cartesian frame set square in the
rack, the shaft of the wheel is tilted through a combination of two angles. See
first that the apex of the cone of Wildhaber lies in the central plane of the
rack and that the relevant angle for this is κ; see figure 5B.02(b) and refer to

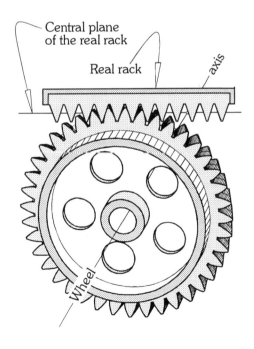

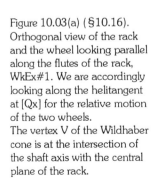

Figure 10.03(a) (§10.16).
Orthogonal view of the rack
and the wheel looking parallel
along the flutes of the rack,
WkEx#1. We are accordingly
looking along the helitangent
at [Qx] for the relative motion
of the two wheels.
The vertex V of the Wildhaber
cone is at the intersection of
the shaft axis with the central
plane of the rack.

359

§5B.04. See next that there is another angle involved, namely λ, and recall that the ramifications of this are mooted already (§10.07). These two angles — κ and λ — are important. They are yet to be the subject of much closer attention.

Length-measured pitches of teeth

17. I am referring throughout this paragraph to the various pitches that are spoken of as being arcuate, normal, chordal etc., measured between the successive teeth of wheels. These well known, length-measured pitches are often used. They are usually abbreviated in the current literature by means of acronyms such as CP (for circular pitch) and NP (for normal pitch), and they are, most often, measured and quoted simply in units of length (mm). This is because they are, most often (ignoring pure ratios such as DP the diametral pitch), either arcs of circles or segments of straight line. The aforementioned, *screw* pitches of this book, such as that of the unique screw residing in the unique pitch line of a gear set (mm/rad), or the screw pitch of such screw-symmetric surfaces as those of the involute helicoids that form the flanks of the teeth in all involute gearing (mm/rad), are not under consideration here. Whereas the terms CP, NP and the like are accordingly within the scope of this paragraph, the terms p, p_{23}, p_{FLANK} and the like are accordingly beyond it. Having said that, I wish to begin by defining, now, *the circumferential pitch of the teeth of a wheel or of a pinion* — the CPW — as the arcuate distance along the key circle, which is the E- or the F-circle as the case may be, from any one profile to the next identical profile. In WkEx#1 the CPW for the wheels was 14.023 mm approximately (§5B.27), the same for both wheels; but, as we shall see, the equality of the two CPW does not obtain with polyangularity. Refer again to the figures 5B.12, 6.06 and 6.08(a)(b), and pay attention to the rack. It contains the straight line running through [Qf] perpendicular to the flutes of the rack and across the said key-circle plane (also called by me the central plane). The same line may be seen at figure 10.03 where it is labeled *best drive*; it may also be called the *key-circle axis* of the rack. By direct analogy here, *the circular pitch of the teeth of the rack* — the CPR — may be defined as the straight-line distance measured from flank to corresponding flank along the said key-circle axis of the rack. The new matter that needs to be sorted out now however is that, contrary to the planar case where the CPR and the CPW are always identical, the CPR in these spatial cases *is always less than* either of the CPW. This can be seen preliminarily, in a primitive way, by seeing the simple fact that the central plane of the rack is in no way parallel with the axis of the wheel. Provided equiangularity obtains and the two angles δ are the same (as they are for example in WkEx#1), *the normal pitches NPR and the NPW are the same however*. I wish to list first for consideration now some numerical results from WkEx#1. These will be followed by similar lists of results for subsequent worked examples. I will thus reveal, step by step, some of the points I wish to make.

MATHEMATICS OF THE MACHINING §10.17

Item (a) for the oblique equiangular set at WkEx#1:
CPW $= [2\pi e]/N = 14.02309870$ mm, the same for both wheels;
NPW $= [2\pi a \cos \alpha]/N = 12.802945$ mm, the same for both wheels;
NPR $=$ the same as the NPW, namely 12.802945 mm;
Angles δ are 20°, set to be the same for drive and coast;
Angle κ is 21.12199160°, the same for both wheels;
Angle λ is 13.69172117°, the same for both wheels;
CPR $=$ NPR sec $\delta = 13.624609$ mm;
CPR $=$ CPW cos $\lambda = 13.624609$ mm, the same, OK;
See the size and shape of the rack at figure 10.03(b), item (a).

I wish to list next a parallel set of numbers for the polyangular WkEx#3. Other things remaining equal (including the angles δ), the point [Qf] has been moved at WkEx#3 from the original E (at WkEx#1) to a newly chosen F, the ratio j becoming not 0.6, as before, but 0.7. This has caused the appearance of two different CPW.

Item (b) for the oblique polyangular set at WkEx#3:
$[CPW]_2 = [2\pi f_2]/N_2 = 13.198205$ mm;
$[CPW]_3 = [2\pi f_3]/N_3 = 15.397914$ mm, which is different;
NPW $= [2\pi a \cos \alpha]/N = 12.3756$ mm, the same for both wheels;
NPR $=$ the same as NPW $= 12.3756$ mm;
Angles δ are 20°, set to be the same drive and coast;
Angle κ is 21.12199160°, the same for both wheels;.
Angle λ_2 is $-3.8317°$, which is [$\Lambda/2$ namely $\lambda -$ the β shift];
Angle λ_3 is $+31.2152°$, which is [$\Lambda/2$ namely $\lambda +$ the β shift];
CPR $= [CPW]_2 \cos \lambda_2 = 13.169$ mm;
CPR $= [CPW]_3 \cos \lambda_3 = 13.168$ mm;
CPR $=$ NPR sec 20° $= 13.170$ mm; all three check OK;
See the size and shape of this rack at figure 10.03(b), item (b).

Generally peruse the figure 10.03(b) and remember just here that it is not the equality of the angles δ that determine equiangularity; it is the equality of the angles τ; and the angles τ, as we know, are the half-angles at the apices of the asymptotic cones (§1.46). Recall also that the half-angles τ are not the half-angles κ at the apices of the Wildhaber cones, which angles are different angles and which, in involute gear sets, are equal to one another in any event [90]. For equiangular sets, the CPW is the same for the teeth on both of the wheels because the key radius of the smaller wheel is exactly k times that of the larger. For polyangular sets on the other hand the CPW for the wheels are different, because j differs from k, and there is unequivocal circumferential sliding at the key circles at [Qf]. This means that, although the same rack still cuts both wheels and its rack angles δ (which have, hitherto, been equal here) remain as before, its CPR is diminished. See this discussed more fully, in connection with the kinematics of rack action, at §10.41. Does

§10.17 GENERAL SPATIAL INVOLUTE GEARING

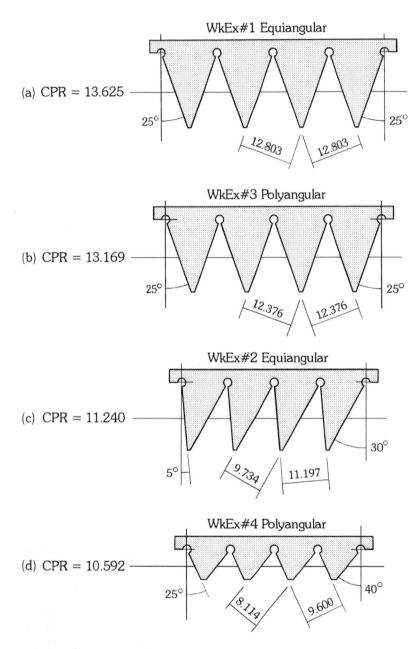

Figure 10.03(b) (§10.17). Rack sizes and shapes for the four worked exercises of the chapters 5B and 6. They appear here the altered order as follows: (a) WkEx#1, (b) WkEx#3, (c) Wk Ex#2, and (d) WkEx#4. All dimensions are in mm.

MATHEMATICS OF THE MACHINING §10.17

the rack and one or other of its mating wheels at the polyangular WkEx#3 constitute a polyangular set? No, it is an equiangular set with $k = 0$ and $R = \infty$! Next I wish to list results from the equiangular WkEx#2, which was an equiangular square set where k was relatively (perhaps even dangerously) small and the two angles δ (for drive and for coast) needed to be different.

Item (c) for the square equiangular set at WkEx#2:
$[CPW]_2 = [2\pi e_2]/N_2 = 11.7883$ mm; this is the wheel;
$[CPW]_3 = [2\pi e_3]/N_3 = 11.7883$ mm; this is the pinion, the same;
Angle δ_D and δ_C are 5° and 30° respectively, the making of
these choices having been aided by the ellipses of mobility;
$[NPW]_{2D} = [2\pi a_{2D} \cos \alpha_{2D}]/N_2 = [2\pi/41][73.063910] = 11.1969$ mm;
$[NPW]_{2C} = [2\pi a_{2C} \cos \alpha_{2C}]/N_2 = [2\pi/41][63.517702] = 9.7340$ mm;
$[NPW]_{3D} = [2\pi a_{3D} \cos \alpha_{3D}]/N_3 = [2\pi/9][16.035439] = 11.1969$ mm;
$[NPW]_{3C} = [2\pi a_{3C} \cos \alpha_{3C}]/N_3 = [2\pi/9][13.942910] = 9.7340$ mm;
$[NPR]_D = 11.1969$ mm, as per D for drive, calculated twice above;
$[NPR]_C = 9.7340$ mm, as per C for coast, calculated twice above;
Angle κ is 42.1304°, always the same for both wheels;
Angle λ is 17.4584°, the same for both wheels (equiangularity);
$CPR = [NPR]_D \sec \delta_D = [11.1969][1.00382] = 11.2397$ mm;
$CPR = [NPR]_C \sec \delta_C = [9.7340][1.154701] = 11.2399$ mm;
$CPR = CPW \cos \lambda = [17.7883][0.953935] = 11.2453$ mm;
all three of which — except for the minor errors — are identical;
See the size and shape of this rack at figure 10.03(b), item (c).

We have seen here that, in the combined event of equiangularity and a chosen set of unequal angles δ, the circumstance obtains where we need to quote, for (a) the rack, and (b) both of the two wheels (the wheel and the pinion), a dual value for the normal pitch. What this means is made clear at the figure. As has always been the case so far, the one rack cuts both wheels, but the circular pitch of the rack is not the same as either of those of the two wheels. For the sake of completion now, I go on to examine the results that come from the square polyangular set at WkEx#4.

Item (d) for square polyangular set at WkEx#4:
$[CPW]_2 = [2\pi f_2]/N_2 = [2\pi/39][67.0059] = 10.795141$ mm; wheel;
$[CPW]_3 = [2\pi f_3]/N_3 = [2\pi/11][26.8024] = 15.309495$ mm; pinion;
Angle δ_D and δ_C are 25° and 40° respectively, the making
of these choices having been aided by the ellipses of mobility;
$[NPW]_{2D} = [2\pi a_{2D} \cos \alpha_{2D}]/N_2 = [2\pi/39][59.586955] = 9.5999$ mm;
$[NPW]_{2C} = [2\pi a_{2C} \cos \alpha_{2C}]/N_2 = [2\pi/39][50.365155] = 8.1142$ mm;
$[NPW]_{3D} = [2\pi a_{3D} \cos \alpha_{3D}]/N_3 = [2\pi/11][16.806603] = 9.5999$ mm;
$[NPW]_{3C} = [2\pi a_{3C} \cos \alpha_{3C}]/N_3 = [2\pi/11][14.205549] = 8.1145$ mm;
$[NPR]_D = 9.600$ mm., as for D twice calculated above;
$[NPR]_C = 8.114$ mm., as for C twice calculated above;
Angle κ is 42.1304°, the same (as always) for both wheels;
Angle $\Lambda/2$ is, by calculation, 17.5440° (see appendix A at 6);

§10.18   GENERAL SPATIAL INVOLUTE GEARING

Beta shift = 28.6726°, from the CAD;
Angle λ_2 = [17.5440 − 28.6726] = − 11.1245°;
Angle λ_3 = [17.5484 + 28.6726] = +46.2210°;
CPR = $[NPR]_D \sec \delta_D$ = [9.600] [1.103378] = 10.592 mm;
CPR = $[NPR]_C \sec \delta_C$ = [8.114] [1.305407] = 10.592 mm;
CPR = $[CPW]_2 \cos \lambda_2$ = [10.79514] [0.9812] = 10.592 mm;
CPR = $[CPW]_3 \cos \lambda_3$ = [15.309495] [0.6919] = 10.593 mm.

What these go to show is that, even in the realm of plain polyangularity where as well as this the angles of obliquity are obliged to be different, the same rack will cut both wheels of the set. See figure 10.03(b) item (d), for the size and shape of this particular rack. See now the totality of figure 10.03(b) to gain an impression, if you can, of the envelope surrounding the overall geometric possibilities (§11.28). By way of accounting for the mixed accuracy of the numbers being quoted here, they come variously from computer driven spreadsheet, CAD, and a hand held calculator. Although after cross checking they are given here to the best-estimated micron, serious gear designers will calculate their own numbers. They will go (by means of computer driven spreadsheet) to whatever accuracy they consider to be a sufficient challenge for the quality of the cutting machines available.

The flat-action phenomenon

18. With some justification this paragraph might be seen as being a mere unimportant interlude that is interesting but dead-ended, a blind alley. But its contents require attention (a) because they are spectacular, and (b) because they are in certain circumstances useful. *The material applies however only for equiangular sets of wheels.* With back reference to the equiangular WkEx#1, which is illustrated at figure 5B.10(a), and using its data, the two different flank slope angle γ_D and γ_C are clearly revealed in the special planar construction laid out at figure 10.04. It comes from the CAD. *This figure is not a visual display of any real geometrical relationship between the wheels. Avoiding Σ, and drawn flat in the plane of the paper, it is a gross distortion of the real picture. It has however a useful purpose, the details of which are about to be explained.* In figure 10.04, each wheel has been cut normal to its own shaft axis across the plane of its own E-circle. The sections thus obtained (both of which should be seen as flat laminae) have next been collapsed into the same plane namely the plane of the paper and slid into tight mesh with one another. They might thus be called the *paper teeth* at the key circles. The respective base circles (which are sections through the base cylinders of the respective base hyperboloids) are shown, and the paper teeth will be seen to be planar involute teeth which, asymmetrical, are directly related to these base circles. Figure 10.04 is a 2D figure bearing little resemblance to the 3D circumstance from which it has come. The extraordinary thing to notice, however, is that the teeth of the laminae are meshing satisfactorily with one another and that the meshing reveals apparent angles of obliquity which are, indeed, the flank slope angles γ of the

MATHEMATICS OF THE MACHINING   §10.18

real relationship. Preliminarily now, and remembering that this applies for equiangularity only, it is easy to see in figure 10.04, for both of the γ, that the following relationship obtains:

$$\cos \gamma = [\text{the relevant radius } a]/[\text{the relevant radius } e].$$

We already have (as we know) independent expressions for both a and e, written in terms of C, Σ, R and δ; see §5A.27 and §5B.16. It might be said that an important part of this chapter is the presentation and reconciliation with one another of these formulae. They do reconcile. Please notice also in

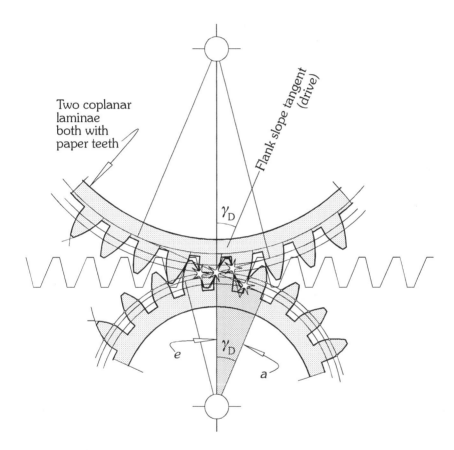

Figure 10.04 (§10.18). This picture I have called the 'flat action' picture. Plane slices of zero thickness (laminae) are taken which contain respectively the different E-circles of the two wheels, and these are lain, in tight mesh, flat with one another in the plane of the paper. This figure (which is not real) is carefully contrived to show the shapes of the planar involute sections meshing with one another in the plane of the paper. These lead us in turn to suspect that the real teeth can be cut by hobbing. What appear to be the angles of obliquity δ in this figure are indeed the flank slope angles γ. It is clear to see in the figure that for each γ, $\cos \gamma$ is the ratio of the relevant a and the relevant e. Because the two γ are not the same, we find that the unreal imaginary rack (drawn here) is asymmetrical.

§10.19 GENERAL SPATIAL INVOLUTE GEARING

passing that, independently of all else, the tangent to the involute at [Qf] (the flank slope tangent) must be perpendicular to the path of Q at [Qf], which path is set by our choice of δ. See figure 10.01. To complete this paragraph I lift out, for consideration later, some old numerical results from WkEx#1. Referring to §10.17, item (a), [CPR = 14.0321 mm.] [Also cos γ_D = (81.796761) / (89.273819) = 0.916245792, so γ_D = 23.6167°; and cos γ_C = (86.366250) / (89.273819), which gives γ_C = 14.6631°]. This pair of γ is the same for both of the two wheels because equiangularity prevails. In the event of polyangularity, the two CPW would be different, and this would obviously preclude the flat-action phenomenon. This flat-action phenomenon might be found to have some relatively simple analogy in the polyangular circumstance, but the matter is pursued no further here.

First algebraic look at gamma in terms of delta

19. Go to the figures 10.05(a)(b). While figure 10.05(a) is drawn directly from figure 5A.02 and is thereby self explanatory, figure 10.05(b) is an unfolded version of the same geometry drawn according to the principles of descriptive geometry. These support the derivation here of a somewhat awkward relationship between the angles γ and δ which has the advantage however of being couched in terms of the original fixed variables of the equiangular set. But the algebra is awkward, and it makes for dull reading. One might, accordingly, and without loss, skip this and the next three paragraphs, unless the tortured line of argument being followed here is for some reason beguiling. More refreshingly and effectively, the same material comes up again at §10.23.

20. I leave aside, just yet, the polyangular. If we say by convention that the CDL is vertical, we say by the same convention that the plane DRS (which contains the E-line at D) is horizontal. Locate the vertical plane JRKS containing the transversal JK. Locate also the two vertical planes containing the two E-circles. These two planes are (a) inclined to one another along Olivier's line at angle Σ, and (b) separately inclined to the polar plane at Olivier's line at angles Σ/2. These angles are shown at true size in figure 10.05(b). Locate, in their several vertical planes the three angles Φ_T, Φ_N and Φ_B; these are marked at their true sizes in figure 10.05(b). Angle Φ_T is the slope of the transversal. Angle Φ_N is the orthogonal projection of Φ_T taken parallel to the FAXL and onto the plane of the E-circle which plane is of course normal to the shaft axis; so Φ_N is the slope of the E-circle radius which extends from the axis of the relevant shaft to [Qx]. Locate also the angles δ_D, δ_D' and $\delta_D'^*$. Angle $\delta_D'^*$ is the orthogonal projection of δ_D' taken parallel to the relevant shaft axis and onto the plane of the E-circle. The same needs to be said for the angles δ_C, δ_C' and $\delta_C'^*$, complete aspects of which latter are not shown at figure 10.05(a). Recall the contents of §10.09.

21. Notice in figure 10.05(a) the two lines marked ℓ_1 and ℓ_2. These are, respectively, (a) the unique line in the plane of the relevant E-radius which is perpendicular to the drive path, namely the tangent to the planar involute

MATHEMATICS OF THE MACHINING §10.22

at [Qx], namely the cutting edge of the hob cutter, and (b) the common perpendicular between the drive path and the relevant shaft axis, namely the throat radius of the relevant base hyperboloid. Because the lines l_1 and l_2 both cut the drive path perpendicularly, they reside in the pair of parallel planes thus defined, each of which is normal to the drive path. But because the lines l_1 and l_2 both cut the relevant shaft axis perpendicularly, they are also contained within the pair of parallel planes thus defined, each of which is normal to the shaft axis. It follows that l_1 and l_2 are parallel with one another. This fact is an important aspect of the argument which follows. In conjunction with the construction at figure 10.05(b) it can easily be used to prove independently what was evident at figure 10.04, namely that γ = arcos $[a/e]$. The actual values which continuously accompany the algebra which follows come from WkEx#1; checked by numerical calculation and by CAD construction, they regularly appear within the usual square brackets.

22. The material of this paragraph applies for equiangular sets only. Refer to figure 10.05(a). Notice first the angles Φ_T and Φ_N. They are in the vertical planes respectively of the transversal and the E-radii (or F-radii). These angles are fixed variables in the current analysis because they depend, as can be seen, solely upon the fixed variables C, Σ and R. Both were mentioned at §10.06. Notice next in the polar plane at [Qx] the angles δ_D and δ_C; these two are the angles of obliquity; they are the independent variables here. Notice also their counterparts δ'_D and δ'_C which are measured not from the best drive (as δ_D and δ_C themselves are) but from Olivier's line. The numerical values for the two δ have been set to be the same, namely 20°. It will be clear that

Φ_T = arcot $[(2R/C)$ tan $(\Sigma/2)]$;
[31.4965°]
and, considering both δ, that

$\delta' = \Phi_T \pm \delta$.
[11.4965°]
[51.4965°]

See also in the figures 10.05 that

Φ_N = arcot $[(2R/C)$ tan $(\Sigma/2)$ cos $(\Sigma/2)]$
= arcot $[(2R/C)$ sin $(\Sigma/2)]$.
[34.0610°]

Given that, for both δ, δ'^* is the parallel projection of δ' taken in the direction of the relevant shaft axis and onto the plane of Φ_N,

δ'^* = arcot $\{$cot $[\pi/2 \pm \delta -$ artan $[(2R/C)$ tan $(\Sigma/2)]]$ sec $\Sigma/2\}$
= arcot $\{$tan $[\delta +$ artan $[(2R/C)$ tan $(\Sigma/2)]]$ sec $\Sigma/2\}$.
[10.4442°]
[48.7241°]

The flank slope angle γ is, by difference, $[\Phi_N - \delta'^*]$, so we may write that

§10.22   GENERAL SPATIAL INVOLUTE GEARING

$\gamma = + \text{arcot} \{(2R/C) \sin (\Sigma/2)\}$
  $\text{arcot} \{\tan [\pm \delta + \arctan [(2R/C) \tan (\Sigma/2)]] \sec \Sigma/2\}$,

which may better be written [52]:

$\gamma = + \arctan \{\tan [\pm \delta + \arctan [(2R/C) \tan (\Sigma/2)]] \sec \Sigma/2\}$
  $\arctan \{(2R/C) \sin (\Sigma/2)\}$.

[+23.6167]
[−14.6631]

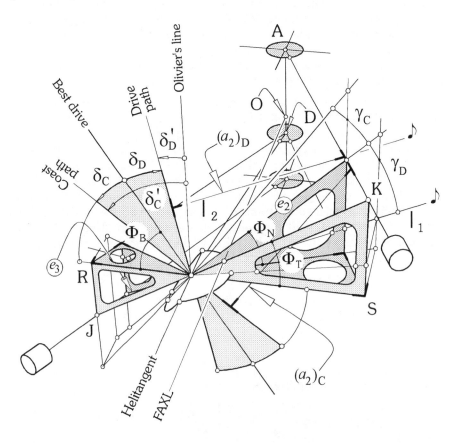

Figure 10.05(a) (§10.19). I am trying to show more clearly here the simple geometrical construction, given δ, for the flank slope angle γ. As shown, we rotate the best drive (with its best facet rigidly attached) through δ in the polar plane. The new location of the best facet becomes the plane of adjacency at the point of contact. This plane cuts the unique plane that contains the relevant E-radius and is parallel with the CDL. We thus determine (for each of the two paths) the hob angle γ. We know that $\cos \gamma = a/e$, but see the awkward algebra for $\gamma = f(\delta)$ at §10.22. Is this inescapably complicated? Or is there a simpler way? Details of the central portion of this figure, where each of the the two planes of adjacency are cut in half for the sake of clarity, is shown more clearly at figure 7.04.

MATHEMATICS OF THE MACHINING §10.22

These are the two angles γ exactly as calculated previously, one for drive and the other for coast. The pairs are the same on both wheels if the architecture is equiangular (as it is in this case), but not the same otherwise. So we have, neglecting sign, that

γ (drive) $= \gamma_D = 23.6167°$,
γ (coast) $= \gamma_C = 14.6631°$.

In the special case of the crossed helicals where, all else remaining the same, R becomes zero, we get the same flank slope angle γ for both sides of both

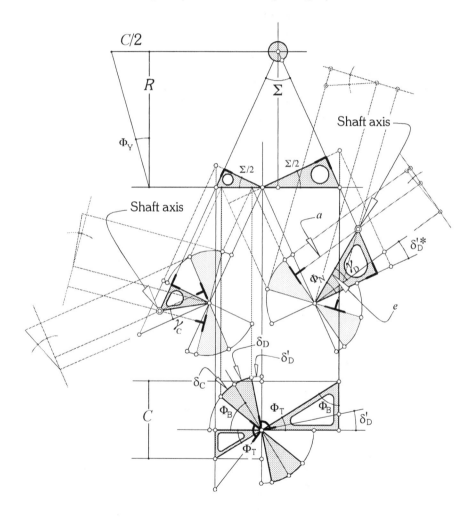

Figure 10.05(b) (§10.20). Descriptive geometry of the scene at figure 10.05(a). I have set out here at true size and shape the relationships between the angles δ, γ and the helix angle θ. For the text regarding θ, please go to §10.26.

369

§10.23   GENERAL SPATIAL INVOLUTE GEARING

sets of teeth. We get that, for $\delta = \pm 20°$, $\gamma = \pm 21.8802°$. But we also know, as has already been seen by mechanical intuition at figure 10.04, that

$$\cos \gamma = [(\text{the relevant } a) / (\text{the relevant } e)];$$

and for these a and e there are expressions, written in terms of C, Σ, R and k, already available. Written directly from the architecture by Killeen and Sticher [54] [52], and appearing together at §5B.16, these expressions are complicated. After some difficulty, but not unexpectedly, they reconcile with the ones derived here [52]. It might be mooted with justification, either that this difficult reconciliation points to the extraordinarily complicated geometrical interrelationships existing between the architecture of the set and the consequent shapes of the teeth, or that some very simple aspect of the matter has, as yet, been overlooked.

Second algebraic look at gamma in terms of delta

23. The above derivation of an expression for γ has shown algebraically but awkwardly that, for equal angles δ, the flank slope angle γ is greater on the drive side of the teeth than on the coast side. See figure 10.04. This phenomenon, which gives the appearance that the teeth are 'leaning over backwards' when in drive, but 'leaning over forwards' when in coast, calls for a more popular explanation now. Accordingly I have drawn figure 10.06. It shows a rectilinear piece of phantom rack (with equal rack angles δ) cut by an oblique sectioning plane in such a way that the angles κ and λ are appropriately involved. The said sectioning plane is called the γ-plane in the figure. The numerical values chosen for the figure are taken directly from WkEx#1. Go to WkEx#1, or to the TNR, for details. The mentioned angles, which are the starting data now, are as follows:

[$\delta = 20°$ exactly]
[$\kappa = 21.12199160°$]
[$\lambda = 13.69172117°$]

See in figure 10.06 that, properly according to figure 10.02, the γ-pane is slanted precisely (albeit relatively) to be parallel with the plane of the key circle (which on account of the equiangularity is the E-circle in this case). The only essential difference between the figures is that, whereas in figure 10.02 the plane of the key circle is rectilinearly arranged (namely square in the figure) and the rack aslant, in figure 10.06 the rack is square and the key circle is aslant. The two figures, in other words, correspond with one another and we can proceed. Paying attention to the essential geometry of the rack and the said oblique plane of the key circle in figure 10.06, it is relatively easy to show by ordinary trigonometry that

$$\tan \gamma = \tan \delta \sec \lambda \cos \kappa \pm \tan \lambda \sin \kappa;$$

and this, as will be seen, is a simpler expression for the two angles γ than the one produced at §10.22 immediately above. It checks numerically. Using however the various relationships mooted as being of possible use at §10.09

MATHEMATICS OF THE MACHINING §10.24

— these involved, remember, (a) the various fixed and obvious angles Φ_T, Φ_N and Φ_B, and (b) the other, obvious, fixed variables C, R and Σ — it will no doubt be possible to make the necessary reconciliation.

An incidental second look at alpha also

24. It can easily be measured furthermore, in the CAD for figure 10.06, that the angles α_{2D} and α_{2C} between the direction of the shaft axis (which is indicated) and the planes of the two flanks of the rack are 4.8318° and 19.3130° respectively; see the lines in the planes of the rack marked star (*), obtained by the dropping of normals to the planes, and see the angles α which, due to lack of space, are marked but not labeled. See these same angles otherwise determined in the CAD and by awkward formulae at §5B.16. In the light of the theorem presented at §9B.02, this exact correspondence between the measured angles is of course highly confirmatory of the theory that the correct flanks of the teeth can be generated by the straight sides of the required phantom rack. I should in the circumstances here seek a simple transcendental formula for α (like the one for γ given immediately above); but, *pro tem.*, I leave this as an exercise for the reader.

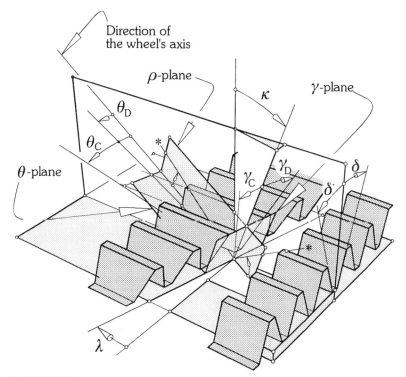

Figure 10.06 (§10.23). Relationships between δ and the other angles, γ, κ, λ and θ, compressed into a single diagram. Find here the angles α also.

371

§10.25   GENERAL SPATIAL INVOLUTE GEARING

An already established fact not to be overlooked

25. Much of chapter 9A was written to highlight the fact that the involute helicoid, the surface exclusively used for the profiles of teeth in spatial involute gearing, is not only by virtue of its geometry screw symmetric about its own axis, but also by virtue of design put with its axis collinear with the axis of the physical wheel. It is important to understand in involute gearing, in other words, *that even in the most unusual of cases, the geometric axes of the shapes of the flanks of the teeth are collinear with the axis of the physical wheel, and that this circumstance can be achieved by means of relatively simple machining processes, despite the occurrences of the angles κ and λ which would appear to deny this simplicity.*

The helix angles θ at the opposite profiles of a tooth.

26. I wish to mention prematurely now but significantly that figure 10.06 will be used to demonstrate, not only the formula for γ at §10.23, but also the formula for θ at §10.27. The *helix angles θ* at the teeth are about to be defined. Please refer again to figure 9A.01 which shows a five-fold set of gear tooth profiles; there are five teeth on the wheel but each tooth is showing one flank only. The figure is showing not only the general shape of the single profile of each identical tooth but also the four different sets (continua) of curves inscribed upon the said single profile. The four continua are (a) the *rulings,* which are straight lines tangent to the core helix, (b) the *planar involutes,* which are identical planar curves drawn in planes normal to the axis, (c) *the slip tracks,* also identical, of which there is a continuum, but only one single one of which is shown, and (d) the *helices,* which collectively signify the *screw symmetry* of the surface (§9A.03). These helices are the twisted curves we need to look at now. If the surface is cut by a continuum of circular cylinders coaxial with one another at the axis of the surface (and this continuum is suggested by a couple of sample cylinders shown at figure 9A.01), there will emerge a corresponding continuum of the helices. Between any tangent drawn to any one of the helices and the axis of the surface there is an angle θ; this is the *helix angle* of that helix. As the helices increase in radius from zero to infinity, the helix angle θ increases from zero to 90°; see figure 9.01 in [1]. *Angle θ does not remain constant with radius. Given the pitch p_{FLANK} (mm/rad) of the flank surface, which is a constant for the surface, the magnitude of θ depends upon the radius, say h (mm) measured from the shaft axis. The formula is:* $\theta = \arctan(h/p_{FLANK})$.

27. Take a toothed wheel now, say the one shown in figure 10.02. Go to the shown origin of coordinates in that figure and see the flank of that tooth clearly exposed at the centre of the figure there. We are about to consider the shape of its intersection with the unique circular cylinder which contains the E-circle; see figure 10.07. The curve of intersection will be, as we have said, a helix; the helix will contain [Qf]. See figure 10.07 and return for reference to figure 10.06. Now in the same way as the conventionally known

transverse plane (§10.30) has revealed by intersection with the sloping sides of the rack the two angles γ_D and γ_C (§10.21), the *key-circle cylinder* reveals the two angles θ_D and θ_C. They are the helix angles at the flanks of the teeth. *The helix angle at the flank of a tooth may be defined as the angle between the tangent to the relevant helix at [Qx] and the generator of the cylinder of the key circle there.* I will not explain at length the simple geometrical constructions made at figure 10.06; their correctness may be seen to be self evident. I do however wish to write the following expression for θ — for both θ — which springs (by trigonometry) directly from these constructions:

$$\tan \theta = \tan \delta \sec \lambda \sin \kappa \pm \tan \lambda \cos \kappa.$$

This is the expression foretold at §10.26. I bears, as I said that it might, a striking resemblance to the corresponding expression for γ which, already found, is set out at §10.23.

28. Figure 10.07 is a rehash of figure 10.02. The perspective has been somewhat altered, the phantom rack has been partly cut away, the unique cylinder of the key circle has been indicated, that part of a chosen tooth that protrudes beyond the cylinder of the key circle has been removed, and the cut cylindric section at the stump of that tooth has been revealed. One might imagine that the truncation had occurred in a lathe, that the wheel had been turned cylindrically, that the tops of the teeth had been removed down to the diameter of the key circle, but that (for convenience in the figure) only one of the teeth has been shown as having been truncated. The shown helix through [Qf], which is drawn upon the surface of the cylinder has been constructed by CAD, and its spatially twisted shape provides the edge, as can be seen, of the cylindrical cut section at the stump of the tooth. The other profile of the tooth in question is of course not at [Qf] just now, but its cut edge can with equal ease be constructed by CAD and the relevant helix is shown. Notice that, although the helices are both left handed and have the same radius, they are of different pitch. The two helix angles θ are thus different. See the $\pm$ sign in the formula for $\tan \theta$ at §10.27; and be aware that the numerical values for the current example turn out to be: $\theta_D = 5.2711°$, $\theta_C = 19.9128°$. These calculated values check exactly with measurements made at the CAD construction for figure 10.06. *It is accordingly nonsense to speak about the helix angle of the teeth of a wheel. The term can only apply to one or other of the profiles of a tooth. The helix angle must moreover (by definition) be measured at the key radius of the wheel.*

29. *Special cases of the angles θ.* Let me mean by the cylindrical helicals not only the parallel and crossed-axes cylindrical helicals but also the doubly special parallel cylindrical helicals where the angle λ is zero namely the case of ordinary spur gears. In all of these special cases the angle κ is zero. The formula for $\tan \theta$ at §10.27 breaks down to explaining that $\tan \theta = \pm \tan \lambda$. *This means that for any one wheel the two angles θ take the same value, namely λ.* There is a single value of λ for each of a pair of equiangular wheels, and two values for the two wheels in the case of polyangularity

§10.29   GENERAL SPATIAL INVOLUTE GEARING

(§10.07). In any event the algebra has confirmed what I have been saying. *This is that, in general spatial involute gearing we must distinguish between the swivel angle λ of a wheel on the one hand and the two different helix angles θ of the flanks of its teeth on the other.* It is only with ordinary cylindrical helical gears with parallel axes and spur gears, where the special geometry causes these characteristic angles of a single wheel (namely the λ and its two companions θ) to be the same, that we can afford to be slipshod. It was at §10.06, remember, where I spoke about the two opposing flanks of

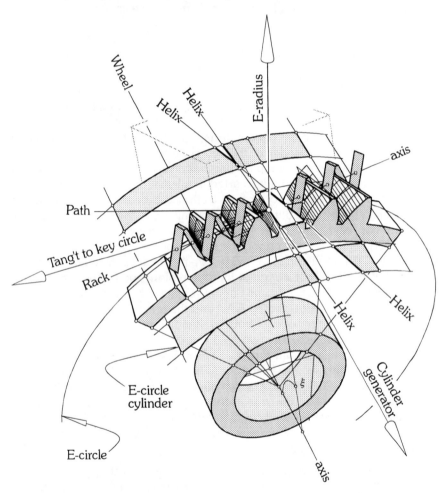

Figure 10.07 (§10.28). This figure (from WkEx#1) shows one whole tooth decapitated by the E-cylinder at the E-circle. The shape of the shown exposed surface (plateau) is not planar, but cylindrical. The two helices bordering this plateau of a tooth determine the helix angles θ of the tooth, and all teeth of any one wheel are identical of course. There are thus in general two helix angles θ for a given wheel, one for drive and the other for coast. See here again the oblique disposition of the phantom rack and the rack triad ghosted in.

these special kinds of teeth being the two opposing branches of the same involute helicoid. Let me mention again, also, that in these special cases the linear complex characteristic of sets in general is a special linear complex where the pitch of the complex is zero and where, as a result, γ and thus θ may take any value (§1.47 *et seq.*)

The various triads

30. See at [Qf] in figure 10.07 the rack triad still ghosted in. Also there, distinguished by means of three prominently drawn arrowheads, is a new set of Cartesian axes. This new set infers of course a new triad of planes mutually perpendicular at [Qx]; *but this last addition to the list of triads is the one most commonly familiar to all gear workers*. While the three axes are named as I have labeled them at figure 10.07, the three resulting planes are commonly known as (a) the *transverse plane,* which is normal to the shaft axis, (b) the *radial plane,* which contains the shaft axis, and (c) the *peripheral plane,* which is tangential to the cylinder of the key circle at [Qf]. For the want of a better name, and to distinguish it from the immensely important, but somewhat concealed, *fundamental triad* of earlier chapters (§3.44, §5B.09, §10.24), I will call this latest of the triads, *the conventional triad.* Collecting for convenience now and listing the various triads for comment, we can see in retrospect: (a) the original, geometric, *curvature triad;* go back to §3.21 for an explanation of this and see figure 3.05; (b) the somewhat elusive *fundamental triad;* study §3.44 for this, and see figure 10.03; (c) *the two rack triads;* recall the discussion at §10.04 for the gist of these; and (d) the practically oriented, obvious, c*onventional triad* of the present paragraph. This last may be detected also at figure 10.06. I have collected and separately mentioned the various triads together just here because we are moving, just now, into the context of the machining of teeth; and this requires by its nature a new attitudinal perspective. Go back to item (d) above. If not already, it will be clear very soon that, although the cylinder exposed by the truncation at [Qf] will continue to be called by many the pitch surface (§P.05), that name will be, in my view, a misnomer.

The helix angles in Beveloid gearing

31. Beam [59] and Smith [61], in their works on Beveloid gearing, pay attention to these somewhat unexpected helix-angle phenomena. Putting four of his consecutive paragraphs together for the sake of brevity here, I quote directly from Smith as follows.

> *Quote.* Applying a taper to cylindrical *spur* involute gears provides an additional degree of freedom and results in a complex involute helicoid surface on the tooth flanks. Opposite flanks will have equal but opposite hands of helix and a common lead. See figure 6. The cylindrical spur gear thus may be considered a special case of the involute helicoid with zero taper, just as the cylindrical spur gear may be considered a special case of the involute helicoid

with zero helix, i.e. infinite lead. Applying a taper to a cylindrical *helical* gear also provides an additional degree of freedom to a gear which is initially a simple involute helicoid with equal and parallel helices of the same hand and common lead and results in a complex involute helicoid of compound structure. The helix resultant of the taper is additive to the original helix on one flank and is reductive to the helix on the opposite flank, there are thus two entirely different helix angles and differing leads on opposite flanks. Relative magnitudes of helix and taper determine whether the flank hands are the same or opposite.

Smith's figures 2 and 6 (1989), which relate to straight Beveloid teeth as Beam explains in his earlier paper (1954), namely teeth where my centre distance C and my angle λ are both zero, are copies of Beam's figures 1 and 2. Beam attributes the figures to an earlier interested worker, Vogel [73], who wrote in 1945. Inspired by the figures, I have borrowed, not the very figures themselves (which relate to the special *straight* case of Beveloid gears as has in effect been said), but the general idea of them. I borrow the general idea of Beam's figures 1 and 2 for my discussion of interference between general spatial involute teeth at §10.42. My composite figures 10.08(a) and 10.08(b), which are of non-straight teeth, have these two figures as antecedents. See also figure 15 in Beam [59]. It resembles my various figures. However, except for its inset portion, the meaning of which is obscure to me, Beam's figure 15 (in conjunction with his table 1) is used to explain machining terminology. Mostly, also, it deals with what he calls the straight Beveloid gear where my angle λ is zero, and where my p_{24} (see later) is also zero. Beam avoided screws. He preferred, characteristically, not of him but of his time, to explain matters otherwise. Next I wish to say the following. Whilst I do not agree that any one involute helicoid can be more 'complex' than another, it becomes clear to me from these and other remarks about Beveloid gearing that the shapes of the teeth of all of the various wheels of the Beveloid sets are satisfactorily enclosed within the more general theory being outlined here. Beam and Smith and other writers bear witness to the fact that Beveloid gearing works (in the sense that its axes may move about), and that it will do that smoothly and effectively without excessive wear. They do not explain, however, how a designer, on being given the requirements C, Σ, R and k for a gear set, might design the set.

Aspects of the motion of the real rack cutter

32. All of the material under this sub-heading might be omitted upon a first reading. It deals with some important conceptual matters nevertheless. It is important first of all to observe that, *as soon as machining becomes the subject matter for discussion*, there are, all told, *four* rigid bodies to be considered. They are the frame 1, the wheel 2, the wheel 3, and the rack body 4. This applies whether the rack is some real rack-shaped tooth-forming instrument (as often illustrated but seldom actually employed) or a mere

MATHEMATICS OF THE MACHINING §10.33

geometrical abstraction representing the overall cutting result of some other tool or tools. With the frame fixed and both wheels steadily turning, the phantom rack will be seen (a) to remain with its flutes aligned with the helitangent at [Qf] namely with the contacting pair of rulings of the naked wheels, (b) to remain with its reference plane coplanar with the transnormal at [Qf], and (c) to translate with a constant linear velocity in the direction of the best drive. Refer in advance to figure 10.08(b); see there that the phantom rack is oriented square with the fundamental triad. While the conventional (the unhidden) triad of §10.30 is marked XYZ in the figure, the fundamental (the hidden) triad of §3.44 is marked with the less obtrusive, spelt labels that are nevertheless quite obvious. It is important also incidentally to see that, for most purposes of argument, any sliding of the rack which may occur in the direction of its own flutes is irrelevant. We might, nevertheless, risking minor loss of generality, revisit *the key circle axis of the rack,* already called by that name at §10.07. This is the line in the *central reference plane of the rack*, initially collinear with the best drive, which we may, if we wish, hold thenceforth fixed in the moving rack but parallel with the best drive.

33. Go to figure 10.08(a) and see there that, for straight rack cutting of teeth (or for plastic forming and thus also for milling, grinding, hobbing etc.), we could set the axis of the rotating wheel in relation to the horizontal flutes of the translating phantom rack by executing, in proper order, the two key angular displacements away from square as shown: *tilt angle κ, and swivel angle λ*. In astronomy, surveying, artillery etc., the corresponding displacements are the *altitude* and *azimuth* angles respectively. But in those fields the relevant mechanisms for orienting the cylindrical member have a built in cruciform piece (the intermediate link of the Hooke's joint which joins the main frame to the said member) which obviates the need for proper ordering [1] [§2.58-59]. If we can *imagine* a Hooke's joint with its cruciform piece arranged with its axes along AA (in the frame) and BB (in the wheel) in figure 10.08(a), there will be no need in that figure, either, to worry about the question of order. We tilt through κ about AA, then swivel through λ about BB. The line B-B in the wheel has been mentioned already at §10.07.

34. Finite angular displacements about fixed axes do not commute [1] [§2.58-59]. The diagram made up of circular arcs at the RHS of figure 10.08(a) is not a spherical rectangle. The arc marked κ is an equatorial great circle about the polar axis AA. The arc marked λ and its partner opposite are both meridian arcs. They are cast about two successive locations of the axis BB, and they are of equal length. The arc opposite the arc marked κ is not great circular; it is the relevant segment of a circle of latitude; it subtends its angle κ, not at the centre of the sphere, but at the polar axis; it and the arc marked κ are not of equal length. There is however a true spherical triangle at the lower RHS of figure 10.08(b); its sides are marked κ, λ and τ; and there is a right angle at the apex $\kappa\lambda$. This reflects the somewhat surprising fact that $\cos \kappa \cos \lambda = \cos \tau$. See appendix C at chapter 6.

§10.35

35. Notice also in figure 10.08(a) that, by employing the angles κ and λ, I have set the central reference plane of the rack to cut the hyperboloidal sheet of the naked wheel along its two intersecting generators at [Qf]. Unless this were so, the imaginary revolute axes AA and BB would not have been, as they are, horizontal and vertical respectively. But these matters are not the whole of the matter. Depending upon the kind of machine tool and the mechanical arrangement of its working parts, various complicated sets of speeds and feeds become involved, and this needs to be understood. How, exactly, when seen under inversion (§1.12), the rack *screws* with respect to the wheel, and how undercut and interference occurs, are also yet to be understood.

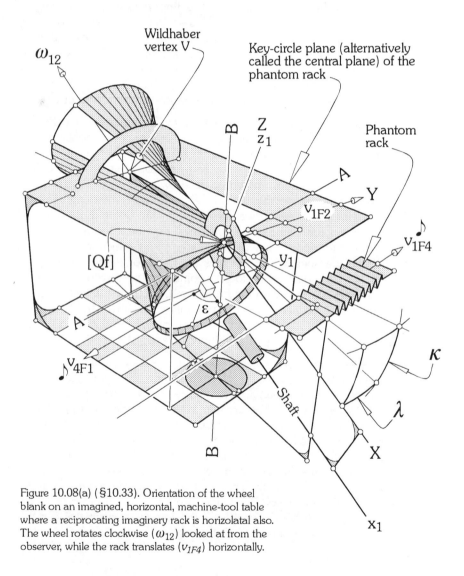

Figure 10.08(a) (§10.33). Orientation of the wheel blank on an imagined, horizontal, machine-tool table where a reciprocating imaginery rack is horizolatal also. The wheel rotates clockwise (ω_{12}) looked at from the observer, while the rack translates (v_{1F4}) horizontally.

Some misconceptions that must be rejected

36. Go back to §1.31 *et seq.* where the contrasting concepts of (a) *the spline-line teeth upon the axodes*, and (b) *the smooth rulings upon the naked wheels*, were first introduced; read also §6.29 again where axodes and naked wheels were more thoroughly compared and contrasted; and wonder now about the following beguiling question. *Is there in general in these offset spatial involute gear sets, somewhere or somehow, a flat, naked rack that rolls respectively upon the tapered peripheries of the naked wheels to act in some way as a correct geometric base for generation of the teeth?* It needs to be said at outset that, because the underlying notions implicit in this question are false, the question itself is nonsensical. The false notions need to be examined however, and the first thing to understand is that, while for two rigid bodies in relative motion the pair of axodes for the motion is unique, the pair of naked wheels for two meshing gears in relative motion is not only not unique, it is a secondary phenomenon relating not to bodies but to various geometric abstractions that are irrelevant. *Sets of naked wheels, which are simply selected portions of matching departed sheets, are not axodes.*

37. The misconceptions implied in the above paragraph arise understandably from the special cases for example of cylindrical crossed helicals with fixed skew axes and a given k where the so called pitch circles (or, better, the pitch cylinders or surfaces) touch at the pitch point and enclose between them the pitch plane containing the central reference plane of the phantom rack, and where, when the helix angles of the two wheels depart from being equal, that is when the throat radii r_2 and r_3 of the axodes must be replaced by the throat radii c_2 and c_3 of the naked wheels (§1.34), the said plane of the phantom rack simply moves along the CDL to suit. In these cases the naked phantom rack *does* roll respectively upon the naked wheels and the teeth of the real wheels are (in actual machining practice) thereby satisfactorily generated. This can be seen explained by Baxter, but only partly, in *Dudley's Gear Handbook*, first edition, 1962, then later by Dudley himself (but again only partly) in Townshend's second edition, 1992; see [96] [97]. Please study also in this connection the doubly special cases outlined at §1.47 *et seq.*; these are the planar spur and the cylindrical helical sets (both with parallel axes), where the naked wheels and the axodes are locked inextricably together, identical; the reasoning here is briefly explained at §1.49 and more fully explained at [1] [§22.30 *et seq.*]. The secret here is to see in the general spatial cases that it is not the tapered hyperboloidal surfaces of the naked wheels that are at issue, but the question of where the real axodes for the relative motions, those of the rack bodies moving relative to the real wheels in the tooth-cutting machine tools, actually are. We shall see, in general (where the offset R is non-zero), that these axodes are *not* at the key circle cylinders (ambiguously called by many the pitch surfaces), but elsewhere. Much of the remaining part of this chapter is now given over to this vexatious question, which is, if we wish to cut teeth, unavoidable.

38. Because, as we have seen, *the axes of the involute hyperboloids of the flanks of the teeth of a wheel reside always upon the axis of the wheel*, it is inescapable that the ISA for the screwing of a rack upon its wheel will always be somewhere parallel with the axis of the wheel. *No properly cutting rack, in other words, will ever be found to be screwing relative to its wheel about an axis oblique to the axis of the wheel.* This does not preclude the possibility that the said ISA, while remaining stationary within the fixed link namely the box, will *move* with respect to both the wheel and the rack; and by symmetry it is clear to see that such movement will only occur circularly, with the ISA, moving steadily and circularly with respect to the wheel, remaining always parallel with the axis of the wheel. Thus we begin to see that the fixed axode for the motion of any rack relative to its wheel will always be some circular cylinder coaxial with the wheel and that the moving axode for the same motion will always be some plane in the rack that is tangential with the said cylinder. It is because these things are inescapable that the question put at §10.36 reduces itself to absurdity.

Rack velocity

39. Among the confusion (which, I hasten to add, will be resolved), one of the kinematic concepts appears at this stage to be clear however. It is the linear velocity relative to frame — it might be called the *absolute* velocity — of the pure translation of the rack in the unique direction of its own key axis, which axis is in the central plane of the rack perpendicular to the direction of the flutes (§10.07). According to my terminology [1] [§12.08], and in view of the rack having been designated link 4 of the current 4-link mechanism (§10.32), the said velocity may be written correctly v_{1F4}. In the context here F_4 is that point F (which is at [Qf]) in body 4 which is instantaneously coincident with another point F in body 1 namely F_1 [1] [§12.08]. Figure 10.08(a) is drawn for the dimensions of the bigger wheel at the equiangular WkEx#1 where, of course, F is at E. The wheel has 40 teeth. Indicated at the RHS of the figure, the absolute velocity v_{1F4} (which is the same for all points in the rack) may be seen and, at the top LHS of the same figure, the angular velocity ω_{12} (clockwise looking from where we stand) may also be seen. Going to figure 6.08(a), which is an orthogonal view along the transversal of that same set, we can see that the said linear velocity of all points in the rack takes place in the direction of the best drive, the best drive being in the plane of the paper and along the vertically arranged edge of the set square shown. If v_{1F2} be the velocity relative to frame of the point F_2, that is the peripheral speed of the big wheel 2 at its F-circle (which has in general neither the same magnitude nor the same direction as the peripheral speed v_{1F3} of the small wheel 3 at the same F), by symmetry the velocity of the rack may be written

$$v_{RACK} = v_{1F4} = v_{1F2} \cos \lambda.$$

Numerical values from WkEx#1 show that, because $e_2 = 89.2738$ mm and $\lambda = 13.6917°$, the rack velocity (for ω = unity) is 86.7369 mm/sec. Looking next at figure 6.06(b), which is drawn for the polyangular WkEx#3, we see that there too, although polyangularity prevails and [Qx] is at F and the naked wheels are slipping circumferentially and the two angles λ are not the same but different, the same formula applies for both wheels.

We may write, as a general statement applicable under all circumstances, that the magnitude of the rack velocity will be the peripheral velocity of the wheel at its key circle multiplied by the cosine of the swivel angle λ pertaining. The direction of the rack velocity will be, moreover, always in the direction of the best drive.

In WkEx#3, for a second numerical example, we measure that $f_2 = 84.0224$ while $f_3 = 58.8157$. The ratio of these is $j = 0.7$, which checks OK. From §6.40, $\beta_{SHIFT} = 17.5234°$, and we know that $\lambda = 13.6917°$, so $\lambda_2 = 13.6917 + 17.5234 = +31.2151°$, while $\lambda_3 = 13.6917 - 17.5234 = -3.8317°$. We have that $\cos \lambda_2 = 0.99776$ and that $\cos \lambda_3 = 0.85523$. So, calculating from each of the two wheels separately (for ω_{12} = unity), we get the same result, namely that $v_{RACK} = 83.835$ mm/sec. What this paragraph has shown is that, within a given set, the rack velocity is a constant for the set (which is not surprising); it is independent of which wheel we start the calculation from. This applies for polyangular sets (which is also not surprising); but when polyangularity increases in a given set (as it does when, other things remaining equal, we crank up the β_{SHIFT}), the rack velocity diminishes. Notice in the two examples given here (and see figure 6.08 for a combined illustration of them) that the rack velocity, while diminishing with the radius f_2, does not vary directly with it. We would need to incorporate the variable j into the writing of an algebraic expression for this.

The analogous oblique-rolling vehicular wheel

40. I wish to mention parenthetically here the case of the oblique-rolling vehicular wheel. Refer to figure 10.16 at [1] [§10.67]: we have (a) the rough but level agricultural ground, (b) the rigid hub of the pneumatic wheel rolling with its angles of camber and drift, and (c) the chassis of the vehicle. In the case of the gear-rack set, on the other hand, we have, correspondingly, (a) the rigid rack, (b) the rigid 'conical' skew-cut wheel with its angles κ and λ, and (c) the gear box or, in the case of tooth generation by controlled cutting, the main frame of the machine tool. In the case of the vehicular wheel we entertain the idea that the ground (the rack) is fixed while the chassis moves, but with the gear-rack set we often entertain the opposite idea that the gear box (the chassis) is fixed while the rack moves and the wheel rotates. In agriculture the 'circumferential slip' is determined by the mechanics of the crumpling of the ground, which in turn depends upon the torque (which is either driving or braking) at the hub. With the rack-gear set, however, that slip (by virtue of there being actual teeth) must be argued somehow to be a calculable quantity. We see in any event that the rack will screw with some finite pitch with respect to the gear wheel about an axis fixed in the box in the same way as does the vehicular wheel screw with some finite pitch with respect to the agricultural ground about an axis fixed in the chassis.

§10.41   GENERAL SPATIAL INVOLUTE GEARING

How the rack moves to cut the teeth of a wheel

41. For the machining of any one gear wheel we can seek — and this will encapsulate all questions of the machining such as milling, grinding or hobbing and all questions of speeds, feeds, and so on — one important thing. *It is the location of the axodes for, and the pitch of the ISA for, the relative motion of the rack body and the wheel.* None of the previously mentioned fanciful material about some kind of mythical rack rolling upon the naked wheels alters the continuously constant nature of the instantaneous relative motion of the four rigid bodies in a set of gears — the frame, the two wheels, and the real rack body (§10.32). In this connection *the theorem of three axes applies*. It applies to whichever set of three bodies we care to choose from the said four [1] [§13.12]. So, for the machining of either one of the wheels, say wheel 2, let us look at the three screw axes which, in the absence of wheel 3, obtain. Refer to figure 10.08(a). Fixed in frame 1 namely the box, we have the following: (a) for the wheel 2 relative to the box 1 an axis of zero pitch along the axis of the wheel; this is ISA_{12}; (b) for the rack 4 relative to the box 1 an axis of infinite pitch in the direction of the best drive (along AA) which is inclined at angle λ to the tangent to the key circle at [Qf]; this is ISA_{41}; the pitch of this axis is infinite and the relevant linear velocity v_{1F4} (or its opposite v_{4F1}) has direction but no specific location [1] [§5.50]; and (c) for the rack 4 relative to the wheel 2 (or *vice versa*) the appropriate combination of (a) and (b); this is the sought for, ISA_{24}.

42. Refer again to figure 10.08(a) which, for the sake of the argument here, and although drawn originally for the equiangular WkEx#1, can legitimately stand for the whole range of plain polyangularity. Accordingly and without loss of relevance I can call the special key circle radius e of the wheel, and the special point E which is at [Qf], by their more general names f and F. For a full explanation of the following go to [1] [§10.49 *et seq.*]. Let the angular velocity of the wheel 2 relative to frame 1 be ω_{12} (rad/sec), whereupon the linear velocity v_{1F4} of all points in the rack (omitting the subscripts) is $\omega f \cos \lambda$ (mm/sec), where f is the key circle radius of the wheel. Locating the basepoint of the relevant dual vector at O (not shown) at the centre of the key circle, we have, at that point, (a) the angular velocity vector of magnitude ω_{12} along the shaft axis, and (b) the linear velocity vector v_{4F1} of magnitude $\omega f \cos \lambda$ in the direction of the quaver ($\flat$) as shown. Note the pair of parallel but opposite directions marked with a pair of the said quavers; both lines are set parallel with the line AA (the best drive). The shown angle ε between (a) and (b), which is acute, and which is not within the plane $O\text{-}x_1y_1$ of the triad $O\text{-}x_1y_1z_1$ (see later), is a simple trigonometrical combination of κ and λ. We can find very easily that

$$\cos \varepsilon = \sin \lambda \cos \kappa.$$

For a full explanation of the following go to [1] [§10.52]. The component of the above mentioned linear velocity which is destined to become the least

MATHEMATICS OF THE MACHINING §10.43

velocity v* upon the *shifting* of the basepoint (with its fixed vector ω) to the home screw is

$$v^* = \omega f \cos \lambda \cos \kappa \quad \text{(mm/sec)}$$
$$= \omega f \cos \lambda \sin \lambda \cos \kappa.$$

Next, because ω (rad/sec) is the relevant angular velocity, the magnitude (mm) of the *shift* is

$$\text{Shift} = f \cos \lambda \sin \varepsilon \quad \text{(mm)}$$
$$\text{where } \cos \varepsilon = \sin \lambda \cos \kappa.$$

Substituting numerical values first from the larger wheel 2 at WkEx#1, we find that the shift turns out to be 0.9476 times e_2 namely 84.5963 mm. This is somewhat less than the radius e_2 of the key circle namely 89.2738 mm; and, due to the newly found, odd inclination of the plane containing ε, the direction of the found shift is not the direction we might (intuitively but wrongly) have expected, namely the radial direction O-z_1, which is within the vertical plane O-x_1z_1 which contains the shaft axis. Go to figure 10.08(c) to see there, not only *the true length of the shift*, but also *the true magnitude of angle ι (iota) which is the inclination of the shift*. Iota, measured from the plane O-x_1z_1 at figure 10.08(a), lies within the plane O-y_1z_1 there, namely the plane of the key circle which is, there, in the plane of the paper. [Taking κ = 21.12199160° and λ = 13.69172117°, we find ε = 77.24425°. We have f = 89.2738, sin ε = 0.975320, and cos λ = 0.971583, so the said shift is 0.9476 times f namely 84.5963 (mm).] Numerical results for the other wheel at WkEx#1 and the pairs of wheels at other worked examples may be seen at the TNR. I will explain these newly found phenomena at §10.45 *et seq.*, but please pay attention just yet to figures 10.08(a) and 10.08(c).

A mating pair of naked wheels and its phantom rack

43. Refer to figure 10.08(b). To arrive at this I have done no more than extend the flutes of the phantom rack in figure 10.08(a) until they are clearly seen to be intersecting the previously unmentioned plane XY in that figure. The plane XY is, indeed, and as labeled, the peripheral plane for the shown wheel (§10.30). I mention this because the plane XY is not the peripheral plane for the mating wheel. The said rack cuts the said plane in a jagged line, the jagged line containing [Qf]. Only parts of this jagged line can be seen, however, because, in order to reveal [Qf] in its local context, the continuity of the plane XY has been interrupted in the figure. With [Qf] at its origin, the axes XYZ are indicating the conventional triad (§10.30). The other orthogonal set of axes at [Qf] (with the named directions) are indicating the fundamental triad (§3.44). The cylinder coaxial with and containing the key circle is shown grey-shaded, and the plane tangential to it, which is (as has been said) the peripheral plane of the conventional triad, is also grey-shaded. Please study next the other rack-shaped jagged line, not the one in the XY-plane but the one in the YZ-plane, spread out along the Y-axis. This is the planar trace of the phantom rack in the transverse plane of the said

§10.43   GENERAL SPATIAL INVOLUTE GEARING

conventional triad. Whereas in this example (WkEx#1) the two angles δ of the rack are equal, the two corresponding angles of this last-mentioned planar trace are not. They are indeed the tooth flank angles γ_D and γ_C whose two different values are already listed at §10.22. *Now despite previous ex-*

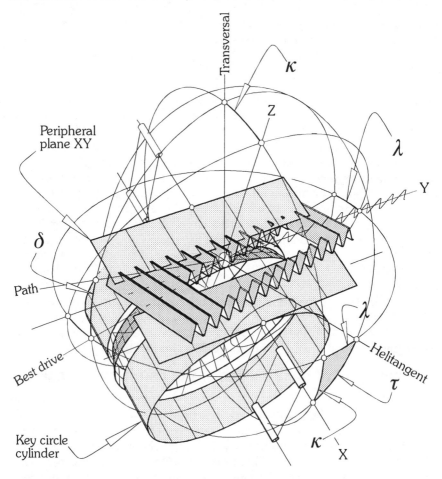

Figure 10.08(b) (§10.43). This is a re-jigged version of figure 10.08(a). The central (key) plane of the rack is shown inclined, as before, at the angles κ and λ to the wheel blank. The two grey-shaded surfaces, the key-circle cylinder and the peripheral plane XY of the wheel, are tangential with one another along the axis X, but they are somewhat more than what their labels tell that they are. They are not the axodes for the relative motion of the rack and the wheel (as we shall see), but, by virtue of their locations, they are the naked wheels for that relative motion. Notice for attention later, the two jagged lines which are the traces of the shown rack, firstly in the plane XY, which is grey-shaded, and secondly in the plane YZ, which is not. The plane XY is the plane of the key circle of the wheel, and the jagged trace of the rack therein displays the unequal angles γ shown at true size at figure 10.08(c).

MATHEMATICS OF THE MACHINING §10.43

perience with the simpler involute gears (the planar spur gears, the parallel helicals and the crossed helicals), and despite figure 10.08(d) about to be mentioned soon, it would be a mistake to claim intuitively that these two grey-shaded surfaces at figure 10.08(b) are the fixed and the moving axodes

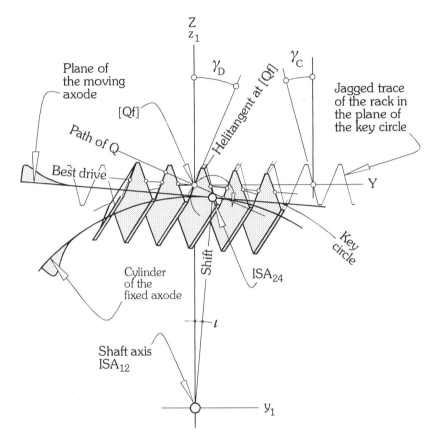

Figure 10.08(c) (§10.44). This is an orthogonal view of the same scene. It is taken directly along the shaft axis of the wheel and thus directly along axis X and orthogonally against the plane YZ of the key circle. The shown front edge of the rack, cut square with the flutes, comes directly from figure 10.08(b); so, except at the point [Qf], it is not in the plane of the paper. The jagged line of the trace of the rack in the plane XY is in the plane of the paper however, and this trace can be seen exhibiting its unequal angles γ_D and γ_C. Most importantly in this view we see the found ISA_{24} which, parallel with the shaft axis, appears as a point. Having calculated the pitch of this particular ISA, which axis remains fixed in the fixed space while moving with respect to both the wheel and the rack, I have constructed the sliding velocity at Q between the rack and the wheel when Q is at [Qf], and have found, not surprisingly, that it occurs exactly along the relevant helitangent (the one for the rack and the wheel) which is parallel there with the flutes of the rack. The fixed and the moving axodes for the motion of the rack relative to the wheel are prominently labelled. Please note that they are a cylinder and a plane respectively, rolling and sliding without circumferential slip and having their imaginery spline-line teeth as axodes properly do.

§10.44   GENERAL SPATIAL INVOLUTE GEARING

for the motion of the tooth-forming rack-body with respect to the wheel blank. It is very important to understand that they are not. On careful reflection we will find, in fact, that they are no more than the *naked wheels* for the relative motion of the rack and the wheel. I wish to say it again: *the grey-shaded surfaces at figure 10.08(b) are not the axodes for the relative motion of the rack and the wheel.*

Finding precisely the location and pitch of ISA$_{24}$

44. Go back now to continue from the end of §10.42. See figure 10.08(c). As before in these figures 10.08, we are at the wheel 2 (the bigger wheel) of WkEx#1. It needs to be understood that the rack is truncated in the foreground here by means of the polar plane at [Qf]. This is square with the flutes; the truncated edge is not in the plane of the paper. The path of Q and the best drive (both of which are within the polar plane) are accordingly not in the plane of the paper either. The shift (§10.42) does not occur along the radius from the F-circle centre to [Qf], but along another line through the same centre somewhat clockwise to the right of it. The plane of the shift is normal to the plane of the paper namely the plane of the F-circle, and normal to the plane of ε. The line of the shift and the apparent direction of the best drive, as seen in this orthogonal view at figure 10.08(c), are perpendicular. This accords with the geometry of the dual vector as explained at [1] [§10.52]. It fits also the fact that the ISA must be somewhere parallel with the shaft axis. Calling this angular displacement of the shift say ι (iota), it can be shown — see appendix A at 10 — that

$$\tan \iota = \tan \lambda \sin \kappa.$$

Using this formula, the value for angle ι (at wheel 2 in WkEx#1) turns out to be 5.0171°. This checks exactly with the CAD drawn for appendix A at 10 (§10.67). In general, angle ι will be different for the different wheels of a gear set, but in equiangular cases the F-circle is of course the E-circle, both κ and λ are the same for both wheels, so ι is the same for both wheels. So we can say in general and from the formula that, with equiangularity prevailing, ι will always be the same for both wheels. I come to the geometric significance of the angles ι later, at §10.49.

45. Evidently now the pitch of the screwing at ISA$_{24}$, namely (v*/ω), namely p_{24} — which is *not* the fundamental pitch of the gear set (namely p namely p_{23} at the pitch line) — is

$$p_{24} = f_2 \cos \lambda \sin \lambda \cos \kappa \text{ (mm/rad)}.$$

[In WkEx#1, given that f_2 is 89.2738 mm, this works out to be 0.21452 times f_2, namely −9.1511 mm/rad.] The sliding velocity between the rack and the flank at [Qf], which is along the helitangent and oblique to the ruling in contact just there, checks with the newly found ISA$_{24}$ and its pitch. It also appears correctly in both of the figures 10.08(c)(d); it comes up in line with the helitangent which is in turn parallel with the flutes of the rack. The shift (§10.42) and the pitch (§10.45) vary directly with f, and they are both independent of ε; this shows that we spoke correctly at §10.32 where it was argued that any uncontrolled component of the travel in the direction of the flutes of the rack would be irrelevant.

Finding precisely the location and pitch of ISA$_{34}$

46. The two formulae presented immediately above have general application. I mean by this that they apply for a wheel and its rack independently of whether the circumstance is equiangular or plain polyangular. What needs to be noted just here is that, if we modify some given equiangular set by doing no more than changing the poly-ratio j, we oblige the two otherwise equal angles λ to differ, while allowing the two otherwise equal angles κ to remain equal. To check here please compare the comparable figures 5A.02(b) and 6.07. See in those figures that, while the equality of the two angles λ has been destroyed by the change in j from 0.6 to 0.7 (all other things remaining equal), the equality of the two angles κ has not. What this means just now is that we can apply the two formulae to the wheel 3 of the present equiangular set (WkEx#1), and obtain without difficulty the following results. [Because tan ι_{34} = tan λ sin κ (§10.44); ι_{34} = 5.5.0171°. Because $p_{34} = f_3 \cos \lambda \sin \lambda \cos \kappa$, where $f_3 = k f_2$ (§10.44), p_{34} = – 11.4907 mm/rad. Because (in general) *Shift* = $f \cos \lambda \sin \varepsilon$, where cos ε = sin λ cos κ (§10.42), we can find (for this wheel 3) that the shift = 50.7578 mm.] These *rack axes*, ISA$_{24}$ and ISA$_{34}$, are of course different axes. In the fixed space 1 of the gear box they do not intersect and, because each is parallel with its relevant shaft axis, they are not parallel with one another: they are skew.

47. Using these results, and looking ahead, I have numerically located and accurately plotted the two rack axes (for the relationships 24 and 34 of the interacting bodies) into the panoramic picture at figure 10.09(a). Because their directions, each shown there with a star (*), are parallel respectively with those of the wheel axes, it is not surprising to find that the common perpendicular between them is vertical (parallel with the CDL). It is however surprising to find that the common perpendicular itself falls exactly upon Olivier's line (§3.66). Noting just now that Olivier's line is in this way relevant for equiangularity only, I take this matter somewhat further at §10.51.

Some well known but misleading phenomena

48. Refer now to figure 10.08(d), which is essentially planar, and be aware that the teeth of the apparent wheel and of the rack depicted there are paper-thin. They are, in the sense introduced at §10.18, sets of *paper teeth*. Both of them are the mere traces upon the YZ-plane (the transverse plane) of the real three dimensional bodies suggested at figure 10.08(b). The relevant (oblique) trace of the phantom rack has already been seen, spread out along the Y-axis at figure 10.08(b). The parts are held to mesh with one another nevertheless, and the paper rack may be seen to have been rolled, *without slip upon the paper key circle radius f*, to generate by means of envelopment

§10.48 GENERAL SPATIAL INVOLUTE GEARING

both flanks of the teeth. This envelopment of the flanks was graphically done step-by-step in the figure. All residual construction lines have been removed. But notice also the generation of each flank separately, not by means of an enveloping line but by means of a traveling, path-tracing point at the end of

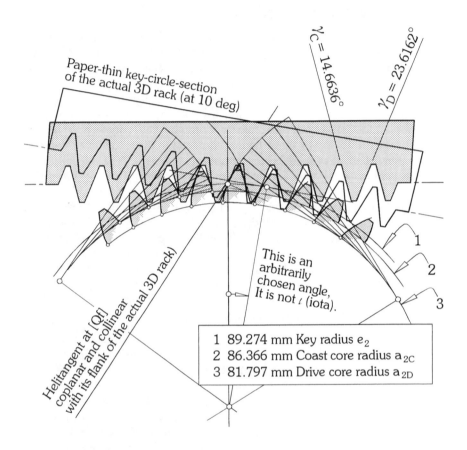

Figure 10.08(d) (§10.48). Based on the data for WkEx#1, this figure shows various correctly presented phenomena that are misleading nevertheless. The toothed, grey-shaded rack-like shape here is not the real rack, but a paper-thin section of it, cut at the plane of the key circle; it is the asymmetrical jagged line in the plane YZ at figure 10.08(b); by rolling without slip with its central line upon the key circle, it generates the paper-thin teeth of the flat-action picture at figure 10.04. The flanks of these paper thin teeth may also be generated by taut cords unwinding (a) clockwise from a circle of the drive core radius a_{2C}, and (b) counterclockwise from a circle of the coast core radius a_{2D}. None of this belies the fact that the real ISA$_{24}$ for the screwing motion of the real rack 4 relative to the wheel 2 occurs parallel with the shaft axis at a radius equal to the shift (§10.41). [This radius is, numerically here, 84.596 mm]. See figure 10.08(c); the cylinder whose radius is the shift is the fixed axode, while a plane (tangential at the ISA and sloping there as shown) is the moving axode; both of these have parallel sets of infinitesimally thin, spline-line teeth (§1.33).

a cord unwinding, *this unwinding occurring not from the key circle but from the relevant core cylinder of radius a_D or a_C* [96] [97]. It is clear from the graphics that this somewhat complicated mechanics being worked out here is correct, that, by considering successive paper-thin slices in this way, we could go on to reproduce the known spatial shapes of the teeth by erecting step by step their twisted continua of parallel planar involutes (§9A.03). *But the mere planar traces of the surfaces enclosing the bodies here are not in themselves bodies, and the mechanics of the real tooth-forming rack (seen as a rigid body traversing its twisted track in the fixed wheel-blank) is better dealt with directly, by means of a clean-cut, once-applied piece of screw theory.* And this latter is precisely what we are in the midst of doing here. The simple fact is that a given real rack, seen as a rigid 3D body, *must be not only rotating about but also sliding along and thus screwing with a finite pitch about its ISA with respect to the wheel-blank, while that ISA is itself migrating to generate the fixed axode which is a cylinder concentric with the shaft axis and the moving axode which is a plane* (§10.38, §10.44, §10.46). There are of course in a gear set not only one but two wheels; and thus there are, for the one single rack, an ISA_{24} and an ISA_{34}. These latter are not the same but different axes. But as all this occurs, as it will, in future, in the reality, and as it did, albeit graphically, here above, the mere planes and ghostly traces, which are translating laterally in directions normal to themselves and moving planarily in directions parallel within themselves, *move with respect to their parent 3D bodies*; and this reduces to absurdity the notion that we are here in this complicated way dealing satisfactorily with the whole of a real rack-cutter and its real wheel-blank. I return now to the central thread of the main argument.

The angles iota

49. Refer to figure 10.09(a). An important aspect of the screw mechanics underlying the continuously instantaneous relative motion among the four bodies in the circumstances here, quite unexposed at the somewhat spurious and thereby inadequate figure 10.08(d), is the question of the angles iota. One of them appears in its true magnitude at figure 10.08(c); the figure comes from WkEx#1, where the two iota are (as we know) equal incidentally (§10.44). We have that $\tan \iota = \tan \lambda \sin \kappa$ (§10.44). The specificity alone of this formula for the angle ι should lead us to suspect that the angle bears some mechanical significance; and it does. Its significance relates, however, not to the generation of teeth around the periphery of some wheel alone by itself, isolated from its mate in some gear cutting machine. It relates instead (a) to the relative locations of the two wheels within the architecture, which determines the single angle κ for the set, and (b) to the choice we make for j, which determines the two angles λ. Notice in figure 10.08(c), with the wheel fixed, that the plane of the moving axode, seen edge-on as a line, has come up parallel with the best drive. This is simply indicative of the fact that the plane normal to the shift line at the ISA, which contains the ISA, is the plane of the said axode of the rack, and this is (as we know) attached to the rack.

§10.50

Attached to the rack as well, however, is another planar axode. In any closed loop there are always two axodes unchangingly attached at every link, because each link is connected by means of a joint to each of its neighbors, and in the current case the two joints are of the same kind and the two neighbors of the rack are the two wheels. The planes of the two axodes intersect in a line, and that line, called for the want of a better name Line X in figure 10.09(a), is parallel (we find in the CAD) with the best drive. This is not surprising because, with the frame fixed and both wheels rotating, the rack is indeed translating in this particular direction (§10.39). We might in due course locate this line algebraically within the fixed architecture of the set, for it might be important (or at least convenient) for us to know as much as we can about this line. In the meantime however we can say this: *the line of intersection of the planes containing the rack-cutting ISAs that are normal to the shifts respectively is parallel with the BDL.* We can easily see now a construction for each of the angles iota: we drop a perpendicular from the centre of the F-circle — see figure 10.08(c) — onto the orthogonal projection of the BDL. We look next along the BDL to see the transversal in the plane of the paper; thus we look parallel with line of intersection X. Refer for further clarification of this to §10.62.

The cylindroid at the meshing

50. Recall the theorem of three axes in kinematics [1] [§13.12]. Pay attention now to another one of the four cyclic arrangements of three bodies taken from among the four bodies 1, 2, 3 and 4 identified here, namely the cyclic arrangement [34 – 42 – 23]. This is the one that excludes body 1. Because I wish to argue from the particular to the general just here, I wish to take for a first example WkEx#1, the example being particular by virtue of its being equiangular. Unremarked at §5A.23, let us belatedly recognize here another special feature of equiangularity. It is that Olivier's line (which is vertical) and the pitch line (which is horizontal) *intersect.* No such intersection occurs in the general arena of polyangularity; but it does occur in the special event of equiangularity; and that is why we need not have been surprised at the end of §10.48. Looking now for the details of the relevant cylindroid [1] [§13.12], and arranging the subscripts cyclically, we have the following: we have the already located ISA_{34} and ISA_{42} and their pitches (§10.46, §10.44); and we have, of course, ISA_{23} and its pitch too; this third of the three resides at the pitch line (§1.09), where p_{23} has been known hitherto simply as p.

[We have (WkEx#1) that
$p_{42} = -19.1511$ mm/rad (§10.44),
$p_{23} = -17.2521$ mm/rad (§5B.09), and
$p_{34} = -11.4907$ mm/rad (§10.46)].

Note incidentally that all three of these are negative and that, although their directions are determined by those of the two shaft axes and the pitch line, their distances apart (along Olivier's line) are determined somewhat otherwise. Refer to figure 10.09(a) where the three axes under consideration are

MATHEMATICS OF THE MACHINING §10.50

mounted at their found locations in the fixed space. The two *rack cutting screw* axes (24 and 34) which are parallel respectively with the *wheel rotation* axes (12 and 13), are each delineated by a star (*), while the pitch line is la-

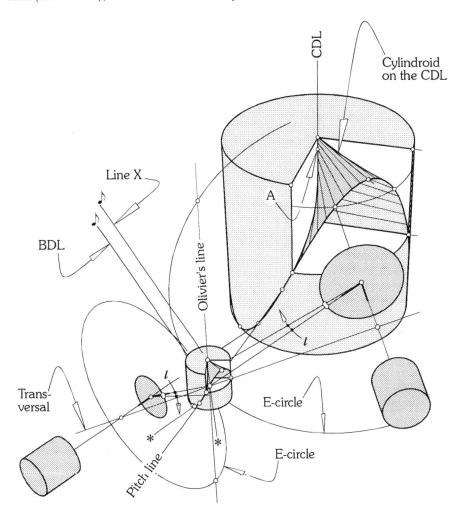

Figure 10.09(a) (§10.50). The well known cylindroid on the CDL, showing also its newly found, shorter, companion cylindroid which is, in this particular case involving equiangularity (WkEx#1), on Olivier's line at the meshing. In the event of plain polyangularity, this other cylindroid, its axis remaining parallel, finds itself at some other location, but always intersecting the pitch line; see figure 10.09(b). These two cylindroids (indeed there are four) exist by virtue of there being not three but four bodies involved in the continuous motion here, frame 1, the two wheels 2 and 3, and the phantom rack 4. For each combination of three of these bodies, the theorem of thee axes applies, and each application engenders a cylindroid [1] [§13.12]. The line of intersection of the planes containing the rack-cutting ISAs that are normal to the shafts respectively (line X) is parallel with the BDL.

beled, as usual, simply with its own name. These three are shown enclosed within the *cylindroid at the meshing*, which is in fact defined by any two of them. The calculations are not given here, but I have put successively into equation (11) at [1] [§15.20] the found locations and pitches of the three pairs of interrelated screws to determine thrice the same length $2B$ of the said cylindroid, thus checking that the numerical results are self consistent, namely that all three screws reside, as they should, upon the same cylindroid at the meshing. The central point, and thus the axial location of the cylindroid, has been determined by means of equation (14) at [1] [§15.20]. On the basis of this graphical work and aided by the said imported algebra, I state this:

> In the event of equiangularity the axis of the cylindroid at the meshing is collinear with Olivier's line.

In figure 10.09 the original cylindroid for the set as a whole is shown in its location coaxial with the CDL for comparison with this one (§1.11). Whereas the length $2B$ of that cylindroid is 104.436 mm approximately, the length $2B$ of the newly found cylindroid is only 21.164 mm approximately. Both cylindroids are presented *square*, with the diameters of their truncating cylinders (which are open to choice) set equal to their respective lengths. They share a common screw, the screw at the pitch line. The existence of this newly found cylindroid should not be surprising. It exists by virtue of the theorem of three axes where there are not, as usual, only three bodies under consideration, but four. The different angular orientation of its principal axes and the different distribution of its pitches are however interesting.

51. If we depart from equiangularity (by putting j either greater or less than k), Olivier's line will no longer cut the pitch line, and conditions will alter. The cylindroid at the meshing will remain as before, superabundantly defined by the three inter-related screws at the meshing, but its central axis, while remaining parallel with Olivier's line, will not necessarily continue to be at E. Whether the central axis will remain at E, or migrate along the pitch line, in one or other (or both) of the two directions from E, as the value of j draws away from that of k, can wait for future investigation. In any event, however, it will continue to cut the pitch line and be vertical, for the simple reasons (a) that the pitch line will always be, by definition, one of its generators, and (b) that the rack axes, always respectively parallel with the wheel axes, will always be horizontal. In the meantime, at least, we can say this:

> Across the range of plain polyangularity, taking all j for a given k within an otherwise fixed architecture, the axis of the cylindroid at the meshing, while remaining always parallel with Olivier's line namely vertical, remains also within the unique vertical plane that cuts the pitch line.

In due course the internal geometry and the general location of this discovered, second cylindroid might be a welcome addition to the armoury of methods at our disposal for finding an even quicker understanding of the

MATHEMATICS OF THE MACHINING §10.52

mechanics of the imaginary rack and its associated real tools (grinding wheels, milling cutters, rotating hobs and the like) for all of the kinds of involute gears that will, in due course, be cut in the tooth-cutting machines.

The other, as yet unmentioned cylindroids

52. We might notice in general that, given the four links of the mechanism here, there are four different combinations of three links. Thus there are four cylindroids all told. The other two — they arise from the combinations [1-2-4] and [1-3-4] of the links — are, in the circumstances here, cylindroids of the fourth kind [1] [§15.55]. The existence of these other cylindroids was touched upon during the argument at §10.41 but not dwelt upon because the cylindroids are degenerate, being infinitely long. The angle ε introduced at §10.41 and shown at figure 10.08(a) is the same as the angle ζ which was mentioned in conjunction with the somewhat difficult matter of *degeneracy and the pitch gradient* at [1] [§15.52] incidentally.

Cutting velocity across the flanks of the rack

53. Figure 10.10 shows a panoramic view of the ISA_{24} in the same radial location upon its fixed cylindrical axode as it was at figures 10.07 and 10.08(c). This is that location where the rack is in contact with the drive flank of the tooth not only at [Qf] but along the whole length of the ruling that contains [Qf]. We know that the flat profile of the rack and the involute-helicoidal, curved flank of the tooth remain in contact along a continuously changing straight line namely the relevant ruling (§10.04), but here under this sub-heading I propose to show by means of the CAD that, given the found ISA and its fixed and moving axodes, the relative motion of the two bodies (the rack and the tooth) is indeed properly represented by the said ISA and its axodes. It will be not only instructive but a further check upon the correctness of the geometry pursued so far.

54. I show first of all (in figure 10.10) that the helitangent at [Qf], found by using the pitch of the screwing at ISA_{24} and the radial distance from the said ISA to the said [Qf] in the reference configuration, falls, as it should, (a) collinear with the same helitangent at [Qf] which originates from the system of helices surrounding the pitch line namely ISA_{32}, and consequently (b) within the plane of adjacency there. This line is labeled in the figure as being not just *the* helitangent at [Qf], of which there are two in the plane of adjacency [1] [§21.22], but the *already known* helitangent at [Qf] which relates directly to the pitch line of the set. That there are two helitangents in the plane of adjacency is a subtle point, and to understand it one needs to study chapter 21 in [1], referring in particular to figure 21.05 where it is shown that, through any point Q in a moving body there is not only the helitangent originating at Q but a single infinity of other helitangents originating at other points in the body. This single infinity is arranged not as the generators of a plane containing Q but as the generators of an elliptical cone whose vertex is at Q, and the reason for this is that the complex of helitan-

393

gent lines in a moving body is not simply a *linear* complex (§2.15), but a *quadratic* complex [1] [§21.22]. The said cone in the case here cuts the plane of adjacency in two lines, (a) the already known, first helitangent mentioned above, and (b) the *unforeseen, second helitangent* along the ruling at [Qf].

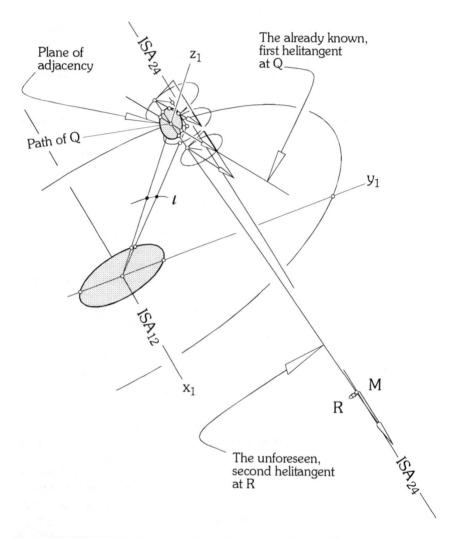

Figure 10.10 (§10.53). Given the unique ISA_{24}, whose location and direction is shown at figure 10.08(c), there are at [Qf] and in the plane of adjacency there, not only one but two helitangents obtaining. This figure shows where they are and why it is that, along the whole of that ruling upon a tooth which is in contact along that ruling with the rack, the sliding velocity at all points is in the plane of adjacency. This could not be otherwise of course, because, otherwise, the rack (at all points except one) would be either drawing away or digging in.

55. To confirm that this second helitangent is, as claimed, along the relevant ruling, we project the ruling itself (as shown) and construct the shortest distance, the common perpendicular (R–M), between it and the ISA$_{24}$ which occurs as shown at the lower RHS of the figure. It is a simple matter now to check with the CAD that this projected ruling is indeed the helitangent at R associated with the ISA$_{24}$ and the unique helix of the relevant pitch surrounding it. We can next check (and we do, as can be seen) that the linear velocities at all points along this helitangent line, which are by definition coplanar [1] [§21.16], are coplanar within the plane of adjacency. This process may be repeated with similar success at neighboring locations of the rack. We can thus show numerically at least if not algebraically that, during the generation of the profile of a tooth by the screw motion of the rack, the rack slides transverse to the instantaneously relevant ruling along the entire length of that ruling while the linear velocities at all points of contact are (as might be expected) confined within the plane of adjacency.

A confirmation by means of inversion

56. The question now arises of whether the above-described screwing of the found rack about its found ISA$_{24}$ will indeed reproduce the earlier-found shapes of the flanks of the teeth. This task needs to be undertaken here lest the overall argument of this book might be seen to be incomplete, open ended. Refer to figure 10.11. Here, to begin, two neighboring rectangular flanks of the rack body (which together form a tooth upon the rack) are drawn at a series of successive locations to straddle the reference location. Without loss of generality, the larger wheel from WkEx#1 is taken for example. Given 40 teeth, the angle subtended by one circular pitch at the axis of the wheel is 360/40 namely 9°, and in the figure the screwing is taking place, for convenience, at the relatively small intervals of 3°. The rack is being not only circumferentially rolled at intervals of 3° (without overlooking the arc-chord adjustment), but also axially slid through successive intervals corresponding to the known value of p_{24}, namely –19.1511 mm/rad (or –1.00275 mm/3°), along the ISA$_{24}$ which is, of course, itself moving relative to the fixed wheel. These successive, finite, small screw displacements, made in both directions away from the reference configuration, are made in those directions consistent with left handed screwing.

57. The straight-line intersections of successive pairs of locations of the flanks of the moving rack in figure 10.11 are giving, of course, only approximations to the actual locations of the rulings upon the flanks of the generated teeth. Although the said lines of intersection lie quite close to the actual generated surfaces of the flanks (unlike flanks of the teeth being generated on both sides of the tooth of the rack), it is only when these intersections are infinitesimally close that they become the rulings themselves. [Checking for accuracy in this respect, I found that, whereas the exact CP = $(2\pi e)/N$ = 14.023 mm, the arcuate tooth width in figure 10.11 measures 7.062 mm, which is somewhat more than half-CP, but acceptable.] Notice the two differ-

ent orthogonal axis systems, $x_1 y_1 z_1$ for the wheel blank, and $x_2 y_2 z_2$ for the rack body. These are simply related, of course, via the angles κ and λ. One can

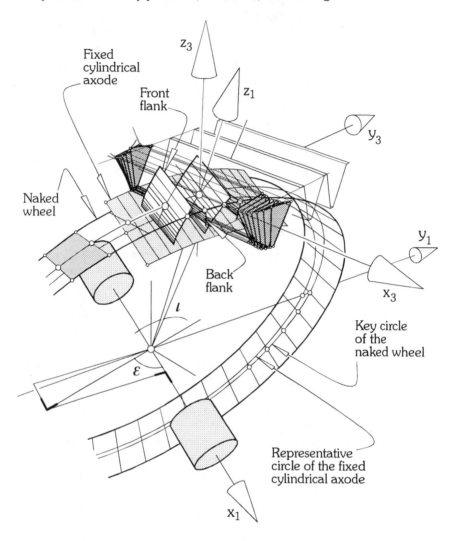

Figure 10.11 (§10.57). Continuous screw motion of the rack. Here the wheel has been held fixed while the rack has been screwed about its ISA $_{24}$, which axis has itself been caused to migrate around the fixed cylindrical axode. The moving axode (a plane not shown) has been rolled around the fixed axode (the cylinder grey-shaded) while at the same time the said plane has been caused to slide axially. Recall that axodes have their infinitesimally thin, spline line teeth, and know that these teeth are here all parallel. The flank shown at foreground is the front flank of the representative tooth whose correct shape, overall, has been reconstituted here. Repeated graphical construction has been used to simulate the simultaneous generation of both flanks of all teeth by the rack but, because of the difficulties involved in drawing this, the picture here is not entirely clear.

see in the figure some completed detail of the generated shape of the representative tooth, conveniently truncated by the cheeks of the blank for the sake of clarity, which satisfactorily mimics the original synthesized tooth of WkEx#1 at chapter 5B. Both flanks of the teeth are generated at the same time by the one rack, and there are as well, for the sake of clarity in the figure (which clarity is somewhat obscure it must be admitted), some neighboring flanks of the neighboring teeth ghosted-in. The result here is hardly amazing, but it is a comforting confirmation that our long circumnavigation of this tooth from its first synthesis from first principles at chapter 5B, through its various practical design troubles, and back to this retrospective end-checking here, has been achieved without apparent error.

Application here of the theorems at chapter 9B

58. We have come this far in this chapter without recourse to the theorems of chapter 9B; and for my part this has been deliberate. On looking back, one sees more clearly now that the elusively moving planes of chapter 9B are not bodies, that the multi-valued and movable screw axes there are not the *unique* screw axes we customarily see in the less forgiving context of the spatial kinematics of concrete mechanism. The material of chapter 9B is dealing with the pure geometry of the involute helicoid when seen not as the outside surface of some real body with recognizable markings upon it, but as a ghostly (albeit definable) surface that can screw about its own axis without our seeing the motion (§9B.09), that can be generated by an unmarked plane that can translate in all directions and rotate within itself in a similar manner, and that, because of these phenomena, is tantalizingly mercurial. The material of chapter 9B is not spurious but, given the overconstrained nature of our real gear boxes, and given the degrees of freedom uncontrolled in the pure geometry, it is difficult to apply in actual design practice. Figure 10.12, although unwieldy, does however straddle the gap between (a) the direct instantaneous screw kinematics of the fully defined rigid bodies of the gear set as pursued so far in this chapter, and (b) the equally sound but pure, invisible, ghostly geometry of the involute helicoid as collected into the theorems of chapter 9B. The two approaches, (a) and (b), are awkward to reconcile but, unless one or other (or both) of them is false, reconciliation must be possible. What follows is my somewhat convoluted job of achieving that.

59. Figure 10.12 is an expanded, parallel version of the perspective figure 10.11. In figure 10.12, where the rack remains at its original reference location, I have located by construction the two points marked 𝓕 and 𝓑 where the axis of the wheel cuts the planes of two neighboring flanks of the rack; for convenience and without loss of generality the chosen pair of neighboring flanks of the rack constitute here a valley in the rack, not a tooth; and the said points are in the foreground and the background of the figure respectively. Due to the relatively large values of the pitches of the flanks of the teeth in this WkEx#1 — the larger is almost one metre per radian — the said points of intersection are relatively far apart. Next by construction I find

§10.59   GENERAL SPATIAL INVOLUTE GEARING

(and not surprisingly) that the angles α between the two planes and the wheel-axis, at 7 and 8 in the figure, are 4.8318° and 19.3130° respectively. These are the same of course as the α_D and α_C at WkEx#1 (§5B.16). Two coaxial circles of radii a_D and a_C are mounted at 7 and 8 respectively and the planes of these circles are accordingly normal to their common axis which is the axis of the wheel. The two planes of the flanks of the rack are next seen

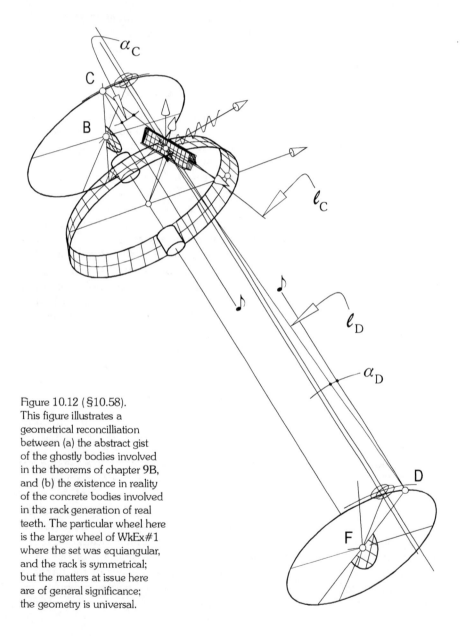

Figure 10.12 (§10.58). This figure illustrates a geometrical reconcilliation between (a) the abstract gist of the ghostly bodies involved in the theorems of chapter 9B, and (b) the existence in reality of the concrete bodies involved in the rack generation of real teeth. The particular wheel here is the larger wheel of WkEx#1 where the set was equiangular, and the rack is symmetrical; but the matters at issue here are of general significance; the geometry is universal.

to cut the planes of the two circles in the lines 𝒥𝒟 and ℬ𝒞 respectively. Next I have mounted, at 𝒟 (for drive) and 𝒞 (for coast), lines inclined at the angles $α_D$ and $α_C$ to the shaft axis. I have then produced these into the zone of the actual tooth at the wheel (figure 10.12) to find that they arrive there from opposite directions exactly collinear with the rulings currently determined by the reference location of the rack (§10.05).

60. Next the angle between the radial lines 𝒥𝒟 and ℬ𝒞 (both of which cut the axis of the wheel perpendicularly) is measured to be, not 40°, which is the angle between the planes of the flanks of the flutes of the rack, but 38.2798°, which is the angle between the *traces* of the those planes that are cut obliquely by the sectioning plane O-y_1z_1, which latter is, of course, the plane of the key circle of the wheel. The planar zigzag trace of the said flutes of the rack is shown at all three of the figures 10.08(b)(c) and (d). We now see that the two lines labeled $ℓ_D$ and $ℓ_C$, in collaboration with their shortest distances (𝒥–𝒟) and (ℬ–𝒞) between them and the wheel-axis, along with their respective pitches p_{FLANK}, are the generating lines respectively of the two involute hyperboloids that form the opposite flanks of the actual tooth. Refer to theorem 1 inset at §9B.02. One thereby sees that this argument is, so far, a satisfactory confirmation that the two approaches under consideration reconcile with one another.

61. We have yet to show however how, given the dimensions of the gear set, we can use the theorems of chapter 9B independently to find the locations and pitches of the two screw axes at the meshing, namely ISA_{24} and ISA_{34}. Refer to theorem 3 at §9B.24, and see the relevant figure 9B.03. Now refer to figure 10.13. This takes for example the finding of the ISA_{24} at WkEx#1. Based on theorem 3 being applied twice, once for the involute helicoid for drive and once for the involute helicoid for coast, the construction at figure 10.13 involves the intersection of two planes. The negative values of the known two pitches p_{FLANK}, [–967.6486 mm/rad for drive and –246.4442 mm/rad for coast], and the negative value of the result [–19.1511 mm/rad], together account for the circumstance that the whole of the graphical action in figure 10.13 takes place below the XY-plane. The two lines each marked by a pair of stars (*—*) correspond with the single line marked similarly at figure 9B.03; both of them within the XY-plane, they are the lines of zero pitch p_s for the two mutually displaced involute helicoids. The mutual displacement is angular about the Z-axis, and this is evidenced by the angles shown, 14.6636° clockwise and 23.6162° anti-clockwise away from O-Y about O-Z. These are precisely the angles at the oblique-cut *planar trace* of the rack that is shown twice (see the jagged line) at figures 10.08(b) and 10.08(c). The angles — they are indeed $γ_C$ and $γ_D$ — add to 38.2798°, which is of course not 40° the angle between the *planes* of the flanks but the angle between the *traces* of the flanks to be seen at the jagged line.

62. In clarification of the matters made evident at §10.49, we can now go on to say the following. The said two planes at figure 10.13 intersect in a

§10.62 GENERAL SPATIAL INVOLUTE GEARING

line, and this line (which runs oblique across the Cartesian space) contains all those points that represent the locations and pitches of the ISA_{24} that may alone be used to generate both of the involute helicoids that are the shapes of the two different flanks of the teeth. At *only one of these points*, however, will the ISA_{24} have the capacity, not only to generate both involute helicoids, *but also to generate them properly spaced,* that is around the periphery of the key circle at exactly half the CP apart. This is where the angle ι (iota) comes into play. See figure 10.08(c).

Iota is the angle at the wheel-axis between (a) the radius of the wheel at [Qf], and (b) the shift line (which line is also radial). Along with the magnitude (namely the length) of the shift, angle ι_2 locates the ISA_{24} within the fixed space of the architecture. Independently of the method of screws, and independently of the current figure, iota may be determined as follows. Without figure,

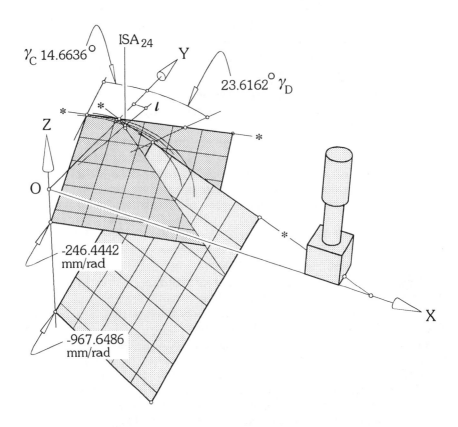

Figure 10.13 (§10.61). Using the theorems of chapter 9B to determine graphically the values at ISA_{24} in WkEx#1. We see here the solution that (a) the angle iota is 5.0171°, (b) the shift is 84.5963 mm, and (c) the pitch p_{24} is minus 19.1511 mm/rad.

let us note (a) that the shift line must be a member of the planar pencil of lines made up of all of the radii of the key circle, and (b) that the same shift line must also be a member of the planar pencil of lines that, emanating from the same key-circle centre, is made up of all those lines perpendicular to the BDL. Clearly the shift line is the line at the intersection of these two planar pencils, namely the unique line that is a member of both pencils. The magnitude of the shift, namely the length of the intercept shown in the figure, may be found as follows. Locate that unique plane containing the BDL that is normal to the found shift line, thus finding at the same time the length of the said intercept.

[In WkEx#1, ι_2 and its shift are 5.0171° and 84.5963 mm respectively.] These have next been marked off, in figure 10.13, from O-Y in the counter-clockwise direction about O-Z, and radially outwards respectively. Next to be noticed — and this requires close attention at the detail — is that the two constructions coincide. Iota has located the unique ISA_{24} at radius 84.5963 mm from the central axis, where the pitch of the screwing (measured in the Z-direction) is –19.1511 mm/rad. Taken here to six significant figures, this result is precisely the same as the result discovered at §10.42.

Speeds and feeds at the machine tool

63. Having done that I wish to say this. To set against the elegance of the theorems at chapter 9B we have the ethereal nature of the ghostly surfaces and the multiplicity of the many axes involved, and this combination of qualities in the theorems appears not to lend itself for easy application. Accordingly I take the view that the elements of screw theory that dominate at §10.42 provide more useful tools for the finding quickly of what we need to know in the present circumstance. We need to know, in machining practice, not so much the nominal values of the two pitches of the flanks of the teeth (which are often quite large), nor the angle iota (which is a function not of an isolated wheel being cut but of the layout of the gear box), but the unique location and the unique pitch (which is often much smaller) of the screwing about a fixed axis required of the imaginary rack in direct relation to the rotating wheel whose teeth are being generated by the cutting action. Independently of the type of machine tool employed, however, this next needs to be said.

64. There are practical arrangements that need to be made at the real mechanical interface between the chosen gear cutting machine (of whatever type) and its output namely the generated involute teeth of the finished wheel. These arrangements will depend of course upon the particular machine, but they need not daunt the machinist; simple combinations of the continuous rotation of axi-symmetric components and the straight-line sliding of otherwise bulky parts is at the root of much screw motion; and here in this benign arena of general spatial involute gearing such sets of simple combinations are customarily the case. Three, four, and five-axis milling and

hobbing machines are in common use in the workshops of today, and these and/or their derivatives will be adequate for the gear cuttings of which I speak [11]. Other than these no special machines are required.

65. I intend to continue here, not with the details of these arrangements (for they reside in the longer and deeper experience of machine-tool designers more knowledgeable than I), but to make some general comments. We are dealing here with the machine-generation of spatial involute teeth. We are not dealing with the machining of the non-involute, hypoid teeth (§2.25), which machining is quite different. Study carefully the difference between the two graphical treatments of the cutting by generation of spatial involute teeth illustrated at the companion figures 10.11(a) and (b). They are simply inversions of one another. Whereas the arrangement at figure 10.11(a), designed for the usual conventional kinematic argument where one of the two relatively moving bodies is held to be fixed while the other moves, the arrangement at figure 10.11(b), where the screw axis remains stationary within a fixed external frame while *both* of the bodies move, is the arrangement more suited to understanding the most likely action within machine tools. In machine tools, the wheel to be cut will most likely be driven to rotate about its own axis of symmetry (its spindle) fixed within the frame of the machine. At the same time the imaginary rack (which is itself a cutting mechanism variously constructed) will be caused to screw about its said screw axis (fixed in space) relative to the rotating wheel. There will be a wholesale rate of rotation of the rack mechanism about the said screw axis along with a wholesale rate of sliding along it, and the ratio of the two will be, as we have shown, p_{24} or p_{34} (mm/rad), according to which of the two wheels is being cut. This said screw axis will of course be parallel with the spindle of the wheel. The ratio of the rate of rotation of the rack (clockwise or otherwise) and the rate of rotation of the wheel (counterclockwise or otherwise) will need to be set. This ratio will come from the simple equations of this chapter. Please notice also the somewhat strange fact already mentioned that, once the necessary data for the cutting are gathered in, the angles iota are irrelevant. Finally remain aware that, while the required exactitude of the shapes of the generated teeth will not be difficult to achieve by the mentioned conventional methods, unnecessary exactitudes in the assembly of the finished wheels in their box can, as we know, be ignored.

Overall strategy of the found mathematics

66. Looked at from the design point of view (and this is the view we have largely adopted, chapter by chapter, throughout this book), we begin our analysis with the given overall architecture, we next proceed step by step through the forrest of decisions that need to be made, and we finish finally with the shapes of the meshing teeth and the found mechanics of the rack. It must be said of course that there are other analyses of other kinds of gears that begin, by and large, with the required nature of the meshing, and which end in due course with an expected architecture. Accordingly, however, we

can say here the following. Given C, Σ, R and k, we can (and we do) locate the pitch line, its pitch p, and the point E upon the equilateral transversal, and thus determine the two E-radii e. Choosing j, we locate F, and thus the two F-radii f, the so-called key radii of the wheels. Knowing the location and pitch of the pitch line and the position of [Qf] (which is at F), we determine the direction of the sliding velocity at [Qf]. There are three lines mutually perpendicular at F: the equilateral transversal, the best drive line (BDL), and the so-called helitangent. It is along this helitangent that the aforesaid sliding velocity lies. Next comes (on either side of the BDL and within the found polar plane at F) the two angles of obliquity δ which, our having been helped in the making of our choices for these by the ellipses of obliquity, are not necessarily equal. Swinging the two thus established paths of the two points Q (for drive and for coast) about the two shaft axes, we locate the four (the two pairs) of the base hyperboloids whose throat radii a and their angles of inclination α thus become established. Thus we get the four (a, α) which, collectively at this early stage, characterize the set. Swinging the vector representing sliding velocity at [Qf] respectively about the two wheel axes, we erect next the doubly-departed sheets, namely that pair of circular hyperboloids that share the helitangent as a common generator. The so-called naked wheels are axially truncated portions (narrow slices taken normal to the axes) of these departed sheets. The naked wheels are of arbitrarily chosen widths, but they do straddle and thus contain respectively the two key circles. The throat radii of the departed sheets (which sheets are ordinary circular hyperboloids and are in turn the naked wheels), are designated t, while the half-angles at the apices of the relevant asymptotic cones are designated τ [90]. Until now in this long paragraph I have been extracting material from the earlier chapters where the basic material will by now have been understood, but next I move on to later chapters and especially to this chapter 10. Two other angles determined at this stage (one of each for each wheel) are (a) the so-called swivel angle λ of the generators of the naked wheel, and (b) the half angle κ at the apex of the so-called Wildhaber cone. Earlier derived equations for these in terms of the fixed variables C, Σ and R of the architecture may be found at §10.08, and later, a quite elegant relationship between the two may be found at appendix C at 6. It reads: $\cos \kappa \cos \lambda = \cos \tau$. Having got thus far, we allocate the numbers of teeth on the wheels, whose ratio must of course accord exactly with the originally given ratio k. Next we proceed to erect the profiles (the shapes of the flanks) of the teeth, there being in general four different profiles, there being the fronts and the backs of the two sets of teeth. Known to be (as they are inescapably) involute helicoids with their axes collinear with the axes of the relevant wheels, the pitches p_{FLANK} of these is given by $p_{FLANK} = a \cot \alpha$. Next we find that the slopes, in the axial plane and at [Qx], of the flanks of the cut teeth need to be studied, and for the relevant angle here we choose the symbol γ. The angles γ, although closely related to the angles δ, are not the same as the angles δ. Indeed they are (at §10.23) given by: $\tan \gamma = \tan \delta \sec \lambda \cos \kappa \pm \tan \lambda \sin \kappa$. We next hypothesize, on the basis of the geometry clearly seen intuitively

§10.67   GENERAL SPATIAL INVOLUTE GEARING

and produced for visualization by the computer-driven graphics, that an imaginary straight-sided rack, built directly upon the two angles δ, will mesh correctly and continuously with the teeth of the wheel when that rack is (a) tilted away from the axis of the wheel through the angle κ, and (b) rotated about a certain related line — see figure 10.08(a) — through the angle λ. This hypothesis is subsequently proven; see the related figures 10.10 and 10.11 and the related argument from §10.53 to §10.57 inclusive. The proven hypothesis lets me say, at this next stage, that the general predictions made at the beginning of the book were thus shown to have been sound. It is true, in other words, and generally speaking (excepting for exoticism), that a single straight-sided rack, suitably screwing about its correct axis somewhere parallel with the axis of the wheel, and with its correct pitch with respect to that wheel, will generate in a steady manner the discovered teeth. We note here of course that the said imaginary rack is itself a geometrical abstraction standing in for the more complicated movements of a hobbing tool or of a suitably shaped milling cutter suitably driven in a multi-axis gear-cutting machine. It remains now in this brief outline of the mathematics to summarize the algebra of the so-called shift of the ISA for the rack and the wheel, and the shift angle ι (iota). There is seen to be an angle ε involved (§10.42), which is of passing importance, and a simple formula for this ε is as follows: $\cos \varepsilon = \sin \lambda \cos \kappa$. Now the shift, which is the radial distance from the centre of the wheel to the ISA for the rack and the wheel, is given (§10.42) by: shift $= f \cos \lambda \sin \varepsilon$. This means that the radius of the fixed axode (which, indeed, is the shift) is somewhat less than the key circle radius. In planar gearing, of course, the radius of the fixed axode is indeed the key circle radius exactly, and this is the so called 'pitch radius' (in planar parlance) of the wheel. We showed also at §10.42 that there is an angle ι (iota) involved; this relates, not to the rack and its wheel when together isolated in the tooth-cutting machine, but to the rack and the wheel when together imagined in the assembled gear box. Mainly for the record (not for the sake of ongoing argument), I wish to mention here that a simple formula for ι (given at §10.44 and proven at appendix A at 10) may be written. It reads: $\tan \iota = \tan \lambda \sin \kappa$. This last mentioned equation accentuates not only the overall ubiquity of the angles κ and λ in the mathematics of the machining, but also the crucial significance of them and their numerical values in the practical business of setting up for operation the gear cutting machines. See some related material at §12.14 *et seq.*.

Appendix A at chapter 10

67. This appendix deals with two matters raised as separate but related issues at §10.42 and §10.44. The matters are (a) that $\cos \varepsilon = \sin \lambda \cos \kappa$, and (b) that $\tan \iota = \tan \lambda \sin \kappa$. These two algebraic statements may be proven by using the same figure. Refer to figure 10.14. This figure, paying attention to the directions of lines at the expense of their altered locations, is more like a velocity polygon [1] than a configuration diagram. It has moreover an enlarged portion displayed separately at inset that needs

MATHEMATICS OF THE MACHINING §10.67

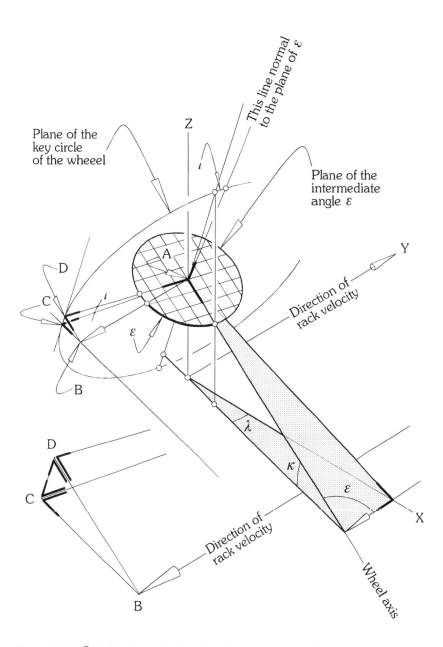

Figure 10.14 (§10.67). This is the figure for the trigonometrical derivations set down at appendix A at 10. The figure is, substantially, a velocity polygon for the mechanism made up of the wheel and the rack at figure 10.08(a). For explanatory information about the general idea of velocity polygn, refer to chapter 12 in [1].

405

to be noticed. Dealing only with the angles, however, and trigonometrically, we can proceed legitimately as follows.

Referring to the two shaded triangles at the RHS of the figure, which are joined by a common side, and, taking the side in the direction of the wheel axis as being of length unity, we see that the following may be written.

The joining side = $\cos \varepsilon$, and the same joining side = $\cos \kappa \sin \lambda$; so it follows that

$\cos \varepsilon = \sin \lambda \cos \kappa$,

which is what was required for us to prove at (a) §10.42. ... (1)

Referring next to the tetrahedron ABCD at the LHS of the figure and its expanded, right-angled triangular face BCD at inset, note the fully illustrated locations of all of the right angles. As well as those shown in the main body of the figure, namely at $\angle ACB$ and $\angle ADB$, there are right angles also at $\angle BDC$ and $\angle ACD$, there being four all told. Although the correct locations of these right angles is a confusing matter to grasp, they are of course vital for the veracity of the algebra. The line CD, incidentally, is parallel not with the line normal to the plane of ε but with the most steeply sloping diameter of the key circle of the wheel.

Taking the side (A—B) of the tetrahedron as being of length unity, there are two ways for writing the short length (C—D). We find, first via (A—D), and next via (A—C),

that the length (C—D) = $\cos \varepsilon \tan \kappa$, and
that the same length (C—D) = $\cos \lambda \tan \iota$.

It follows that

$\cos \lambda \tan \iota = \cos \varepsilon \tan \kappa$;

but, from (1) above,

$\cos \varepsilon = \sin \lambda \cos \kappa$,

so

$\cos \lambda \tan \iota = \sin \lambda \cos \kappa \tan \kappa$,

which reduces to

$\tan \iota = \tan \lambda \sin \kappa$,

which is what was required of us to prove at (b) §10.44 ... (2)

CHAPTER ELEVEN
ASPECTS OF THE PHYSICAL REALITY

Contents

01. This chapter contains a miscellany of matters that are nevertheless related. First I present a few ideas about the inevitable question of simple interference among the teeth of wheels and ways for avoiding it. Associated with this *simple* interference, which always pertains until, as an ordinary matter of course, we avoid it by truncation, there is however a more serious phenomenon. This I have called, for the want of a better name, either the *excessive pointedness of teeth*, or the *mutual decapitation of flanks*, which, when it occurs in design, does so as a kind of accident. It occurs, by definition, when the otherwise full, unity-plus of one or other of the contacts ratio, is reduced to below unity. Such an event, of course, unless especially looked for for some particular reason, is catastrophic. Next there is a kind of *envelope*, ill defined as yet, within which we are safe against this dangerous phenomenon, but beyond which we are not. I try to come to terms with this. Finally I describe an apparatus [54], free of these faults and faithfully based upon the results of WkEx#1, which confirms that the velocity ratio k of a properly designed, spatial involute gear set is insensitive to all small errors of assembly; we can see with the help of this apparatus that if the errors are not small, but large, the troubles outlined in this chapter will obtrude.

Interference at teeth and among teeth

02. In the area of tooth generation, we have seen that, while the parallel spline-line teeth of the planar (moving) axode of the rack are parallel with the parallel spline-line teeth of the cylindrical (fixed) axode of the wheel, and while both sets of them are thus parallel with the axis of the wheel, *the central plane of the rack (which is not parallel with the spline line teeth) is not parallel with the axis of the wheel.* The rack is rigidly attached, obliquely, upon the plane of the rolling axode, a combination of the angles λ and κ being involved. Refer for this, not to figure 10.08(b) where the shaded surfaces there are not the axodes, but to figure 10.08(c) where the axodes are shown, as they need to be in that view, end-on. With the wheel fixed, let the

rolling at ISA$_{24}$ be clockwise as seen in the figure. The ISA$_{24}$ will, accordingly, and as the rolling occurs, be moving parallel with itself clockwise around the cylindrical fixed axode. Because the screwing is left handed — p_{24} is negative — the accompanying sliding along ISA$_{24}$ will be, accordingly, towards the foreground of the figure. Notice now the points upon the surface of the rack that are 'forward moving' (towards the right) and those that are 'backward moving' (towards the left), and think about the implications of this for rack interference with the roots of the teeth. The forward moving points are above the rolling, axially sliding plane of the planar axode while the backward moving points are below it. The kinematic implications for the properly generated portions of the flanks of the teeth above the plane would seem to be satisfactory but, for those below the plane (towards the rear of the figure), there is interference brewing probably.

03. Look back at the sketchy figure of the pinion at 5B.16(d) and see in advance the more carefully drawn figure 11.01(a). Based upon the same numerical data — that for WkEx#2 beginning at §5B.45 — figure 11.01(a) shows the intersecting working surfaces of the flanks of all nine teeth of the said pinion. The teeth are of course identical. All flanks are wholly convex (§3.18, §3.24). Except at the intersections of the surfaces of their flanks along the apparently twisted curves of their crests and roots (§11.07), the teeth are *untruncated*. The diameters of the two different core cylinders are very close to one another incidentally, the difference being within the thickness of the thicker lines, and the drawing cannot cope with the nicety. The drawing shows only the greater diameter of the two; and so, when I say that the whole of the active profiles of both flanks from the core helices outwards are shown, I speak to this extent approximately. The angular velocity ratio of the set (9/41) is low; and the shaft angle Σ (90°) is large. Conditions are accordingly severe. I wish to leave aside just now the usual difficult questions of manufacture, statics, friction, strength and wear, and look, to begin, at the equally difficult, but most important practical question of the real kinematics. Can this gear set, *geometrically*, be made to work? Basic requirements of the spatial involute theory determine the smooth geometric shapes of the contacting surfaces, and these are not open to question, *but we the designers must determine the truncations*, and the question is as follows. Can we avoid by judicious truncation, in this particular case and in general, unwanted clashings on the one hand and unwanted clearances on the other between the two different pieces of the *enclosed body material* of the wheels? This enclosed body material might be seen, *pro tem,* to fill the enclosed, truncated shapes of the two sets of real teeth; the exact meaning of the term is however defined more carefully at the next paragraph.

04. *Interference in general.* During the preliminary geometric layout of a proposed gear set, the flanks of the teeth are seen to be algebraically defined, as yet unbounded, rigid surfaces which, when separately pivoted to frame at their axes and made to mate with one another, produce the required kinematic action. Although separately considered, the fronts and the

ASPECTS OF THE PHYSICAL REALITY §11.04

backs of the teeth are both involved. While the concave in-sides of the mating surfaces are imagined to be supported by *body material*, the convex out-sides are imagined to be exposed to open space. Upon the intersection of the surfaces on any one wheel, it is only those spaces *filled twice by the body material* that survive to become the interior of the body of the wheel; all parts of the surfaces other than those that are thus seen to enclose body material cease to exist. They disappear. See the self explanatory nature of this at figure 11.01(a). Each of the wheels must accordingly be seen (in the long run) as a tidily bounded rigid body enclosed within its own surface in such a way that all teeth are enclosed within their own flanks, ends, troughs etc., and the wheel itself similarly enclosed. If then pointed teeth, upon mat-

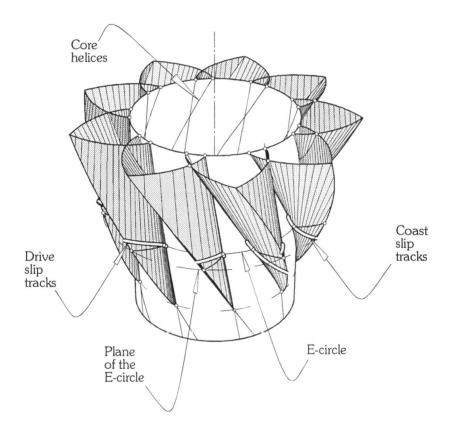

Figure 11.01(a) (§11.03). The untruncated teeth of the pinion at figure 5.16(e). The figure there and the figure here both come from the equiangular WkEx#2 of chapter 5B. Note the slip tracks (front and back), the E-circle, the plane of the E-circle, and the central planar involutes cut by that plane. Drive and coast being different, there are of course two sets of core helices, but the magnitudes of the radii of these (and of the E-radius of the wheel itself) are so close to one another they are indistinguishable here.

ing, fail to connect properly, allowing backlash to occur in either drive or coast, or if any portion of one of the toothed wheels contemporaneously occupies a space already occupied by a portion of the other, trouble occurs. Whether or not these various troubles occur depends upon the chosen basic geometry and upon our policies for the truncation, but it also depends upon the method of manufacture.

05. *Generation by rack.* By way of recapitulation I wish to say again that, whereas exotic polyangular designs may require the imagination of two phantom racks working in turn at different set-up angles κ and λ for the fronts and the backs of the teeth, plain polyangular and equiangular designs require the imagination of only one phantom rack doing its job in one single operation, even though the angles of obliquity at the sides of the flutes of the rack (for drive and for coast) may be different. The latter is the case for the example under consideration here (WkEx#2). Refer to the TNR for numerical values. The set is equiangular, $\kappa = 21.1220°$, $\lambda = 13.6917°$, the angles of obliquity δ for drive and for coast are 5° and 35° respectively, and we need to imagine only one phantom rack. Now in the unlikely event of actual generation of the teeth by means of a real rack reciprocating, we might note to begin that, whereas the working surfaces of the flanks (the involute) are the envelopes of a series of locations of the *plane surfaces of the sides* of the rack, the sides and the bottoms of the troughs between the teeth are generated by the *straight lines of the leading edges* of the rack. It is not so easy to see, however, where the one leaves off and the other begins, namely to envisage the borderline between the two. Will the borderline occur exactly at the core helix upon the base cylinder, and even if so, is the transition smooth or interrupted there? Whether in the event of cutting by means of reciprocating rack, or milling, hobbing, grinding etc., all of which will have their different effects, we are dealing here with the complicated question of *geometric interference and subsequent actual undercut,* and this is the subject matter of the following three paragraphs.

06. *Pointed teeth.* In the sense that the unbounded algebraic surfaces of the two flanks of any one tooth intersect along the *sharp crest* of that tooth, as shown at figure 11.01(a), all teeth are pointed. *But there must be enough of the slip tracks left extant on the thus decapitated surfaces of the flanks to ensure that the contact ratios (for drive and for coast) remain greater than unity.* I digress to define the *inside face* of a wheel as that bounding plane (or concave end) of the wheel which, normal to (or coaxial with) the axis of the wheel, faces towards the CDL. I define the *outside face* of a wheel correspondingly. Let us accordingly speak of the *inside end* and the *outside end* of a tooth. If, at the inside end of a tooth, the intersecting flanks *mutually decapitate one another*, thus (by definition) chopping the slip tracks off in such a way that the contact ratio is reduced below unity, we have *troublesome pointedness*. In practice the said sharp crests of the teeth are often intentionally blunted by conical (or other) subsequent truncation. It is clear in any event that at the inside face of a wheel the flanks must enclose within the

body material of the wheel all necessary parts of the relevant slip tracks. We see in particular that the inside ends of the teeth must not be cut too deep in the axial direction.

07. *The essential planarity of the sharp nose.* I wish to comment here incidentally that the curve of intersection of the unbounded flanks at the crest of a tooth (this curve constituting the sharp nose of the tooth) appears, surprisingly, whether the teeth be symmetrical as in the straight Beveloids, or asymmetrical as they are here, not to be *twisted* (as we might expect), *but to be planar* (§11.03). This intelligence, which has accidentally revealed itself as a product of the computer-driven graphics here, betrays no doubt an important theorem about the intersection of involute helicoids that can be proved somehow by formal argument. By way of enlargement next upon this, I report that the curve of intersection of the unbounded flanks between the roots of teeth (this curve constituting the sharp V-shaped crotch between teeth) appears, surprisingly, whether the teeth be symmetrical as in the straight Beveloids, or asymmetrical as they are here, not to be twisted either, but to be planar. *Indeed the nature of the geometry appears to be such that the curve at the crotch is precisely the same as the curve at the nose, and the endless repetition of the geometry as we go outwards and endlessly in the axial direction of the wheel is therefore intriguing.*

08. *Occluded troughs.* Whereas *troublesome pointedness* at the inside ends of the teeth can occur due to lack of attention being paid to axial truncation of the wheel itself there, *troublesome occlusion* can occur at the outside ends of the teeth due to similar lack of attention to detail there. The two flanks facing one another across the trough between a pair of neighboring teeth can cause at the outside ends of those two teeth the trouble of which I speak. See figure 11.01(a). As we go along any one trough towards its narrow end (at the outside ends of the teeth), intersection of the facing flanks becomes in due course inevitable. This (as before) excludes doubly filled body material from consideration (§11.04), and it may reduce below unity the contact ratio for any relevant pair of slip tracks operating there (§11.10). It is clear that the outside face of a wheel must enclose within the body material of the wheel all necessary parts of the slip tracks. The outside face of the wheel must also be cut appropriately.

09. *Undercut.* It seems to me that this is a separate, subsequently occurring phenomenon. Whereas the fundamental geometrical questions discussed above must be decided during the initial processes of design, undercut occurs later, during the actual cutting. All teeth are in due course cut, not by a phantom rack, but by a real one (or its equivalent), and the extent to which the sharp crests and troughs of an unblunted real rack are smoothed or actually cut off, determines whether or not *undercut* occurs during the generating process. The cutting process wipes out the question of the double filling of space by body material (§11.04), but it might, when it occurs, cause interruption to the smooth kinematic action for the same fundamental reason. *There*

must be enough of the slip tracks left extant on the truncated (and subsequently undercut) surfaces of the flanks to ensure that the contact ratio for both pairs of slip tracks (for drive and for coast) remains greater than unity. All three of these effects — (a) *troublesome pointedness at the inside ends of the teeth,* which comes from wrong or impossible geometry, (b) *troublesome occlusion at the outside ends of the troughs,* which stems from the same source, and (c) *undercut,* which occurs later, by virtue of a poorly chosen generating process — play their interrelated roles together in the complex business of interference.

Kinetostatic evaluation of the result at WkEx#2

10. I am referring here to the gear set as so far synthesized at WkEx#2 and illustrated at the series of figures 5B.16. The worked-out pinion of this set has appeared already at figure 11.01(a), but now it appears again at figure 11.01(b). Leaving questions of strength aside, I reckon the set is acceptable from the kinematical and statical points of view. Studies at chapter 5B have checked that the two paths of the point of contact (for drive and for coast) have contacts ratio well in excess of unity (§5B.34); and I see no difficulties with problems of friction. Refer to my various remarks about wedge-film lubrication at §5B.24 and elsewhere. Figure 11.01(b) shows the beginnings of the likely shapes of the teeth of the real pinion before final cropping of the

Figure 11.01(b) (§11.10). The pinion of the previous figure truncated more severely than is probably necessary between the teeth and less than is probably possible at the crests (and at the heels) of the teeth. This is an exercise in kinematics; cosmetic considerations and considerations of strength are not yet at issue here.

sharp edges and adjustments to the conicality of its inside and outside cheeks which might yet be altered to take account of strength and cosmetic considerations. Its capacity in collaboration with its mating wheel to adjust for mislocated axes is of course inherent in the design, and this needs no further mention here.

11. Figure 11.02(a) is interesting. The wheel and pinion (and the rack) of the equiangular, square set at WkEx#2 are here untruncated, with raw-cut teeth that are still sharp-nosed. They are standing independently of one another upon the imaginary rack, which, having generated them, is common to them both. In the absence of mass and the presence of friction, they might just stand there, but in the absence of friction, the presence of mass and the force of gravity, they might lean to new locations, or simply fall over. For the sake of the argument here, however, please imagine the massless scenario,

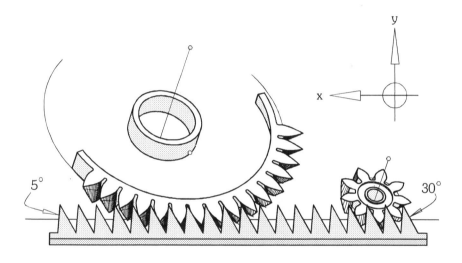

Figure 11.02(a) (§11.11). The wheel and the pinion of the equiangular WkEx#2. This is a parallel view looking along the flutes of the relevant rack and thus orthogonally against a plane cut normal to the flutes of the rack. Notice the rack flank angles 5° and 30° which replicate of course the angles of obliquity δ that were chosen. The roots and the crests of the rack and the crests of the teeth on the wheels are left untruncated here; the mechanics of undercut might thus be better studied. Both wheels are meshing here with the same rack (which is not surprising), and, because the set at WkEx#2 is equiangular, not only are the angles κ equal, but the angles λ are equal too. To be in mesh, both wheels are rotated from the plane of the paper, firstly about the x-direction right handedly through angle κ [24.1304°], and secondly about the y-direction right handedly through angle λ [17.5484°]. Accordingly the wheels are resting here with the planes of their E-circles and thus their mechanical axes parallel. This equality of orientation against the rack will be seen to be a defining characteristic of equiangularity.

§11.11 GENERAL SPATIAL INVOLUTE GEARING

and that the wheels are in their reference locations, with their slopes determined by κ and λ intact, and with their Q at [Qf]. Refer to §7.15 for some detailed remarks about the plane of adjacency and perceive their relevance here. The plane of adjacency between the wheel and the rack is precisely the

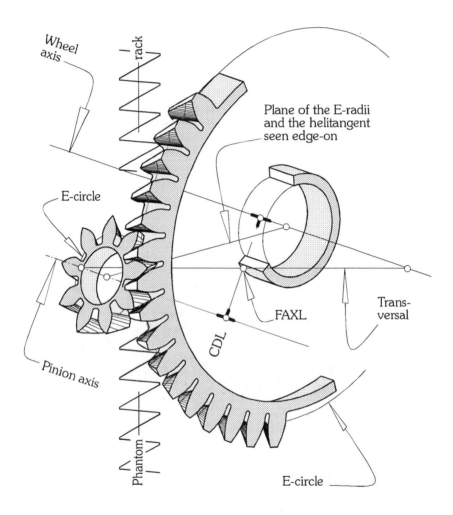

Figure 11.02(b) (§11.12). An orthogonal view taken along the helitangent at [Qf] of the wheel and its pinion at WkEx#2, in mesh. The set is a square set, and equiangular. The transversal is in the plane of the paper, and the FAXL is normal to the plane of the paper. The shaft axes appear parallel in this view because we are looking along the helitangent which is, due to the equiangularity, here parallelel with the FAXL The teeth of the wheel (but not yet those of the pinion) have been truncated to avoid the otherwise obvious interference. We are looking end-on at the phantom rack. All four flanks are divided here @ five rulings per CP. Kinetostatically, the set appearsto be satisfactory.

plane of the flank of the rack itself. Both the straight line of contact between the rack and the tooth at [Qf] and the helitangent there (which is a different line) reside in the plane of adjacency. The same conditions would apply, moreover, at other flanks of the tooth or teeth were the point of contact Q not at [Qf], that is, in other words, if the wheel were not at its reference location. Recall my mention of pronged teeth at §5B.53. But I wish to point out, here and above all, that, because the pair of angles κ and λ is the same for both wheels, the wheels stand (in their reference locations in so far as these two slopes are concerned) equally inclined to the rack. *This stance of the wheels is a characteristic of equiangularity: the wheels of a polyangular set could not be put to sit in this particular way, neatly parallel with one another.*

12. Refer now to figure 11.02(b). This is an orthogonal view of the equiangular, square set resulting from WkEx#2. The necessary choices having been made by the author within the attendant computer-driven graphics, the teeth of the wheel have been truncated without serious damage to the contacts ratio. Both of these remain at well above unity (§11.06), and the set as it stands appears to be a going proposition.

13. Finally, though, I wish to make some remarks about the statics. Refer to §5B.17 for the forces **F** at the two Q, and to §6.41 for the counterpart in statics of that kinematical law, *the law of the speed ratio* (§6.41). It seems to me (in the absence of friction) that, for a given, unit set of opposing torques at the shafts, the smaller force **F** at the relevant Q might be chosen to indicate, by definition, that that chosen set of directions of the opposing torques is indicating *drive*, while the greater force **F** at the relevant Q might indicate by definition that the opposite set of directions of the opposing torques is indicating *coast*. Using these criteria we find, in this WkEx#2, that the ratio of the two **F**, extracting an appropriate pair of values for *a* cos α from the TNR, is $\mathbf{F}_{COAST}/\mathbf{F}_{DRIVE} = [64.608359]/[63.293066] = 1.0208$, which is barely greater than unity. We have at all points of contact, of course, due to the teeth being involute, kinematic correctitude; there is, in the normal course of events, no subsequent adjustments to be made after a first and final cutting of the teeth to compensate for kinematical problems stemming from wrong geometry and/or overconstraint. It follows that in properly cut involute gears such geometrically-based vicissitudes do not obtrude upon the question of which is which, drive or coast. Given friction we might say however that the proportion of the power at input which escapes dissipation due to friction at the meshing (namely the efficiency of the gear set) should be used to determine which is the drive and which the coast, drive by definition being the more *efficient* of the two. But knowing that the sliding velocities at the points of contact and the friction forces in the directions of those velocities — both of which are changing across time with the changing configuration of the mechanism (§3.50, §4.59-64) — are both involved in estimations of power loss at the meshing, I forebear to comment further here about this complicated question: *which is which between drive and coast?*

Commenting more widely now but inconclusively

14. For the others of the worked examples, similar truncations have been mooted or actually made, the author having checked the feasibility of the truncations on all or most of those occasions. But troubles with mutual decapitation at the flanks of teeth (soon to be dissected more thoroughly at §11.25) have occurred, and my remarks about statics, rubbing velocities, friction and the likelihood of irreversibility continue to apply. I am acutely aware — see for example the late-occurring accident at the offset worm (the generic spiroid) at WkEx#6 (§7.19) where the oops took place at figure 7.08(b) — that there is, at present, no cast-iron security against the likelihood of sudden, catastrophic trouble. More needs to be known. One asks, for example, what causes a blunt-bottomed slip track that goes wide across a tooth, thus causing likely trouble with the mentioned mutual decapitation? This blunt-bottomness is caused, of course, by large angles α! If movement towards the polyangular will make the two alphas of a set more equal, as I think that it often might, then might this not be a good thing? We need to look more deeply into matters such as these. I currently reckon, also, that a full discussion of the actual *avoidance of interference* on the one hand and *irreversibility* on the other is beyond the scope of this book; but please find what you can of this big subject elsewhere. In any event I can always finish here, in due course, by saying that we can always set up the gear-cutting machine (of whatever kind) by knowing the angles κ and λ for the relative orientation of the rotational axes and the required ratio cos λ for the peripheral speeds of the wheel and the rack. Looking from the pragmatic point of view, this is, perhaps, all the machinist needs to know. But the κ and the λ come directly from the a and the α and thus directly from the core helix. Thus the slip tracks (of which there is an infinity) are also determined early. When contact is made between the mating wheels (and not before) the *specific* slip track is determined. Ultimately I would wish succinctly somehow to trace this whole investigation full circle, showing how we can get, *via* the phantom rack and the two angles κ and λ, back to where we started from; but this brings us, collectively, back to the *Manual* of course (see Preface).

Ideas about the envelope

15. At the risk of creating a grossly enlarged, and thus unacceptable sub heading here, the heading might well have been written as follows: 'Probing from the inside the outside envelope of geometric practicability'; or, better, 'On stepping into the geometrical wilderness once we leave the safe haven of geometric practicability'; or, even better, 'Independently of questions relating to friction (non-back-driving etc.), and independently of physical difficulties that might arise in connection with the machining of teeth, we study here the purely *kinematic* limits to the proper, smooth working of the set'. Were this book to be a study in planar involute gearing only, we would be including in this chapter such matters as the minimum number of teeth permissible upon a wheel, the limits (both high and low) to be set upon the angle of obliquity

(the pressure angle), addenda and dedenda, the adequacy of contact ratio, and other such matters. Here I wish to emphasize in general that, unless the kinematics is initially correct, all other matters (such as lubrication, wear, strength, dynamics etc.) will never be entirely satisfactory. Unquestionably the kinematics is the main matter at issue always, but this might, as we know, either suddenly or gradually, *become geometrically impracticable.*

16. I have tried by making such comments here to emphasize that the wilderness of uncertainty outside the restricted envelope of geometric practicability is vast. What I have said applies, of course, to all gearing of whatever kind; and the general spatial involute gearing of this book is no exception. Accordingly I try here. by way of example, at least to draw attention to some of the puzzling phenomena that occur when we tackle a new problem in design that is at — or, should I say, within — the vague interface between what is geometrically practicable and what is not. When the matter is well beyond the envelope, it should be understood, the situation is hopelessly impossible. Having worked it before, I know the troubles to be met in the example that follows; I consider it to be at (or near or within) the said envelope, namely the vague interface between the possible and the impossible, but to expose the troubles, I need to work the problem. So here is the problem.

A worked example previously tried but discontinued

17. This example for investigation is a severe modification of WkEx#2. We are looking here for a low-ratio, square, plain polyangular set where $k = 9/41$ as before, but where j is much larger than that. We try here with $k = 9/41$ (which is only 0.2195 approximately) and $j = 0.4$. We suspect, with these requirements, that a satisfactory solution may not be easily available. The data is the same as that for WkEx#2, except that we nominate our movable point F not at E but at another point on the transversal where the radius-ratio j becomes 0.4. This j has already been explained at §6.38. It might be said by way of comparison now — and this is a rough comparison — that the amount by which the ratio j/k differs from unity is a measure of polyangularity. As we jumped from the equiangular WkEx#1 into the polyangular WkEx#3 (other things remaining equal), the ratio j/k was (0.7)/(0.6) say roughly 1.2. We also jumped successfully from WkEx#2 to WkEx#4. Here, however, as we jump directly from the equiangular WkEx#2 into this un-numbered example (other things remaining equal), the ratio j/k is substantially greater, namely (0.40)/(0.2195) = say roughly 1.82. I refer again to figure 6.09, already explained. This purports to illustrate in a useful way the question of the distribution among the variables S, T and ε upon the R-cylinder (§6.20) of the various worked examples of this book. Does it reveal that with this un-numbered exercise we are in trouble? Once again I offer the deeper aspects of figure 6.09 as an exercise for the reader.

18. Having thus seen this un-numbered exercise as posing yet another challenge to the theory, as being even more difficult than previous examples for various reasons, and perhaps impossible to bring to completion satisfacto-

§11.18   GENERAL SPATIAL INVOLUTE GEARING

rily in any event, let us carry on. It is in the general nature of the problem that, in the absence of experience and without a strong mathematics, and thus without a well worked out spread-sheet for quickly achieving trial solutions, we must, somewhat blindly, simply explore. Refer to the series of figures 11.03 (a)(b)(c). Using the chosen data for WkEx#2 and this one, and using the geometrical principles already elucidated and understood for the construction of the relevant sets of naked wheels, the computer drawn figures here (like the ones at the figures 6.10) compare and contrast the architectures of the said two examples. We need to remember in the figures that we are jumping here from an equiangular set ($k = 9/41$) directly to a polyangular set where k remains at 9/41 while j is caused to jump from 9/41, namely 0.22 approx., to 0.40 exactly. Thus j/k has jumped (as already explained above) from unity to 1.82. As said before, the fundamental geometry of the contact-

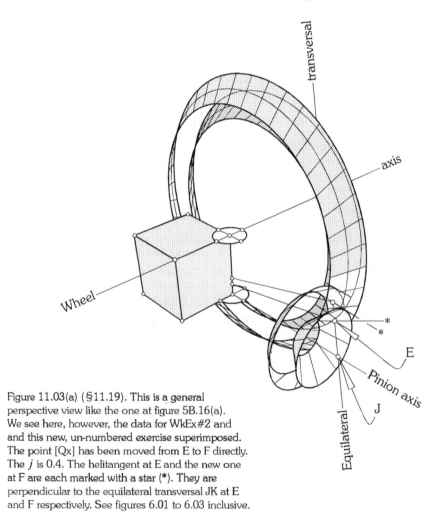

Figure 11.03(a) (§11.19). This is a general perspective view like the one at figure 5B.16(a). We see here, however, the data for WkEx#2 and and this new, un-numbered exercise superimposed. The point [Qx] has been moved from E to F directly. The j is 0.4. The helitangent at E and the new one at F are each marked with a star (*). They are perpendicular to the equilateral transversal JK at E and F respectively. See figures 6.01 to 6.03 inclusive.

418

ing surfaces of the chapters 3 and 4 is sound, but what is at issue here is that, in actual mechanical reality, the necessarily truncated edges of the said surfaces may interfere with one another, either within the outside shape of one self-standing tooth or between the sets of meshing teeth. So let us see.

19. Figure 11.03(a) is a general perspective view closely corresponding with that of figure 5B.16(a). We see straight away that by moving from E to F at the new $j = 0.4$, we have increased the size of the pinion and decreased the size of the wheel, thus making the gear set overall more compact. It is not easy to see in the figure but it can be seen that the points E and F are upon the same equilateral transversal at $R = 40$ mm, and that the *angular disposi-*

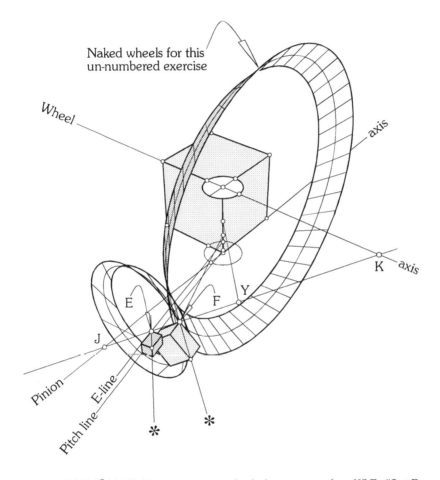

Figure 11.03(b) (§11.19). Here we see more clearly that, in getting from WkEx#2 at E to this un-numbered exercise at F directly, [Qx] has been moved along from E to F. The helitangents at E and F have each been marked with a star as before, and the five points JEFYK in line along the transversal may now be easily seen. The small rectangular boxes are indicting the helical relationship of the rubbing velocities with the pitch line.

tions of the generators of the naked wheels — these remain undefined as before (§6.42) — have been changed by the sudden switch from WkEx#2 to this one considerably. Figure 11.03(b), another perspective view of the two sets, does show the transversal and the two points E and F clearly; they are at the intersections with the transversal of the two different helitangents respectively indicated by the two stars (*) (*) which may be seen. Figure 11.03(c) is special; it is the unique orthogonal view looking along the transversal where (a) the angle Λ which applies for both sets (§6.46), and (b) the so called beta shift which applies between them (§6.46), appear at their true magnitudes. [By formula, $\Lambda = 35.09680°$, and by measurement, $\beta_{SHIFT} = 42.6542°$.] Notice that the beta shift, which is a clear measure of our departure from equiangularity in going from WkEx#2 to this one, is of such magnitude that the handedness of the teeth on the wheel (41 teeth) has changed from being moderately left handed to being moderately right handed. The handedness of the teeth on the pinion on the other hand (9 teeth) has remained left; this has become more severe however. One might fear at first that this severity of the handedness at the pinion might be a source of trouble; but (and look for this at chapter 7) what if the pinion were a worm?

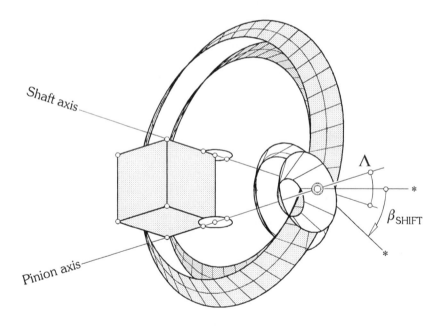

Figure 11.03(c) (§11.19). Yet another perspective view of the same scene, this time, however, looking directly along the transversal. This is the special view that was introduced incidentally at figure 5B.16(d). The specialty of the view is that it reveals the magnitude of the angle Λ which, given a fixed C, Σ, k and R, is a fixed function of the architecture.

ASPECTS OF THE PHYSICAL REALITY §11.20

20. Next, as usual, we must allocate within the polar plane the two angles of obliquity. This is a tricky job. If we set them too small we run the risk of setting one or other or both of the Z_l too short (shorter than the NP); if we set them too great we run the risk of decapitating teeth by mutual interference of the flanks thus of interrupting a slip track at a sharp nose; and, as we have seen already at WkEx#2, unless we choose their relative magnitudes sensibly, the two lengths Z_l will be markedly unequal (§5B.49). We are trying to avoid these hazards.

21. We need, once again, the ellipses of obliquity (§5B.40); and, at the risk of repetition, I wish to explain these ellipses again. As we know, all the legitimate straight-line paths for a moving point Q at a chosen F are confined within the radial array of lines that constitutes the polar plane at that F; the polar plane is normal to the velocity vector (the sliding velocity) at that F, and its central point is at the basepoint of that same vector. With polyangularity here prevailing, the polar plane is no longer normal to the FAXL (§5A.24). If we construct the shortest distances between any one of the legitimate paths in the polar plane at F and the two shaft axes, we construct in effect the input and output links RS and SR of the fundamental RSSR of figure 2.02. If we do this for all of the legitimate paths within the polar plane we will locate two points S on each of the possible paths, and the loci of these become the ellipses of obliquity. *Notice that, independently of the two paths chosen, the ellipses of obliquity are a function of the architecture of the set; the ellipses are not a function of the chosen paths.* Now refer to figure 11.04.

22. Figure 11.04 is an orthogonal view along the helitangent of the data of the current example (the un-numbered one). The view is accordingly taken directly against the polar plane which appears in the figure as a circle; and, as usual now in this book, the picture has been arranged to set the transversal (already in the plane of the paper) horizontal across the page. The ellipses of obliquity are quite clearly *intersecting* at the centre of the polar plane and, in comparison with figure 5B.16(b) where equiangularity obtained (§5B.40), they are no longer arranged with their minor axes parallel with the CDL. The CDL appears in the figure, oblique and in the background. This latter is a natural result of the new set of conditions, namely the new polyangularity occasioned by the fact that the ratio *j* is no longer the same as the ratio *k*.

23. Trying judiciously now to set the angles of obliquity to optimize an outcoming result which is of course unforeseen as yet, I give in figure 6.09 my first considered choices for the locations of the rods SS. The angles of obliquity are set in the figure — whether they are for drive and coast respectively or for the opposite is unanswered as yet — at 40° and 24°. While setting these angles it is both possible and desirable to envisage in space the shapes and dimensions of the two equivalent RSSR, with their less than hemispherical sockets at the joints SS (§2.14), and to gather from these the likelihood of higher than wanted transmission forces, of unhappy effects of friction due to high sliding velocities at low angles χ (§5B.37), of unwanted

§11.23 GENERAL SPATIAL INVOLUTE GEARING

(or indeed wanted) absences of reversibility namely occurrences of non-back-driving, or of other unwanted troubles or special requirements. Without exceptional geometrical insight, or quickly available computer-calculated predictions based on a completed algebra, or long experience in practical fly-by design, this is not easy to do however.

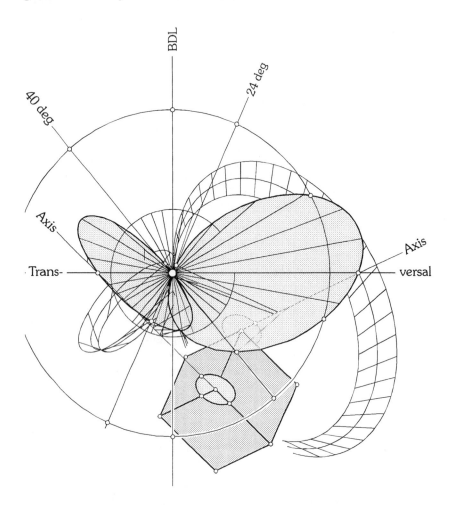

Figure 11.04 (§11.22). The ellipses of obliquity for the current (un-numbered) worked example. The basic data here is the same as it was for the equiangular WkEx#2, centre distance = 40 mm etc., except that ratio j is changed from 9/41 to 0.4. Recall figure 5B.16(b). To show the changed detail more clearly here, a new scale magnifies the scene there by a factor of almost two. The larger circle in the plane of the paper here signifies the polar plane. Here, also, I have chosen cautiously some first trial values for the angles delta, 24 deg. and 40 deg., keeping a concerned eye open for imagined, unknown trouble if these angles were chosen too large.

ASPECTS OF THE PHYSICAL REALITY §11.24

24. Next in the business of design we can follow the procedures already outlines at §5B.16 *et seq.* for a long way towards completion. Ignoring as yet the finer points of undercut, interference and contact ratio, we can come up with some preliminarily drawn pictures of the shapes of the flanks of the teeth and the actual meshing between teeth. Refer to the views at figure 11.05 for a first visual impression direct from the CAD. See §11.25 next for some further explanation, and to appendix A at 11 for selected numerical values.

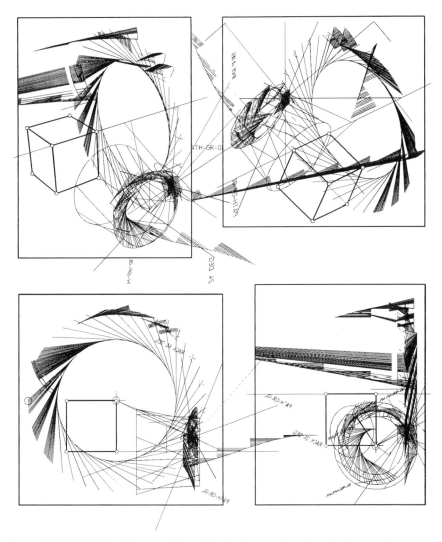

Figure 11.05 (§11.24). Glimpses of the unfinished, un-numbered, worked example of this chapter. These pictures are extracted raw from the wire-mesh of the CAD.

§11.25 GENERAL SPATIAL INVOLUTE GEARING

25. In figure 11.05 there are some interrelated views which variously illustrated at an early sage of the work my fist rough look at the teeth of the wheels in this current, un-numbered, worked exercise. Note the teeth of the worm-like pinion beginning to appear at top right. This group of views is instructive, perhaps, in that it shows the kinds of graphical activity that have led (in the long run) to most of the regular figures of this book. But also it give a clear impression in this event of the fact that, unless we take special steps to do so, we fail to detect in the early stages of any piece of design work the more obscure of the interference troubles that may be looming.

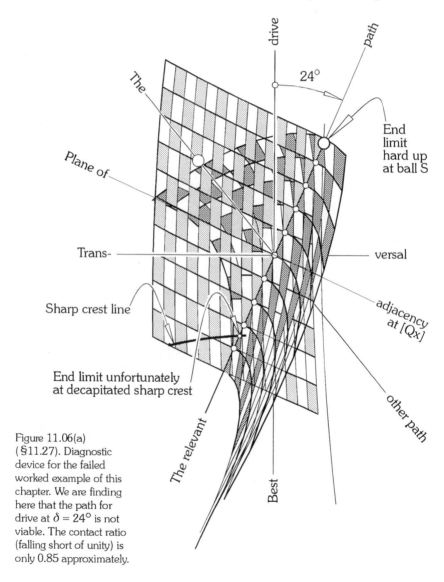

Figure 11.06(a) (§11.27). Diagnostic device for the failed worked example of this chapter. We are finding here that the path for drive at $\delta = 24°$ is not viable. The contact ratio (falling short of unity) is only 0.85 approximately.

424

ASPECTS OF THE PHYSICAL REALITY §11.26

26. I wish to try out, here, some new words that will become necessary soon. In connection with the forthcoming *slip-track locations matrix* (which is itself, collectively, a new word), I wish to note here some suggestions before I alter them, enlarge the list, or forget. The *ideal* lengths of the zones of contact Z_I are terminated at the ball and socket joints SS of the equivalent linkages RSSR; their lengths Z_I (there are two of them) are indeed the lengths of the two rods SS; this idea and the words for it are already well enough estab-

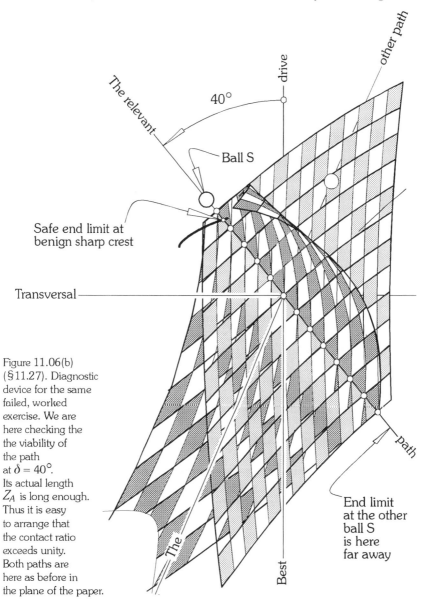

Figure 11.06(b) (§11.27). Diagnostic device for the same failed, worked exercise. We are here checking the the viability of the path at $\delta = 40°$. Its actual length Z_A is long enough. Thus it is easy to arrange that the contact ratio exceeds unity. Both paths are here as before in the plane of the paper.

425

lished (§4.51). The *actual* zones of contact, of lengths Z_A, on the other hand, are determined by either (a) one or other of the already mentioned terminators SS, which would be poor practice (5B.24), or (b) *generated terminators* such as the edge of a sharp crest or the cleft of a sharp trough, see figure 10.09(a), or (c) *subsequently superimposed terminators* such as one or other of the edges of a *truncated crest* usually conically truncated, or one or other of the edges of a *previously excavated trough* usually cut in a helical manner like a square thread with a cylindrical tool. The lengths Z_A must be greater than the normal pitches NP (of which there are two) measured between successive relevant flanks.

The slip track locations matrix

27. Figures 11.06(a) and 11.06(b), which are drawn to the same scale, refer respectively to the paths of (a) one of the points of contact Q in drive, and (b) the other point of contact Q in coast. What follows now is an explanation of these two figures. As each gear wheel rotates about its own shaft axis, each of the two slip tracks sweeps out a *torus* in the fixed space. The tori to be conjured up in the imagination here are not the simple, right-circular tori that look like an anchor ring and are generated by a mere circle set in a plane containing the axis of revolution. The tori here are more general tori. The generating curves here are not circles coplanar with the respective axes of rotation, but general twisted curves in space, and the swept out tori themselves are neither circular in right-cross-section nor closed surfaces. In both of the figures the view is an orthogonal view of the zone of the meshing taken directly along the helitangent. This is the same as the view taken at figure 6.11. Accordingly we are looking in both figures directly at the polar plane within which reside the two paths of the points of contact Q. In each figure each of the two slip tracks are presented in a series of circumferentially displaced locations distributed at intervals of one tenth of the circular pitch of the relevant teeth, and they are seen to intersect at points regularly distributed along the paths of the points of contact. The fact that in each picture the pairs of tori intersect in a straight line indicates unequivocally that the tori themselves are ruled surfaces and that they can be none other than hyperboloids. It is not surprising next to find that they are, indeed. the base hyperboloids from which the slip tracks themselves originally sprung (§4.76, §5B.23). This curious circularity of the geometrical argument is not some ghastly mistake: it may be seen as a useful device with which the graphical constructions (or the numerical calculations based upon the relevant algebra) may be checked. The shown matrices of the successive locations of the slip tracks plotted at intervals of one tenth of circular pitch across the whole region of the meshing may be used moreover to study the real limits to the actual paths of Q (namely the intercepts Z_A), and the contacts ratio. Please be aware of course that, while the straight-line paths of Q (the lines of intersection of the tori) are in the plane of the paper, the curved surfaces of the tori themselves (the base hyperboloids of the flanks of the teeth) are not.

ASPECTS OF THE PHYSICAL REALITY §11.28

The closed envelope of kinematic possibilities

28. This book would be quite incomplete unless, at or near its end, it contained some such sub-heading as this. Having accordingly written it, I find it difficult now to write some intelligible text to support it. But what I can say is this. It is well known in the arena of planar, involute gear-design that (given the restrictions placed upon the angles of obliquity, for example, with which

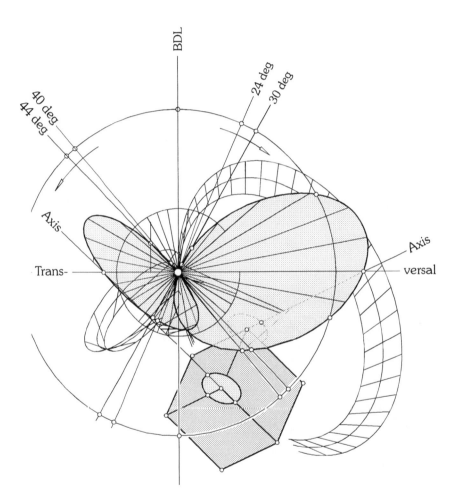

Figure 11.07 (§11.29). These are the same ellipses of obliquity as the ones for the un-numbered. Following now upon figure 11.04, however, and in order to improve the intersections of the paths with the said ellipses, and thus to improve the lengths of the paths (at the expense no doubt of causing trouble with other aspects), both of the angles delta have been increased in this new figure, the one from 24° to 30°, and the other from 40° to 44°. Now we can try for success again. Might this for the reader be a useful exercise at the relatively unknown edges of the envelope?

§11.28   GENERAL SPATIAL INVOLUTE GEARING

restrictions most designers agree) there is a minimum number of teeth to be found on the pinion below which mechanical action is known to be impossible. Some of the similar impossibilities in the planar arena (of which there are many) relate to the pure kinematic geometry of undercut, contact ratio, truncation etc., while others relate to questions of statics, sliding velocities, and the various estimated values of the coefficients of limiting friction and so on. Dudley [47] has written well (but mostly kinematically) about such matters, and all practicing engineers in the field of gearing well understand the com-

Photo-figure 11.08(a) (§11.30). Prototype comprised of two gear bodies based upon the dimensions discovered at WkEx#1. Notice the triangular, fabricated frame (link 1) supporting the skew, live shafts (2 and 3) within the free-to-be-set, self-aligning, bearing housings bolted on to the milled plates set horizontal and vertical as shown. Liberal clearance holes in the lugs of the bearing housings ensure that, in the absence of care, the assembly will suffer some or all of the errors to which an architecture might be prone.

428

plexity of the matters at issue. Figuratively I would like to say that, here in this wider and much more complex field of *general spatial* involute gearing, we can work now with some tried tools but within a closed, but transparent *envelope*, whose general shape, special local convolutions, various degrees of penetrability, and overall whereabouts, together produce an irritating mystery; we are imprisoned within a tightly-drawn transparent fence in the middle of a maze of wide proportions without knowing exactly where we are; clearly we need (and I speak for myself) to find our way more confidently

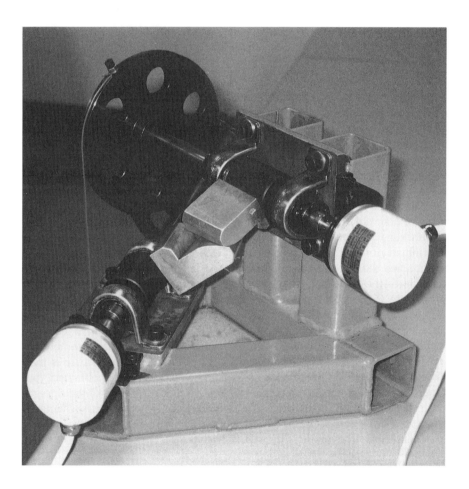

Photo-figure 11.08(b) (§11.30). Another view of the same prototype showing the teeth (held in mesh by means of the opposing counterweights) and the electrical encoding devices which are capable of accurately measuring the angular displacements of the shafts. The flanks are milled (by straight-line generation in a suitable NC machine) to the correct shapes of their respective involute helicoids, and correctly with respect to the shaft axes. The teeth being relatively soft, faint wearings at the slip tracks can be seen.

about. Exploratory work needs to be done. More theory is wanted, not only in the arena of the kinematics and the truncations (which arena I have somewhat rashly claimed to be the subject matter of this book), but also in the arena of the statics, limiting friction, the sliding velocities (when limiting friction permits the slidings to occur), and the efficiency, namely the efficiency with which power is transmitted at the meshing.

An exercise for the reader

29. The reader might refer, for exercise, to figure 11.07. It relates to the possibility of trying again to find a satisfactory solution for the un-numbered exercise which has failed to submit, so far, to my efforts (§11.17). The figure, taken along with its caption, is self explanatory.

Early confirmation with a first prototype

30. I come now to the prototypical work of Michael Killen [54]. Armed with the workings and numerical results that were the product of WkEx#1 (§5B.09), he designed and supervised the construction of the mechanical apparatus illustrated at the photo-figures 11.08(a) and 11.08(b). Refer to the figures 3.09 and 3.10 and see §3.35 *et seq.* for the concept of *gear body*. Killeen set out to test — by means of two carefully machined gear bodies in appropriately controlled, flank to flank contact with one another — one of the principal hypothesis of this book, namely that, independently of all small errors of assembly, any set of two, properly machined gear bodies will steadily transmit rotational motion at a predestined velocity ratio, the ratio being determined not by the precise architecture of the set, but by the precisely cut gear bodies themselves. In his case the velocity ratio k was set by the parameters of the exercise; and this was, of course, $k = 0.6$ precisely. Notice, on reading the captions to the photos, that the apparatus is constructed in such a way that, without very careful adjustment (which was purposely not provided for), the actual dimensions of the architecture, both linear and angular, will not accord with the nominal dimensions; errors at assembly are accordingly endemic; and the main point to be mentioned of course is that, despite this, the intended velocity ratio $k = 0.6$ not only obtained, but remained constant across the whole range of useful movement of the apparatus. It must be said in conclusion here that Killeen, having variously estimated his experimental errors, included in his work a careful analysis of the reliability of his results. This analysis is not reported here.

Appendix A at chapter 11

31. Listed here are some of the numerical values of the results obtained at the un-numbered, unsuccessfully worked example attempted in this chapter. These, along with a few included notes, are provided here for the ongoing benefit of the rare working-reader who might actually tackle the exercise suggested at §11.29.

ASPECTS OF THE PHYSICAL REALITY §11.31

Original data carried over from worked example #2:

$C = 40$ mm
$\Sigma = 90°$
$R = 60$ mm
$k = 9/41 = 0.219512..$
$p = -8.3768445..$ mm/rad
$r_3 = (B-P) = 1.838820..$ mm
$\psi_3 = 12.3807569..°$

and, judiciously,

$\delta_2 = +5°$ (clockwise from best drive)
$\delta_3 = -30°$

New data for this exercise:

With trepidation, let $j = 0.4$, and then, judiciously,
$\delta_2 = +24°$ (clockwise from ditto)
$\delta_3 = -40°$

Lengths (by CAD) of the two rods SS:

$[Z_j]_D = 53.9514$
$[Z_j]_C = 28.8662$

Details (by CAD) of the two naked wheels, the axial locations S_t of their throats, their throat radii t, and their twist angles τ:

$S_{t2} = -25.5439$ mm
$S_{t3} = +56.7989$ mm

$t_2 = 38.3695$ mm
$t_3 = 25.0198$ mm

$\tau_2 = -47.8127°$ RH
$\tau_3 = +68.3746°$ LH

The axial locations S, by CAD, of the throats of the four base hyperboloids:

$S_{a2D} =$
$S_{a2C} =$
$S_{a3D} =$
$S_{a3C} =$

The four sets of parameters $[a, \alpha]$ (by CAD) of the four base hyperboloids:

$[a_2, \alpha_2]_D = 47.3440$ mm, 10.9615 deg
$[a_2, \alpha_2]_C = 66.9213$ mm, 34.0770 deg
$[a_3, \alpha_3]_D = 26.7153$ mm, 67.5476 deg
$[a_3, \alpha_3]_C = 12.8206$ mm, 18.3568 deg

The four products $[a \cos \alpha]$, one for each set of the flanks:

§11.31   GENERAL SPATIAL INVOLUTE GEARING

$[a_2 \cos \alpha_2]_D = 46.4802$
$[a_2 \cos \alpha_2]_C = 55.4299$
$[a_3 \cos \alpha_3]_D = 10.2030$
$[a_3 \cos \alpha_3]_C = 12.1682$

The two ratios of the $[a \cos \alpha]$, one for drive and one for coast:

$[10.2030]/[46.4802] = 0.2195 = k$, OK
$[12.1682]/[55.4299] = 0.2195 = k$, OK

Pitches $[a \cot \alpha]$ mm/rad of the helicoids: [In square brackets, for comparison, are the corresponding values at example #2]

2D = 244.441 mm/rad   [445]
2C = 98.9279 mm/rad   [120]
3D = 11.0398 mm/rad   [97]
3C = 38.6374 mm/rad   [26]

The tooth-interval angles are (a) for the pinion 40°, and (b) for the wheel 8.780848°. These are, in radians, (a) 0.698132, and (b) 0.153248.

So the pitches p_{FLANK} of the helicoids expressed mm/(one tooth interval) are as follows. [The mentioned colours were the colours in my CAD.]

2D = 37.4601   Olive
2C = 15.1605   Orange
3D = 7.7072   Green
3C = 26.9740   Yellow

These pitches no longer occur pairs, as they did at WkEx#3, where equiangularity was prevailing.

The wheel circular pitches CPW are by calculation:
(a) for equiangularity at WkEx#2, using the f at §5B.27, the E-radii are 76.9228 mm and 6.8855 mm; the ratio of these (namely k) = 0.2195, OK; so the circular pitches are 11.7883 mm and 11.7883 mm; and
(b) for polyangularity at this un-numbered example, the F-radii are 67.0059 mm and 26.8024 mm; the ratio of these (namely j) = 0.4, OK;. so here the CPW are 10.268548 mm and (b) 18.711605 mm.]

Note. In going from #2 to this un-numbered example we have, roughly speaking, decreased the radius of the wheel by 13%, but increased the radius of the pinion by 59%. Rewriting (with an eye to §10.17) to suit the newly occurring polyangularity, we have

$[CPW]_2 = [2\pi f_2]/N = 10.268548$ OK
$[CPW]_3 = [2\pi f_3]/N = 18.711605$ OK

NPW: There are two of these, not one for each wheel, but one for each mating pair of the flanks. The normal pitches are:

ASPECTS OF THE PHYSICAL REALITY §11.31

$[2\pi a \cos \alpha]_D/N$ = 7.123 mm both wheels
$[2\pi a \cos \alpha]_C/N$ = 8.495 mm both wheels

These are also the two NPR for the rack.

But the CPR, the rack circular pitch, measured from flute to flute in a straight line along the infinitely large E-circle of the rack, is a measure which has only one value. As before the same rack cuts both wheels.

CPR = NPR sec δ:
CPR = 7.123 sec 40° = 9.298 mm
CPR = 8.495 sec 24° = 9.298 mm
It is the same, as expected, for both δ.

Angle κ: This, as per example #2 (and as per always), is independent of $k, j,$ and the two subsequently chosen δ; it depends solely upon the fixed variables of the basic architecture namely C, Σ and R. See the formula at §10.08 which I feel sure is correct. It gives κ = 42.1304..°. This checks with the CAD. Naturally, and as usual for offset skew gears, angle κ (always somewhat less than $\Sigma/2$) is the same for both wheels. Like Λ, it is a basic characteristic of the set, depending only upon C, Σ and R.

Angle λ: I have no formula from first principles for finding the two τ in terms of the fixed variables including j and the two δ, but I can check from the measured values of the two λ the validity of the formula for cos λ just derived, namely $\cos \lambda = \cos \tau \sec \kappa$. Measured from the cone generator towards the naked generator looking from the outside, the two λ, by CAD and by calculation, are:

[λ_2 = 25.1058° RH by CAD = 25.1038° RH by calculation.]
[λ_3 = 60.2026° LH by CAD = 60.2020° LH by calculation.]

CHAPTER TWELVE
THE SCREW SYSTEMS
AND GEAR DESIGN

Explanation

01 This short chapter is comprised of two main themes. I try first to give an overview of the fundamental geometry of the general spatial involute gear set expressed in the mixed terminology of classical line geometry and modern screw theory. This begins at §12.05. I try next, beginning at §12.14, to summarize the imaginative steps that one might take to encompass in the mind's eye the logic of the design argument. This second theme might be said to be a summary of the geometry, made in ordered retrospect, of the contents of this book; and this relates, of course, to the contents of §10.66.

Introduction

02. I attempt, firstly, to write a logical account of the abstract, line-geometric elements embedded within the actual mechanical reality of the spatial involute gear set (§12.05). These line-geometric elements are not only embedded in the set, but interwoven in such a way that I am obliged, in attempting to describe them, not to follow the chronology of the text as already written (unless it might, in places, be appropriate), but to concentrate directly upon the whole result as it now presents itself. I wish to bring into focus here, not the incremental discoveries made as we went consequentially, but the fundamental relations of the underlying geometry. It seems to be that this geometry is a tightly woven identity that has been there for our finding. It is not, I believe, the mere product of our wish to invent.

03. The kinematical aspects of the task undertaken have been to achieve smooth transmission of motion from one rotating shaft to another where (a) the shafts, each carrying a coaxial wheel, are skew, (b) the meshing elements that make contact are evenly distributed sets of identical teeth upon the wheels, (c) the paths in space of the point (or points) of contact between teeth are not required to pass through the common perpendicular between the shafts, (d) the motion is to be possible in both directions, and (e) the mobility must be unity. This latter means that, while all necessary constraint is to be incorporated, overconstraint must remain absent [1].

04. Looked at kinetostatically, we may say that the analyses of the book have been made on the assumptions (a) that the continuous motion must be steady, (b) that there are no errors made in the shapes of the contacting, relatively sliding links, (c) that friction is absent, and (d) that thus the gear box, although a *complex joint* between the input and output shafts, is nevertheless a *workless joint* [1] [§14.13]. This means that under the said circumstances the power consumption (the power lost at the meshing) is zero and that the efficiency of the transmission is 100%. This also means that under the said circumstances the following may be said: *the overall action screw and the overall motion screw are reciprocal* [1] [§14.20]. In the presence of steady friction, the complex joint becomes no longer workless, and the efficiency becomes no longer 100%. The statics (although altered) remains independently of the kinematics however, and the problem remains kinetostatical [100]. It might be said indeed that, except for start-up and sudden changes of speed and shut-down, there is no place in the design of conjugate gear sets for any serious consideration of the dynamics.

Intersecting screw systems inhabiting the gear set

05. The notion that all of the tangents to a uniform helical curve, a helix (§9A.01), form a ruled surface called an involute helicoid is relatively old; but the fact that any two involute helicoids, placed back to back at a single point of contact between the convex sides of their surfaces and with their axes set to be rotatable at two lines fixed in space, will constitute a constant-velocity, cam-like coupling between the axes is relatively new [68]. It is this that might be used to begin the description of an overall geometric picture. If we take the two axes (the two shaft axes) here introduced, and see them first as the two screws of zero pitch within a cylindroid, we see that the said cylindroid is fully defined by the said two axes [1] [§15.05]. Its central axis is the common perpendicular between the said two screws of zero pitch, and the single infinity of generators are of course each perpendicular to this central axis [1] [§15.13].

06. Given the cylindroid that is determined by the given C and Σ of a gear set, the pitch lines for all possible k are generators of this cylindroid. Surrounding each one of these pitch lines there is a linear complex of lines whose pitch is the pitch p at that pitch line, the complex constituting all the ∞^3 of the legitimate paths for any Q, given that k. The respective central axes of this single infinity of different complexes all cut the same line (the CDL) perpendicularly, and each one of the complexes (as we prove next) intersects all of the others in the same congruence. It is known that any pair of the complexes intersect in a congruence [1] [§11.40], and that this occurs in such a way that the directrices of the congruence of intersection (if real) are a pair of conjugate lines in both of the complexes [1] [§11.40]. Refer back to the 'important observation' noted at §6.15 where it was proven that, in a gear set with given C and Σ, and independently of k, the velocity vectors at J and K while skew with are at right angles to the shaft axes at K and J respectively.

It follows from this that, independently of k, the polar planes at all points along either one of the shaft axes intersect in the one line of the other shaft axis. As already explained, however, the shaft axes constitute a pair of conjugate lines in every one of the single infinity of complexes [1] [§11.36], so the proposition that all of the complexes intersect in the same congruence whose directrices are the two shaft axes is hereby proven.

07. The said congruence of intersection accordingly consists of lines which are legitimate paths for Q, whatever the k, namely wherever the pitch line may happen to be. These 'universal' legitimate paths are however not only special but also from the practical point of view useless. Referring to the equivalent RSSR explained at §2.12, they are all rods SS where the cranks RS are of zero length (§6.54). They are, indeed, all of the transversals (across the shaft axes) that may be drawn. Because they are right lines in the moving body (the other body remaining fixed), and because they are legitimate paths also, the velocity vectors at all points along all lines of this congruence are perpendicular to the line. If we apply to these transversals the extra restriction that they must cut the FAXL perpendicularly, we become reduced to the line-series of transversals (the regulus) that is indeed the parabolic hyperboloid of the F-surface. If, as has been said, the F-surface is swept out by the transversals that cut the FAXL (and all of these cut the FAXL perpendicularly), what is the other family of its generators? The answer is that the other family, the straight lines upon the F-surface that cut the CDL perpendicularly, are the E-lines for the various k. From this discussion the reader might deduce some further useful information about the distribution of the best paths in the fixed space for the two Q and the distribution of the related rubbing velocities. I leave this as an example of the many such questions that might be asked (and answered in this and subsequent essays) by myself and/or the reader.

08. It needs to be made clear that, although (as explained above) the two are intimately related, the F-surface and the cylindroid are not the same (§1.31). Nor is the F-surface, illustrated at figure 5B.01, the same as the surface marked (d) at figure 1.07 (§1.37). Whereas the linear distribution of the generators of the cylindroid are related *sinusoidaly* with their angular locations, the linear distribution of the generators of the F-surface and the surface marked (d) at figure 1.07 are related by means of the *tangent function* with their angular locations. The latter two surfaces are both parabolic hyperboloids, of which there are more than a few extant in the gear set (§4.78).

09. As has been said, the population of the ∞^1 of legitimate paths that cut the FAXL perpendicularly defines the F-surface. Whereas *all* of the ∞^2 of the special legitimate paths that cut the two shaft axes somewhere form an hyperbolic congruence, the F-surface comprises only a sub-set of these. The F-surface is indeed a special regulus, a parabolic hyperboloid. One could also argue that the F-surface is the intersection of two congruences: (a) the hyperbolic congruence which is all the transversals across the shaft axes, and (b) the special congruence which is all the lines in space that cut the FAXL per-

pendicularly. What is the mechanical significance of these latter, these lines that cut the FAXL perpendicularly? Is there one?

10. We know, and have often spoken about, the helicoidal field of linear velocity vectors (the sliding velocities) surrounding the pitch line, where each vector has its basepoint exclusively at one of the points in space (§10.54). There are an ∞^3 of such vectors. We know in other words about the field of coaxial helices and the velocities tangential to the helices and so on. Litvin [5] mentions this. But what we can see more clearly and deeply here is that, at each point in space there is not just one helitangent belonging to the velocity vector based at that point, but a single infinity of them based upon other points in line with this particular one (§10.54). There is indeed, as has been shown, a *quadratic complex* of helitangent lines surrounding the pitch line. Refer [1] [§21.19] and study, in [1], figure 21.05. For each given k, this quadratic complex of helitangents is coaxial (at the pitch line) with the linear complex of legitimate paths. There must be a congruence of intersection somewhere between these two which will contain all those lines that can (for a given k) be both (a) a legitimate path, and (b) the line for some sliding velocity somewhere.

11. Chapter 6 in Litvin [5] is interesting but to readers such as myself with close attachment to the classical screw systems and the proven geometrical concept of the interacting involute helicoids generally incomprehensible. Paragraph 6.10 of chapter 6 — beginning at page 135 — deals with the concept called by that writer the *axes of meshing*. These axes are held by him to occur (in spatial gearing in general) along straight lines rigidly connected to frame through which passes the common normal to the mating surfaces at any point of contact of the surfaces. He goes on to argue that in certain non-parallel and/or non-intersecting cases (namely in some of the cases where the shaft axes are skew) there is not only one but two of these axes. This appears to imply that there is a hyperbolic congruence of contact normals, determined by the said two axes acting as directrices, each of which normals emanates from some one point of contact among those distributed along the curved line of contact, but it seems unlikely that this was Litvin's intended meaning. Chapter 6 of Litvin [5] does appear to show however that there are tenuous convergencies between his ideas and mine.

12. *Some jotted down, but as yet ill-considered reminders to myself.* All the tangents to all the helices of pitch $a \cos \alpha$ about a shaft axis is a quadratic complex (§5A.08); but all the tangents to one of the said helices is a line series (which is not a regulus). Whereas a regulus merely repeats itself as its generating line continues to rotate about the central axis of the regulus, the involute helicoid 'winds along' its central axis as well. Can the *double F-surface* (§8.15) be properly presented as a classical entity when considered against a background of the general screw systems? Peruse chapter 23 of [1]. I should mention here somewhere the parallel cylindrical helicals where the *special linear complex* [1] [§9.41] obtains (§5.40). Find at appendix A at 5A a

proof that the tilted triangles there are similar, and that this is an interesting condition applying for all points E along the E-line. At §2.16 *et seq.*, there are two proofs by means of self-manifestation of the fundamental law; each of them contain elements of screw theory involving the idea of reciprocity; and that material, too, should find its way into this general essay. There is not enough included here about the mechanical significance of the various screw systems and their reciprocal screw systems. Look, for example, into the mechanics of the pure transmission force **F** (which is a 1-system) and the relevant system of all of the linear velocities that can exist (which is a reciprocal 5-system), and so on. There are for general consideration also (a) the geometry of the two cylindroids (one for each shaft axis) determined by the orientation upon the transversal of the polar plane at the chosen [Qf], (b) the ellipses of obliquity which are the intersections of these cylindroids with the said polar plane, (c) the relationship of the cylindroids to one another, and (d) our separate choices for the paths of the two Q made with an eye to the shapes of the equivalent linkages RSSR. It remains true overall (it must be said) that, unless I can find some strong central theme for this ongoing essay, it will be difficult to pull these disparate bits of classical geometry together into an intelligible whole.

13. It might be wise, therefore, to say no more, just here, than this: except to wonder at the sheer complexity — or the dense simplicity — of the interwoven relationships of the screw systems within the 3-link, involute gear set, it is a useless subject for serious study by all practical persons. The same thing has been said, however, about the Bennett mechanism [1] [§20.52]. This mechanism, within its own domain of the overconstrained yet mobile, 4-link, 4-revolute linkages, is reduced by some to being a mere conundrum. It is nevertheless a wonder to behold, and the here-mentioned, 3-link, involute gear set of mobility unity deserves, it should be said, a similar magical status. We do go about the practical job of *designing* specific sets for specific purposes however, and the *logic* of this (although dispersed) has been one of the main subjects matter of this book. Accordingly, I conclude here with the following, hopefully intelligible, ordered remarks.

The overall design strategy seen now in retrospect

14. As prospective designers we need to become aware firstly that two involute helicoids in external (convex to convex) contact with one another, each pivoted to frame along its central axis, the two pivot-axes being generally disposed in the frame, will steadily transmit rotary motion with a fixed angular velocity ratio, and that that velocity ratio will be independent of the locations of those axes in that frame. The angular velocity ratio k depends solely upon the characterizing dimensions (a, α) of the two contacting surfaces [66] [67] (§6.40). The frame here is of course the gear box (link 1 of the mechanism), and the pivot-axes are the axes of the two gear wheels (namely the links 2 and 3), the three links forming a closed kinematic loop whose mobility is unity.

§12.15   GENERAL SPATIAL INVOLUTE GEARING

15. Without knowing as yet the characterizing dimensions, namely the (a, α), of the contacting surfaces, namely the shapes and sizes of the flanks of the meshing teeth, we can select, to begin, the chief dimensions of the architecture. These are (a) the shortest distance C between the two skew shaft axes; (b) the angle Σ between these axes; and (c) the length R of a certain distance called the *radial offset*. Very roughly speaking — there being no precise definitions for the mentioned terms as yet — the radial offset is the distance from the CDL to the *zone of the meshing*.

16. Needing a frame of reference now we erect the *F-surface*. This is indeed a *double* surface consisting of the unique pair of parabolic hyperboloids (there being two of them) defined by the CDL, the two shaft axes, and a limiting condition placed upon a certain line. The limiting condition is that a *transversal*, any line cutting both shaft axes, shall be called an *equilateral transversal* if its points of intersection with the two shaft axes are equidistant from the CDL. Each of the separate F-surfaces is a ruled surface swept out by the relevant equilateral transversal. In conventional practice the zone of the meshing is arranged to occur within the acute angle Σ, and the F-surface that occurs there (as distinct from the other one that occurs at the obtuse angle Σ) is the one that belongs to conventional practice (§8.16).

17. Given the architecture and the ratio k, the *pitch line* can be determined (§1.05). This is the instantaneous screw axis (which remains fixed in the space 1 of the gear box) for the relative motion of the two wheels 2 and 3. The pitch line is under all circumstances one of the generators of the cylindroid uniquely determined by the two shaft axes. Whereas the two shaft axes form the unique pair of generators of zero pitch of the cylindroid, the pitch line is one of the ∞^1 of generators whose pitches are non-zero [1] [§15.15]. The pitch line and its pitch, our having already employed the *theorem of three axes* [1] [§13.12], can easily be determined algebraically (§1.14). It must however be mentioned that (according to the two different relative directions of rotation of the wheels) there are two different pitch lines; these relate to the two different aspects of Σ, the *acute* and the *obtuse*; and there is, accordingly, a *conventional one* of the two pitch lines to be preferred under most normal circumstances (§1.11).

18. Unless the design is to be *exotic*, in which case there will be some special problems peculiar to the consequent *exoticism* (§8.02), we need next to choose the important single point [Qf] somewhere upon that equilateral transversal, which transversal is at the chosen distance R from the CDL. The distance R is measured of course along the common perpendicular between the CDL and the said transversal, namely along the FAXL (§1.43). The point D upon the CDL divides the CDL according to the ratio k (§1.41), and the line D-e, *drawn parallel with the pitch line*, cuts the said transversal in E. If we put [Qf] at some point F on the said transversal, which F is not at E, the set will become a *plain polyangular set*. If however we put [Qf] exactly at E, the set will become an *equiangular set*.

19. Be aware that in general the pitch line is not a generator of the F-surface. *The E-line, however, parallel with the pitch line, is a generator of that surface.* All points F upon the said transversal (which include the special point E) are possible positions for [Qf], but the point [Qf], it must be said, is defined not as some point along the transversal but as that point upon the straight-line path of some point of contact Q where that path cuts the F-surface (§4.27). Any point [Qf] belongs, in other words, not to the F-surface but to the straight-line path in the fixed space of some point of contact Q. Any one Q will occur by definition at the point of contact between mating involute helicoids of the flanks of meshing teeth; and neither the paths of the two Q (for drive and for coast), nor the shapes and locations of the contacting involute helicoids, is known as yet.

20. It is here at this juncture that we must (at last) find the paths for the two Q. *It is here, moreover, and for the first time, that the fundamental law must be seen to be obeyed.* The law states, in effect, that the path in the frame 1 of any point of contact Q between the mating involute helicoids must be a line of the linear line-complex surrounding the pitch line, the pitch p of which complex is the same as the pitch p at the pitch line (§2.02). It follows that the two paths (for drive and for coast) must both be a member of the planar pencil of lines inhabiting the polar plane of the complex at [Qf] [1] [§11.33]. We can locate the polar plane of the complex at [Qf] as follows: from the basic formula of the screw — equation (25) at §1.51 — we determine the direction of the sliding velocity at [Qf], then erect the required polar plane at [Qf] normal to that direction. The polar plane, it will be found, contains (a) the equilateral transversal, and (b) the best drive line (the BDL); thus the line of the sliding velocity, namely (c) the so-called helitangent, is mutually perpendicular with (a) and (b); and the three directions (a), (b) and (c) constitute the *fundamental triad* at [Qf].

21. But we need to *choose* the two paths, each with its own angle of obliquity δ measured away from the BDL, and each with its own judged effectiveness with respect to contact ratio and so on (§6.62). This is done with the help of the *ellipses of obliquity* (§5B.40), as follows. Taking account of Cayley's theorem (§5B.67), we draw all of the shortest distances between each of the shaft axes and the rays of the polar plane. This erects the cylindroids whose generators show at each shaft axis the directions of the different cranks RS of the linkages RSSR that are legitimate. Each shaft axis is the central axis of its own cylindroid. Each cylindroid intersects the polar plane in its relevant ellipse [1] [§15.26], and with these as guides the paths SS of the two chosen RSSR can be chosen most intelligently.

22. Having done this the shapes of the profiles of the teeth are next determined. To begin — and we do this with one or the other of the pairs of the mating profiles meeting, *pro tem.*, precisely at [Qf] — we note that the mating pair of rulings that intersect at [Qf] can be found as follows: knowing the radii a of the relevant core cylinders (and we take for these one of the

two alternative core cylinders at each of the two wheels), we drop straight lines from [Qf] to be tangential respectively to the two core cylinders in such ways that the required angles α are achieved. What we do, in effect, is this: we construct (twice) the triangle BWA at figure 3.03, where point B is [Qf], where angle BWA is α, and where angle ABW is a right angle. The twin figures 5B.07(a) and 5B.07(b) are relevant here. See also figure 3.08.

23. Next we see straight away that each of the two mating profiles is generated by its own unwrapping roll of paper, the axis of each roll being at the axis of the relevant wheel, the radius *a* of each roll being the relevant core radius *a*, and the generating edge of each cut paper being set at angle α to the axis of the roll. See figure 5B.08. Having done this for one point of contact Q (say drive) beginning at [Qf] and rolling both inwards and outwards, we do it again for the other point of contact Q (say coast) beginning at the same [Qf] and rolling both inwards and outwards, but this second maneuvre, for the obvious reason that the backs and the fronts of the teeth are different, is executed one half-CP of circumferential displacement later. See figure 5B.09 for the fronts and the backs of the teeth on a single wheel, and see there the usual half-CP of circumferential displacement along the key circle between successive flanks.

24. Having thus *drawn* the teeth, namely made drawings of them as distinct from actually *making* them, we might next, in the most usual of circumstances, begin to think about the *truncations* that will be necessary to avoid interference, and *undercut* that might or might not occur according to the method chosen for manufacture. But this, as has been seen, is a field of endeavour requiring, in the absence physical trial, a geometrical imagination beyond the ordinary.

25. Next it needs remarking that the actual machine-cutting of teeth brings us onto a new, apparently different arena. The arena does lend itself to rational analysis however, and the same or similar methods of analysis bear similar fruit. The new matter pertaining is the pure geometrical material of chapter 9B. This relates firstly to the known fact that an involute helicoid is generated by an enveloping plane which is screwed about a fixed axis with a fixed pitch in space. This piece of pure geometry is not limited to the boundaries set by its own apparent scope however. It is capable of much wider interpretation. The wider facts of the matter include the fact that, as we relax the necessity for the said instantaneous screw axis to remain fixed in the fixed space, as we allow it to become located further and further away from (but remaining parallel with) the central axis, and as we allow it to migrate about the surface of the coaxial circular cylinder there (namely its fixed axode), and as and we allow its pitch to vary to suit the radius of that axode, there is a continuous range of pairs of axodes (comprising the successive fixed circular cylinders and the successive rolling-and-crosswise-sliding planes parallel with the central axis) that can be employed towards the same result, *namely the generation by envelopment of the same involute helicoid*. This obviously

bears upon the possibility that one of the flat flanks of a rack continuously screwing with respect to the wheel might be arranged to generate one of the two curved flanks of a required tooth.

26. Among the single infinity of ways that one flank of a relatively screwing rack might be arranged to generate one flank of a required tooth, however, *there is a single, unique way in which the same rack (employing both of its flanks) might be arranged to generate both flanks of a required tooth contemporaneously*; and herein lies the crux of the machining matter. The material of chapter 10 not only explains qualitatively why this assertion is true, but provides the necessary algebra, surprisingly simple, for finding quantitatively the machine-tool settings for achieving any required, non-exotic, polyangular result. The required radial displacement from the central axis of the wheel to the unique ISA for the screwing motion of the single, double-flanked rack relative to the wheel, namely the so called *shift* of that particular ISA, namely the radius of the fixed cylindrical axode for the required screwing motion of the rack, *is neither the core radius nor the key radius of the wheel. It is only in those special cases of the parallel cylindrical helical sets and the planar spur-gear sets that the said fixed cylindrical axode for the generation of teeth by rack coincides with the relevant key cylinder.* In those most special of cases, and (dare I say) those the least acceptable, considering the troublesome overconstraint inherent there, it is of course a pity that the said axodes (coinciding as they do with the key cylinders) are loosely known by the somewhat unfortunate term *pitch cylinders* (§10.48) [96] [97]. May I modify the bald severity of that last remark by referring the reader to chapter 2 of my [1]. I see no practical alternative to the gigantic, herring-bone, parallel, cylindrical, helical, but flexible gears used in the propulsion of big ships for example. Constraint and overconstraint are the yin and the yang of machine design (§3.66) [40].

TABLE OF NUMERICAL RESULTS

	WkEx#1 Equi acute	WkEx#2 Equi square	WkEx#3 Poly acute	WkEx#4 Poly square	WkEx#5 Poly worm 1	WkEx#6 Spiroid
Centre distance C mm	80	40	80	40	63	60
Shaft angle Σ deg	50	90	50	90	90	90
Offset radius R mm	140	60	140	60	0	60
Tooth ratio k vulgar	[24:40]	[9:41]	[24:40]	[11/39]	[1:18]	[1:20]
Tooth ratio k decimal	0.6	0.21951220	0.6	0.282051	0.0555	0.05
Pitch p_{23} mm/rad	-17.25208	-8.3768445	-17.25208	-10.450670	-3.489231	-2.992519
Distance (J-K) namely L	153.125824	126.491106	153.125824	126.491106	63.000000	134.164079
Distance (D-E) namely E	140.948102	71.235946	140.948102	68.767434	ZERO	80.913156
Poly ratio j	NA	NA	0.7	0.4	0.5	0.4
Circular pitch CP_2 mm	14.023099					
Circular pitch CP_3 mm	14.023099					
Distance (A-M), S_{e2} or S_{r2}	119.985636	69.579307	122.014299	49.913420	ZERO	60.609153
Distance (B-L), S_{e3} or S_{r3}	133.780545	15.273506	131.751882	34.939394	1.172876	24.243661
Radius r_2 mm	52.01119406	38.161180	52.011194	37.052375	0.193846	0.149626
Radius r_3 mm	27.98880594	1.838820	27.988806	2.947625	62.806154	59.850374
Radius e2 mm	89.273819	76.922818	NA	NA	NA	NA
Radius e3 mm	Calc 53.564291	16.885497	NA	NA	NA	NA
Radius f2 mm	NA	NA	84.022418	67.005939	42.000000	29.692300
Radius f3 mm	NA	NA	58.815692	26.802376	21.000000	74.230749
Angle ψ_2 deg	31.64935163	77.619243	31.64935163	74.248826	3.179830	2.862405
Angle ψ_3 deg	18.35064837	12.380757	18.35064837	15.751174	86.820170	87.137595
Delta drive δ_D deg	20	5	20	40	15	40
Delta coast δ_C deg	20	30	20	25	15	-40
Distance S_{a2D}	116.962403	21.349246	126.697573	38.083443	66.584013	52.066632
Distance S_{a2C}	127.905218	12.058795	123.878665	30.791216	-66.584013	27.607398
Distance S_{a3D}	131.966605	70.913006	120.938401	68.768354	1.172876	55.074404
Distance S_{a3C}	138.532294	68.873639	139.068209	56.367202	-1.172876	135.853000
Radius a_{2D} mm	81.796761	64.164934	80.114201	65.433526	8.001911	70.196896
Radius a_{2C} mm	86.366250	76.136596	78.944005	53.190600	8.001911	46.076825
Radius a_{3D} mm	49.078057	14.084986	50.091424	17.159588	40.552120	2.339865
Radius a_{3C} mm	51.819750	16.712911	57.899290	26.284983	40.552120	29.086220
Angle α_{2D} deg	4.83181	9.455995	10.476164	24.404699	73.743472	49.057565
Angle α_{2C} deg	19.31299	31.941922	3.707951	18.758885	73.743472	3.307850
Angle α_{3D} deg	4.83181	9.455995	19.331418	11.641626	6.123314	10.591148
Angle α_{3C} deg	19.31299	31.941992	35.276941	57.286162	6.123314	85.464582

continued overleaf

GENERAL SPATIAL INVOLUTE GEARING

	WkEx#1 Equi acute	WkEx#2 Equi square	WkEx#3 Poly acute	WkEx#4 Poly square	WkEx#5 Poly worm 1	WkEx#6 Spiroid
p_{FLANK} (a cot α)$_{2D}$ mm/rad	967.648556	385.251951	433.263388	144.215967	2.333333	60.897517
p_{FLANK} (a cot α)$_{2C}$ mm/rad	246.444193	122.119299	1218.150393	156.614660	2.333333	797.216764
p_{FLANK} (a cot α)$_{3D}$ mm/rad	580.589138	84.567504	142.787604	83.287749	378.000000	12.513653
p_{FLANK} (a cot α)$_{3C}$ mm/rad	147.866516	26.806602	81.843922	16.883625	378.000000	2.307228
(a cos α)$_{2D}$	81.506076	63.293066	78.778748	59.587026	2.240042	46.000057
(a cos α)$_{2C}$	81.506076	64.608359	78.778748	50.365130	2.240042	46.000057
(a cos α)$_{3D}$	48.903646	13.893600	47.267248	16.806596	40.320755	2.300003
(a cos α)$_{3C}$	48.903646	14.182312	47.267248	14.205549	40.320755	2.300003
[NPW]2D = 2	12.802945	11.196942	12.3756			
[NPW]2C	12.802945	9.733988	12.3756			
[NPW]3D	12.802945		12.3756			
[NPW]3D	12.802945		12.3756			
Angle κ deg	21.121992	42.130415	21.121992	42.130415	90.000000	39.2315205
Angle β_{SHIFT} deg	0	0	17.5234	28.67		
Angle λ_2 deg	13.691721	17.4584	-3.8317	-11.12		
Angle λ_3 deg	13.691721	17.4584	31.2151	46.22		
Angle ε_2 deg	77.244425		86.4261			
Angle ε_3 deg	77.244425		61.0901			
Angle ι_2 deg	5.0171		-1.3826			
Angle ι_3 deg	5.0171		12.3182			
Shift$_2$ (prop'n f_2)	0.9476		0.9958			
Shift$_3$ (prop'n f_3)	95%		75%			
Pitch p_{24} (prop'n f_2)	0.21452		-0.0622			
Pitch p_{34} (prop'n f_3)	0.21452		0.4134			
p_{24} mm/rad	-19.1511		5.2261			
p_{34} mm/rad	-11.4907		-24.3151			

NOTES AND REFERENCES

[1] Phillips, Jack, *Freedom in machinery*, in two volumes, Volume 1 *Introducing screw theory*, Volume 2 *Screw theory exemplified*, Cambridge University Press, 1984-90. [Fist mention §P.04, thence often.]

[2] *Note*. There is a dichotomy associated with *rigidity and point contact* on the one hand, *and flexibility and consequent compressibility* on the other. The term *Hertzian* is avoided in the context of gears because Hertzian theory refers to spheres and planes in contact only. It is accordingly preferable to speak of a *contact patch* or a *patch of contact* between the contacting flanks of flexible teeth under compressive load. [§1.03]

[3] *Note*. The terms *drive* and *coast* are used to distinguish the two modes of operation of a gear set whereby (a) *drive* relates to one pair of the meeting flanks of mating teeth being in contact to carry the load, and (b) *coast* relates to the other pair. The distinction is made in respect of the direction of the torque being transmitted, the directions of rotation of the wheels being irrelevant. [§1.05]

[4] Steeds, William, *Mechanism and the kinematics of machines*, 2nd ed., Longmans Green, London (1940). [§1.08, 14, 15, 35, 50, 53; §2.10]

[5] Litvin, Feydor L., *Gear geometry and applied theory*, Prentice Hall (1994). See also [81]. [§1.08, 15; §2.28, 31]

[6] *Note*. The cones of Wildhaber loom large and consistently. Wildhaber E., Basic relationships of hypoid gears, *American Machinist,* this is the first of eight papers serially presented, Vol. 90 et seq., 1946 et seq., pp. 108-111, 150-152, etc. [§1.08, 15; §2.12, 25, 28, 29, 31; §4.36]

[7] *Note*. Numerical results from the various worked examples (WkEx#1, #2, #3 etc.) are collected into the Table of Numerical Results (TNR). See page 445. Both in text and in the TNR, quoted values are given accurate to no more than 5 or 6 significant figures, often less. [§1.11]

[8] Reuleaux, Franz, *Theoretische Kinematik Grundzüge einer Theorie des Maschinenwesens*. This is the title given to a collected works published in conjunction with a technical journal *Berliner Verhandlungen* during the period 1874-75. That journal had previously published some of the chapters of the collected works serially. There is also the Kennedy translation of 1876 [1]. [§1.12; §9B.12]

[9] Konstantinov, Michail, et al., *Teoria na Mechanismite i Mashinite,* this book is in Bulgarian, Technika, Sofia (1980). [§1.14]

[10] *Note.* Dr John Gal of the University of Sydney made an algebraic reconciliation of these formulae in 1978. [§1.14]

[11] Moncrieff, A. Donald, Gear cutting. This is chapter 16 of *Gear H'book second edition*, see figure 16.10 on page 16.14 for his diagrammatic explanation of a six-axis hobbing machine. [§10.64]

[12] Willis, Robert. See his book *Principles of Mechanism, second edition*, Longmans Green, London 1870. See also Roth on Willis, *Dictionary of Scientific Biography*, Vol 15, pp. 403-4, Scribners New York 1976. The first edition of the book was published by Parker, London 1841. Various mentions of the fundamental law (for planar action only) may be found in the mentioned second edition: article 31 on page 21; article 116 on page 84, article 174 on page 129, which gives mention also of Euler's early ideas and an exact reference to Euler's important paper (in Latin), namely Teeth of wheels (*Nov. comm. Petr.* ix, 209, 1767); same article page 130 for a reference to his own earlier paper (1838) which (upon inspection) is not yet clear about the fundamental law; article 120 for various historical data and mention of Camus (1733, 1806, 1837); article 123 for who copied who and without due reference, most interesting;. Also his preface at page xvi accurately cites his earlier paper, On the teeth of wheels, *Trans. Inst. Civ. Engs.*, Vol 2, 1838. [§2.03, 07, 24, 46; §3.01; §4 84; §5.03]

[13] Buckingham, Earle, *Analytical mechanics of gears,* McGraw Hill, first edition, New York 1949. See also Dover 1962. Buckingham was professor at MIT, and was chairman of an AGMA committee that wrote this book. Buckingham defines base pitch as I do, the normal distance between the successive involutes (along the tangent to the base circle). [§2.07, 28, 30, 46; §3.01]

[14] Phillips, J.R., Letter to the Editor, 08 Dec 1993, *Mechanism and machine theory*, Vol. 29, 1994, p. 905. In this letter the mentioned law was first enunciated by the present author. [§2.02, 21, 46; §4.84]

[15] Baxter, M.L, Basic geometry and tooth contact of hypoid gears, *Industrial Mathematics,* Vol. 11, pp. 19-42, Detroit (1961). [§2.25, 28, 29]

[16] Baxter, Meriwether L. Jr., Basic theory of gear-tooth action and generation. This is the opening chapter of *Gear Handbook first edition*, editor Darle Dudley, McGraw Hill, New York 1962. [§2.25, 28, 30, 38]

[17] [void]

[18] Mac Cord, C.W., *Kinematics or Mechanical Movements,* John Wiley, New York, 1883. [§2.28, 37; §3.61; §4.84; §5.03]

[19] Dyson, A., *Kinematics and Geometry of Gears in Three Dimensions,* Clarendon Press, Oxford 1969. [§2.28, 31]

[20] Shtipelman, Boris A., *Design and Manufacture of Hypoid Gears,* John Wiley 1978. [§2.28, 31, 32]

[21] Minkov, K., Abadgiev, V. and Petrova, D., Some aspects of the geometric and technological synthesis of hypoid and spiroid gearings, *Proceedings of the Seventh world congress IFToMM, Seville Spain,* 1987, Vol 3, 1275-78, Pergamon 1987. [§2.28, 32]

[22] *Note.* Author Valentin Abadjiev [101] and some others appropriately employ the term, *first equation of the meshing.* [§2.32]

[23] Korestilev, L.V., Instantaneous velocity ratios in spatial gearing (in Russian), *Theory of Gearing and Machines,* a collection published annually by the in-house press of the Machine Building Institute, now STANKIN, in Moscow, pp. 39-40, Moscow 1970. [§2.28, 34]

[24] Dooner, D. and Seireg, A., A generalised geometric theory for toothed bodies as function generators, *Proceedings of the Eighth world congress IFToMM, Prague Czechoslovakia,* Vol 1, 85-88., Soc. Czech. math. & phys., Prague 1991. [§2.28, 36]

[25] *Note.* I wish to acknowledge the software packages (a) *Caddsman of Australia,* and (b) *Coral Draw of Canada,* for (a) the underlying geometrical work of the drawn figures, and (b) the subsequently applied labels, art work, captions, and colour. [§1.06, §2.15]

[26] Ball, Robert Stawell., *A Treatise on the Theory of Screws,* Cambridge University Press, Cambridge England 1900. [§2.19, 22].

[27] *Note.* The second and third quoted vector proofs for the law, provided by Dr Ian Parkin of Sydney and Professor Karl Wohlhardt of Graz respectively, were received by means of direct personal communication during 1994. [§2.31]

[28] Karsai, Géza, A felületkapcsolódás "els[o"]törvénye", (A "first law" of surface-contact), the Hungarian technical journal *Gép,* Vol. XLV, March 1995, pp. 26-29. [§2.21]

[29] Rosenauer, N. and Willis, A.H., *Kinematics of Mechanisms,* Associated General Publications, Sydney (1953), also Dover (1967). [§2.38]

449

[30] Skreiner, M., A study of the geometry and kinematics of instantaneous spatial motion, *Journal of Mechanisms,* Vol. 1, pp. 115-43, Pergamon (1966). [§2.38]

[31] Parparov, I., Bogdanof, A. and Minkof, K., CAD for spatial gear drives, *Proceedings, Eighth world congress IFToMM, Prague Czechoslovakia,* Vol 2, pp. 379-82., Soc. Czech. math. & phys., Prague (1991). [§2.44]

[32] Phillips, J.R., Computational geometry in the synthesis of skew gear teeth, *Fourth ARK international workshop IFToMM, Ljubljana Slovenia,* Editors Jadran Lenarcic and Bahram Ravani, Kluwer Academic Publishers, 1994. [§2.44]

[33] Drago, Raymond J., for example, in his Chapter 2 of *Dudley's Gear Handbook second edition*, 1991, editor Dennis P. Townsend, p. 2.18, for a clear description of and a useful commentary on the crossed helicals. See also Drago, R.J., *Fundamentals of Gear Design*, Butterworth, Stoneham, Massachusetts 1988. [§3.01, 03]

[34] Houser, Donald, Gear noise. This is chapter 2 of *Gear Handbook second edition*, 1991, see page 14.11. [§3.03]

[35] Derek Smith, J., *Gears and their Vibration,* Macmillan, New York 1983. [§3.06]

[36] Opitz, H., Noise in gears, *Phil. Trans. Royal Soc.,* Vol 263, 1969, pp. 369-380. [§3.06].

[37] *Note.* For algebraic reasons associated with the concept of length, I regularly distinguish between the mere segment of line (A−B) and the infinitely long line AB of which the segment (A−B) is a part. [§3.08]

[38] *Note.* The book by Porteous, Ian R., *Geometric Differentiation*, Cambridge University Press, 1994, for example, gives a pure mathematical account of the involute helicoid. [§3.15]

[39] *Note.* In the same context here I have referred to (a) an equal and opposite pair of rubbing velocities, and (b) an un-delineated rubbing velocity having direction and no sense. Although for technical reasons I prefer (a), (b) is acceptable. Although the bare term *rubbing* velocity conveys its own unclear meaning, it is a useful term for ordinary discussion nevertheless. I use it uncritically throughout. [§3.45; §6.33]

[40] Olivier, Théodore, *Théorie géométrique des Engrenages destinés à transmettre le mouvement de rotation entre deux axes ou non situés dans un même plan*, Paris 1842. [§2.37; §3 65, 66; §4.84; §5.03]

[41] *Note.* The models are engraved *Aachen Geselschaft Politechnisches Arbeits Institut*, and the maker is mentioned, J. Schröder of Darmstadt. These models are spoken of also by Mac Cord. [§5B.12]

NOTES AND REFERENCES

[42] *Note.* Dr Sticher has made a number of contributions that have been relevant. Here is the title etc. of his doctoral thesis. *Contributions to geometric interpretations and methods in the study of spatial linkages*, University of Sydney, 1972. [§4.28, 32; §5B.67; §10.22]

[43] Drago, Raymond J., Gear types and nomenclature. This is chapter 2 of G*ear Handbook second edition.*, 1991. See in particular section 2.1.4 entitled, *Nonparallel, noncoplanar gears (nonintersecting axes)*, and go the figures 2.14 (on page 2.17) and 2.42 (on page 2.49). [§5A.30]

[44] *Note.* This is another, poorly regarded by some, Greek-Latin hybrid; but we already have, mixed, and mixed into English, *polynomial, television, terramechanics*; so why not another? [§5A.03]

[45] Salmon, George, *Analytical Geometry of Three Dimensions, seventh edition*, Vol. 1, Longman Green & Co., London 1928. [§4.22, 25, 27]

[46] Lagutin, S.A., English version. The meshing space and its elements, *Mashinovedenie*, No. 4, 1987, pp. 69-75. [§2.28, 35]

[47] Dudley, Darle W., Gear tooth design. This is chapter 4 of *Gear Handbook second edition,* 1991. See the sub sections entitled, (a) *Zones in which involute gear teeth can exist*, and (b) *Pointed teeth*; and see in particular his figures 4.6 and 4.7 on pages 4.13 and 4.15. [§6.09]

[48] Phillips, Jack, Involute skew gears for flexible joints in machinery, *Procedings of the 9th world congress IFToMM, Milan Italy,* August 1995. Vol. 2, pp. 1609-1613. [§3.01]

[49] Dooner, David B. and Seireg, Ali A., *The kinematic Geometry of Gearing: a Concurrent Engineering Approach,* John Wiley, New York 1995. [§2.18, 38]

[50] Hotchkiss, Robert G., McVea, William R., and Kitchen, Richard L., Bevel and hypoid gear manufacturing. This is chapter 20 of *Dudley's Gear Handbook second edition,* Editor Dennis P. Townsend, Mac Graw Hill 1991. [§2.01, 33]

[51] First paper: Bregi, Ben F., Gear finishing by shaving, honing, and lapping, *Chapter 18 of Gear H'book first edition*, 1962; see picture of crowning on page 18-13. Second paper: Dugas, John P., Gear finishing by shaving, rolling, and honing, *Chapter 18 of Gear H'book second edition*, 1991; see this for additional information. [§2.06]

[52] *Note.* Dr F.C.O. Sticher, of the University of Technology, Sydney, has made a number of algebraic suggestions and improvements to the contributions of Killen [54]. These are marked with this reference number [52] wherever appropriate. [§5B.16; §10.09, 22]

[53] Bär, Gert, Curvatures of the enveloped helicoid, *Mechanism and Machine Theory,* Vol. 32, pp. 111-120, Pergamon 1997. [§3.24]

[54] Killeen, Michael, *The mathematics of equiangular spatial involute gearing*. Much but not all of Killen's contribution to the algebraic work was contained within this Undergraduate Thesis completed at the University of Technology, Sydney, 1966. [§5B.16]

[55] Rankine, W.J.M. See [18] [§368 et seq., pp. 242 et seq.] for Mac Cord's sceptical views on the possibility of skew wheels being cut for straight line contact. At §369 he criticises both of the two professors Willis and Rankine — for Rankine he quotes *Machinery and mill work*, page 146 — for cautiously but erroneously thinking in terms of cones, hyperboloids, and the possibility of straight line contact. See my §1.53 for a clarification of what Mac Cord meant in his time (the same as I do in mine) by the term *skew* as applied to wheels. Rankine wrote, among other books, *Applied mechanics* (1858), and *Manual of machinery and millwork* (1869). Rankine and Willis were contemporaries, they were professors in Glasgow and Cambridge respectively. [§1.53]

[56] Minkov, Kolyo, A new approach to basic geometry and classification of non-orthogonal gearing, *International power transmission and gearing conference, ASME,* Chicago, Vol 2, 1989, pp 593-598. See also [21]. [§2.12, 28, 32].

[57] *Note.* Due to the shortage of available symbols, common symbols such as N have sometimes been ascribed more than one meaning. The context will most often be enough for clarity, but a consultation of the Symbols Index, beginning at page 458, will definitively clear all misunderstandings I hope.

[58] Merritt, H.E., *Gears*, third edition, Isaac Pitman & Sons, 1954. [§0.01; §2.37; §5A.14, 27; §9A.06]

[59] Beam, Albert S., Beveloid gearing, *Machine Design,* Vol. 26, December 1954, pp. 220-238,. [§2.37; §3.01; §10.31]

[60] Mitome, K., Table sliding taper hobbing etc., (in two parts plus another part), *Journal of Engineering for Industry,* Trans ASME, Vol 103, November 1981. pp. 446-455. [§2.37]

[61] Smith, Leonard J. AGMA technical paper No. 89 FTM 10 (1989), *The involute helicoid and the universal gear,* ISBN 1-55589-539-5. For this paper (or a version of it) published again later; consult the references at Innocenti [62]. [§2.37; §3.01; §10.31]

[62] Innocenti, Carlo, Analysis of meshing of beveloid gears, *Mechanism and Machine Theory,* Vol. 32, 1977, No. 3, pp. 363-373. [§2.37]

[63] Abadjiev, V. and Petrova, D., Testing of the kinematic conjugation of the flanks active surfaces of gear-pairs of type hypoid, *Mechanism and Machine Theory,* Vol. 32, 1997, No. 3, pp. 343-348. [§2.31]

NOTES AND REFERENCES

[64] *Note.* With my use of the word 'ghostly', I intend to mean something different from the accepted meaning of the word 'imaginary'. Whereas *imaginary* might apply to the mental construct of the imaginary rack, because that thing, although unreal, is a *bounded body* in the kinematical sense, I reserve the word *ghostly* for application to geometric constructs such as lines, planes, algebraically defined curves surfaces and the like, which are not bodies in the kinematical sense and can freely interpenetrate one another. Refer also note [95]. [§1.29, 33] [§9B 09]

[65] *Note.* Whereas the idea *naked wheel* is a carefully defined geometrical abstraction, a *wheel blank*, as is well known, is a roughly hewn piece from which a finished, toothed wheel is to be cut. [§9A.08, 30].

[66] Giovannozzi, R., Meccanica.— Intorno alla transmissione del moto fra assi schembi mediante ruote elicoidali, *Atti della Academia Nazionale dei Lincei,* 1947, Serie Ottava, Rendiconti, Classe di Scienze finiche, mathematiche e naturali, Volume II, 1° semestre 1947, pp 586-595, Published in Roma by Dott. Giovanni Bardi etc. [§3.01, 51]

[67] Poritsky, H. and Dudley, D.W., Conjugate action of involute helical gears with parallel or inclined axes, *Quarterly of applied mechanics,* Vol VI, No, 3, October 1948, pp 193-214. [§3.01, 51].

[68] Giovannozzi, Renato. Giovannozzi appears to be the author of two books in Italian. Both contain an explanation of the phenomenon, by that time apparently well understood, of the consequences of there being two involute helicoids in contact at a point. Both were published by Casa Editrice Prof. Riccardo Patron, Bologna: (1) *Construzione di Machine, Vol II,* (first edition *circa.* 1964); (2) *Corso di Meccanica applicata alle Macchine, Vol. I,* (second edition *circa.* 1972). [This information (possibly corrupted by me by way of poor translation) was kindly offered by Dr Carlo Innocenti, of U Bologna.] [62]. [§3.01].

[69] *Note.* Whereas I use the words *parabolic hyperboloid*, others use *hyperbolic paraboloid*. Please find at §4.70 an explanation of my personal view on this not very hotly, but disputed matter.

[70] Euler, Leonard, (as quoted by Willis), his 'second paper' on the Teeth of Wheels, *Nov. Comm. Petr.*, ix 209, 1767. [§2.46]

[71] Sticher, F.C.O., see Appendix 2 in: *Contributions to geometric interpretations and methods in the study of spatial linkages,* his doctoral thesis, University of Sydney, 1972.

[72] *Note.* Recognising that the term *Hypoid* is the registered commercial property of the Gleason Company at Rochester, New York, I have used the term *hypoid* generically in this book to describe all offset skew gear sets where the shaft angle Σ is a right angle. It is a tribute to Gleason that nobody seems to have improved upon this word.

[73] Vogel, W.F., *Involutometry and Trigonometry*, Michigan Tool Company, 1945. This reference is extracted from Beam [59]. [§10.25].

[74] Shoulders, Walter L., High ratio right-angle gearing manufacturing. This is Chapter 21 of *Dudley's Gear H'book, second edition 1991*, McGraw Hill, New York etc. 1991. [§2.33]

[75] Zhang Wen-Xiang and Xu Zhi-Cai, Algebraic construction of the three-system of screws, *Mechanism and Machine Theory*, Vol. 33, No. 7, pp. 925-930, October 1998. [§4.62]

[76] DIN 868, 1976, *Algemeine Begriffe und Bestimmungsgrössen für Zhanräder, Zahnradpaare und Zahnradgetreibe*, see figure 24. [§P.01]

[77] Disteli, Martin. These references and the different cities of origin of this series of papers, which appeared in *Zeitschrift für Mathematik und Physik* during the period 1901-1911 approximately, are listed here as follows: (1) Band 46, 1901, pp.134-181, Karlsruhe; (2) Band 51, 1904, Heft 1, pp. 52-88; Strassburg; (3) Band 56, 1908, Heft 3, pp. 233-258, Dresden; (4) Band 59, 1911, Heft 3, pp, 243-298, Karlsruhe. The original German titles of the papers are not included here, but my free translation of them into English appears for the English reader's convenience within the text at §4.84. [The papers were kindly drawn to my attention by Prof Dr Helmuth Stachel, of TU Wien.] [§4.84].

[78] [void]

[79] [void]

[80] Simon, Vilnos, Transmission errors in loaded hypoid gears, *Proceedings of the 10th world congress IFToMM*, Oulu Finland, 1999, Vol. 6, pp. 2239-2244. [§2.25, 33]

[81] Litvin, F.L. and Gutman, Y., Methods of synthesis and analysis for hypoid gear drives of formate and hexiform, *ASME Journal of Machine Design*, Vol. 103, 1985, pp. 83-113. See Litvin [5] page 716 for details. [§2.25; §3.05]

[82] Bricard, Raoul, *Leçons de cinématique*, Gauthier-Villars, Paris, 1927. [The relevant passage is separated from the main chapters, copy obtained from Michel Fayet] [§2.37; §3.74].

[83] Hauser, Donald, Gear noise, This is Chapter 14 of *Gear H'book, second edition*, [§3.06].

[84] [void]

[85] Routh, E.J., *A treatise on analytical statics, with numerous examples*, Analytical statics, 2nd edition, in two volumes, Cambridge University Press, 1896. [§4.56],

[86] Litwin, Faydor L., Theory of gearing and application. This is the first chapter of *Gear Handbook, second edition*. [§5B.13]

NOTES AND REFERENCES

[87] *Note.* It is said that a book by author Woodbury, *History of the gear cutting machine*, 1958, contains mention of the reverse proposition of Bricard. Also, by the same author, there is *Studies in the history of machine tools*, MIT, 1972.

[88] *Note.* The alternative terms *phantom rack* and *imaginary rack* might be seen by some to be synonymous but, as I wish to use the terms, they are not exactly synonymous. The term *phantom* conjures up for me and well describes the elusive nature of the said rack in the imagined case where the two wheels are in mesh; for it can with truth be said in this case that the rack, running as it does between the two wheels, is invisible. The term *imaginary* on the other hand implies an imaginary reciprocating cutter with a material body and a fluted surface; it is imaginary in the sense that it may be standing in (in the imagination) for a rotating hob cutter or the cutting tool of a milling machine. [§5B.35]

[89] Drago, Raymond J., Gear types and nomenclature. This is chapter 2 of *Gear H'book, 2nd ed.* See page 2.18, where the possibility even of seizure is mentioned. See photo there also. [§5B.60].

[90] *Note.* If the symbols α and τ, both of which are shown at figure 3.01, are both used to describe the same circular hyperboloid, the angles (whatever their values) will clearly be complementary; they will sum to $\pi/2$. If the symbols are however used — as I use them — separately to designate different kinds of hyperboloids, this will not be so. Whereas I will on the one hand be using the throat radius *a* and the inclination angle α, written (a, α), to designate a *base hyperboloid* (§3.67), I will on the other be using the throat radius *t* and the twist angle τ, written (t, τ), to designate the *hyperboloidal shape of a naked wheel* (§5B.02). There is no underlying logic here; the circumstance is simply an historical accident of the algebra; it does however make certain distinctions clear, and I have kept the convention. It is important to be clear about this. [§3.04; §6.02].

[91] *Note.* The words *equiangularity* and *polyangularity* do not lend themselves to easy contraction. Please be aware that "with [Qx] at E" and similar phrases are shorter but nevertheless legitimate ways of saying "with equiangularity prevailing". Similarly, "with [Qx] at F" and "when [Qx] is not at E" are shorter but nevertheless legitimate ways of saying "with polyangularity prevailing". [§5B.11]

[92] Litvin, Faydor L., in *Gear H'book 2nd ed.*, 1991, see §1.1.5–§1.1.6, beginning page 1.8, and other references to the unwrapping of the slant-cut paper. Baxter, Meriwether L., in *Gear H'book 1st ed.*, 1962, see §1-10 at page 1-12 for a clearer explanation, but see the so called surface of action (not so clearly drawn) at figure 1-15.

[93] Prudhomme, R. and Lamasson, G., *Cinématique,* École nationale supérieure d'Arts et Métiers, École d'Ingenieurs, Dunod, Paris 1966. Go to section 3.7 headed *Enveloppe d'une surface dans virition.*, then go to sub section 3.7.3.0. [This book was kindly drawn to my attention by Dr Michel Fayet, of INSA Lyon.] [§9B.01]

[94] Grant, George B., *A treatise on gear wheels*, 1st edition 1899, 20th edition 1958, Philadelphia gear works Inc., Philadelphia 1899. Library of the Rose Hulman Institute of Technology USA. [§P.02]

[95] *Note.* See the idea *tick* at [1][§4.04], peruse note [64], and see that these earlier mentioned matters are relevant also here. [§9B.09]

[96] Baxter, Meriwether L., in *Gear H'book 1st edition*, see the sub section 1-7 headed, *Involute tooth profiles*, and see the two diagrams well explained at figures 1-11 and 1-12 on page 1-9. But see also Darle Dudley, in *Gear H'book 2nd edition*, discussing pointed teeth, page 4.15 at figure 4.7. See also items [16] and [97]. [§10.08, 37].

[97] Paul, Burton, *Kinematics and dynamics of planar machinery*, see Proof of the generating rack principle, p. 119. Prentice Hall, Englewood Cliffs, NJ, 1979. Here we find a much better explanation of the finer points of the matter. [§10.37].

[98] Fayet, Michel, On the reverse of one property of involute gears, *ASME Journal of Mechanical Design*, in press for Vol 24, June 2002. [5B.62].

[99] Baxter, Meriwether L. For a clear explanation of octoid teeth, go to *Gear H'book first edition*, 1962, page 1-15. [§5B.62].

[100] *Note.* Delivered by the present author at the IFToMM sponsored Second International Symposium on Cams and Cam Mechanisms (Kurvengetriebe), Berlin, by invitation in March 1989, a paper entitled "Screws and jamming in the general spatial cam" was not published among the Proceedings of that Symposium. Due for publication otherwise soon, it is available just now for inspection. [§12.04]

[101] Abadjiev, Valentin, On the helical tooth flank synthesis of skew-axes gear pairs. *Mechanism and Machine Theory*, Vol. 36, No. 10, 2002, pp.1135–1146. [§2.32]

[102] Olivier, Théodore, see [40], Olivier's earlier paper of 1839, *Bulletin de la soc d'Encourragement etc.*, Vol. xxviii, page 430. This reference is lifted from the footnote of Willis's book [12], on his page 151. [§4.84]

[103] Brauer, Jesper, Analytical geometry of straight conical involute gears, *Mechanism and Machine Theory*, Vol. 37, No. 1, 2002, pp 127-141. [§2.37]

SYMBOLS INDEX

ACRONYMS USED AND THEIR MEANINGS

BDL	Best drive line
CAD	Computer aided design
CDL	Centre distance line
CP	Circular pitch.
CPR	CP of teeth of rack
CPW	CP of teeth of wheel
DP	Diametral pitch (not used)
FAXL	F-axis line
HDL	Hydrodynamic lubrication
HTG	Helitangent
LH	Left handed
LHS	Left hand side
NP	Normal pitch
NPR	NP of teeth of rack
NPW	NP of teeth of wheel
PLN	Pitch line
PTH	Path
RH	Right handed
RHS	Right hand side
RSSR	Revolute spherical spherical revolute
SAX	Shaft axis
TNR	Table of numerical results
TVL	Transversal

GENERAL SPATIAL INVOLUTE GEARING

SYMBOLS FOR POINTS
ARE GENERALLY
UPRIGHT UPPER CASE ROMAN

A
    upper terminator of the CDL:
        [1.06]
    lower terminator of the
        rotating segment (A–B):
        [3.08] [3. 27]

B
    lower terminator of the CDL
        [1.06]
    upper terminator of the
        rotating segment (A–B):
        [3.08] [3.21,24,26,27]

C
    general point on the CDL:
        [1.37] [1.37,39-41]
        [7.03,08],
    lower terminator of the
        rotating segment (C–D):
        [3.09]

D
    special point on the CDL:
        [0.16]
        [1.40][1.36,40]
        [5A.20,22]
        [5B.09,12]
        [6.08]
    upper terminator of the
        rotating segment (C–D):
        [3.09]

E
    any point on the E-line
        [3.09]
        [4.64]
        [5A.21-22]
        [5B.09,22]
        [6.14]

F
    any point on the
        equilateral transversal
        [6.04]

G
    any point that is not
        upon the F-surface

H
    central point of the
        rectilinear array
        [4.38] [4.38-39]
        [6.13,17,34]

[Ht]
    that unique point H upon the
        common generator of the naked
        wheels where the smooth surfaces
        of the wheels are tangential
        [4.38-39,66]
        [5B.10]

J (also j)
    lower terminator of the transversal
        [4.47]
        [5A.16] [5A.16-17]

K (also k)
    upper terminator of the transversal
        [4.47]
        [5A.16] [5A.16-17]

L
    foot of the perpendicular
        dropped from E (or F)
        onto the shaft axis BJ
        [5A.21]

M
    foot of the perpendicular
        dropped from E (or F)
        onto the shaft axis AK
        [5A.21]
    the freely chosen 'middle point' in
        the open G-space mentioned
        and made use of in the hypoid
        literature.
        [2.29]
        [5B.13]

N
    foot of the perpendicular (of length h)
        dropped from a general Q onto the
        pitch line (see figure 1.08)
        [1.06]
        [3.52]

O
    origin at O_1
        the mid-point of the CDL
        [1.43]
    origins at O_2 and O_3:
        [5B.38]

SYMBOLS INDEX

P
    the pitch point:
        [S.02,07]
        [1.09]
        [1.09,16-22,26-30,41-42]
        [4.04,11]
        [5A.13]
        [5B.05,?]

Q
    the point of contact:
        [1.05] [1.05-07]
        [2.12-13]
        [3.26,56]
        [4.70]
        [5B.02]

[Qc]
    that point on the path of Q
    that bisects the intercept Z_l
    [3.57] [3.30,55,57]
    [5A.12-13]

[Qf]
    that point on the path of Q
    where the path
    pierces the F-surface
    [3.39]
    [4.27]
    [4.27,36,38-39,47,58,61,65,69]
    [5A.xx]
    [5B.13]
    [6.24,29]
    [10.15]

[Qm]
    that point on the path of Q
    where the algebraic sum of
    the changing radii q_2 and q_3
    are at a local minimum
    [3.56] [3.37-39,56,69-71]
    [4.01,06 15-18,29,40-42,43-56]
    [5A.13,22,23,25]
    [6.25-26]

[Qp]
    that point on the path of Q
    closest to the pitch line
    [3.55]
    [5A.09[
    [5A.11,13]
    [6.14,34]

[Qs]
    that point on the path of Q
    which is at the line of striction
    of the girdling hyperboloid
    [4.60,61,64],
    [10.13]

[Qt]
    that point on the path of Q
    where the intersecting base
    hyperboloids are tangential
    [3.71]
    [4.05,11]
    [5B.10]

[Qx]
    that point on the path of Q if
    such a point exists where the path
    cuts the path of the other Q
    [4.73,74]
    [5A.03,11,13,32]
    [5B.02,13,20,22,29,32,34,50]
    [6.03-10]
    [7.01,02,07]

R
    see Fig 9.03.

S
    see Fig 9.03.

T
    [4.05]

U
    at a slip track's U-turn
    [10.07]

V
    vertex of a
    Wildhaber cone

W
    that point on the core helix directly
    above point A in figure 3.03 allowing
    the angle ABW to be a right angle
    [3.05,07]
    [5B.38]
    [7.05]

Y
    the central point of the
    equilateral transversal
    [5A.17]
    [5B.02]
    [6.19]
    see also figure 6.05

Γ
: the general point that is free to roam the common generator of two generator-joined circular hyperboloids [4.04-24]

[Γd]
: [4.08]

[Γf]
: [4.27]

[Γm]
: [4.16,29]

[Γt]
: [4.07] [4.04,07,20-23,30-31] [5B.03,10].

SYMBOLS FOR ANGLES
ARE GENERALLY
REGULAR LOWER CASE GREEK

α
: in general the inclination angle of an hyperboloid but in particular the inclination angle of a base hyperboloid [90] [2.12] [3.04] [3.08-19,28,31,62] [4.75] [5A.23] [5B.04,17-18, 24,29,52] [6.39] [7.05] [10.01,24]

β
: angle turned through by the sliding velocity vector as we migrate from the centre H outwards along the transversal of the rectilinear array [6.35-39]

β_{SHIFT}
: angle turned through by the sliding velocity vector as we migrate from the point E outwards along the transversal of the rectilinear array. [6.44, 58] [10.39]

γ
: flank slope angle of the flank of a tooth [S.14] [7.16] [9.02] [10.14] [10.01,11,18,23,27]

δ
: angle of obliquity previously called the pressure angle [S.07] [2.10] [3.50,54] [4.45,58,59,61,64,75] [5B.15,19-21,34,40,43,49,53] [6.01] [10.01,11,18]. [11.05,20,23]

SYMBOLS INDEX

MISCELLANEOUS SYMBOLS
FOR RATIOS ETC. ARE GENERALLY
ITALIC LOWER CASE MIXED

c_H
    [3.17]
    [5B.30]

j
    the so called radius-ratio
    $[j = f_2/f_3]$
    [6.42]
    [10.39,46,51]
    [11.22]

k
    angular velocity ratio
    $[k = e_2/e_3]$
    [S.02,03,07]
    [1.07,10-11,21,36,38-45]
    [3.33,36,51]
    [4.35]
    [5A.23]
    [5B.17,45]
    [6.01]
    [11.03]

p
    pitch of the gear set (mm/rad)
    [S.01,02]
    [1.10,15-17,20]
    [2.02]
    [5B.09,58]
    [6.38]

p_{FLANK}
    pitch of the involute helicoid
    at the flank of a tooth (mm/rad)
    [9A.08]
    [9B.01]
    [10.06]

v
    linear velocity (mm/sec)
    [1.09,51]
    [4.26]
    [5A.09]

w
    a useful ratio:
    [5B.17]
    [6.68]

ω
    angular velocity (rad/sec)
    [1.09,51]

SUBJECTS AND AUTHORS INDEX

AUTHORS NAMES, MARKED WITH AN ASTERISK (*), APPEAR ALPHABETICALLY AMONG THE GENERAL ITEMS LISTED HERE. ALGEBRAIC SYMBOLS ARE ALSO LISTED ALPHABETICALLY, BUT THEY APPEAR UNDER THE SEPARATED SUB-HEADINGS POINTS ANGLES RADII LENGTHS AND RATIOS IN THE PRECEDING INDEX, THE SYMBOLS INDEX. ERRORS DILLEGENTLY SOUGHT FOR BUT UNDETECTED BY THE AUTHOR ARE REGRETTED.

Abadgiev*
 [2.16]
Accuracy
 unnecessary at assembly
  [1.01]
  [2.09,25]
  [3.03,33-34,35-36,72]
  [5A.06]
  [5B.65]
  see also Errors
 of cutting
  [1.01]
 and rigidity
  [4.54,56]
Addenda and/or dedenda
 [3.39]
Algebraic relationships
 [1.09,14,17,37,42,47,51]
 [2.02,12,17,19,22,23,38]
 [3.17,23,33,53]
 [4.02]
 [5A.27-28]
 [5B.05.17]
 [6.19-21,33-34,38-41,44-50,66-68]
 [7.17,33]
 [10.09,19-29,42,44,67]
Altitude and azimuth
 [10.33]

Analysis-v-synthesis
 [1.13]
 [4.01]
 [5A.02,15]
 [5B.08]
Angle of inclination α
 [3.04]
 see also synonym Angle α
Angle of obliquity δ
 [S.07,14]
 [6.59]
 [10.11]
 [11.05,20]
 see also synonym Angle δ
Angular pitch
 of teeth
  [5B.26]
Angular velocity polygon
 [1.14]
 [5B.06]
Angular velocity ratio
 see Ratio k
Annular discs
 [5B.13]
Apices
 of the W'haber cones
  [5B.04]
Applications
 [2.05]
Architecture
 [2.32,34]
 [5A.19]
 [6.37-39]
 [12.15]
Asymptote
 at the curved intersection of
  two circular hyperboloids
  that share a generator
  [4.31]
 at the discontinuity in the
  surface at figure 1.04
  [1.19]

Asymptotic cone
  of an hyperboloid
    [4.02,32]
  of a naked wheel
    [5B.05,11,13]
Axes
  O-xyz
    [1.06,41]
    see also Triads
  Q-uvw
    [5B.06-07]
    see also Triads
Axial
  edges
    [8.12]
  location
    [8.14]
  offset
    [3.33]
    [8.14]
  withdrawal
    [8.17-20]
Axis Axis and Line
    [4.05] [4.05-09,29]
Axodes
  for wheel-wheel
    [1.26-30,31-33,39-40]
    [4.01,04,10-14,35,39]
    [5A.30-31]
    [5B.02]
    [6.29-32]
    [10.36]
  for wheel rack
    [10.38,43,48]
    [11.02]
Backhandedness
    [8.01,12-16]
Backlash
    [2.05]
    [3.34]
    [9A.33]
Bald wheels
    [6.54]
Ball in tube
    [4.66,68]
    [6.11,13]

Bär*
    [3.24]
Base
  circle
    [3.20]
  cylinder
    [3.15]
  hyperboloids
    [S.06]
    [3.15,67-71]
    [4.35,39,74]
    [5B.16,23,29-30,36,52]
    [7.12]
    [10.03]
    [11.27]
Basic
  rack
    [2.27]
  ruling
    [3.60]
Basic equation
  of meshing
    [1.13]
    [2.32]
  of the screw
    [1.09,51]
    [3.53]
    [12.20]
Baxter*
    [2.28-29]
    [5B.13,40]
Beam*
    [P.02]
    [2.37]
    [3.01]
    [7.17].
Bed spring
    [3.13-16]
    [5B.25]
    [7.04,23,26].
Beer mats
    [3.43]
Bending
    [1.03]

SUBJECTS AND AUTHORS INDEX

Benign
  curvatures
    [2.07]
    [5B.19,20-22,39]
  region
    [1.25]
Bennett mechanism
  [12.13]
Best
  drive line BDL
    [3.48,49]
    [4.63]
    [5B.40]
    [6.38]
    [10.49]
  facet
    [3.03,48]
    [5A.23, 24-26]
    [5B.09,36]
Beta shift
  [6.44]
  [10.39]
  [11.19]
  see also Symbols β_{SHIFT}
Bevel sets
  both straight and spiral
    [1.04,50]
    [2.06]
    [4.31]
    [5B.13,61-62]
    [6.62]
Beveloid sets
  known also as conical
    [P.08]
    [1.50]
    [2.37]
    [3.01]
    [5A.30]
    [5B.14]
    [6.61
    [9A.07]
    [10.31]
Binormal
  [2.39]

Black bag
  [2.19]
Blooming
  of the slip track
    [3.20]
  of the single point of contact
    [5B.58]
  at all stations
    [4.25]
Blunt-bottomness
  [11.14]
Body material
  [11.04] [11.03,04-06]
Box (gear box)
  [1.01,06]
Branches
  of the involute helicoid
    [3.15,59-64]
    [8.04]
    [10.06]
  of the slip track
    [3.59-64]
    [5B.52]
Bricard*
  [2.37]
  [3.74]
  [5B.62]
Buckingham*
  [2.28,30]
  [3.01]
Bumpiness
  [2.13]
Buttons
  [9B.16]
CAD
  [11.25]
Cartesian equation
  for the hyperboloid
    [4.02,25,35]
Cayley*
  [5B.42]

CDL
  [S.01]
  [1.06,39-46]
  [4.26,29]
  [5A.23]
  [5B.05,21]
  [6.22]
  [11.22]
Central
  plane of the rack
    [11.02]
  point of the ISA
    [4.13,14]
  reference location
    [9A.18-20]
Centre distance C
  [S.01]
  [1.06,37]
  [4.26]
  see also Length C
Chapter 22 in [1]
  [5B.05]
  [6.01]
Chasles*
  [4.25]
Circle
  of intersection
    of the coaxial hyperboloids
    namely the key circle
      [4.76]
  of throat location
    [5B.11-12]
Circular
  appearance of an
  ellipse of obliquity
    [5B.44]
  contact patch
    [5B.39]
  hyperboloids
    [3.04]
    [4.03-24,74-76]
  pitch (CP)
    [5B.27]
    [5B.24-33,35,36]
Circumferential slip
  [6.31]

Clearance, backlash
Closure equation
  [1.13]
Coaxiality
  of the circular hyperboloids
    [4.74,76]
    [5B.23]
  of the involute helicoids with
  the axes of the wheels
    [5B.55]
    [9A.01-06]
Collision
  [5A.10]
Comfort discomfort
  [6.03]
Common generator
  of the base hyperboloids
    [3.29] [3.67-69]
    [5B.18]
  see also Path
Common perpendicular
  between the two shaft axes
    [S.01]
    [1.06,37]
    [4.04]
    see also CDL
  between a path and either
  one of the shaft axes
    [5B.16]
    see also Radius a
  between the helitangent at [Qf]
  and either one of the shaft axes
    [5B.10-11]
    see also Radius b
Common tangent plane
  at generator-joined hyperboloids
    [4.01,07,20-23]
    [5B.03,10-11]
    see also Plane of adjacency
Complementary angles
  [3.04]

SUBJECTS AND AUTHORS INDEX

Cone-hugging
  of the naked wheels
    [4.36]
    [5B.02,24]
    [6.29]
Cones of Wildhaber
  [4.36]
  [5A.23]
  [5B.02,04,11,13-14,37]
  [10.07,16]
Congruence
  of the fifth kind
    [5A.24]
    [5B.21]
Conical
  [9A.06]
Conical blanks
  [9A.12]
Conicality
  [10.05]
Conicity
  [5B.05]
Conjugality
  [1.04,07]
  [2.01] [2.01,32]
Constant
  linear velocity of Q
    [2.04]
    [4.71]
  transmission force
    [4.50]
  sum $[F = f_2 + f_3]$
    [6.06]
    [10.08]
Constraint and/or overconstraint
  [12.03]
Construction
  for angle χ
    [5B.37]

Contact
  normal
    [1.07,13]
    [2.12]
    [3.36]
    [5B.50]
  patch
    [S.05]
    [1.03]
    [2.07,25]
  ratio
    [S.17]
    [2.18]
    [3.34]
    [5B.34]
    [5B.25,27-34,49,53,55]
    [7.25]
    [9A.05]
    [11.12]
Continua
  [9A.03]
  [10.04]
Conventional
  paths
    [4.26]
  triad
    [7.15]
    [10.30,43]
Cookie-cutter
  [10.05]
Core
  cylinder
    [3.14]
    [11.03]
    see also synonym Throat cylinder
  helix
    [3.12]
    [3.12-14,35]
    [5B.25,29-30]
    [7.04-05]
    [8.02,08]
    [9A.02]
    [10.01]
    [11.05]
Corrugations
  [6.30]

CP
    see Circular pitch
C-radii
    see Radii c
Crests
    [11.03,06-07]
Crossed helicals
    [P.01]
    [1.39-46]
    [2.27]
    [3.01,16,64]
    [4.68]
    [5A.12,30-31]
    [5B.60]
    [5B.05,12,19-25,37,60]
    [6.01,11,23,27]
    [7.03,09]
    [9A.06,17]
    [10.22,29,37]
Crowning
    [S.04]
    [2.06,09,25]
    [3.05]
Cubical box.
    [6.61]
Cucumber shaped silhouette
    [S.05]
Curvature
    of the slip track
        [3.24]
    of the tooth profile
        [0.14]
        [3.03,19,24]
        [5B.20,22]
Curvature triad
    [3.21,24]
        see also synonym Triad at B
        see also Triads various
Curved
    line contact
        [P.03]
        [2.25]
        [9A.35]
    surface of action
        [2.25]
        [3.05]

Cuspidal edge
    [3.15]
    known also as
    Line of regression
Cutting
    [1.03]
    [5B.27]
Cutting velocity
    across the flanks of the rack
        [10.53-55]
Cycloidal form
    [1.04]
Cylinder
    elliptical
        [4.25]
Cylindrical helicals
    with crossed axes
        see Crossed helicals
    with parallel axes
        [3.01]
        [5B.04,58-59]
        [10.29]
        see also Parallel cylindrical helicals
        see also Planar gearing
Cylindroid
    at the CDL
        [1.11] [1.11,15,37-38,44]
        [2.35]
        [4.04,12,68]
        [5B.42]
        [12.06]
    at the meshing
        [10.41,47,50-51]
    the unmentioned
        [10.52]
    of cranks RS.
        [12.12]
Dancing
    [5B.39]
    [8.38]
de la Hire*
    [2.10]

SUBJECTS AND AUTHORS INDEX

Decapitation.
[6.59]
[11.01,06,14]

Dedendum
see Addendum

Dedicated algebra
[6.37]

Degenerate slip track geometry
[3.16,22,58-64]
[5B.24-25]

Departed sheets
see Singly, Doubly departed

Design
[2.26-28]
[5A.14-15]

Developable non-developable
[3.11]
[6.30]

Dichotomy
[1.13]

Difficulty of assembly
[3.03]

Digging in
[5A.08]

Disability angle δ
[2.10]
[3.48,54]
see also synonym Angle δ

Disjunction
[2.28,32]

Displacement
by half the CP
[5B.32]

Distance symbols
see Length and
distance symbols

Disteli*
[P.02]
[2.37]
[4.84]

Dooner & Seireg*
[2.36]
[3.35]

Double point
at [Qx]
[5A.11,16]

Doubly departed sheets
[1.31]
[3.39]
[4.36]
[5B.11]
[5B.02-04,10-13,16,24]
[6.01,29]

DP not mentioned

D-radii
see Radii d

Drago*
[3.01]
[5A.30]

Drawing work
[5B.27]

Drive and/or coast
[S.03]
[1.05]
[2.13-14]
[3.30]
[4.72,76]
[5A.03,10]
[5B.15,16,19,21-24,49]
[9A.13]
[11.13]

Dual vector
[10.44]

Dudley*
[2.28,30,33,46]
[6.09]

Dyson*
[2.28,31-33]

E-circles
[3.41]
[4.76]
[5A.21,23]
[5B.02,23,32,34,36,59]
[6.32]
[7.07]
[10.03]

Eclectic workers*
[2.34-36]
Edges
[3.40]
[11.05]
Efficiency
[2.25]
[11.13]
[12.04]
Elastic deformation
[S.05]
[3.01]
E-line
[S.07]
[5A.03,21-22,23]
[6.07,21]
[12.19]
Ellipses of obliquity
[S.07]
[5B.16,21,40-44,47,49,52]
[6.01,60-62]
[7.22]
[11.21-23]
[12.12]
Elliptical cylinder
containing the locus of the points of intersection of the perpendicularly intersecting generators of an elliptical hyperboloid
[4.17,18]
supporting the curved part of the quartic at the intersection of two hyperboloids sharing a common generator
[4.32]
Elliptical hyperboloid
in general
[3.04]
[4.02,07-10]
girdling the axes
[4.07-11,40,25,40]
Embryo gear set
[5B.02]
Endless travel
[3.56]

Ends of Z_i
[3.23,32,57-59]
see also Degenerate slip track geometry
Energy
Enumerative geometry
[2.25]
[6.08]
Envelope
surrounding the kinematic possibilities
[3.08,09]
[11.01,03,15-16,28]
of planes enveloping the involute helicoid
[9B.02]
[10.48]
E-point
see Point E
Equation of meshing
[2.32]
Equiangular
architecture
[0.16,20-21]
[1.41,46]
[4.38,64]
[5A.19-29] [5A.16]
[5B.09]
[6.02,08,14]
rack
[5B.35]
Equiangularity
[5A.02-04,16,20]
[5B.02,11,48]
[6.01,27]
[7.08]
[11.11]
Equilateral transversal
[4.26] [4.70-71]
[5A.16-17,24,32]
[5B.48,53]
[6.04,05,07,11]

SUBJECTS AND AUTHORS INDEX

Equivalent linkage
  four-bar
    [**2.12**] [2.12-14]
    [3.28,40]
    [4.24]
  three-bar
    [1.05-08]
    [2.12]
    [3.37]
E-radii
  see Radii e
Errors
  permissible at assembly
    [S.03]
    [2.04]
    [3,03,33-34]
    [5B.65]
    [11.10]
    see also Accuracy
Euler*
  [P.02]
  [**2.46**] [2.26,46]
Exotic architecture
  [S.10]
  [4.72-73]
  [5A.13,19]
  [5B.61]
  [8.15,33]
Exoticism
  [S.10]
  [4.72-73]
  [6.09-10,28]
  [8.01-05,20-21]
  [9A.15]
  [11.05]
  [12.18]
External and/or internal
  axodes
    [1.30]
  meshing
    [1.10,42]
    [4.14,49]
Facet at B
  [3.09,11,21]

Families of generators
  [3.04]
  [4.02,18]
FAXL
  [S.08]
  [**1.43**][1.44]
  [3.39]
  [4.26,27,58]
  [5A.17-19,23]
  [5B.20,41]
  [6.22]
F-circles
  [3.41]
  [4.76]
Field of helices
  [1.15]
  see also Helices
Finite screw displacements
  [9B.17]
  [10,34]
First and higher orders
  of tangency
    [4.18]
Five planes
  [5A.26]
Fixed parameters
  [S.01]
Fixity of the paths
  [9A.16]
Flank slope angle γ
  [10.10,43].
  see also Symbols Index
Flank slope tangent
  [10.41]
Flanks of teeth
  [1.07]
  [5B.23,28-34]
  [6.45,63]
  [7.14,20,23,26,30]
  [9A.02-05]
  [10.57]
  see also Profiles
Flat action
  [10.18]

475

Flexibility of the box
[2.04-05]
[9A.34-35]

Flush configuration
[9A.18,30]

Force
[1.03]
[2.21]
[5B.17]
[6.02]
[12.12]

Forehanded and backhanded design
[8.16]

Formulae
see Algebraic relationships

Fortuitous facts
[4.38]

Four
kinds of rectilinear
parabolic hyperboloids
[4.78-83]

F-radii
see Radii f

Friction
[2.11]
[3.23,31]
[6.02]
[12.04]

Frictionless
cylindrical sliders
[4.15]

Frilly garters
[4.36]
[5B.13]
[10.07]

Fronts and/or backs
[2.13]
[5A.10-11]
[5B.32]
[9A.08-10]
[11.04]
see also Drive and/or coast

F-surface
[S.08]
[1.43,44]
[4.26,27,58,65]
[5A.16] [5A.31-32]
[6.07-11,22,18-24]
[8.13,15]
[12.07,12,16]

Fundamental law
vector proofs of
[2.16-23]
manifestations of
[2.27-20]
otherwise
[S.02-03]
[1.14]
[2.02]
[2.01-08,16-23,25-26,33,39-46]
[3.02,05,07,52]
[4.40-42]
[5A.09,11]
[5B.14]
[6.01]
[10.20]

Fundamental triad
[3,44]
[7.10]
[10.43]
[12.20]
see also Triads various

Gashes
[5B.34,36]

G-circles
[1.07]
[3.37,41]
[5A.21]
[6.25]

Gear bodies
[2.12,36]
[3.35-39]
[4.39]
[5A.33];
[11.30]

Gear box
[1.06]
[2.20]

476

SUBJECTS AND AUTHORS INDEX

Generated teeth
  in hypoid practice
    [P.03]
    [2.01]
  in involute planar practice
    [10.48]
Generated terminator
  [11.26]
Generation
  of planar teeth with
    an enveloping line
    and/or by a point
    [10.48].
  of spatial teeth with
    an enveloping plane
    and/or by a rack
    [10.57]
    [11.05]
Generation by a plane
  of the involute helicoid
    [9B.01 *et seq.*]
    [10.58]
Generator-joined hyperboloids
  [4.03-24]
Generators
  of the girdling hyperboloid
    cutting one another
    perpendicularly
    [4.30]
  of the cylindroid
    [4.13]
  of the cones of Wildhaber
    [5B.04]
  of the naked wheels
    [5B.04]
Geometrical/physical Q
  [3.56]
Ghostly surfaces
  [1.29,33]
  [6.56]
  [7.09]
  [9A.05]
  [10.58]

Giovannozzi*
  [2.37]
  [3.01]
Girdling hyperboloid
  in general
    [4.21]
    [4.17,21-23,58,62,64]
    see also Hyperboloid girdling
  for a given path
    [4.58-63,70-71].
G-radii
  see Radii *g*
G-space
  [6.07]
  [6.07-09]
Gutman*
  [3.05]
Gyratory motion
  of the triad at Q
    [3.50]
    [4.61,63,70]
Half measure
  [4.77]
Half the CP
  [5B.32,50]
HDL
  hydrodynamic lubrication
    [S.05]
    [1.03]
Helibinormal
  [3.50,52]
Helicality
  [5.10]
Helices
  surrounding the pitch line
    [1.15,42]
    [2.31,39]
    [4.26]
    [5A.08]
    [6.31]
  surrounding a shaft axis
    [3.12] [3.12-14,35]
    [5B.28-29]
    [9A.09]

Helicoid
see Involute helicoid
Helicoidal
surface
[3.14]
velocity field
[1.09,51]
[4.26]
[5A,08]
[6.12,16]
[7.09]
[12.10]
Helicopter
[2.05]
Helinormal
[3.50,52]
Heliradius h
[1.38,52]
[5A.23]
Helitan view
[6.48,60]
[11.22]
Helitangent
[1.06]
[3.45,47,52]
[4.21]
[5A.23(a),(g)-(h), 24]
[5B.02,05,09-11,25,35-36,41,48]
[6.31]
[7.22]
[10.54]
Helix
[9A.01-02]
Helix angle
τ at an hyperboloid
[3.04]
see also Angle τ
θ at the flanks of teeth
[1.15,39,46,49]
[5B.08,36,58-59]
[7.01,08,16]
[10.26] [10.10,26-29,31]
see also Angle θ

Hob angle γ
at the cutter
at the tooth
see Angle γ
Hobbing
[S.14]
[3.02,03]
[5A.05]
[5B.36]
[10.03,06]
Hobbing cutter
[3.45]
Home points
for B and Q
[3.26]
Hooke's joint
[10.33]
Horizonal (sic) edges
[4.76]
[5B.13]
[5B.23,30-31]
Horizontal-vertical
[1.06]
[5A.29]
[10.20]
Hugging
[4.21]
[5A.04]
[5B.13]
see also Cone hugging
Hyperbolic congruence
[12.11]
Hyperboloid
elliptical
[4.02,21-23]
circular
[3.04,14]
[4.02].
see also Hyperboloids
girdling
see Girdling hyperboloid
Hyperboloidal conical
[0.01]

SUBJECTS AND AUTHORS INDEX

Hyperboloids
  intersecting
    along a common
    generator
    [S.06]
    [1.27,28]
    [3.29,62-65]
    [4.01-24,34]
    [5B.03,04,10-11,18,30]
  intersecting
    in circles coaxially
    [4.30-31]
    [5B.23]
Hypoid gearing
  [P.03]
  [1.04]
  [2.01,25-33]
  [7.19]
Hypotheses
  [3.02],
Imaginary rack
  see note [64].
Inaccuracy
  [0.03]
Inclination
  angle of
    see Angle α, β
  of the slip track
    see Angle σ
Inertia
  [2.05]
Inevitability of irregularity
  [1.05]
Infinity
  in general
    [4.31]
  of slip tracks
    [3.34]
    [9A.21].
Innocenti*
  [2.37]
  [3.31]

Inside and/or outside
  [11.06]
Inspection
  [2.45]
Instantaneous screw axis, axes
  [1.08-09,11]
  see also synonym ISA
Interference
  between the naked wheels
    [5B.03]
  between teeth
    [5B.17,55-57]
    [6.02,63]
    [11.02,04]
Internal
  meshing of teeth
    [1.10]
    [3.32]
  rolling of the axodes
    [1.30]
  forces
    [4.54]
Intersecting
  axodes
    [1.26-30]
    [4.14]
    [8.13]
  base hyperboloids
    [3.67]
  coaxial circular hyperboloids
    [4.74]
  ellipses of mobility
    [11.22]
  non-coplanar circles
    [3.41-43]
    [6.32]
  travel paths
    [6.03]
    see also Travel paths
Intersection of
  base hyperboloids
  and involute helicoids
    [5B.52]
  the ellipses of obliquity
    [5B.40]

Inversion
    [1.12]
    [4.26,70]
    [10.56]
Involute
  action
    [0.13]
    [1.04]
    [2.01]
    [3.01,02]
    [5B.13]
  gears
    [2.04]
    [5B.13]
  helicoid
    [S.03]
    [3.01,03,08-19,34-35]
    [5B.29-30,52]
    [7.04]
    [9A.01-05]
    [9B.01]
    [12.05,14]
  spirals
    [3.18]
Iota ι
    [10.42]
    see also Symbols Index
Irreversibility
  of gear sets
    [2.11] [6.02]
    [11.23]
    see also Reversibility
ISA
  pinion-rack
    [10.46-47]
  wheel-rack
    [10.05,37-38,44-47]
    [11.02]
  wheel-wheel
    see Pitch line
Jamming
    see Friction Reversibility
Joints
    [1.05]

Junction station
    [5B.27]
    see also [Qx]
Karsai*
    [2.21]
Key circle
    synonym for E-circle F-circle
    [4.76]
    [7.02]
Killeen*
    [5B.17]
Kinematic
  chain (loop)
    [1.01,05-06]
  equivalence
    [3.40]
Kinetostatics
    [3.01,03]
    [4.17]
    [11.10]
    [12.04]
Knurled knobs
    [1.12]
Konstantinov*
    [1.13-15,47]
Korestilev*
    [2.34]
Lagutin*
    [2.28,35]
Lane changing
    [9A.21-22]
Law of the speed ratio
    [6.40] [6.53]
    [11.13]
Least
  distance q from the pitch line
    [2.02] [2.02-03,05]
    [3.53]
    see also Distance q
  velocity of crosswise sliding
    [1.14]

Legitimacy
    [2.17,36]
    [3.07,38]
    [4.35,40-42]
    [6.17]
    see also Fundamental law

Length-measured pitches of teeth
    [10.17]

Line
    of action
        [2.12]
        [3.27]
        [4.48,51]
        [5A.12]
        [5B.19,20]
        see also Length Z_L
        see also Zones of action
    of contact
        see Straight line of contact
        see also Curved line of contact
    of striction
        [3.38]
        [4.13,14,60,61,64]
        of the girdling hyperboloid
        [10.13]

Line geometry
    [12.02,05-13]

Line symmetry
    of the lines of action
    about the CDL
        [5B.20-22]

Line X
    [10.49]

Linear
    array of velocity vectors
        [6.33,36,66]
    complex
        [1.49]
        [2.15] [2.17,20,35-36]
        [3.47]
        [5A.08-09]
        [5B.40,58]
        [12.08,20]
        see also Special linear complex
    congruence
        special of the fifth kind
        [5A.24]
        [5B.21]
        see also Special linear congruence
    velocity vectors   [4.26]
        see also Linear velocity
        see also Rubbing velocity

Links 1, 2, 3 (and 4)
    [1.08]
    [10.39]

Literature surveys
    [2.25-38]

Litvin*
    [1.08,15]
    [2.28,31,32]
    [3.05]
    [5B.13]
    [12.11]

Load bearing
    [7.32]

Load distribution
    [2.03]

Location of the meshing
    [2.27]

Locus
    of perpendicular
    intersections
        [4.22-23,25,60]
    of [Qm]
        [4.43-49]

Loop
  of three links
    [1.05-06]
    [3.51]
  of four links
    [2.12,14]
Lubrication and wear
  see Wear and Lubrication.
Mac Cord*
  [P.02]
  [1.38]
  [2.37]
  [3.61]
  [4.84]
  [5B.12,39]
MacCullach*
  [4.25]
Machining
  in general
    [S.14-15]
    [5A.05]
    [5B.28,36]
    [6.65]
    [10.25,32-35,57]
    [11.05,09]
  at worms
    [7.18]
Manifestations
  of the fundamental law
    [2.17-20]
Manual
  [11.14]
Mating involute helicoids
  [3.25]
  [5B.28]
Mean contact point
  [P.05]
  [2.29]
Mechanical
  computer
    [1.13]
  significance of [Qf]
    [4.65-67,69]

Merritt*
  [P.02]
  [2.37]
  [5A.14]
  [8.06,14]
  [9.xx]
Meshing
  [3.67]
  [4.35]
Meta-statics
  [4.55-56]
  [6.26]
Middle point
  of the path
    [3.37-39,46]
    see also Point [Qm]
Milling
  [3.02-03,11]
  [5A.06]
  [5B.36]
  [10.03]
Minimum
  length of changing transversal
    [4.58,70]
  length of Z_L
    [5B.20
Minkov*
  [1.14]
  [2.28]
Misalignment
  of hypoid shafting
    [P.03]
Misnomer conical
  [9A.06-07]
Mitome*
  [2.37]
Mobility
  [1.01]
  [9A.23-25]

SUBJECTS AND AUTHORS INDEX

Naked wheels
[S.06]
[1.34-36]
[4.01,20-23,36-39,66,73]
[5A.04]
[5B.02] [5B.02,08-12,23,36-48]
[6.01,12-13,29-32]
[7.09]
[10.07,35,43]

NC tools
[2.24]
[3.03]

Neutral stability
[4.56]

Noise and vibration
see Vibration and noise

Non-departed sheets
namely axodes
[1.32,35,41]

Non-equiangular cones of W'haber
[5B.13]

Non-intersecting paths
[4.72-73]
[8.17]
see also Travel paths skew

Non-involute teeth in gearing
[P.03]
[2.35]
[3.05]

Non-reversibility
of gear sets
[3.23]
[6.02]
see also Reversibility

Non-square hypoids
[1.25]

Normal pitch NP
or synonym Base pitch
[5B.27]
[5B.25,27-34,35,51]
[6.48]
[7.01]

Nose-end and other end
[7.25]

NP
see Normal pitch

Numbers N of teeth
[5B.27-33,34]

Numerical values
[1.06,11,43]
[3.68-69]
[5B.17,19,20,26,28,30,33]

Oblique
[6.01]
[11.02]

Oblique-rolling vehicular wheel
[10.40]

Obliquity
angle δ of
[5B.15]
see also synonym Angle δ
synonym for Pressure angle

Obtuse Σ
[1.25,10-11,16-25]
[2.15]
[4.34]
[8.12-13]

Occluded troughs
[11.08]

Offset
[P.03]
[1.34,51-54,55]
[3.33]
[5B.12,20,37,53]
[8.12]
see also synonym Radial offset

Offset involute worms
[S.14]
[7.19]

Olivier*
[P.02]
[2.37]
[3.66]
[4.84]
[5B.21]
[8.07]

Olivier's line
  [3.66]
  [4.84]
  [5A.29]
  [10.47,50-51]
One-start two-start etc.
  [7.01,06]
  [9A.03]
Oops
  [7.29]
Opposite branches
  of the same helicoid
  [3.15]
  [10.06,29]
Optimisation
  [P.09]
  Origin O
  [1.06]
Orthogonal projections
  of the linear array
  [6.16]
Other laws
  [2.38]
Overall strategy
  of the found mathematics
  [10.66]
  [12.14-20]
Overconstraint
  [P.03]
  [1.01,49]
  [2.05,18,25]
  [3.03]
  [5B.26]
Paper rack
  [10.48]
Paper teeth
  [10.18,48]
  see also Flat action

Parabolic hyperboloids
  considered together
  [4.77]
  [6.23]
  considered separately
  [S.08]
  [1.37-38]
  [4.77] [4.58,68,77]
  [5B.21]
  [6.13,17]
  of linear velocity vectors
  [6.11-21]
Parallel cylindrical helicals
  [1.47-49]
  [3.02,10]
  [4.25]
  [5B.58-59] [5B.40-41]
  [9A.04]
  [10.29]
  see also Cylindrical helicals parallel
Parallel planes
  delineating the CDL
  [1.06]
  as used by Korestilev
  [2.34-35]
Parameter of distribution
  [4.13]
Parametric equations
  [4.02]
Parkin*
  [2.22]

SUBJECTS AND AUTHORS INDEX

Path
  of B
    [3.08,09,19]
  of Q
    [1.07]
    [2.25]
    [3.05,07,26]
    [4.27,35,40-42]
    [5A.32]
    [5B.07,15-16,33]
    [6.38,54]
    [8.14,17]
    [12.20]
  dependency of [Qm]
    [4.46,69]
    [6.15]
  independency of [Qf]
    [4.38,69]
    [6.15]
  upon the transversal
    [6.17]

Peripheral plane
  of the conventions triad
    [7.15]

Perpendicular intersection
  of the generators of the naked
    wheels with the generators
    of the base hyperboloids
    [4.76]
  of the generators of
    the girdling hyperboloids
    [4.22-23,25,60]
  of tangent planes
    [4.01,09]

Petrova*
  [2.16]

Phantom rack
  [5B.35,51]
  [6.64]
  [7.15]
  [9B.23]
  [10.03,37,43]
  see also Imaginary rack.
  see also Note [88]

Pinion
  in general
    [7.20]
    [8.09]

Pismanik*
  [2.32]

Pitch
  complex
    8.01
  helices
    [3.51]
    [4.21]
    [5A.08-09]
    see also Helices
  line
    [P.06]
    [S.01]
    [1.09] [1.05-12, 13,14-15,16-22]
    [2.37,41-43]
    [3.02,51]
    [4.10,35]
    [5A.03]
    [5B.05]
    [6.06,21,38]
    [8.13]
    [12.17]
  of the gear set (mm/rad)
    [P.05-07]
    [S.01]
    [1.10,15-17,20]
    [2.02]
    [5B.09,58]
  of the core helix (mm/rad)
    [3.03]
    [10.01]
  of the flanks of teeth (mm/rad)
    [8.08]
    see Symbols Index for p_{FLANK}
  point
    [P.05]
    [1.09] [1.45]
    see also Point P
  of teeth (mm)
    [5B.26]
    see also CP, DP, NP, etc.
    see also Length-measured pitches

Pitch (in terms of Wildhaber)
  pitch cones
    see Cones of Wildhaber
  pitch surface
    [P.05]
  pitch vertical
    [5B.04]
Pitches of teeth
  [10.07]
Pitting
  [1.03]
Plain polyangularity
  [S.08]
  [6.28][6.10-11,28,38]
Planar
  spur gearing
    [S.02,16]
    [1.01,04,10,15,24,47-49]
    [2.03,06,10,40,46]
    [3.02,16,20,61]
    [4.39,65]
    [5A.10]
    [5B.18,35,53,58-59,64]
    [6.09,15]
    [9A.04,17]
    [10.01,11,29,37]
  involutes
    [S.02]
    [2.07]
    [3.18,20]
    [5B.36,58]
    [10.01,03,48]
Planarity of the crests
  [11.07]
Plane of adjacency
  [2.07,12]
  [3.25] [3.25,27-28,35-36]
  [5A.26]
  [5B.24,32]
  [7.15]
  [10.14]
  [11.11]
Plane of the E-circle
  [10.03]

Plane of the E-radii
  see Radiplane
Plane of Wildhaber
  [3.46]
  [5B.04]
Plateau
  [5A.03]
Plücker*
  [4.02,25]
Plurality of Q
  [2.28]
Point contact
  between the cones of Wildhaber
  [5A.04]
Point of contact
  [5B.35]
  [9A.32}
Pointed teeth
  [S.17]
  [6.09,28]
  [11.01,04-06]
  see also Troublesome pointedness
Points O_2, O_3
  [5B.36]
Polar plane
  [3.03,47,49]
  [5A.10,11,24]
  [5B.09,15,20,40-44,59]
  [11.21]
Pole
  [1.26]
Polodes
  [1.26]
Polyangular architecture
  [S.08-10]
  [1.46]
  [4.68]
  [5A.05,31].
  [5B.04,23]
  [6.02,04]

SUBJECTS AND AUTHORS INDEX

Polyangularity .
  plain
    [4.63,68]
    [5A.06,30-31,32]
    [5B.53]
    [6.02,04,07]
  exotic
    [8.01-22]
Polygon of angular velocities
  [1.04]
Poly-ratio
  see Radius-ratio
Poritsky*
  [2.37]
  [3.01]
Power
  [1.02-03]
Power weight ratio
  [6.02]
Practical limits
  [3.58]
  [4.49]
Pressure angle
  [2.10,20]
  see synonym
  Angle of obliquity δ.
Primary and secondary axodes
  [9B.12]
  see also Reuleaux*
Prime numbers
  [5B.45-46]
Principal curvature
  [3.24]
Profiles
  of the flanks of teeth
    [S.03]
    [1.02,04,07]
    [2.09]
    [3.11]
    [4.72]
    [5B.29-33]
    [10.06]
  see also Flanks

Prognoses
  [5A.15]
  [5B.47]
Prototypes
  [S.18]
  [11.01,30]
Pulsation
  [1.02]
Quadratic complex
  [5A.08]
  [10.54]
  [12.10]
Quartic
  [4.28]
Rack angles analysis
  [10.24]
Rack triad
  [10.04]
  see also Triads various
Rack velocity
  [6.50]
  [**10.39**] [10.16,39,67]
Radial
  offset R
    [5A.17]
    [5A.15,17-18,28,30]
    [5B.09-11,16]
    [6.06]
  planar section
    through the flank of a tooth
    [7.15.16]
  load
    [4.52-53]
  plane
    of the conventional triad
    [7.15.16]
Radii c
  [7.08]
  [10.37]
  see also C-radii
Radii e
  [7.07]
  see also E-radii

Radiplan
   [5B.38-39]
   [6.47]
Radiplane
   [5A.23,25-26]
   [5B.10,38-39]
Radius ratio
   namely the ratio j
   [4.59]
   [6.42]
Rapid wear
   [S.05]
Ratio j
   see Radius ratio
Ratio j/k
   [11.17-18]
Ratio k
   the angular velocity ratio
   [S.01-02]
   [1.07,10-11,21,36,38-45]
   [3.33,36,51]
   [4.35]
   [5B.39,42]
   see also Miscellaneous other
RCP rack circular pitch
   [10.05]
R-cylinder
   [6.18-21]
Real and unreal
   branches of the slip track
   [5B.52]
Real rack
   [10.05, 16]
Reciprocity
   the principle of
   [2.16,19]
   [12.04]
Reconciliation
   of the methods
   [10.59-62]
Rectangular array
   of linear velocity vectors
   [6.12-16]

Rectilinear parabolic hyperboloid
   [4.26]
Referencing
   the methods of
   see Syntax
Region
   of the meshing
   [S.01]
   [1.35,38] [5B.30]
   see also Zone of the meshing
   of the charts
   [1.20]
Regulus
   properties of
   [4.02]
Relocation at the axes
   [9A.26-29,31]
Reuleaux*
   [9A-12-13]
Reversal
   of function
   [2.13]
   of travel
   [2.13]
   of sliding velocity
   [5B.64]
Reversibility
   of the gear set
   [2.09,11]
   [3.23,31]
   [6.02]
Rho-plane and/or rho-curve
   [7.17,33]
   [10.24]
Right lines
   [4.26,70]
   [6.15]
Rigidity
   and the inevitability
   of irregularity
   [1.05]
   [2.11]
   [3.01]
Rigour
   [3.02]

Rods SS
   [2.12]
   [3.28]
   [4.48]
   [5B.18,40-41]

Rolling
   and sliding contact
     [3.25]
     [4.38]
   external internal
     [4.30]
   rocking and boring
     [5B.07]

Routh*
   [4.56]

R-radii
   see Radii r

RSSR
   regular
     [2.12-14,15]
     [3.28,40]
     [4.72]
     [5B.16,18]
     [6.62]
     [7.11,13,22]
     [11.21,23,26]
     [12.12
   special
     [6.54]

Rubbing velocity
   the sliding velocity undirected
     [1.09,51-52]
     [3.45] [3.31,38,45]
     [4.36]
     [5A.08]
     [5B.02,24,25]
     [8.21]
   see also Sliding velocity v

Ruled surface
   [3.10] [3.10,11]

Rules
   for avoiding interference
   and ensuring continuity
     [5B.56]

Rulings
   of the cylindroid
     [1.38]
   of the involute helicoid
     [S.03]
     [3.08,09,21,35]
     [5B.24,25,30,40]
     [9A.02,04]
     [10.55]

Salmon*
   [4.25]

Schröder models
   [4.11]
   [5B.12]

Scoring
   [1.03]

Screw axes
   [1.51]
   [6.32]
   see also Instantaneous ditto

Screw motion of the rack cutter
   [10.32,35,41,48,58]

Screw symmetry
   of the involute helicoid
     [3.03,13,41].
     [5B.36]
     [9A.01-02, 05-06]
     [10.01]

Screw systems
   [12.05-13]

Screw theory
   mentions of
     [P.04-05]
     [S.01]
     [1.51]
     [2.16-20]
     [4.84] [4.01,84]

Separated panoramic views
   of the two wheels
     [6.49]

Severality of Q
   [2.18]
   [4.53]

Shaft angle Σ
   [S.01]
   [1.06]
   see also Angle Σ
Shapes
  of flanks
    [10.25]
    [11.03-04,24-25]
  of the slip tracks
    [5B.24-25,52]
    [11.10]
  of the teeth
    [5B.34,36,50,55]
    [11.10]
Sharp nose
   [7.20,26]
   [11.20]
Shift
   [10.42,44,46]
Shtipelman*
   [2.28,31-32]
Sign
  the general question of
    [1.22-25]
  of Σ
    [1.24]
  of k
    [1.21]
Simon*
   [2.33]
Sine-curve
  on an R-cylinder
    [6.08]
Single
  point contact
    [S.03,05]
    [1.05]
    [2.04,06-09]
    [3.03,25,34]
    [5B.07,25,34,39]
  point of tangency
    [6.29]

Singly departed sheets
   [1.34-36]
   [5B.05]
   [6.29]
Size sizes
  of the obliquity angles δ
    [5B.14]
    [6.60-62]
  of the wheels
    [5B.20]
  of the parabolic hyperboloids
    [4.77]
    [6.22]
  (relative) of the wheels
    [6.02,05]
Skating
   [S.11]
   [5B.26]
   see also synonym Lane changing
Skew paths
   [4.72]
   [5A.11,30]
   [6.03,10]
   see also Non-intersecting paths
Skreiner*
   [4.13]
Skull of a person
   [5B.55]
Sliding (rubbing) velocity
   [1.09,20,38,51-52]
   [2.21]
   [3.45]
   [4.63,66,68,70]
   [5A.24,25]
   [5B.10,36-37,39]
   [6.11-12]
   [7.09]
   [10.03]
   [11.13]
   see also Rubbing velocity
Slip
  at the naked wheels
    [4.36]
    [5B.11]
    [6.29]

SUBJECTS AND AUTHORS INDEX

Slip tracks
  in general
    [S.05]
    [2.12]
    [3.16]
    [3.03,05,08,12-14,16-19,
    21-23,29,34,35,57-60] check
    [4.70]
    [5A.06]
    [5B.23-26,29-30,34,36,52]
    [6.59]
    [7.05,12,14,23]
    [9A.21-22]
    [10.03]
    [11.06,09]
  unsupported
    [3.65]
  crossing
    but nearly tangential
    with the rulings
      [3.22,60-61]
      [5B.25,33]
    specific
      [11.14]
Slip-track locations matrix
  [11.27]
Slots at the troughs
  [7.27]
  [10.03]
Small angles χ
  [5B.37]
  [7.32]
Smith*
  [2.37]
  [5B.61]

Smooth
  elastic cord
    [4.15,49]
  naked wheels
    [1.35]
    [4.36,38]
    [5B.03,08]
    [6.29,32]
    [10.36]
Sockets
  less than hemi-spherical
    [2.12-13]
    [3.30]
    [5B.40]
Soggy
  [4.66]
Spatial involute action
  [2.07]
Special
  congruence of the fifth kind
    [5A.24]
    [5B.21]
  equiangulars
    [5B.05]
  linear complex
    [1.49]
    [5B.40-41,58]
    see also Linear complex special
Spectrum
  of core helices
    [3.12]
    [5B.29]
Speed ratio k
  [S.01-02]
  [1.07]
Speeds and feeds
  [10.35,63-64]
Sphere
  of the two loci
    [4.25]

Spherical
   involute geometry
      [5B.62]
   joints
      [2.22,24]
      [4.23]
      [5B.16,40]
   truncations
      [5B.38]
Spline-line teeth
   of the axodes
      [1.31-33,35]
      [4.36]
      [5A.25]
      [5B.02,03]
      [6.30]
      [10.36]
      [11.02]
Spoons
      see Two spoons
Spur gears
      see Planar gearing
Square
   sets
      [S.03,07]
      [1.14]
      [5B.45-53]
      [6.01]
   brackets
      [5B.50]
   prismatic rods
      [6.11-12]
Standard cutters
      [1.03]
Star of lines
      [5B.40]
Statics
   in general
      [5B.17] [?]
      [6.41]
      [11.13]
      12.04]
   at [Qm]
      [4.17,50-54]

Stations
      [5A.04,13]
Steeds*
      [1.08,15,35,42,47,53]
      [5B.62]
Steps of one tenth
      [5B.30-31,51]
Sticher*
      [5B.17,42]
Straight Beveloids
      [5B.61]
      [10.08]
      see also Bevel sets
Straight bevels
      [5B.62]
      [10.08]
      see also Bevel sets
Straight line contact
   between teeth
      [1.49]
      [2.06]
      [4.84]
      [5B.40,58]
   between the transnormal and
      the cones of Wildhaber
      [5B.13]
      [6B.04]
Straight path of Q
      [S.03-05]
      [2.07]
      [3.29][3.01-07,29,36,68]
      [4.20]
      [5A.05-06]
Strain energy
      [4.15,55-57]
Strategy
      [10.66]
      [12.14]
Strength
   in general
      [1.03]
   of teeth
      [5B.26,38]

Stress
 [1.02]
Structure
 of this book
  see Syntax
Stumpy elements
 [7.23-25]
Subtle switch
 from B to Q
  [3.26]
Surface of action
 [2.25]
 [5B.40]
Swept volume
 [5B.54-56]
Swivel angle λ
 [**10.07**] [10.07,29,33]
 see also synonym Angle λ
Synthesis
 in general
  [2.05]
  [5B.09-18,26-35]
 against analysis
  [1.13]
  [5A.02,15]
  [5B.08]
Tambourines
 [5B.50]
Tan-curve
 on an R-cylinder
  [6.08]
Tandem
 points Q in
  [2.09]

Tangency
 of the slip track
  at the throat circle
   [3.15]
 of the axodes
  along the pitch line
   [4.04,10-14]
 of the naked wheels
  at [Qf]
   [4.20-23,36,39]
   [5B.10]
 of the ellipses of obliquity
  [5B.40-41]
Tangent plane
 at the point
  of contact Q
   see Plane of adjacency
 at intersecting hyperboloids
  [4.01,09]
  [5B.10-11]
  see also Point [Γt]
Tangent planes
 perpendicularly intersecting
  [4.01,09]
Tapered
 gears
  [P.08]
 teeth
  [5B.36]
Terminology
 [P.03-05]

493

Theorems
   of the phantom rack
      [S.12]
      [9B.02,19,22]
      [10.58]
   of the pitch helices
      [5A.08]
   of the tangency of
     the naked wheels
      [5B.10]
   of the three axes in kinematics
      [1.08] [1.08,11-14]
      [10.41]
   of two circles intersecting
     at a single point
      [3.42-43]
      [4.19]
   of Plücker
      [4.12]
   of two hyperboloids sharing
     a common generator
      [4.03-11,14]
   of equiangular architecture
      [5A.21-29]
   of the hyperboloid
      [3.04]
      [4.02]
Thick teeth
   [9A.11]

Three
   kinds of circular hyperboloids
      [4.74]
   kinds of pairs of
    intersecting hyperboloids
      [4.37]
   links
      [1.01,05,06]
      [3.25].
      [9A.26]
   paths of Q
      [2.12]
      [3.25,29,65]
      [4.70]
      [5A.05]
   rigid bodies
      [1.08]
   hyperboloids intersecting
    in the key E-circle
      [5B.23]
Throat
   circle circles
      [3.15]
      [3.08,09,13,15]
      [4.05]
      [5B.05,11-12,18,23]
      [7.05,11]
   cylinder
      [3.08,15]
      [5B.18,30]
      [10.01]
   location circle
      [5B.11] [5B.11-12,24-25]
   plane
      [3.15]
   radii a, b
      [3.08],
      see also Radii a, b
   in general
      [4.14]
      [5B.24]
Tight mesh
   [3.34]
   [5B.32]
   [9A.13,17]

SUBJECTS AND AUTHORS INDEX

Tilt angle κ
  half angle
    of the cone of Wildhaber
      [5B.05]
      [10.08,33]
      see also Angle κ
Tip relief
  [2.03,09]
Tooth
  frequency
    [5B.43]
  ratio k
    [0.02]
    [6.42]
    see also Angular velocity ratio
Torque
  [4.50,55]
  [5B.17]
Transient mobility
  [4.65,66]
Transmission
  angles
    [3.31]
  error
    [P.03]
    [1.02]
    [2.25,33]
    [3.03]
    [5B.65]
  force
    [4.50,55]
    [5B.17,25]
Transnormal
  [3.03,3.42-43,45,49]
  [4.05]
  [5A.23,26]
  [5B.04,09,13]

Transversal
  JFK the equilateral transversal
    [3.42,43]
    [4.26,58,59,68,70]
    [5A.16,23]
    [5B.03,09]
    [6.04-05,11-13,15]
    [11.19]
  jQk the general transversal
    [3.39,42,43,45]
    [4.20,58,59,62,66,70]
Transverse plane
  of the conventional triad
    [7.15]
Travel path
  see Path of Q
Travel paths
  intersecting
    [4.72-73]
    [5A.03,11]
    [5B.15,16,18]
  skew
    [4.72-73]
    [6.28]
    [8.01-22]
Triad
  at [Qx]
    [5B.05,09]
    [7.10]
  at B
    [3.21] [3.08,21].
  at Q
    [3.03,44-49]
    [4.37,70]
  of axes Q-uvw
    [5B.06,07]
Triads
  various.
    [10.30]
Troublesome pointedness
  [11.06]

Truncation of teeth
 [3.36,40]
 [5B.29-31,33,36,53,54-56]
 [7.20,31]
 [8.22]
 [9A.05]
 [10.03,28]
Twiddle in the middle
 [4.15]
Twist angle τ
 [3.04]
 [5A.23]
 [5B.05]
 see also synonym Helix angle τ
Twisted paths
 of Q
 [2.25]
Twistedness
 [6.34]

Two
 branches of the involute helicoid
  [3.15] [3.15,62-64]
 branches of the intersection
  [4.28,30]
 intersecting circles
  [3.41] [3.37,41-43]
  [4.12,18,19]
 intersecting planes
  locating the transversal
  [4.70]
 linkages RSSR
  [5B.16]
 pairs of naked wheels
  of different gear sets
  [5B.24]
 pairs of naked wheels
  of the same gear set
  [4.73]
  [8.18-19]
 paths of Q intersecting
  [5A.10-14]
  [5B.14,49]
 paths of Q skew
  [4.23-24,27]
  [5A.11]
  [8.01-22]
 pitch lines
  [1.11] [1.11,23]
  [4.04]
 points [Γt]
  [4.04-10,20,25]
 points Q
  [2.24]
  [4.23,24]
 points [Qm]
  [4.18]
 screw axes
  [3.03]
 sets of wheels in unison
  [5B.26]
 special plain polyangulars
  [6.27]
 spoons
  [1.07,13]
  [2.12]
  [3.40]

SUBJECTS AND AUTHORS INDEX

Undercut
  [S.17]
  [1.29]
  [5B.23]
  [6.02,09]
  [11.05,09]
Unequal angles of obliquity
  [5A.06]
Unit mobility
  [4.68]
Unsupported slip tracks
  [3.60]
Unwrapping
  [3.10]
  [5B.61]
U-turn
  [5B.23]
Vault
  jump sideways over
  [3.62-64]
  [7.05]
Vector polygons
  [1.13-14]
Velocity vector at F
  [6.35,43,67]
  see also Sliding velocity
Vertical-horizontal
  see Horizontal-vertical
Vertices
  of the cones of Wildhaber
  [5B.11]
Vibration and noise
  [2.01,04,09,13,35]
  [3.06]
  [4.54]
  [5A,06]
WCP wheel circular pitch
  [10.05]

Wear and lubrication
  [P.09]
  [S.05]
  [2.07,12]
  [3.01,22]
  [5A.06]
  [5B.07,19,24-25,37,46,64]
  [6.02]
Weight
  [6.02]
Wetted surfaces
  [5B.64]
Wheel blanks
  [5B.33]
  [7.30]
Whole geometric construct
  [**3.19**] [3.08-19,27-28,35]
  [10.01]
Widths of teeth
  [5B.33]
Wildhaber cones
  see Cones of Wildhaber
Wildhaber*
  [1.08]
  [2.25,28,31]
  [3.46]
  [10.07]
Willis*
  [0.02]
  [1.11]
  [**2.46**] [2.03,24,46]
  [3.01]
  [4.84]
Windows of opportunity
  [S.09]
  [11.01-28]
Wohlhart*
  [2.23]

Worked examples
  #1
    [5B.01,09-17,27-34]
    [10.02]
  #2
    [5B.45-56]
    [11.10-12]
  #3
    [6.42-52]
  #4
    [6.55-64]
  #5
    [7.06-14,33]
  #6
    [7.19-31]
  un-numbered
    [11.17-26]

Worm
  definition of
    [7.01]

Worms
  [S.14]
  [1.14]
  [2.29]
  [4.75]
  [7.06,19]

Zone
  of contact (actual)
    a line of limited length Z_A
    [3.58]
    [4.53]
    [5B.18,20,24-25,34,43]
    [11.26-27]
  of contact (ideal)
    the coupler (a line of length Z_I)
    of the equivalent RSSR
    [4.37]
    [5B.07,16,19-25,32-34,40,43,49]
    [11.20]
  of the meshing
    a vague region
    [0.01] [0.01,06-07]
    [1.44]
    [2.12]
    [5A.22]
  of single point contact
    [4.53]

Printing (Computer to Plate): Saladruck Berlin
Binding: Stürtz AG, Würzburg